Cellular Signaling in Health and Disease

BIOLOGICAL AND MEDICAL PHYSICS, BIOMEDICAL ENGINEERING

The fields of biological and medical physics and biomedical engineering are broad, multidisciplinary and dynamic. They lie at the crossroads of frontier research in physics, biology, chemistry, and medicine. The Biological and Medical Physics, Biomedical Engineering Series is intended to be comprehensive, covering a broad range of topics important to the study of the physical, chemical and biological sciences. Its goal is to provide scientists and engineers with textbooks, monographs, and reference works to address the growing need for information.

Books in the series emphasize established and emergent areas of science including molecular, membrane, and mathematical biophysics; photosynthetic energy harvesting and conversion; information processing; physical principles of genetics; sensory communications; automata networks, neural networks, and cellular automata. Equally important will be coverage of applied aspects of biological and medical physics and biomedical engineering such as molecular electronic components and devices, biosensors, medicine, imaging, physical principles of renewable energy production, advanced prostheses, and environmental control and engineering.

Martin Beckerman

Cellular Signaling in Health and Disease

Springer

Martin Beckerman
Y-12 National Security Complex
602SCA MS 8225
Oak Ridge, TN 37831-8225
USA
b3m@y12.doe.gov

ISBN 978-0-387-98172-7 e-ISBN 978-0-387-98173-4
DOI 10.1007/978-0-387-98173-4
Springer Dordrecht Heidelberg London New York

Library of Congress Control Number: 2009900943

Printed on acid-free paper

Springer is part of Springer Science+Business Media (www.springer.com)

Preface

Today, noninfectious diseases have largely supplanted infectious diseases as the primary cause of morbidity and mortality in the developed and developing worlds. The diseases that now occupy center stage are age- and lifestyle-associated. They have proven to be remarkably difficult to treat let alone cure. The disorders do not have discernable Achilles heels and to-date few if any magic bullets have been found.

These challenges and difficulties are reflected in the extraordinary efforts now underway around the world to understand the underlying causes and find ways of effectively combating these diseases. The efforts cut across multiple disciplines and have led to the creation of entirely new subfields within the traditional physics, chemistry, biology, engineering, and computer science communities. Many of these newer fields are by now long- and well-established, and their names well known to everyone. Their combined goal is nothing less than a full understanding of how the human body works.

This goal is the greatest of the grand challenges in science today. It requires understanding at multiple spatial and temporal levels from molecules to cells to tissues and organs to systems. It requires that this understanding be quantitative and describe in mechanistic detail how the different parts and components work together. The study of diseases and their disease states are an essential part of the endeavor – the failures and breakdowns in the body reveal a great deal about how the body works and, in turn, understanding the workings provides the needed underpinnings for devising ways to treat and eventually find cures for the diseases.

Three categories of noninfectious diseases predominate. The first are the metabolic disorders. These are associated with changes in lifestyle that lead to obesity. The two most prominent disorders within this category are type 2 diabetes and atherosclerosis. The second major grouping is the cancers. These range from metastasis-prone solid tumors to the leukemias. The third class consists of the neurodegenerative diseases. The most prominent of these are Alzheimer's disease and Parkinson's disease.

Each of these disease categories is associated with a particular group of target tissues. Type 2 diabetes affects tissues that manage nutrients and energy resources. Atherosclerosis deals with endothelial cells that line medium and large blood vessels. Solid tumors are prone to emerge from epithelia that must repair and replicate from time to time. Neurodegenerative disorders emerge, of course, from neurons, but interestingly, each disorder targets a specific type of neuron and not others. The tissue selectivity of these diseases arises from the properties and associated functions of the cell types that render them particularly susceptible to the causal agents. At the same time, and most significantly, these diseases share many common features. Among these are inflammation, altered metabolism, impaired oxidative balance, misfolded proteins, and dysregulated signaling. These manifest themselves in the different tissues in ways that reflect their aforementioned specific functions and properties.

The goal of this text is to present an up-to-date exploration of these three disease categories and their diseases that accurately reflects the multidisciplinary character and findings of ongoing research activities. In doing so, particular attention and emphasis are given to models and explanations that supply mechanistic details at the molecular level of how a certain process or set of events occurs thereby addressing "how" questions. The aim is to let each disease "tell its own story" in a way that brings out the unique character of each disease and in the process reveals something important on how the body works. Once that is done and understood, one can appreciate the common features that ultimately unite the diseases, and may eventually give rise to a unified "theory of the disease state" and how to prevent such states from being occupied for appreciable amounts of time.

The discourse will, by its very nature, encompass a set of models or exemplars on how the diseases under discussion might develop and progress. The models are not intended to be all inclusive, that is, they cannot include everything known or posited as having a possible role in a given disease. Such models by including everything usually end up by explaining nothing. Models that do not provide the mechanistic information on how something is happening are of dubious utility in an endeavor of this type and these have been avoided as well. The selections of what to include and what not to include, are, as always, subject to limitations on the part of the writer. Apologies are therefore advanced to those researchers who feel either neglected or misrepresented. The faults are not theirs but mine.

Knoxville, Tennessee Martin Beckerman

Contents

Part I
Metabolic Syndromes

Chapter 1
Introduction

Claude Bernard (1813–1878) is widely regarded as the father of modern experimental physiology. Writing in 1853 Bernard noted that what makes a living organism different from a nonliving one is its ability to regulate its own internal environment. The internal environment referred to by Bernard meant the bodily fluids, especially lymph and blood that circulate throughout the body and make contact with and bathes all its cells. What characterizes life is not some vital force, but rather the body's ability to maintain the constancy of its *milieu intérieur* in the presence of, and independently from, a sometimes dangerous and always varying external environment

The pancreas and liver have key roles in diabetes mellitus. Claude Bernard explored the roles of these organs in a series of pioneering experiments carried out over several years beginning in the mid to the late 1840s. He began with the pancreas and its secretions during digestion and then moved on to study the liver and glucogenesis. He established that the liver not only secretes bile juices but also synthesizes glucose, which it releases into the circulation. In addition, he discovered that the liver synthesized glucose from a stored precursor glucagon, and in healthy individuals blood glucose levels are controlled and maintained at a constant level by the sympathetic nervous system.

Living organisms actively regulate their internal environments and maintain their constancy in the face of varying external conditions. This is accomplished in the case of glucogenesis by "internal secretions" – the term used by Bernard for what is now known as hormones produced by endocrine cells and released into the bloodstream. The hormones function as long-range signaling proteins that coordinate and regulate the actions of the key organs in the body. The endocrine system helps maintain the constancy of the internal environment, the *milieu intérieur*. The particular set of cells in the pancreas responsible for the internal secretions was discovered a few years later, in 1869, by Paul Langerhans, and named the islets of Langerhans in his honor, and sometime later the connection between this small population of endocrine pancreatic cells and diabetes was established.

The ability of an organism to regulate its internal environment comes about because the organism possesses the machinery that enables it to recover from

M. Beckerman, *Cellular Signaling in Health and Disease*,
Biological and Medical Physics, Biomedical Engineering,
DOI 10.1007/978-0-387-98173-4_1, © Springer Science+Business Media, LLC 2009

disturbances, or perturbations, internal and external. Because compensatory mechanisms are present, the states of the body are stable. The term introduced by Walter Cannon (1871–1945) in 1932 for the kind of stability that is maintained dynamically (actively) rather than statically (passively) is homeostasis.

1.1 The Cellular Signaling Machinery Makes Homeostasis Possible

The nature of Bernard's internal secretions was fully determined in 1902 by Bayliss and Starling who identified the first hormone, secretin, sent by the duodenum to the pancreas. They posited that secretin functions as a "chemical messenger" or "hormone," that is, as a signal sent by cells in one organ or tissue through the bloodstream to cells in another organ or tissue in order to coordinate or regulate their actions. The term "hormone" was coined by Starling in 1905 and was taken from Greek for "I excite or arouse."

There are two types of diabetes. Type 1 diabetes, referred to historically as juvenile-onset diabetes, is caused by the loss of insulin-secreting cells in the islets of Langerhans. Type 2, or adult-onset, diabetes arises from insulin resistance and impaired insulin secretion. The causal relationship between insulin secretion and diabetes was made by Fred Banting and Charles Best working in the laboratory of JJR MacLeod at the University of Toronto in 1921–1922 and assisted by biochemist James Collip. The name "insulin" was proposed by McCleod and taken from Latin for island. The first insulin injections to treat Type 1 diabetes were given in 1922, and the commercialization by Eli Lily and Co. followed shortly thereafter.

The body possesses three super cellular signaling systems – the endocrine system that secretes hormones such as insulin, the nervous system that uses neurotransmitters and neuromodulators as signaling mediators, and the immune system that employs cytokines. The term cytokine is a broad one and encompasses a large family of signaling molecules used by cells of the immune system. This term, coined by Stanley Cohen in 1974, refers to a group of signaling molecules that are used not only by cells of the immune system to communicate with one another but also by other cells in the body. Five families of signaling proteins are regarded as cytokines – Toll and Toll-like proteins that mediate inflammatory and innate immune responses, hematopoietic signaling molecules (e.g., interleukins, growth factors), tumor necrosis factors, chemokines that serve as chemical chemoattractants, and interferons that are important in defense against viruses.

1.2 Inflammation Is Present in Diseases

A particular subset of these cytokines referred to as inflammatory mediators has a prominent role in the disease states identified with metabolic disorders, cancers, and neurodegeneration. Cytokines not only are involved in managing immune responses but also have a large role in the remodeling of tissues that have been subjected to a wound or have become infected or inflamed. The cytokines join the endocrine/hormonal and nervous system neurotransmitters and neuromodulators as the third super signaling system of the body responsible for maintaining homeostasis and health.

Rudolf Virchow (1821–1922) is honored as the father of cellular pathology. Writing at about the same time as Bernard, in 1856, he identified a triad of factors that can produce a venous thrombosis. These factors are, in modern terms, (i) abnormal blood vessel walls (vascular injury); (ii) abnormal blood flow (turbulence, bifurcations, and shear stresses); and (iii) abnormal blood constituents (hypercoagulation) involving platelets and/or other soluble coagulation mediators. In examining atherosclerotic lesions, Virchow discovered in 1858 that inflammatory cells were present in the plaque and posited that atherosclerosis was an inflammatory disorder brought on by a local injury to the intima. This observation has resonated down to the present time with the appearance of the highly influential article by Russell Ross in 1999 positing that atherosclerosis is an inflammatory disorder.

A key feature of inflammatory responses is that they bring on changes in the chemical milieu of the body. The cytokines attract white blood cells to the damaged area and activate the body's immune system. They also convey messages to the liver. Upon receipt of the cytokine messages, the liver secretes a variety of proteins into the bloodstream. These proteins help limit damage to the tissues and, in the case of an infection, neutralize invading pathogens. In doing so, they maintain production of cytokines and alter the body's metabolism. Production of triglyceride-rich lipoproteins is increased, synthesis of total cholesterol is altered, and the HDL/LDL cholesterol balance is shifted. The cytokines stimulate increases in binding of LDL, which serves to neutralize pathogens, and the cells to which the LDL cholesterol molecules attach increase their production of cytokines, thereby sustaining the process.

Coronary artery disease is the leading cause of death in the world today although that distinction may be passing to cancer. Its underlying cause is atherosclerosis. The disease and its major properties were described by Virchow in his 1858 opus on cellular pathology. He coined the term embolism in 1847 to describe the clots and in 1848 introduced the term thrombosis. He further described what is today referred to as the "response to injury" model of how the disease develops. Specifically an initial injury to the arterial wall sets off a set of inflammatory responses that over time lead to a growth of a clot (a thrombosis), an irregularity of the lumen that impairs blood flow and increases coagulation. This plaque intrudes into the arterial lumen and if it breaks free becoming an embolus, it can trigger either stroke if it travels to the brain, a heart attack if it travels to the heart, or a pulmonary embolism if it lodges in the lungs.

1.3 Cholesterol Together with Inflammation Promotes Atherosclerosis

Cholesterol is essential for life. It is not only a constituent of the plasma membranes of all animal cells, but present in large quantities in membrane micro-domains specialized for signaling and transport and serves as raw material for making steroid hormones and bile acids. Cholesterol is synthesized by cells and brought into the body by foodstuffs. It is transported through the circulation and in and out of cells. Correct maintenance of cholesterol homeostasis is crucial. Excessive amounts of cholesterol, or hypercholesterolemia, are prominently associated with a number of disease states, most notably, with atherosclerosis.

Cholesterol was initially discovered by Poulletier de la Salle in 1769 and again in 1815 by Michael Chevreal. The molecule was named cholesterine by Chevreal from Greek for bile (chole) and hard fat (stereos), both individuals having extracted it from gallstones. The name was changed to cholesterol in 1900 to better reflect the presence of an alcohol (OH) group. The association of cholesterol with atherosclerosis was discovered in several stages. It was found that high cholesterol levels led to a buildup of plaque in the arteries, and too much cholesterol led to heart disease. That the plaque studied by Virchow and found in atherosclerosis was composed of cholesterol was established firmly in 1932 by Otto Diels, Heinrich Wieland, and Adolf Windaus. The key question of where the cholesterol was coming from was answered by Konrad Block who together with David Rittenberg used radioisotopes to show that it was generated by animal cells themselves. Along with Feodor Lynen, they uncovered some of the steps in the biosynthetic pathway in the cell responsible for the synthesis of cholesterol.

1.4 Signaling Pathways Responsible for Maintaining Cellular Homeostasis Are Uncovered and Explored

The term "signal transduction" is a fairly new one. The term was coined by Martin Rodbell (1925–1998) and was inspired by the language and concepts of cybernetics. In Rodbell's vernacular, signal transduction involved sequential activation in a signal transduction pathway containing discriminators, transducers, and amplifiers. Signaling begins when a ligand such as insulin binds its receptor, the discriminator. The signal is then transduced across the plasma membrane into the cell and then amplified by the effectors to stimulate the cellular responses. In keeping with this type of terminology, it is customary to speak of the components of signal transduction pathways in terms of their functional roles – as ligands, receptors, second messengers, adapters, anchors, scaffolds, and transcription factors.

In all of these situations, initiating signals arriving at a cell surface, whether they be hormones or neurotransmitters or cytokines, promoters of growth and differentiation or death, are converted into cellular response by signal transduction pathways. These pathways are integrated with intracellular pathways that monitor the internal health of the cell and maintain cellular homeostasis. The meshwork of outside-in and inside-out pathways and intracellular stress pathways has been the subject of intensive investigation for the past 50 years or more.

Insulin and other polypeptide hormones bind to receptors functioning as tyrosine kinases, and this mechanism – tyrosine phosphorylation and autophosphorylation – in the receptor's intracellular region is responsible for transduction. Other types of hormones bind a different kind of receptor and utilize a correspondingly different signal transduction mechanism. The receptors studied by Rothbell are known as G-protein-coupled receptors or GPCRs. Receptors belonging to this large family bind glucagon produced by the endocrine pancreas, adrenaline and noradrenaline produced by the adrenal gland, and hormones produced by the hypothalamus and anterior pituitary. Other agents acting in the nervous system as amine (e.g., dopamine, histamine, and serotonin) and peptide (e.g., oxytocin, substance P, and vasopressin) neuromodulators bind G-protein-coupled receptors, too. A third group of GPCR ligands are those associated with sensory inputs to the nervous system such as light (the rhodopsin receptor), tastants, and odorants. The G-proteins and the attendant signal transduction mechanism were discovered by Rodbell and Alfred G. Gilman in the 1970s.

1.5 Biophysical Techniques Provide Detailed Information on the Three-Dimensional Structure of Macromolecules

There are two preeminent methods for exploring the shape and structure of proteins at the atomic level – x-ray crystallography and nuclear magnetic resonance (NMR). These physical techniques provide detailed three-dimensional information on how the proteins are organized into their functionally distinct folding domains and what happens upon binding. They identify which amino acid residues are critical for catalysis, regulation, and binding, and show the functional consequences of mutations of specific amino acid residues.

A chain of scientific breakthroughs led the way to the successful application of x-ray crystallography to protein structure studies. These milestones begin with the discovery of x-ray by William Conrad Röntgen in 1895 and include the development of the underlying physical basis for the diffraction process by the Braggs, father (William Henry) and son (Lawrence), in the 1907–1914 time period, Max von Laue in the 1907–1914 time period, and Arthur Compton in the years following. The first proteins to be completely characterized by this means were myoglobin by John Kendrew and hemoglobin by Max Perutz in and just before

1960. These breakthroughs followed the pioneering work of Dorothy Hodgkin who solved the structure of several important biomolecules including cholesterol in 1937 and insulin in 1969; the latter effort took 35 years from start to finish.

The determination of the structure of large and irregular proteins such as those encountered in signaling pathways is particularly challenging. Each protein represents a notable achievement. Two recent examples of successful solutions are the potassium channel by the American biophysicist Roderick MacKinnon in and just after 1998 and the insulin receptor by an Australian team led by Colin Ward in 2006. The insulin receptor was first identified in 1969, its amino acid sequence determined in 1985, and the three-dimensional structure of its kinase domain was solved in 1994, but it took another 12 years to completely solve its crystal structure.

The second major method, NMR, has a distinguished set of landmarks, as well. These begin with the molecular beam magnetic resonance experiments of Isidor Rabi in the late 1930s, the discovery of the magnetic moment of the proton by Otto Stern in 1943, and the first successful NMR experiments of Felix Bloch and Edward Purcell in 1946. These were followed by further developments of NMR methods by Norman Ramsey, Richard Ernst, and Kurt Wütrich, while another series of developments led to magnetic resonance imaging (MRI) pioneered by Paul Lauterbur and Peter Mansfield and today widely used as a diagnostic tool in medicine.

1.6 Signaling Pathways Have Been Illuminated Through Intensive Efforts Spanning the Last 50 Years

During the years following the first structural breakthroughs, the basic properties of the cellular signaling pathways were uncovered by researchers working in multiple fields. A short and incomplete list of major milestones includes allostery, second messengers, modularity, posttranslational modifications, and proteolysis. A central concept on how proteins function was put forth by Monod, Changeux, and Jacob in 1963. In an allosteric interaction, the protein itself functions as a signal transducer. Binding of a ligand at a particular site on the protein triggers conformational changes leading to alterations in structure and function at distant sites on that protein. By this means binding to a regulatory site can turn on or off catalytic activities or mask or unmask a binding site. In feedback inhibition, a process that inspired the development of this model, a downstream product of a linear chain of reactions binds to a regulatory site on an upstream element deactivating that component and terminating metabolism or signaling.

Second messengers were first discovered by the American biochemist E.W. Sutherland, in 1955. In his study of hormonal signaling, he found that cyclic adenine monophosphate (cAMP) functioned as a signaling intermediate in the pathways that mediate the hormonal effects of epinephrine on metabolism.

These breakthroughs were followed by those of Rodbell and Gilman dealing with second messengers that underlie G-protein-induced action. Other well-characterized second messenger systems utilize calcium and phospholipids. Most recently nitric oxide has emerged as a legitimate second messenger.

Proteins, especially the ones involved in signaling, are highly modular. The components of the proteins are built up from pieces – domains and motifs – each with a well-defined function and each capable of folding quasi-independently into its physiologically competent state. This modularity characterization applies to transmembrane proteins known because of their highly modular structure as mosaic proteins and to modular intracellular adapters, scaffolds, and anchors that are centrally involved in transducing signals into the cell.

Reversible protein phosphorylation plays a major role in signal transduction. That protein kinases catalyze the addition of phosphate groups to substrate-signaling proteins using ATP as the donor while protein phosphatases do the opposite and catalyze the removal of these groups was discovered by Edmond Fischer and Edwin Krebs in the 1955–1956 time period. These operations play major roles in transducing signals from the plasma membrane, through several intermediaries to the eventual cellular targets such as the gene transcription machinery in the cell nucleus. Protein phosphorylation is not the only kind of posttranslational modification used for signaling. Other posttranslational modifications are acetylation, methylation, SUMOylation, ubiquitination, and S-nitrosylation.

The importance of proteolytic degradation has become increasingly apparent during the past 20 years. It is now recognized to be comparable in importance to reversible protein phosphorylation, working together with the latter in regulating cellular functions in health and disease. Malfunctions in the proteolytic machinery contribute to tumor growth and metastasis, to cystic fibrosis, and to some of the most prominent neurodegenerative disorders. Up until about 1975 the lysosome was believed to be the locale where unwanted proteins were shipped for degradation/and or recycling. This view, along with the notion that cellular proteins were all long-lived, was overturned with the discovery of the cytosol-based ubiquitin-mediated proteolysis by Israeli scientists Avram Hershko and Aaron Ciechanover and American scientist Irwin Rose.

1.7 Mutated, Misfolded Proteins Cause Cancer

The combinations of biophysical techniques for studying the three-dimensional structure of proteins and DNA and the primary structures determined through genetic sequencing provide the underpinnings for understanding how inappropriate growth and proliferation develops. The combination of these and cell biological studies of cancers have revealed the existence of an impressive number of cancer genes. Some of the proteins encoded by these genes when mutated lose one or more of their critical functions while others gain new ones. Not all mutations in a protein lead to disease, and when they do generate a certain disease, not all mutations endow that protein with the same conformational, disease-causing properties. Thus, the molecule-level details are important.

Certain types of proteins are centrally implicated in cancer. Foremost among these are proteins that operate in the signaling pathways that regulate growth and metabolism, survival and death, and DNA repair. The identification of these classes of cancer-causing agents has led to a further understanding that dysregulated growth, survival, and repair signaling are the cellular consequences of the mutated and misfolded proteins. The cancers that arise in this way all exhibit a related set of hallmarks, six or seven or eight depending on who does the enumerating. A landmark paper in this arena was the one by Douglas Hanahan and Robert Weinberg in 2000 in which a core set of hallmarks was presented.

Cancer cells increasingly rely on aerobic glycolysis and proportionally reduced their dependence upon oxidative phosphorylation. This aspect was first noted and remarked upon by Otto Warburg over 80 years ago. This shift in emphasis is named the Warburg effect in his honor. This particular shift in metabolic strategy is associated with alterations in the tumor microenvironment. These environments are oxygen starved, that is, they are hypoxic, and they are also acidic. More generally, metabolic reprogramming contributes to the establishment and maintenance of the different classes of diseases. Nowadays, altered metabolism is seen as playing a considerable role not only in cancers but also in Type 2 diabetes and in neurodegeneration.

Another rather stunning set of findings during the past few years has confounded the issues greatly. Not only are genetic influences and alterations important, but epigenetic ones are crucial as well. Eukaryotic DNA is packaged into chromatin, and chromatin structure can be altered through a variety of protein–DNA and protein–protein interactions and posttranslational modifications. As a result, some genes can be turned on inappropriately and others turned off. Cellular homeostasis can be lost in multiple ways. To further confuse, but maybe illuminate the situation, a host of small noncoding RNA particles have been discovered in the past few years whose existence was not previously suspected. These are encoded by the so-called junk DNA. These RNA particles interact with and alter essential signaling pathways in ways that contribute to cancer and other disorders.

1.8 The Microenvironment Is an Important Ingredient in Cancer Metastasis

The vast majority of cancer deaths, 80–90%, come from cancers that have metastasized. The key pathway of cancer metastasis – the migration of cancerous cells from the primary site through the circulatory and lymphatic systems to far-removed secondary sites – was identified in 1865 in the case of skin cancers by the

German surgeon Carl Theisch. That this is the route taken was confirmed as a general result for all cancers in 1872 by the German anatomist Wilhelm Waldeger.

Local signaling by cells in the microenvironment is a key aspect of both tumor growth at primary sites and metastatic spread to secondary sites. This aspect was recognized by Judah Folkman in a landmark paper published in 1971 that noted that all tumors are angiogenesis dependent. A general, accompanying feature of these environments is that the oxygen levels are low, that is, the tumors grow under hypoxic conditions.

The importance of the microenvironment for cancer metastasis was first noted 120 years ago, in 1889, by Stephen Paget in examinations of a large number, more than 900, of autopsy records of people who suffered from various kinds of cancers. In studying the sites where metastasis took place, he reached the conclusion that specific organs were being targeted not randomly but rather because they had properties that favored further cancer growth. In his vernacular, the "seed" referred to the cells of the primary tumor that were attempting to migrate and the "soil" denoted the sites found by them supporting further colonization. The circulation and lymphatics form the highway for transport of the cells from one site to another. However, the establishment of colonies in organs and tissues far removed from the primary site is site specific, while mechanical factors tied to the transport route, as postulated by James Ewing in 1928, have an important role at the regional level, e.g., at draining lymph nodes.

Cells continually exchange signals with their neighbors and with the supporting extracellular matrix. These signals maintain tissue homeostasis and healthy state and have a direct causal role in the transition to and maintenance of disease states. In examining how atherosclerosis and cancers develop, it is clear that signaling is taking place between several populations of cells including, depending on the specific disease, macrophages, fibroblasts, stem cells, and smooth muscle cells, and also platelets. Microenvironments are also present and influence disease states in the nervous system. Astrocytes and microglia engage in tight associations with neurons, and the accompanying two-way signals have a role in health and neurodegeneration.

In tissue remodeling, and in cancers, new tissue must be generated and damaged tissue and debris removed. The cells that are generated must be supplied with nutrients and oxygen and waste removed. Macrophages and connective tissue-promoting fibroblasts working together carry out these functions by secreting molecules that guide other cells to damage sites. The expression of cell-surface adhesion molecules and signal receptors, and secretion of various factors, attract and guide cells to the site of attachment. Secretion of proteolytic and growth-promoting factors helps tear down and build up the extracellular matrix and create an environment supportive of angiogenesis and repair. Other signals active during cancer metastasis trigger changes in cellular morphology driving a differentiation from an attached, polarized form acting as part of a tissue or organ to a solitary, mobile form able to migrate to remote sites. And still other signal transduction events regulate the metabolic strategy, optimizing it for the particular environment – nutrients and oxygen.

1.9 Neurons Are Cells Highly Specialized for Long-Range Signaling

The first key step in elucidating the core properties of the nervous system was taken by the Spanish anatomist Santiago Ramón y Cajal (1852–1934). He correctly identified through careful examinations in a light microscope that the nervous system was made of up discrete units – cells – and that these neurons consisted of a central soma containing the cell nucleus and two processes, the dendrites serving as input units and the axons as output unit. The key point made by Ramón y Cajal was that in contrast to the prevailing view at the time (the reticular model) that the nervous system consisted of a single continuous and physically attached set of filaments, the neurons did not form a continuum but instead sending axons and receiving dendrites were simply in close proximity but not touching, that is, they were contiguous but not continuous.

The gap formed between the two neurons in communication with one another was named a "synapse" in 1906 by the British scientist Arthur Sherrington. The mystery of just how information was conveyed from one neuron to another across the synapse was solved by Henry Dale and Otto Loewi with their discovery of the first neurotransmitter, acetylcholine, by Dale in 1914 and the demonstration by Loewi in 1921 that this molecule functioned as a chemical messenger across the synapse from the presynaptic, or axonal side, to the postsynaptic dendritic side.

The job of the neurons is communication over distances large and small. They carry out this function through membrane depolarization and action potentials, and neurotransmitter release. The neurons are highly polarized with long and narrow processes extending out from the cell body, or soma. The nucleus is located in the cell body and large numbers of mitochondria are distributed throughout in ways that adjust to and support the large energy demands. The cell's communication abilities are made possible by amazing networks of receptors, ion channels, and intracellular signaling transducers and by an extensive railway trafficking system that conveys material to and from the far-away ends of the neuron. Their extended structure, strong metabolic requirements, and long life render neurons particularly susceptible to age-associated disorders such as Alzheimer's disease and Parkinson's disease.

1.10 Amyloids Are an Essential Ingredient in Many Diseases

Amyloids, or amyloid fibrils, are (mostly) extracellular deposits of misfolded proteins that have clumped together to form long, insoluble, strandlike structures. That such deposits form has been known for some time, having been described for at least 400 years. These deposits have been referred to over the years first as "sago spleen" and as being lardaceous, and then following Virchow as amyloids. These deposits present a waxy or fatty character when found at autopsy in livers and other organs of people suffering from a variety of illnesses including tuberculosis and syphilis.

The use of the term amyloid for these deposits is due to Rudolf Virchow who described these deposits in 1854. The term was coined in 1838 by the German botanist Mathias Schleiden to describe plant starch/cellulose. Virchow used an iodine-sulfuric acid test for starch/cellulose, and receiving a positive response initially identified the deposits as being starchlike. The correct association of these deposits with protein (albumenaceous) was made shortly thereafter by several researchers.

When stained with a Congo red dye and then viewed under polarized light, amyloids exhibit a characteristic yellow-green birefringence pattern. This test was first devised by Divry and Florkin in 1927. The key element in this staining test is a vegetable dye, first synthesized in 1883 and subsequently used in the textile industry. It was given the mysterious-sounding name "Congo red" to assist in its marketing. The dye is no longer used as a textile dye because of its tendency to run and change color, but remains in use today as the preeminent marker for the presence of amyloids.

Fibrils are long, unbranched strings, or fibers, aligned in parallel and held together in a planar fashion by means of hydrogen bonds. A common example of a fibrillar structure is cellulose. As already noted, in contrast to cellulose, which is made from carbohydrates, amyloid fibrils are made from protein. The biophysical connection between the birefringence pattern and the ordered amyloid fibrils was established by means of electron microscopy by Cohen and Calkins in 1959. Since that time amyloid fibrils have been studied with electron microscopy, solid-state nuclear magnetic resonance, and x-ray crystallography. A formal, biophysical definition of amyloids is – amyloids are proteinaceous deposits showing a yellow-green (apple green) birefringence when stained with Congo red and viewed in a light microscope under polarized light, and consisting of nonbranching fibrillar structures 8–10 nm in width and indeterminate length, and with an x-ray diffraction pattern consistent with crossed beta structures.

The deposit of waxy materials puzzled over by Virchow occurs in the spleen, liver, and kidneys. They were most likely due to what is today called AA amyloidosis involving buildup of amyloid A (AA) protein over time as a consequence of long-lasting infections and chronic inflammation and resulting from, for example, tuberculosis and rheumatoid arthritis. Amyloid A protein deposits are derived from an overproduction of serum amyloid A (SAA), an acute phase protein serving as a precursor for the AA protein. These buildups occur most often in the heart, but can occur in other tissues and organs as observed by Virchow.

In systemic amyloid diseases such as AA amyloidosis, the amyloids collect in multiple organs and tissues of the body. In another disease of this type, immunoglobulin light-chain amyloidosis (AL), B cells secrete light-chain peptides that misfold and move through the circulation to collect in the kidneys and also in the heart and liver leading to organ failure and death. Light-chain amyloidosis is associated with B-cell myelomas. The light-chain peptides forming the core of the amyloid deposits partially unfold from their native state and then refold into an alternate conformation that promotes amyloid formation. The formation of amyloids as a consequence of misfolded proteins is the key biophysical feature of all amyloidoses.

In contrast to the above-mentioned systemic amyloidoses, there are a number of diseases in which amyloid deposition is specific to a single organ or tissues of the body. This grouping includes the transmissible spongiform encephalopathies, or prion diseases, and major public health problems such as Type 2 diabetes mellitus and Alzheimer's disease. These diseases are local amyloid diseases, involving a particular organ or tissue. In Type 2 diabetes, misfolded aggregates of islet amyloid polypeptide (IAPP) form in the pancreas; in Alzheimer's disease, the culprit is the amyloid β protein in the brain. In the prion diseases, the disease-causative agents are misfolded prion proteins that not only form amyloid deposits but are also transmissible, that is, these proteins can, by themselves, transmit the disease from one subject to another.

Table 1.1 lists some of the most prominent amyloid diseases. The first group of diseases in the table is the systemic amyloidoses. Among these are several diseases that become particularly prominent in the elderly, especially in individuals older than 80 years. The list of diseases associated with misfolded proteins and aggregation includes diseases in which the aggregation and deposition of misfolded proteins is intracellular rather than extracellular. Parkinson's disease, Huntington's disease, and Lou Gehrig's disease are neurodegenerative disorders in which intracellular deposits of misfolded proteins have a major role. The term "conformational diseases" or, alternatively, "protein-folding

Table 1.1 Amyloid diseases

Disease	Misfolded protein(s)
Light-chain amyloidosis	Immunoglobulin light chain (AL)
Heavy-chain amyloidosis	Immunoglobulin heavy chain (AH)
Transthyretin (TTR)-amyloidsosis	ATTR
Dialysis-related amyloidosis	B_2-microglobulin ($A\beta_2M$)
(Apo) Serum amyloid A	AA
Apolipoprotein-associated	AApoAI, ApoAII, ApoCII, ApoE
Type 2 diabetes mellitus	Human islet amyloid polypeptide (hIAPP)
Prion diseases	PrP^{Sc} protein
Alzheimer's disease	Amyloid beta (Aβ) protein
Parkinson's disease	α-synuclein
Polyglutamine repeat disorders (e.g., Huntington's disease)	CAG repeat-containing proteins
Amyotrophic lateral sclerosis	Copper-zinc superoxide dismutase (CuZnSOD)
Serpinopathies	Antithrombin, Antitrypsin, PAI-1
Taupathies	Tau protein

diseases" encompasses this broader class of disorders of the aging human. With the exception of Type 2 diabetes, which will be discussed in Part I, these diseases will be looked at in greater or lesser detail in Part III of this text.

1.11 Reactive Oxygen and Nitrogen Species Carry Out Signaling in Ways That Contribute to Health and Disease

Reactive oxygen and nitrogen species were initially regarded as acting in two ways – as a killing agent (superoxide) used by phagocytes to kill pathogens and as a set of intracellular damage-causing agents released mainly by the mitochondrial electron transport chain that attack cellular macromolecules – lipids, DNA, and proteins. Far more recently, these immune system adjutants have been recognized as acting in a signaling capacity that promotes cellular health. They do so in two distinct ways – as second messengers and through posttranslational modifications that alter the conformations and activities of signaling proteins. The posttranslational modifications may have either a beneficial or a harmful effect. In the former instance, they act in ways that contribute to homeostasis. In the latter case, they enhance, along with heavy metals and other environmental toxins, the propensity for certain proteins to misfold, aggregate, and form amyloids, alter cellular metabolism, and promote cellular death, apoptotic and/or necrotic.

Reactive oxygen and nitrogen species are associated in multiple ways with all three disease categories. In atherosclerosis, for example, key disease-causing mediators are LDL particles that have been oxidized by factors released by cells residing in the arterial wall. In cancers, these agents are associated with mitochondrial function, and through that connection with the raising or lowering of anticancer barriers such as apoptosis. In neurons their particular susceptibility to oxidative and nitrosative insults and to the damaging effects of misfolded proteins are reflected in the multiple entries appearing in Table 1.1.

Signaling by reactive oxygen and nitrogen species will be explored in Chapter 8. That examination will immediately follow that of inflammation, the subject of Chapter 7. Part II, dealing with cancer growth and metastasis, will follow these two chapters. The remainder of Part I will have as its focus both Type 2 diabetes and atherosclerosis and will contain detailed discussions of cholesterol transport and insulin signaling.

Further Reading

Milieu Intérieur and Homeostasis

Cannon WB 1932 *Wisdom of the Body*. New York: W. W. Norton.

Starling EH 1905 The chemical correlation of the functions of the body. *Lancet* 2: 339–341. (*reprinted in* Henderson J 2005 Ernest Starling and 'Hormones': an historical commentary. *J. Endocr*. 184: 5–10.)

Inflammation

Ross R 1999 Atherosclerosis – an inflammatory disease. *N. Engl. J. Med.* 340: 115–126.
Virchow RC 1858 *Die Cellularpathologie in ihrer Begründung auf physiologische und pathologische Gewebelehre*. Berlin: August Hirschwald.

Signaling by Cells of the Immune System

Bigazzi PE, Yoshida T, Ward PA and Cohen S 1975 Production of lymphokine-like factors (cytokines) by simian virus 40-infected and simian virus 40-transformed cells. *Am. J. Pathol.* 80: 69–78.
Cohen S, Bigazzi PE and Yoshida T 1974 Similarities of T cell function in cell-mediated immunity and antibody production. *Cell Immunol.* 12: 150–159.

Signal Transduction

Baron M, Norman DG and Campbell ID 1991 Protein modules. *Trends Biochem. Sci.* 16: 13–17.
Fischer E 1894 Einfluss der Configuration auf die Wirkung der Enzyme. *Ber. Dtsch. Chem. Ges.* 27: 2985–2993.
Koshland DE Jr, Némethy G and Filmer D 1966 Comparison of experimental binding data and theoretical models in proteins containing subunits. *Biochemistry* 5: 365–385.
Monod J, Changeux JP and Jacob F 1963 Allosteric proteins and cellular control systems. *J. Mol. Biol.* 6: 306–322.
Pawson T 2004 Specificity in signal transduction: from phosphotyrosine-SH2 domain interaction to cellular complexity. *Cell* 116: 191–203.

Cancers

Dvorak HF 1986 Tumors: wounds that do not heal. Similarities between tumor stroma generation and wound healing. *N. Engl. J. Med.* 315: 1650–1659.
Ewing J 1928 *Neoplastic Diseases* (6th Ed.) Philadelphia: WB Saunders.
Folkman J 1971 Tumor angiogenesis: therapeutic implications. *N. Engl. J. Med.* 285: 1182–1186.
Hanahan D and Weinberg RA 2000 The hallmarks of cancer. *Cell* 100: 57–70.
Paget S 1889 The distribution of secondary growths in cancer of the breast. *Lancet* 133: 571–573.
Warburg O 1930 *The Metabolism of Tumors*. London: Constable Press.
Warburg O, Posener K and Negelein F 1924 Über den Stoffwechsel der Tumoren. *Biochem. Z.* 152: 309–344.

Signaling in the Nervous System

Ramón y Cajal S 1911 *Histologie du Système Nerveux de l'Homme et des Vertébrés* (L. Azoulay transl.). Paris: A. Maloine.
Sherrington CS 1906 *The Integrative Action of the Nervous System*. New York: Charles Scribner's Sons.

Amyloids

Cohen AS and Calkins E 1959 Electron microscopic observations on a fibrous component in amyloid of diverse origins. *Nature* 183: 1202–1203.

Divry P, Florkin M and Firket J 1927 Sur les propriétés optique de l'amyloide. *CR Soc. Biol. (Paris)* 97: 1808–1810.

Schleiden MJ 1838 Einige Bemerkungen über den vegetabilischen Faserstoff und sein Verhältnis zum Stärkemehl. *Ann. Physik u. Chemie* 43: 391–397.

Virchow R 1853 Über eine im Gehirn und Rückenmark des Menschen aufgefundene Substanz mit der chemischen Reaktion der Cellulose. *Virchows Arch. Pathol. Anat.* 6: 135–138.

Chapter 2
Energy Balance

The body maintains energy balance throughout the daily feed-fast cycles, across periods where energy demands increase due to physical exertion, and when energy supplies are depleted due to nutrient deprivation. During a typical day, several meals are consumed leading to periods of time several hours long after the meals, the fed state, where there is a net energy gain. This situation changes during sleep and during gaps in meal intake where the net energy balance, that is calories in minus calories out, becomes negative. These are the fast states that lead to refeeding. The ability of the body to maintain energy balance is centered about a set of peripheral organs and tissues that digest, convert, ship, and store nutrients and energy, namely,

- Pancreas
- Liver
- Adipose tissue
- Muscle
- Gastrointestinal tract

In the fed state, nutrients – carbohydrates, fat, and amino acids – are taken in, converted, and stored. In response to the increase in circulating glucose, the pancreas secretes insulin that serves as a fed-state start signal to the other peripheral tissues. The liver takes up and stores the glucose as glycogen. Triglycerides are packaged into large, transportable particles; these are synthesized and stored in adipose (fat) tissue, and muscle rebuilds its supply of protein.

In the fast state, nutrients are no longer taken in but instead energy sources residing in peripheral tissues are mobilized and released to maintain energy homeostasis. This state is signaled by glucagon release from the pancreas. The liver now breaks down and releases the stored glucose/glycogen, initiates gluconeogenesis, initiates fatty acid oxidation, and generates ketone bodies as an alternative/supplement to the dwindling glucose supply. Adipose tissue degrades their triglycerides thereby supplying fatty acids and glycerol to other peripheral tissues, muscle utilizes fatty acid oxidation, and ketone bodies are shipped to the brain.

M. Beckerman, *Cellular Signaling in Health and Disease*,
Biological and Medical Physics, Biomedical Engineering,
DOI 10.1007/978-0-387-98173-4_2,

To properly manage feeding and energy balance, the central nervous system and peripheral tissues communicate continuously with one another. Local signals are sent and received by cells residing in the stomach and intestines, pancreas, liver, adipose tissue, and muscle. Longer ranging signals are sent through the circulation to specific regions in the brain from these peripheral tissues and, in response, feedback regulatory signals are sent from the brain to the peripheral tissues.

The brain exerts considerable influence over feeding. Hormones secreted into the bloodstream by the pancreas, adipose tissue, stomach, and intestines pass through the blood-brain barrier and enter the part of the hypothalamus called the arcuate nucleus. The hypothalamus is a region buried deep in the brain responsible for regulating, for example, eating, drinking, and sleeping. The hormones secreted into the bloodstream by the food-processing and energy-storing organs convey needed information on the amount of stored fat and the energy balance. In response, neurons in the arcuate nucleus send out signals in the form of neuropeptides that instruct other control centers to either stimulate or suppress the appetite.

The central nervous system regulates fat stores to maintain proper energy balance in the body. In order for this to happen, there has to be a feedback mechanism operating between regulatory center(s) and fat stores. The existence of such a mechanism was first postulated by Kennedy in 1953. He noted that in healthy animals, fat stores remain fairly constant in the face of changing conditions and suggested that the hypothalamus might act as a central regulator of energy balance through negative feedback mechanisms much like those described by Norbert Wiener in his 1948 landmark tome on cybernetics. He further suggested that a hypothalamic center might have a key role in the regulation, noting that age-related changes and damage to this region lead to obesity. The discovery that insulin and leptin convey signals to the hypothalamus serves as a confirmation of Kennedy's hypothesis.

Several conditions need to be met for this type of negative feedback regulation to take place. First the signal strengths, as measured by the amounts of insulin and leptin released into the circulation, should be proportional to the amount of fat. Next the amount entering the CNS from the circulation must reflect the amount of body fat, and third in response to increasing signaling there has to be a negative behavioral response – reduced feeding. Under normal healthy conditions in the body, all of these conditions are met for insulin and leptin, but as will be discussed in detail they fail under the unhealthy conditions of the metabolic syndrome.

Alterations in metabolism vary from short-term changes associated with fed-fast states to long-lasting changes in metabolic strategy associated with healthy and disease states. These alterations are made possible by the presence of multiple-layered, overlapping, signaling systems that tightly integrate together growth and metabolism. Some of the regulatory signals arrive at critical control points, typically rate-limiting enzymes, within metabolic pathways. Other signals are conveyed to the nucleus where they induce changes in expression of genes involved in metabolism. The arriving signals encompass not only hormonal and growth factor signals such as insulin sent from cells in other tissues but also local signals indicative of the cell's current nutrient and energy status.

The goal of this chapter is to provide an overview of the signals regulating energy balance sent and received by peripheral tissues and the central nervous system. This overview will illustrate many of the themes mentioned in the preceding paragraph. It will begin with the liver and the reciprocal actions of insulin and glucagon signaling at critical control points for glucose management. It will include a first look at β-oxidation, or fat burning, in muscle tissue, and in doing so examine how this process is influenced by hormonal signals as well as local signals that convey information about the cell's energy status. A critical intracellular sensor of energy status – AMP-activated protein kinase (AMPK) – will be introduced, and the chapter will conclude with an examination of how insulin, leptin, and ghrelin hormonal signals are processed in the hypothalamus.

2.1 Hormonal Signaling by the Endocrine Pancreas

The pancreas consists of two distinct populations of cells. Those belonging to the exocrine pancreas have roles in digestion while those in the endocrine pancreas, the Islets of Langerhans mentioned in Chapter 1, are prominent regulators of glucose and lipid homeostasis and energy balance. These endocrine cells along with populations of endocrine cells strategically situated in other peripheral organs secrete a variety of signaling molecules – hormones and signaling peptides – that act locally to coordinate the actions of the peripheral organs and globally over a long range conveying energy balance information to the central nervous system. These signaling molecules are listed in Table 2.1.

Table 2.1 Pancreatic Hormones

Hormone or signaling peptide	Sending cells	Receiving cells	Physiological function
Glucagon	Pancreatic Islet α-cells	Liver	Stimulates release of glucose into the circulation and conversion of glycogen into glucose (L)
Insulin	Pancreatic Islet β-cells	Liver, muscle, adipose tissue, brain	Storage of glucose (L), uptake of glucose (M, A), storage of fatty acids (A), inhibits breakdown of glycogen (L), satiety signal (B)
Somatostatin	Pancreatic Islet δ-cells	Intestine, stomach, pancreas	Suppresses release of gastrointestinal hormones, inhibits gastrointestinal motility and blood flow (I), secretion of stomach acid (S), inhibits the release of insulin and glucagon (P)

The exocrine pancreas receives signals from the gastrointestinal tract. Recall from the Introduction that secretin was the first hormone discovered. In response to secretin, cells in the exocrine pancreas secrete bicarbonate that buffers low pH conditions, while cholecystokinin (CCK) triggers secretion of digestive enzymes. There are four populations of pancreatic endocrine cells. These are referred to as α-, β-, δ-, and PP-cells. In humans, these cells are scattered throughout the Islet. These cells are associated with other endocrine cells to facilitate paracrine communications and are aligned along the blood vessels. More than half of the cells are β-cells, with α-cells the next most common cell type. These cells respond to changes in glucose levels by either increasing or decreasing secretion of glucagon, insulin, and somatostatin thereby maintaining glucose and lipid homeostasis and energy balance throughout the feeding cycle.

Insulin, a polypeptide hormone secreted by the pancreas is the central regulator of energy balance in the body. It is secreted by Islet β-cells in response to eating as indicated by increased glucose levels released into the circulation from the small intestine. The hormone instructs the liver to take up and store glucose by forming glycogen from it. This hormone keeps blood glucose levels from becoming too high. It does so by increasing glucose uptake by fat and muscle cells and inhibits glucose production by the liver. It also stimulates the synthesis and storage of lipids and inhibits their degradation and release into the bloodstream.

Insulin receptors in muscle and adipose tissue stimulate uptake of glucose by the GLUT4 transporter (Table 2.2). In the liver, insulin stimulates the storage of glucose in the form of glycogen, a storage polymer form of glucose. If the amount of glucose taken up exceeds the storage capacity, the excess is converted to fatty acids that are released into the bloodstream as lipoproteins. Insulin also facilitates the entry of glucose into adipose tissue where it is converted to glycerol. The glycerols together with fatty acids are then used to make triglycerides. In sum, insulin stimulates uptake and burning of glucose (glycolysis) and storage of fatty acids in adipose tissue.

Table 2.2 Glucose transporters and their distribution in the body

Glucose transporter	Distribution
GLUT1	Erythrocytes, endothelial cells of the blood-brain barrier
GLUT2	Liver, pancreatic β-cells, kidney, intestines
GLUT3	Neurons, placenta
GLUT4	Adipose tissue, striated muscle – skeletal and heart

2.2 In Response to Signals from the Pancreas, the Liver Maintains Glucose and Lipid Homeostasis

Once blood glucose levels decline, pancreatic α cells secrete glucagon that instructs the liver to release the stored glucose into the bloodstream. This is accomplished by the conversion of glycogen to glucose (glycogenolysis), by the synthesis of glucose from noncarbohydrate sources (gluconeogenesis), by the inhibition of glycolysis, and by inhibiting the storage of glucose as glycogen molecules (glycogenesis).

Insulin signals received by the liver launch the opposite set of actions. In response to feeding leading to an elevation in circulating glucose, insulin instructs the liver to take up glucose and store it in the form of glycogen and inhibits the release of glucose into the bloodstream. Insulin is a polypeptide hormone and binds to receptor tyrosine kinases that transduce the insulin signals into the hepatocytes. This pathway will be examined in detail in Chapter 3. A brief sketch at the chain of events leading to the conversion and storage of glucose as glycogen in the liver is presented in Fig. 2.1. As shown in this figure, glucose is converted sequentially to glucose-6-phosphate and from there to glucose-1-phosphate. Two enzymes control the interconversions between glucose-1-phosphate and glycogen. One of these, glycogen synthase (GS), catalyzes the formation of glycogen from glucose-1-phosphate. This enzyme is inactivated when phosphorylated and activated when dephosphorylated. Another enzyme, glycogen phosphorylase (GP), when phosphorylated catalyzes the conversion of glycogen to glucose-1-phosphate. Insulin and glucogen exert their regulatory influences by toggling on and off GS and GP at the correct times in fed-fast cycle.

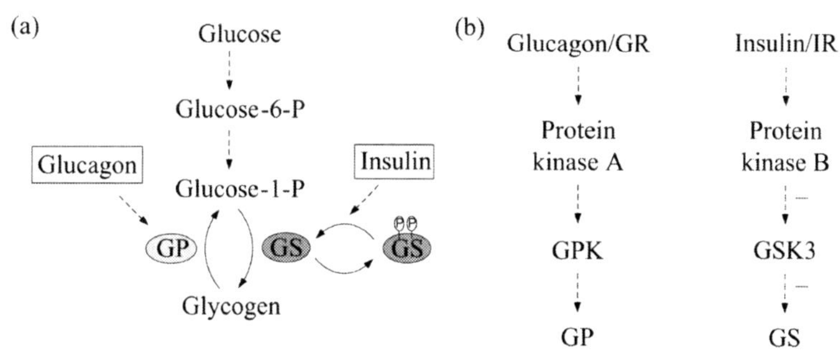

Fig. 2.1 The glycogen synthase shuttle

Phosphorylation and dephosphorylation actions regulate whether glucose is converted to glycogen for storage or whether the converse takes place where glycogen is converted back to glucose for release. Glycogen synthase is the rate-limiting enzyme in the conversion of glucose to glycogen. Insulin promotes the storage process by activating protein kinase B (Akt), which then phosphorylates glycogen synthase kinase-3 (GSK3) thereby inactivating it. GSK3 is the leading kinase that phosphorylates and inactivates GS. Thus insulin promotes the dephosphorylated state of GS leading to activation and glycogen storage. Glucagon exerts its regulatory actions through protein kinase A, which phosphorylates and activates glycogen phorphorylase kinase (GPK), which then phosphorylates and actives glycogen phosphorylase. Glycogen phosphorylase plays a comparable role to GS operating as the rate-limiting enzyme in the breakdown of glycogen to glucose. The insulin/glycogen ratio thus acts on GS and GP to switch back and forth between the two modes of action (Fig. 2.1).

An illustrative example of how hormonal and internal signals come together to regulate a metabolic program is that of gluconeogenesis in the liver. Here again insulin and glucagon operate in opposition to turn on and off a metabolic process, but in this case the critical control points are those of transcription factors and their cofactors operating in the nucleus. The key regulatory targets in this case are the transcription factor known as FOXO1 and the coactivator referred to as TORC2. Insulin signaling during the fed state prevents these factors from turning on transcription of a pair of enzymes critical for gluconeogensis, namely G6Pase and PEPCK. Glucogon, in contrast, promotes activation of the transcription factors during fasting and stress situations leading upregulation of these key enzymes and increased generation of glucose. Most interestingly, the internal energy-status sensor AMPK can override the hormonally derived pro-gluconeogenesis signals when energy levels in the cell are too low. The insulin signaling pathway is the main topic in Chapter 3 and metabolic reprogramming is the chief subject in Chapter 4. The aforementioned signal transducers and transcription factors will be defined and examined in considerable detail in those two chapters.

2.3 Energy in the Form of Lipids Is Stored and Released When Needed in Adipose Tissue

Adipose tissue is composed of fat-storing cells called adipocytes. There are two types of adipose tissue – white adipose tissue (WAT) and brown adipose tissue (BAT). Fuel in the form of fat is stored in white adipocytes and converted to heat in brown adipocytes. These cells grow and shrink as they gain and lose lipids, which are stored in the form of triglycerides within lipid droplets. The trafficking of lipids and glucose in and out of these cells is regulated by the sympathetic nervous system and by hormones in a manner that parallels the movements of glucose in and out of other tissues.

Insulin stimulates the uptake of lipids and glucose. It does so by triggering the translocation of the glucose transporter GLUT4 and several kinds of the fatty acid transporters from intracellular parking locations to the plasma membrane. Insulin also inhibits lipolysis, the hydrolysis of lipids, the central step required for release of the lipids back into circulation, and stimulates *de-novo* free fatty acid synthesis. As was the case in other tissues, insulin signals are transduced into the cell by means of receptor tyrosine kinase activation leading to PI3K activity and the activation of Akt. The latter stimulates the activity of phosphodiesterase PDE3B, which antagonizes cAMP second messengers thereby repressing lipolysis.

2.4 Adipose Tissue Functions as an Endocrine Organ

Adipose tissue serves as an energy store, provides insulation and cushioning for the body, and presents an inflammatory milieu. Adipose tissue also functions as an endocrine organ, secreting hormones into the bloodstream. The obesity (ob) gene product leptin was identified in 1994 by Jeffrey Friedman and coworkers. The name leptin was coined from the Greek word "leptos" meaning thin. This hormone is secreted into the circulation by adipocytes. Leptin receptors were found shortly thereafter in 1995 in the hypothalamus, in the arcuate nucleus, and other nuclei. Leptin is widely regarded as a long sought for lipostatic factor – the amount of leptin secreted by white adipocytes into the circulation tracks the total amount of body fat present as well as the body mass index (BMI). The more fat tissue the greater the amount of leptin secreted into the bloodstream, and this is accomplished in a slow constitutive manner.

Leptin was just the first of a number of adipose tissue hormones to be found. Its discovery in 1994 was followed by that of adiponectin in 1995/1996. These and still other hormones discovered during the past few years are collectively referred to as adipokines, by analogy to the term cytokines. Some of the adipokines act on cells of the adipose tissue itself; others such as adiponectin binds to receptors in the liver or in skeletal muscle, and still others cross the blood-brain barrier to act on the central nervous system as does leptin. A short list of white adipose tissue hormones and their target organs and tissues is presented in Table 2.3.

Leptin acts not only as a hormone but also as a pleiotropic cytokine. The gastric mucosa of the stomach secretes leptin in an exocrine manner into the lumen of the stomach and in an endocrine manner into the circulation. Whereas adipocyte-secreted leptin acts over a long time scale to regulate energy homeostasis, stomach-associated leptin acts on much short time frame. The amount of stomach-secreted leptin goes up when eating and goes down afterwards. Secreted leptin working together with another peptide, cholecystokinin (CCK), secreted by the small intestine, regulates actions by the duodenum, the part of the small intestine into which the stomach first empties. These signaling peptides regulate stomach emptying and also help control meal size.

Table 2.3 Hormones secreted into the bloodstream by cells in white adipose tissue (WAT)

Hormone or signaling peptide	Sending cells	Receiving cells	Physiological function
Adiponectin	WAT	Liver, muscle	Glucose, insulin, and energy management
IL-6	WAT	Liver, muscle	Energy and insulin management
Leptin	WAT	Brain, muscle	Satiety signal (B)

2.5 Ghrelin Released by Endocrine Cells in the Stomach Acts in Short-Term Feeding and Long-Term Energy Management

Endocrine X/A cells in the stomach secrete ghrelin, a 28 amino acid peptide, into the circulation. This peptide ghrelin has two distinct actions. It binds receptors expressed in the anterior pituitary to stimulate the release of growth hormone (GH). Second, it supplies "need-to-feed" or hunger, signals to the hypothalamus, and regulates energy balance by opposing the actions of leptin.

Growth hormones are secreted by somatotroph cells in the anterior pituitary. Their release is regulated by two hypothalamic peptides – growth hormone releasing hormone (GHRH) and growth hormone inhibiting hormone (GHIH), or somatostatin. The release of GHs can also be stimulated by artificial compounds called growth hormone secretagogues (GHSs). Ghrelin is the natural signaling peptide for these receptors. The existence of receptors for ghrelin in both the anterior pituitary and arcuate nucleus of the hypothalamus provides a coordinating link between growth and feeding.

Ghrelin is produced predominately by the stomach. Its concentration in the circulation increases during fasting and decreases during refeeding. The opposite pattern is seen in the stomach where decreases occur during fasting and increases take place during feeding. This inverse response property is a consequence of the continual secretion of the peptide hormone from the stomach during fasting. Thus, ghrelin is a hunger signal sent by the stomach that acts on the arcuate nucleus (ARC).

Ghrelin supplies orexigenic signals to the same hypothalamic nuclei as leptin. It stimulates the synthesis and release of neuropeptide Y (NPY) and agouti-related protein (AgRP) in the arcuate nucleus thereby stimulating increased food intake. The ghrelin receptor is capable of signaling in the absence of ligand binding and does so about 50% of the time. The circuit mediating feeding behavior is depicted in Fig. 2.9 and will be discussed later in the chapter.

2.6 Satiation Signals Are Sent by Cells in the Gastrointestinal Tract

Several populations of neuroendocrine cells in the gastrointestinal (GI) tract secrete hormones and peptides that coordinate the actions of different parts of the digestive system (Table 2.4). These cells also transmit satiation signals that halt feeding. To accomplish this task, biophysical changes indicative of fullness such as pyloric pressure (the pylorus is the region connecting the stomach to the duodenum), stomach motility and muscle relaxation, and gastric filling are sensed. In response, the neuroendocrine cells send out "stop eating" messages that complement the "start eating" signals conveyed by ghrelin. Two key satiation hormones sent out by the neuroendocrine cells are cholecystokinin (CCK) and glucagon-like peptide 1 (GLP-1).

Table 2.4 Hormones/peptides secreted into the bloodstream by cells in the gastrointestinal tract

Hormone or signaling peptide	Sending cells	Receiving cells	Physiological function
Cholecystokinin (CCK)	I-cells of the duodenum, jujunum	Exocrine pancreas	Stimulates release of digestive enzymes
Ghrelin	Stomach (A-cells of the gastric fundus)	Brain	Hunger signal
GLP-1	L-cells of the distal small and large intestine	Endocrine pancreas	Regulates insulin and glucagon release, gastric emptying and secretion
$PYY_{3\text{-}36}$	L-cells of the distal small and large intestine	Brain, I, P	Satiety factor (B), inhibits gut motility (S) and pancreatic secretion (P)
Secretin	S-cells of the duodenum	Exocrine pancreas	Stimulates secretion of bicarbonate

Cholecystokinin is secreted by cells called I cells in response to the presence of fat or protein in the duodenum. These cells are able to sense the presence of nutrients and respond by secreting the hormones. The hormone binds to receptors in the vagal afferents triggering vagus nerve messages to the hindbrain. There are several different bioactive forms of this hormone. The most prevalent forms are designated as CCK8, CCK33, and CCK58. These hormones serve as short-term satiation signals from the upper part of the GI tract. Among their several regional roles is the control of gallbladder contraction; another is the inhibition of gastric emptying.

Glucagon-like peptide 1 functions in a similar role in the lower part of the GI tract (ileum). In response to the presence of carbohydrates and lipids, it adjusts stomach and intestinal motility, thereby contributing to an "ileal brake" to eating. Receptors for this peptide are expressed in the GI tract, pancreas, vagal afferents, brainstem, and hypothalamus. This peptide is also released by neurons in the brain as a feeding inhibitory peptide, in regions involved in energy balance and associated with leptin signaling.

2.7 Brown Adipose Tissue Carries Out Adaptive (Diet-Induced and Cold-Induced, Nonshivering) Thermogenesis

Brown adipocytes are unique to mammals. These cells utilize a protein-uncoupling protein 1 (UCP1) to redirect the electron transport chain to produce heat instead of ATP. Recall that the electron transport chain embedded in the inner mitochondrial membrane (IMM) couples electron flow to the pumping of protein across the membrane from the matrix to the intermembrane space. That is the energy

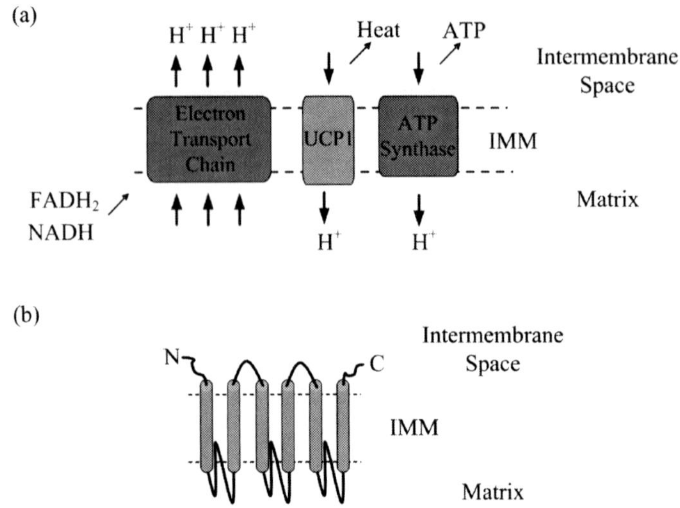

Fig. 2.2 Uncoupling proteins. (**a**) Organization and actions of the electron transport chain, uncoupling protein 1 (UCP1), and ATP synthase in the inner mitochondrial membrane (IMM). (**b**) UCP1 chain topology

derived from the flow of the electrons from Complex I to Complex IV is used to pump protons and establish a proton electrochemical gradient. As shown in Fig. 2.2, the UCP1 protein resides in the IMM and facilitates proton leak in the other direction from the intermembrane space back into the matrix. This step cuts off the last step in oxidative phosphorylation, in which reentry of the protons generates ATP. It bypasses the F0/F1-ATPase generation of ATP in Complex V and in its place produces heat.

Mitochondrial uncoupling proteins (UCPs) belong to a large family of mitochondrial transporters encoded by SLC25 genes. These transport proteins shuttle metabolites across the inner mitochondrial membrane. They form dimers; each chain passes back and forth through the IMM six times with both N and C terminals in the intermembrane space as shown in Fig. 2.2. The founding member of the UCP family, UCP1, is expressed in brown adipose tissue. Other family members, UCP2–UCP5, are broadly distributed and will be discussed further in later chapters.

The signaling route that mediates adaptive thermogenesis is, as follows. Feeding triggers signaling to the CNS, which then responds by sending return signals via the sympathetic nervous system to peripheral tissues such as BAT. These nerves emit norepinephrine, which then binds β-adrenergic receptors on target tissues. In BAT the predominant type of receptor is the β_3-adrenergic receptor. Receptor binding leads to cAMP and PKA activation as it did in WAT. Two sets of changes result from these events in BAT. The first is that CREB gene expression is enhanced leading to expression of UCP1 and increased mitochondrial biogenesis. The second is PKA activation of HSL and perilipin leads to FFA release from lipid droplets. The FFAs then bind and stimulate UCP1 activity. In addition, the FFAs are metabolized in the mitochondria via β-oxidation and the citric acid cycle to produce reducing agents NADH and $FADH_2$ thereby increasing the rate of oxidative phosphorylation leading through UCP1 activity to enhance heat generation.

2.8 Muscle Cells and β-Oxidation

β-Oxidation of free fatty acids is a major source of energy for the body. It is especially important for skeletal muscle, cardiac muscle, and liver. In liver and heart, it provides some 80% of the total energy production. In liver, it is used to produce ketone bodies, that is, acetone, 3-hydroxybutyric acid (β-hydroxybutyrate) and acetoacetate, all of which are used as extra energy sources by tissues such as the brain.

The β-oxidation process begins with the activation of the FFA through formation of a bond between the carbonyl group of the FFA and the sulfhydryl group of Co-enzyme A (CoA) and leads to formation of acyl-CoA (Fig. 2.3).

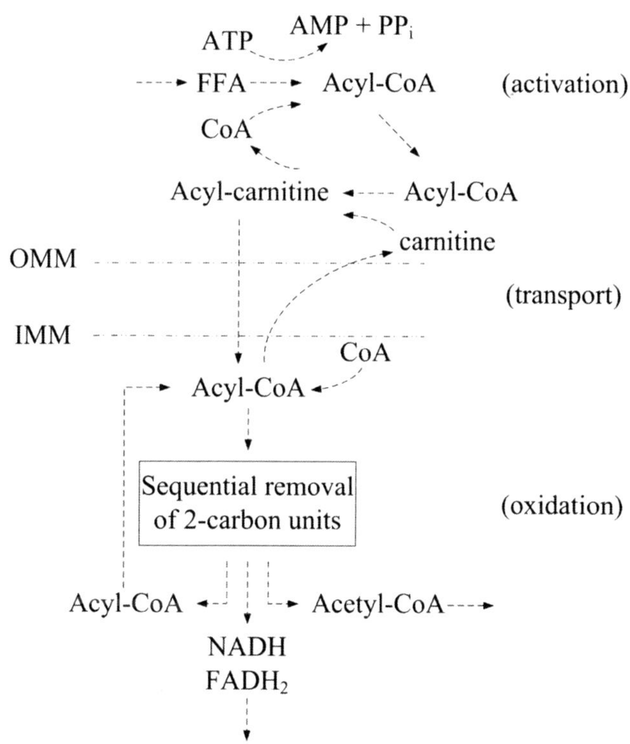

Fig. 2.3 β-Oxidation of free fatty acids illustrating its three stages of activation, transport, and oxidation

This step takes place on the outer mitochondrial membrane (OMM) and is catalyzed by acetyl-coenzyme A synthase (ACS). In the next preparatory step, the acyl-CoA molecules are transported across the OMM and IMM into the mitochondrial matrix by carnitine carrier molecules. This operation is facilitated by carnitine palmitoyltransferase 1 and 2 (CPT1 and 2), the former catalyzing carnitine attachment at the cytosol-OMM and the latter carnitine detachment at the IMM-matrix.

Once inside the matrix, the main steps in β-oxidation begin. In β-oxidation, pairs of carbon atoms are removed from the end of fatty acyl-CoA molecules in a series of enzymatic reactions using NAD^+ and FAD as coenzymes. These generate from each fat molecule one NADH, one $FADH_2$, and an acetyl-CoA. The process takes the name β-oxidation from the observation that the cuts to the acyl-CoA are made at the β-carbon position. The acetyl-CoA molecules output from β-oxidation are used in the Krebs cycle and the electron-rich NADH and $FADH_2$s are used in both the Krebs cycle and in the electron transport chain of oxidative phosphorylation.

The rate-limiting step in β-oxidation is the transport step catalyzed by CPT1. As will be discussed shortly, this step is regulated by intracellular signaling molecules such as AMPK and hormones such as leptin and ghrelin in response to energy needs and nutrient supplies. The output from β-oxidation, acetyl-CoA, is centrally involved in a variety of metabolic processes. It is used by the liver as a precursor in biosynthesis of ketone bodies, and as indicated in Fig. 2.4 can be generated by means of glycolysis as well as through β-oxidation. Acetyl-CoA is also used as the starting point in feedback regulation of β-oxidation. As illustrated in Fig. 2.4, the negative (endpoint) feedback loop terminates at CPT1 – acetyl-CoA is converted to malonyl-CoA, which then inhibits CPT1.

Fatty acid oxidation is regulated in skeletal muscle in response to hormonal input and stress brought on by activities such as exercise. The signaling pathway that responds to these activities is the AMPK pathway that monitors energy status and is activated when increases in energy supplies are needed. As shown in Fig. 2.5, low-energy supplies plus signals from an upstream kinase such as LKB1 activate AMPK. This kinase, in turn, phosphorylates ACC. There are

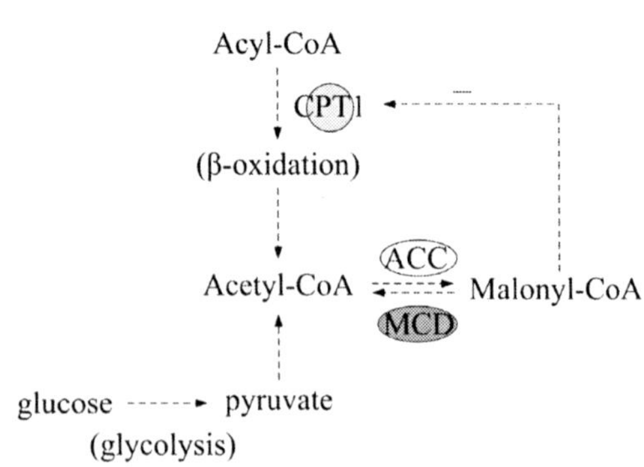

Fig. 2.4 Feedback regulation of β-oxidation. Abbreviations: ACC, acetyl-coenzyme A carboxylase; MCD, malonyl-coenzyme A decarboxylase

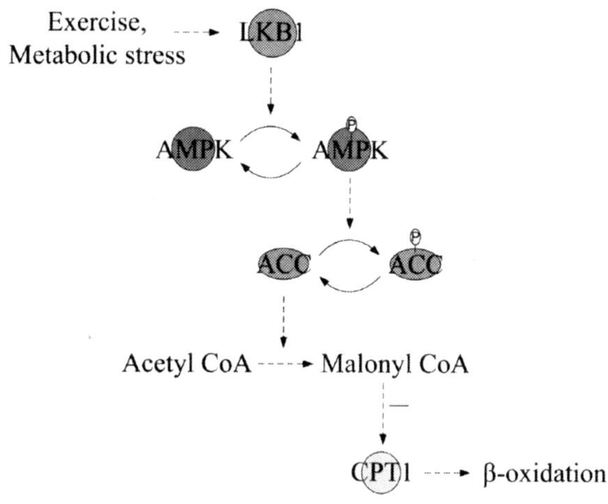

Fig. 2.5 Regulation of beta oxidation by the AMPK, metabolic stress-activated pathway

two ACC isoforms, ACC1 and ACC2. The latter is involved in regulating fatty acid oxidation in skeletal muscle. In the absence of AMPK phorphorylation, unphosphorylated ACC2 stimulates the conversion of acetyl CoA to malonyl CoA, which inhibits CPT1 thereby preventing β-oxidation. This blockage is relieved by AMPK-mediated phosphorylation of ACC2.

Fatty acid oxidation is regulated at the transcriptional level in addition to its regulation at the catalytic level by AMPK and through end-point feedback. Key regulators of fatty acid synthesis are members of the nuclear receptor family that will be examined in Chapter 4. Several members of this family are of particular interest. These include most notably the peroxisome proliferator activated receptors (PPARs) and the associated PPARγ coactivator 1α (PGC-1α) protein. The latter plays a central role in coordinating fatty acid metabolism and the Krebs (TCA) cycle. As will be seen in the next chapters, chronic overfeeding leading to excessive fatty acids and metabolic overload is believed to have a causative role in insulin resistance and Type 2 diabetes (T2D).

2.9 AMPK Is an Intracellular Energy Sensor and Regulator

AMP-activated protein kinase, or AMPK, is the key regulator of energy balance operating at the cellular level. Once AMPK is activated, it acts in several ways to bring energy supply and expenditure back into balance. It both stimulates energy-producing catabolic processes and throttles back energy-using anabolic ones. This is accomplished by phosphorylating key elements of the metabolic pathways and key regulators of the expression of genes involved in metabolism. Representative examples of energy balancing by AMPK in different tissues are presented in Fig. 2.6.

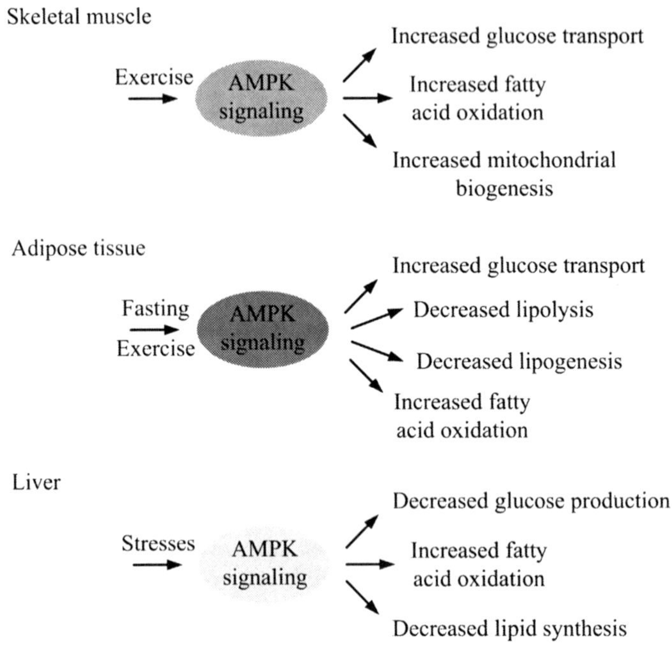

Fig. 2.6 Energy balancing by AMPK in skeletal muscle, adipose tissue, and liver

The first example shown in Fig. 2.6 is skeletal muscle, one of the principal sites of glucose and fatty acid utilization in the body. Stimulation of glucose uptake occurs through AMPK stimulation of the translocation of GLUT4 transporter to the plasma membrane. This occurs through AMPK-catalyzed phosphorylation of AS160, a Rab GTPase activating protein (GAP) involved in GLUT4-bearing vesicle transport (discussed further in Section 3.7). Insulin signaling through Akt also targets AS160, and both routes operate synergistically to promote glucose uptake.

Exercise rapidly depletes the available ATP, and AMPK signal transduction helps maintain ATP supplies in several ways one of which is the stimulation of fatty acid oxidation. This is accomplished through AMPK phosphorylation of the metabolic enzyme acetyl-CoA carboxylase (ACC). This enzyme serves as a key control point in fatty acid oxidation; it catalyzes the conversion of acetyl-CoA to malonyl-CoA, which in turn represses CPT1 activity. This enzyme is the rate-limiting enzyme for transport of long-chain fatty acids into the mitochondria where they undergo β-oxidation as discussed in the previous section. AMPK stimulates β-oxidation by phosphorylating and inhibiting ACC's ability to catalyze the conversion of acetyl-CoA to malonyl-CoA. As indicated in Fig. 2.6, increased fatty acid oxidation is one of the ways that AMPK restores energy balance in other peripheral tissues, not just skeletal muscle.

Another key operation carried out by AMPK is to inhibit protein synthesis, an intensive energy-consuming process. This is accomplished by AMPK phosphorylation of components of the tuberous sclerosis complex (TSC). This complex functions as a key signaling node involved in the signaling pathways regulating cellular growth and metabolism. By phosphorylating TSC2, AMPK prevents activation of the target of rapamycin (TOR) protein, which activates key mediators of mitochondrial biogenesis and protein synthesis. Several signals impinge on AMPK to trigger its regulation of protein synthesis. Leptin activates AMPK as does ischemia, hypoxia, and other stress conditions. In addition to these AMPK relayed signals, the TSC and TOR signaling nodes receive signals from the insulin signaling pathway. The TSC, TOR signaling node, and other key elements of this pathway will be discussed in detail as Chapter 3 where signaling from insulin and its relationship to Type 2 diabetes will be examined.

The liver has a major role in the homeostatic regulation of whole body metabolism and energy balance. As was the case for skeletal muscle, it regulates fatty acid oxidation in adipose tissue and in the liver. Overall, it integrates hormonal and nutrient signals by switching on energy-generating catabolic pathways and turning off ATP-consuming ones. AMPK does so by phosphorylating regulatory proteins and by turning on gene expression.

Metformin is the leading drug in use today against noninsulin dependent diabetes mellitus (NIDDM). Its use is especially recommended for people suffering from obesity and insulin resistance. This drug acts on the liver to reduce sugar levels in the blood and acts on peripheral tissues to increase their sensitivity to insulin. This drug exerts its protective actions in the liver through the AMPK pathway.

2.10 AMPK Is Activated by Upstream Kinases and by Depleted Energy Supply as Indicated by Increased AMP/ATP Ratios

AMP-activated protein kinases, or AMPKs, are serine/theorine kinases. They possess three subunits, a catalytic α-subunit, and a pair of regulatory subunits designated as β and γ. As depicted in Fig. 2.7,the α-subunit contains an N-terminal kinase domain and a C-terminal $\beta\gamma$-subunit-binding domain. The N-terminal domain has a residue, Thr-172, situated in the activation loop, which is phosphorylated by an upstream kinase. Two protein kinases – LKB1 and CaMKK – are capable of phosphorylating AMPK on Thr-172. The CAMKK signaling route is prominent in neurons, while the LKB1 path seems to be the important route in nonneural cells.

The γ-subunit contains a pair of sites adjacent to one another in the C-terminal region that are bound by AMP in a manner that depends on the AMP/ATP ratio. High levels of ATP, indicating plentiful energy supplies, inhibit binding and activation of the AMPK by AMP. In contrast, when energy supplies are depleted,

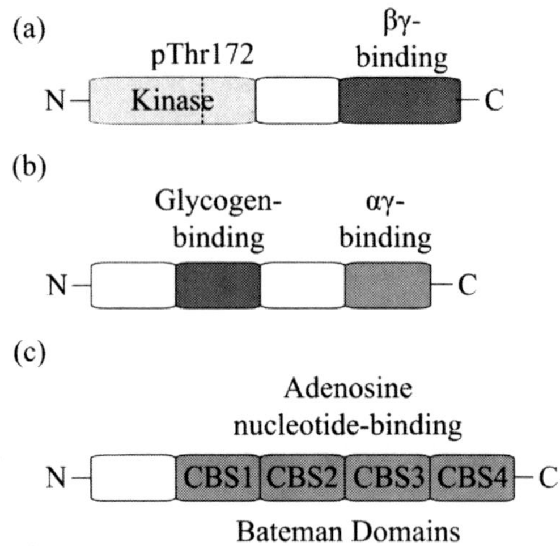

Fig. 2.7 AMPK catalytic and regulatory subunits. (**a**) α-subunit showing the arrangement of the N-terminal kinase domain and a C-terminal βγ-subunit binding domain. (**b**) β-subunit showing the presence of a glycogen-binding domain and an αγ-subunit binding domain. (**c**) γ-subunit. The N-terminal regions vary in length for the three different isoforms. These are followed by four repeats of a sequence referred to as CBS motif. The first two form an AMP/ATP-binding domain and the second pair another AMP/ATP-binding domain

resulting in a higher AMP/ATP ratio, AMP binding can occur. The selection of AMP/ATP rather than ADP/ATP renders the protein highly sensitive to small changes in available energy supply. AMP binding triggers a conformation change that enables the protein to be phosphorylated by upstream kinases such as LKB1. Once this happens the kinase is catalytically active.

The LKB1 protein forms a complex with two regulatory subunits – a kinase-like protein called STE-related adaptor (STRAD) and a scaffolding protein named mouse protein 25 (MO25). These subunits provide cytoplasmic anchorage for the kinase and serve to activate it. The tertiary complex is constitutively expressed but in the absence of AMP binding, dephosphorylation by resident protein phosphatases prevents activation of AMPK. Binding by AMP inhibits dephosphorylation by resident protein phosphatases. Eventually, dissociation of AMP from the γ-subunit leads to dephosphorylation on Thr172 within the catalytic subunit, returning AMPK to an inactive state.

The physical arrangement of the three subunits as determined through X-ray crystallography is depicted in Fig. 2.8. Shown in this figure are results obtained for *Saccharomyces pombe*. Similar findings have been reported for *S. cerevisiae* and human AMPK, but there are also differences in the details among the three AMPK complexes. Overall, one observes that the binding of AMP does not generate large and obvious conformational shift in the complex. Instead AMP binding to the γ-subunit activates the kinase though an allosteric mechanism

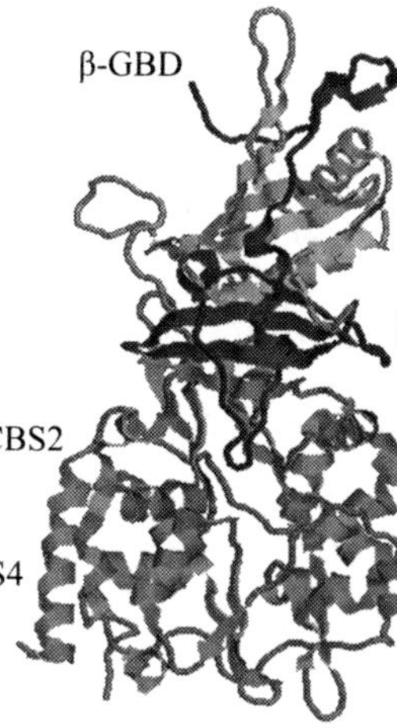

Fig. 2.8 Structure of the AMPK energy sensor determined by means of X-ray crystallography. The figure was prepared using Jmol with atomic coordinates deposited in the PDB under accession number 2OOX

involving inter-subunit interactions that increase the catalytic activities of the α-subunit and inhibit dephosphorylation. It is noteworthy that each of the three subunits is in contact with the other two.

The domain structure of the γ-subunit is quite revealing. The four repeats depicted schematically in Fig. 2.8 are labeled as cystathionine β-synthase motifs 1–4 (CBS1–4). These motifs are arranged as two mirror pairs facing one another. One pair, CBS1 and CBS3, jointly referred to as Bateman domain A faces the other pair, CBS2 and CBS4, termed Bateman domain B. These domains form a deep binding pocket that provides sites for AMP binding, $Mg^{2+}\cdot$ATP binding, and perhaps ADP binding as well. The BCS motifs are also a major site for disease-causing mutations. In these situations, nucleotide binding is disrupted and AMPK signaling is lost.

2.11 The Hypothalamic Network Provides Feedback Signals to Peripheral Tissues

The hypothalamus integrates energy and nutrient signals and based on these signals provides feedback to key peripheral tissues such as the liver. Hormones secreted by cells in pancreas, adipose tissue, stomach, and gastrointestinal tract impinge on neurons in the arcuate nucleus, which function in a sensory role relaying signals from the periphery to second and higher order neurons. Unlike most regions of the central nervous system, the blood-brain barrier of the hypothalamus is reduced so that key hormones and metabolites can pass through.

Insulin and leptin enter by means of saturable transporters; that is, by receptors that chaperone the signaling molecules across the blood-brain barrier. Ghrelin, the third member of the "big-three regulatory hormones" (insulin and leptin being the other two), is a 28 amino reside acid peptide with an unusual, acyl modification. It too passes through the BBB by means of saturable transport. That these three key signaling molecules pass through the BBB by means of saturable transport is significant – this type of mechanism can be regulated by the BBB and most importantly the amount of hormones that passes across the BBB at high concentrations is limited. In contrast, entry by means of passive diffusion is nonsaturable, that is, it is nonlimiting. Metabolites and other peptides pass through in one of several ways. Some enter by passive diffusion as in the case of free fatty acids while others are aided by transporters as in the case of ketone bodies. Many of the signal peptides secreted by the gut (e.g., PYY_{3-36}) rely on passive diffusion for entry into the hypothalamus.

Two populations of neurons residing in the arcuate nucleus of the hypothalamus are the targets of insulin, leptin, and ghrelin signals. These neurons alter their expression and secretion of neuropeptides either positively or negatively in response to binding by these hormones to their receptors. One population of neurons secretes neuropeptide Y (NPY) and agouti-related protein (AgRP), which stimulate feeding, while the second releases pro-opiomelanocortin (POMC) [α-MSH], which acts opposite to the others by inhibiting feeding behavior. The core feeding circuitry is pictured in Fig. 2.9 and operates in the following manner. When energy intake minus energy expenditure is positive as is the case just following feeding, insulin and leptin satiety signals impinge on the two populations of neurons. They reduce the firing and peptide release from feeding-stimulatory NPY cells and at the same time promote the firing and peptide release from stop-feeding POMC neurons.

Ghrelin hunger signals are received by NPY neurons. The NPY neurons respond in three ways to these signals. First, the NPY neurons inhibit signaling by the POMC neurons through release of the inhibitory neurotransmitter

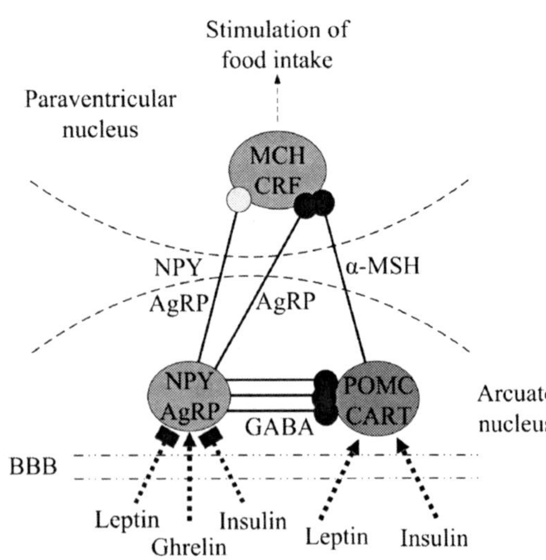

Fig. 2.9 Central hypothalamic regulatory network. (See text for details.)

γ-aminobutyric acid (GABA) at synapses between NPY and POMC. Second, ghrelin receptor activity stimulates release of NPY and AgPR peptides at second-order, MCH/CRF neurons in the paraventricular nucleus (PVN). Third, the NPY neurons send out processes that target the same postsynaptic sites that are used by the POMC neurons and inhibit the effect of α-MSH through release of AgRP. The PVN neurons receive greater input from the NPY stimulatory neurons and reduced input from inhibitory POMC neurons. The PVN neurons, in turn, send out signals that stimulate food intake.

2.12 Leptin Signaling and Regulation of Energy Balance in the Hypothalamus

The 3D structure of leptin as determined by means of X-ray crystallography is shown in Fig. 2.10. As can be seen in this figure, the protein forms four-helix bundle (HA–HD) with a fifth, short helix (HE) serving as a hydrophobic cap. The four-helix bundle is an unusual one. The helices are connected to one another by means of two long loops (L_{AB} and L_{CD}) and one short loop (L_{BC}). These loops give rise to an up-up-down-down topology for the four helices.

This tertiary organization is structurally similar to Class I hematopoietin cytokines, in particular to long-chain helical cytokines such as the pro-inflammatory cytokine IL-6 and IL-11 that signal via receptors sharing a gp130 subunit. Cytokines of this type are pleiotropic in their functions and leptin is no exception to this trend. Leptin receptors have been identified in a variety of central and peripheral tissues. In support of these varied roles, six alternatively spliced forms can be generated. These are designated as Ob-Ra through Ob-Rf. These receptors can be placed into three groups – short, long, and secreted. Short and long receptors have the

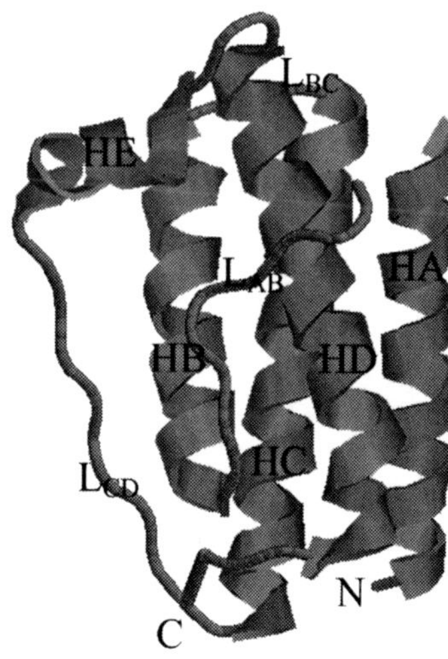

Fig. 2.10 Structure of leptin determined by means of X-ray crystallography. The figure was prepared using Jmol with atomic coordinates deposited in the PDB under accession number 1AX8

same extracellular and transmembrane domains but differ in the length of their cytoplasmic regions. Short forms such as Ob-Ra possess cytoplasmic regions of 30–40 residues while long forms such as Ob-Rb possess 300 residue-long cytoplasmic regions forming a more substantial signaling platform.

Binding of leptin to Ob-Rb, a long form of the leptin receptor, initiates leptin signal transduction leading to alterations in the mix of neuropeptides being released. As is customary for cytokine receptors, the Ob-Rb receptor signals through a Jak-STAT pathway. The Jaks are a family of nonreceptor tyrosine kinases. Members of this family, namely Jak2s, are recruited by the activated receptors, and phosphorylate the receptors at several cytoplasmic sites. These locales function as docking sites for the signal transducer and activator of transcription (STAT) proteins. The STATs, in particular, STAT3s, undergo phosphorylation by the Jaks, dimerize, and translocate to the nucleus where they function as transcription factors (Fig. 2.11). One of the genes upregulated by the STAT3s encodes SOCS3, a negative regulator of leptin signaling that operates through a feedback loop to block tyrosine phosphorylation by the Jak2s and terminate leptin signaling. The appetite-suppressing neuropeptide POMC is upregulated and the appetite-stimulating neuropeptide AgRP is downregulated, both actions supporting leptin's role as a satiety factor.

Leptin signaling is not restricted to Jak-STAT pathway alone but rather operates through multiple pathways. One of the other pathways utilized by leptin is the PI3K pathway. A sketch of this pathway has been included in Fig. 2.11. As indicated in the figure, signals proceed from PI3K to phosphodiesterase 3B (PDE3B) resulting in reduced signaling by cAMP and a corresponding reduction in expression and secretion of the stimulatory neuropeptide NPY. This leptin pathway shares a number of components with the central

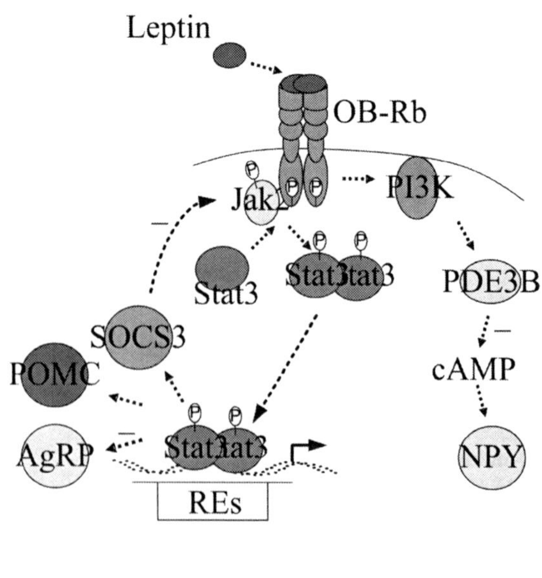

Fig. 2.11 Leptin signal transduction through the Jak-STAT system in the hypothalamus

signaling pathway used by insulin. Because of its importance not only in diabetes but also in cancer, the insulin pathway will be dissected in considerable detail in the next chapter.

2.13 Ghrelin Signaling and Regulation of Energy Balance in the Hypothalamus

Ghrelin modifies the firing rate of NPY neurons, acting via AMPK and β-oxidation-associated metabolites. This regulatory activity occurs in the following manner. Receptor activation in response to ghrelin binding leads to phosphorylation and activation of AMPK. In the next step, AMPK phosphorylates and deactivates acetyl-coenzyme A carboxylase ACC. This enzyme, as shown in Fig. 2.12, catalyzes the formation of malonyl-CoA from acetyl-CoA. Malonyl-CoA is a negative regulator of CPT1 and by deactivating ACC malonyl-CoA is not generated from acetyl-CoA. As a result CPT1 is free to transport acyl-CoA molecules into the mitochondrial matrix where it undergoes β-oxidation. The resulting energy increase supports an increased neural firing rate.

Ghrelin binds growth hormone secretagogue receptors (GHSRs) on NPY neurons in the arcuate nucleus of the hypothalamus. This receptor is a member of a small group of seven-pass receptors that exhibit a high level of constitutive activity. In particular, the GHSR exhibits an activity level in the absence of ligand binding of about 50%. When active it, like other G-protein-coupled receptors, signals via G-proteins. In its inactive form, the G-proteins consisting of a Gα and Gβγ subunits are bound and inactive. Receptor activity leads to dissociation and signaling. Recall that there are four families of Gα subunits. The GHSR utilizes the Gαq family to activate phospholipase C (PLC), which generates the second messengers diacylglycerol (DAG) and inositol 1,4,5-triphosphate (IP_3) from phosphatidylinositol 4,5 biphosphate (PIP_2). In this case, IP_3 stimulates the release of Ca^{2+} from the intracellular stores resulting in activation of calcium-dependent serine/threonine kinases such as CaMKIV.

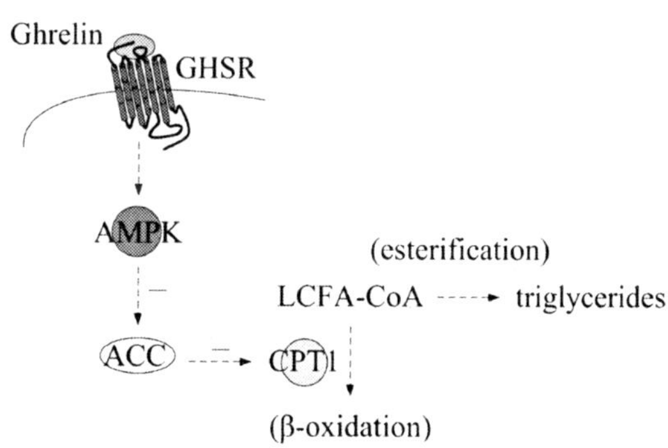

Fig. 2.12 Ghrelin signaling via AMPK to modify the firing of NPY neurons

Fig. 2.13 Signaling and gene expression induced by Ghrelin

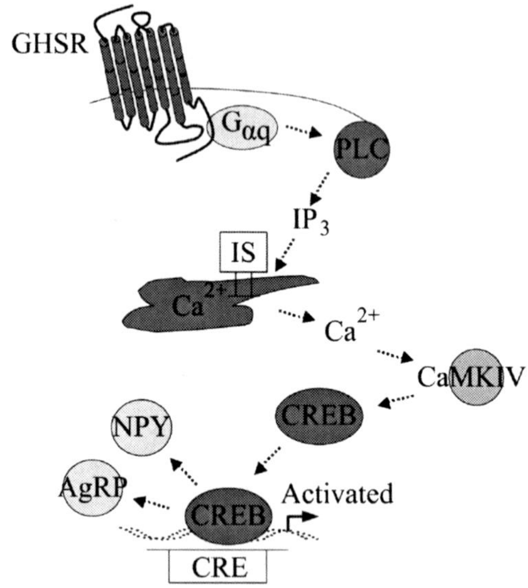

This kinase phosphorylates and activates CRE-responsive binding protein (CREB) transcription factor leading to upregulation and release of the neuropeptides NPY and AgRP (Fig. 2.13).

Further Reading

Benoit SC et al. 2004 Insulin and leptin as adiposity signals. *Recent Prog. Horm. Res.* 59: 267–285.

Flier JS 2004 Obesity wars: molecular progress confronts an expanding epidemic. *Cell* 116: 337–350.

Kennedy GC 1953 The role of depot fat in the hypothalamic control of food intake in the rat. *Proc. R. Soc. Lond.*140: 578–592.

Wiener N 1948 *Cybernetics: Or Control and Communication in the Animal and the Machine.* Cambridge, MA: MIT Press.

Zigman JM and Elmquist JK 2003 Minireview: from anorexia to obesity – the Yin and Yang of body weight control. *Endocrinology* 144: 3749–3756.

Hormonal Signaling by the Endocrine Pancreas

Cabrera O et al. 2006 The unique cytoarchitecture of human pancreatic islets has implications for islet cell function. *Proc. Natl. Acad. Sci. USA* 102: 2334–2339.

Lang J 1999 Molecular mechanisms and regulation of insulin exocytosis as a paradigm of endocrine secretion. *Eur. J. Biochem.* 259: 3–17.

MacDonald PE, Joseph JW and Rorsman P 2005 Glucose-sensing mechanisms in pancreatic β-cells. *Phil. Trans. R. Soc. B* 360: 2211–2225.

Maechler P and Wollheim CB 2001 Mitochondrial function in normal and diabetic β-cells. *Nature* 414: 807–812.

Saltiel AR and Kahn CR 2001 Insulin signaling and the regulation of glucose and lipid metabolism. *Nature* 414: 799–808.

Liver Maintenance of Glucose and Lipid Homeostasis

Dentin R et al. 2007 Insulin modulates gluconeogenesis by inhibition of the coactivator TORC2. *Nature* 447: 366–369.
Jiang G and Zhang BB 2003 Glucagon and regulation of glucose metabolism. *Am. J. Endocrinol. Metab.* 284: E671–E678.

White Adipose Tissue (WAT)

Frayn KN et al. 2003 Integrative physiology of human adipose tissue. *Int. J. Obesity* 27: 875–888.
Friedman JM and Halaas JL 1998 Leptin and the regulation of body weight in mammals. *Nature* 395: 763–770.
Frühbeck G et al. 2001 The adiposite: a model for integration of endocrine and metabolic signaling in energy metabolism regulation. *Am. J. Physiol. Endocrinol. Metab.* 280: E827–E847.
Rajala MW and Scherer PE 2003 Minireview: the adiposite – at the crossroads of energy homeostasis, inflammation, and atherosclerosis. *Endocrinology* 144: 3765–3773.
Zhang Y et al. 1994 Positional cloning of the mouse *obese* gene and its human homologue. *Nature* 372: 425–432.

Brown Adipose Tissue (BAT)

Cannon B and Nedergaard J 2004 Brown adipose tissue: function and physiological significance. *Physiol. Rev.* 84: 277–359.

Stomach and Gastrointestinal Tract

Badman MK and Flier JS 2005 The gut and energy balance: visceral allies in the obesity wars. *Science* 307: 1909–1914.
Chen HY et al. 2004 Orexigenic action of peripheral ghrelin is mediated by neuropeptide Y and agouti-related protein. *Endocrinology* 145: 2607–2612.
Kojima M and Kangawa K 2005 Ghrelin: structure and function. *Physiol. Rev.* 85: 495–522.
Murphy KG and Bloom 2004 Gut hormones in the control of appetite. *Exp. Physiol.* 89: 507–516.
Van der Lely AJ et al. 2004 Biological, physiological, pathophysiological, and pharmacological aspects of ghrelin. *Endocrine Rev.* 25: 426–457.
Zhang JV et al. 2005 Obestatin, a peptide encoded by the Ghrelin gene, opposes Ghrelin's affects on food intake. *Science* 310: 996–999.

Muscle Cells and β-Oxidation

Bartlett K and Eaton S 2004 Mitochondrial β-oxidation. *Eur. J. Biochem.* 271: 462–469.
Eaton S, Bartlett K and Pourfarzam M 1996 Mammalian mitochondrial β-oxidation. *Biochem. J.* 320: 345–357.

Jager S et al. 2007 AMP-activated protein kinase (AMPK) action in skeletal muscle via direct phosphorylation of PGC-1α. *Proc. Natl. Acad. Sci. USA* 104: 12017–12022.
Muoio DM and Koves TR 2007 Skeletal muscle adaptation to fatty acids depends on coordinated actions of the PPARs and PGC1α: implications for metabolic disease. *Appl. Physiol. Nutr. Metab.* 32: 874–883.
Saks V et al. 2006 Cardiac system bioenergetics: metabolic basis for the Frank-Starling law. *J. Physiol.* 571: 253–273.
Stanley WC, Recchia FA and Lopaschuk GD 2005 Myocardial substrate metabolism in the normal and failing heart. *Physiol. Rev.* 85: 1093–1129.

AMPK as an Energy Sensor and AMP Signal Transduction

Amodeo GA, Rudolph MJ and Tong L 2007 Crystal structure of the heterotrimer core of *Saccharomyces cerevisiae* AMPK homologue SNF1. *Nature* 447: 492–495.
Hardie DG, Hawley SA and Scott JW 2006 AMP-activated protein kinase – development of the energy sensor concept. *J. Physiol.* 574: 7–15.
Long YC and Zierath JR 2006 AMP-activated protein kinase signaling in metabolic regulation. *J. Clin. Invest.* 116: 1776–1783.
Scott JW et al. 2004 CBS domains form energy-sensing modules whose binding of adenosine ligands is disrupted by disease mutations. *J. Clin. Invest.* 113: 274–284.
Townley R and Shapiro L 2007 Crystal structure of the adenylate sensor from fission yeast AMP-activated protein kinase. *Science* 315: 1726–1729.
Xiao B et al. 2007 Structural basis for AMP binding to mammalian AMP-activated protein kinase. *Nature* 447: 496–500.

AMPK in the Liver

Shaw RJ et al. 2005 The kinase LBK1 mediates glucose homeostasis in the liver and therapeutic effects of metformin. *Science* 310: 1642–1646.
Viollet B et al. 2006 Activation of the AMP-activated protein kinase in the liver; a new strategy for the management of metabolic hepatic disorders. *J. Physiol.* 574: 41–53.
Zhou G et al. 2001 Role of AMP-activated protein kinase in mechanism of metformin action. *J. Clin. Invest.* 108: 1167–1174.

Hypothalamic Regulation of Metabolism

Bouret SG et al. 2004 Trophic action of leptin on hypothalamic neurons that regulate feeding. *Science* 304: 108–110.
Elmquist JK and Flier JS 2004 The fat-brain axis enters a new dimension. *Science* 304: 63–64.
Horvath TL and Diano S 2004 The floating blueprint of hypothalamic feeding circuits. *Nat. Rev. Neurosci.* 5: 662–667.
Schwartz MW and Porte D Jr. 2005 Diabetes, obesity, and the brain. *Science* 307: 375–379.
Schwartz MW et al. 2000 Central nervous system control of food intake. *Nature* 404: 661–671.
Pinto S et al. 2004 Rapid rewiring of arcuate nucleus feeding circuits by leptin. *Science* 304: 110–115.

Neuronal Glucose Sensing

Marty N, Dallaporta M, Foretz M, Emery M, Tarussio D, Bady I, Binnert C, Beermann F and Thorens B 2005 Regulation of glucagon secretion by glucose transporter type 2 (glut2) and astrocyte-dependent glucose sensors. *J. Clin. Invest.* 115: 3545–3553.

Free Fatty Acid Sensing

Lam TKT, Schwartz GJ and Rossetti L 2005 Hypothalamic sensing of fatty acids. *Nat. Neurosci.* 8: 579–584.

Leptin Hypothalamic Network Plasticity

Aurora R and Rose GD 1998 Helix capping. *Protein Sci.* 7: 21–38.

Bouret SG, Draper SJ and Simerly RB 2004 Trophic action of leptin on hypothalamic neurons that regulate feeding. *Science* 304: 108–110.

Frühbeck G 2006 Intracellular signaling pathways activated by leptin. *Biochem. J.* 393: 7–20.

Kokoeva MV, Yin Hl and Flier JS 2005 Neurogenesis in the hypothalamus of adult mice: potential role in energy balance. *Science* 310: 679–683.

Pinto S et al. 2004 Rapid rewiring of the arcuate nucleus feeding circuits by leptin. *Science* 304: 110–114.

Tartaglia LA 1997 The leptin receptor. *J. Biol. Chem.* 272: 6093–6096.

Zhang F et al. 1997 Crystal structure of the *obese* protein leptin-E100. *Nature* 387: 206–209.

Ghrelin Signal Transduction

Banks WA et al. 2002 Extent and direction of Ghrelin transport across the blood-brain barrier is determined by its unique primary structure. *J. Pharmacol. Exp. Ther.* 302: 822–827.

Chen HY et al. 2004 Orexigenic action of peripheral ghrelin is mediated by neuroprptide Y and agouti-related protein. *Endocinology* 145: 2607–2612.

Crowley MA et al. 2003 The distribution and mechanism of action of ghrelin in the CNS demonstrates a novel hypothalamic circuit regulating energy homeostasis. *Neuron* 37: 649–661.

Holst B and Schwartz TW 2004 Constitutive ghrelin receptor activity as a signaling set-point in appetite regulation. *Trends Pharmacol. Sci.* 25: 113–117.

Holst B et al. 2003 High constitutive signaling of the ghrelin receptor – identification of a potent inverse agonist. *Mol. Endocrinol.* 17: 2201–2210.

Kojima M et al. 1999 Ghrelin is a growth-hormone-releasing aceylated peptide from stomach. *Nature* 402: 656–660.

Pantel J et al. 2006 Loss of constitutive activity of the growth hormone secretagogue receptor in familial short stature. *J. Clin. Invest.* 116: 760–768.

Chapter 3
Insulin Signaling and Type 2 Diabetes

Diabetes mellitus, or diabetes, is caused by dysregulated metabolism resulting in insufficient glucose entry into tissues leading to excessively high levels of blood glucose (hyperglycemia). The dysregulation arises from a failure of insulin-secreting cells in the pancreas to secrete sufficient amounts of insulin and from a failure of target tissues to properly respond to the insulin signals directing them to take up the glucose.

There are two main forms of diabetes – Type 1 diabetes (T1D) also known as insulin-dependent diabetes mellitus (IDDM) and Type 2 diabetes (T2D) or noninsulin-dependent diabetes mellitus (NIDDM). The cause of T1D is destruction of insulin-producing pancreatic β-cells by the immune system's T cells. That is, T1D is an autoimmune disease in which there is a failure to produce enough insulin. Type 2 diabetes arises through a failure of insulin to elicit a large-enough response in the target tissues. That is, T2D is brought on by insulin resistance on the part of the tissues that need to import glucose. In the early stages of T2D, the pancreatic β-cells are able to compensate for the inadequate response to insulin but eventually can no longer do so resulting in later stages of the disease. If not treated insulin resistance and β-cell failure can lead to cardiovascular disease, renal failure, retinal damage/blindness, and poor wound healing/gangrene.

That there are two distinct sets of causes and forms of diabetes was first recognized by Harold Himsworth. In 1936, and later in his Goulstonian Lectures of 1939, he not only made the distinction, but also recognized that the latter was due to insulin resistance, and further noted associations between diabetes, obesity, hypertension, and atherosclerosis. The term "syndrome X" was coined in 1988 by Gerald Reaven in which there was an emphasis on insulin resistance and compensatory hyperinsulemia and the path from these to cardiovascular disease. This event was followed a few years later by the introduction of the phrase "metabolic syndrome" to describe the constellation of factors of disordered energy balance listed in the Introduction, Chapter 1. This expanded list of disease factors is as follows:

- Abdominal obesity
- Atherogenic dyslipidemia

M. Beckerman, *Cellular Signaling in Health and Disease*,
Biological and Medical Physics, Biomedical Engineering,
DOI 10.1007/978-0-387-98173-4_3,

- Raised blood pressure
- Insulin resistance and glucose intolerance
- Proinflammatory state
- Prothrombotic state

In this chapter, the insulin signaling pathway will be examined in detail along with those events that can lead to insulin resistance. The insulin signaling pathway is a central one in regulating metabolism and growth, and will be encountered again in Part II of this text dealing with cancer. Insulin signaling in healthy and diabetic states will be studied in the first part of this chapter; this will be followed with an examination of β cell failure and the increasing loss of β cell mass as it develops in response to insulin resistance.

3.1 Type 2 Diabetes Develops in a Series of Stages from Overnutrition

Diabetes and insulin has received an enormous amount of attention from the time of insulin's discovery in 1921 to the present time. Insulin was the first protein for which a primary amino acid sequence was determined, and the use of insulin to treat Type 1 diabetes was a momentous medical achievement. The exact sequence of events leading to Type 2 diabetes, and the relative importance of the different mechanisms implicated in the disease is a difficult and ongoing challenge. In spite of the challenges remaining in understanding how a persistent energy imbalance can lead to Type 2 diabetes, and how insulin signaling impacts these events, considerable progress has been made.

A sketch of the key steps in the disease progression based on this understanding is presented in Fig. 3.1. In this depiction, overnutrition leads to two sets of consequences. One of these is the release of inflammatory factors brought on by excessive fat storage within adipose tissue. The second consequence is metabolic overload, a term that encompasses the effects of excessive fatty acids on the other peripheral tissues. These two sets of alterations converge on the insulin-signaling network responsible for managing glucose throughout the body. Leptin signaling, which complements the insulin-signaling network, is negatively impacted as well, worsening the situation as does mutations to key elements of the insulin- and leptin-signaling pathways. These factors all contribute to the development of insulin resistance, that is, the inability of insulin signals sent from pancreatic β-cells to elicit appropriate glucose uptake responses in other peripheral tissues.

To compensate for the growing ineffectiveness of the insulin signals, the pancreatic β-cells increase their secretion of insulin. These increases do not abate, with a return to normal levels. Instead, over time, the persistent increases set off a spiral cascade of deleterious events leading to the death of β-cells and the

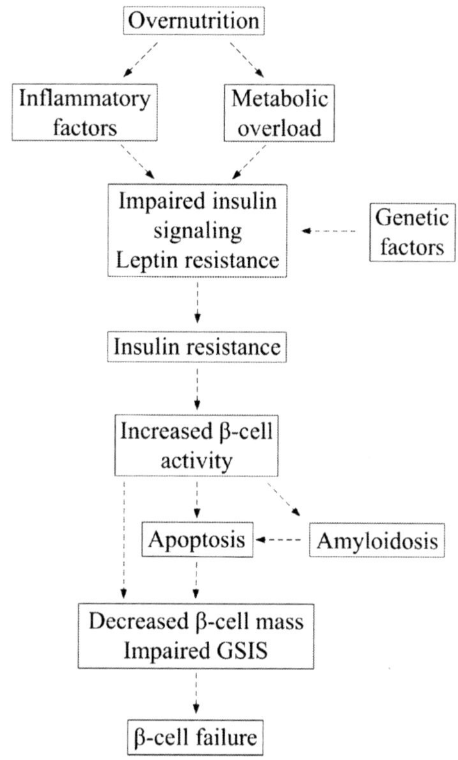

Fig. 3.1 Development of type 2 diabetes as a consequence of overnutrition

progressive inability of the islets to produce an adequate amount of insulin in response to glucose stimulation. Insulin secretion is accompanied by secretion of the polypeptide amylin (also known as islet amyloid polypeptide, or IAPP). Amyloid deposits of this peptide contribute to the progressive deterioration of the Islets. β-Cells die off at an increasing rate, and full-blown Type 2 diabetes ensues.

3.2 Adipose Tissue Functions as an Immune Organ

The microenvironment surrounding adipocytes, referred to as the stromal vascular fraction (SVF), contains several types of cells. The SVF contains endothelial cell, smooth muscle cells, and blood cell along with preadipocytes and a fibroblast-like population of multipotent stem cells. As the adipose tissue increases in mass, macrophages infiltrate into the tissue and become a prominent member of the community of cells. As noted in the last chapter, the size of the macrophage population reflects the amount of fat present in the tissue.

The cells of the adipose tissue microenvironment send out a variety of signaling molecules to one another. These signaling molecules include factors that promote inflammation, vascularization, and proliferation. Among these are pro-inflammatory cytokines such as TNF-α and IL-6 along with the developmental factor TGF-β, nitric oxide, and a variety of polypeptide growth factors and chemoattractants. Many of these factors are secreted by both adipocytes and macrophages. In atherosclerotic lesions, to be discussed in detail in Chapter 6, macrophages take up lipids and become foam cells; there are parallels in lipid storage function and inflammatory signaling between adipocytes and macrophages.

The pro-inflammatory signaling events in the adipose tissue generate the same type of alternations in the body's chemistry as produced by an infection. Ordinarily, the inflammatory responses are short lived. Viruses and bacteria are eliminated, the damage is repaired, and the inflammatory response is ended. This timely cessation of inflammatory activities does not occur when the cause is excessive body mass. Instead, the body experiences chronic low-level inflammation. Worse, inflammation and metabolic overload influence one another creating positive feedback linkages that enhance their negative effects.

As the adipose tissue acquires and stores excess lipids, a point may be reached where the lipid overload can no longer be accommodated. Under these conditions, lipids are stored elsewhere in the body including and especially in liver and muscle. In liver, this buildup has strong parallels with adipose tissue. The liver has its own population of macrophages. These are called Kupffer cells. One of the results of lipid overload in the liver is the generation of a robust inflammatory response in which, like adipose tissue, TNFα, IL6, and other inflammatory mediators are released, and the NF-κB pathway is activated. These effects are pronounced in, for example, nonalcoholic fatty liver diseases.

3.3 Metabolic Overload Occurs in Energy-Responsive Tissues

Insulin signals regulate the uptake, storage, and burning of glucose in muscle and other tissues. Figure 3.2 recapitulates the key points made in the last chapter regarding insulin-promoted glucose metabolism. Recall from Chapter 2 that glucose is stored in liver and muscle as glycogen, long-branched chains of glucose molecules strung together that serve as an energy reserve. The synthesis of glycogen polymers is triggered by insulin acting through protein kinase B. The crucial steps in this storage process were outlined in Fig. 2.1 and are recapitulated in more detail in Fig. 3.2 (upper panel). The starting point for glycogen synthesis is glucose-1-phosphate (glucose-1-P). An energy source is required and this is provided by uridine triphosphate (UDP). The activated form of glucose-1-P is denoted as UDP-glucose. The addition of UDP is catalyzed by the enzyme UDP glucose pyrophosphorylase. The key control point in glycogen storage is evoked at the next step, the addition of glucose

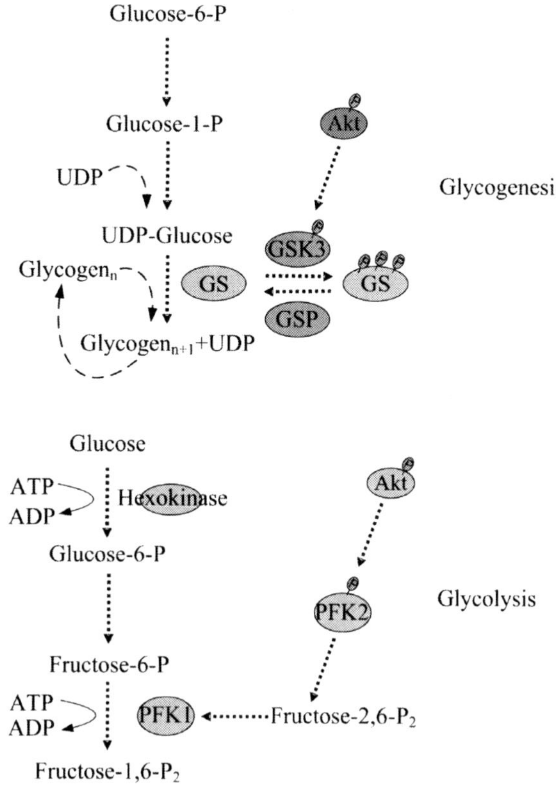

Fig. 3.2 *Upper panel*: regulation of synthesis of glycogen from glucose (Glycogen synthesis, or glycogenesis) by insulin/Akt, and *lower panel*: regulation of glycolysis by insulin/Akt

molecules to the growing glycogen chain. This operation is catalyzed by glycogen synthase (GS), which in turn is regulated through phosphorylation by glycogen synthase kinase 3 (GSK3). Phosphorylation by GSK3 deactivates GS and this state is maintained until a new glucose supply is made available through feeding. Activation of the insulin-signaling pathway upon feeding leads to phosphorylation and activation of Akt, which targets GSK3 as one of its substrates. Phosphorylation by Akt deactivates GSK3 thereby relieving its inhibition on GS, which is then able to catalyze the polymerization and storage of the newly arrived glucose.

Insulin signaling via the PI3K – protein kinase B pathway stimulates glycolysis in adipose tissue and muscle cells. It does so in two ways. One of these is through increased uptake of glucose into the cell by the GLUT4 transporter. The second way is through phosphorylation and activation of phosphofructokinase 2 (PFK2). This enzyme along with its sibling phosphofructokinase 1 catalyzes the transfer of phosphoryl groups from ATP to fructo-6-phosphates. These latter enzymes serve as critical control points in glycolysis and, as shown in Fig. 3.2 (lower panel), act just after glucose is converted first by hexokinase to glucose-6-P, and then (by phosphoglucose isomerase) to fructose-6-phosphate.

The phosphofructokinases are unusual enzymes. They possess an N-terminal kinase domain, a centrally located phosphatase domain, and a C-terminal regulatory domain that contains several regulatory phosphorylation sites. The PFK2 enzyme, in particular, controls the synthesis and degradation of fructose 2,6 biphosphate, an allosteric activator of PFK1 (that stimulates the kinase domain and inhibits the phosphatase domain), which as can be seen in Fig. 3.2 (lower panel), catalyzes the conversion of fructose-6-phosphate to fructose 1,6 biphosphate. The overall consequence upon adipose and muscle tissue of insulin action is to increase the uptake of glucose and its metabolism through glycolysis.

During the past few years, it has become increasingly apparent that the failures of this glucose management system originate in overnutrition, particularly, excessive amounts of lipids, and in the attendant inflammation in adipose tissue and liver. The resulting lipid-handling problems lead to metabolic overload in other regions of the body, especially skeletal muscle and Islet cells of the pancreas, resulting in T2D and other metabolic syndromes.

As discussed in the last chapter, the liver metabolizes fatty acids and stores excess triglycerides. Excess fatty acids affect liver metabolism and signaling. They alter signal transduction pathways and nuclear receptor-mediated gene expression. Too much fatty acids negatively effects the liver's ability to oxidize fatty acids (releasing then as VLDLs and ketones) and increases its production of triglycerides. Skeletal muscle and pancreatic islet cells are exposed to elevated levels of these lipids and also to inflammatory mediators released by over-lipidized tissue. The insulin-signaling pathway will be examined shortly, and this will be followed by a look at how excessive lipids and inflammation can directly disrupt this signaling pathway and generate insulin resistance.

3.4 Signal Transduction Begins with the Insulin Receptor and Its Substrate Proteins

The insulin receptor is a member of the receptor tyrosine kinase (RTK) family of polypeptide growth factor receptors. This discovery, made in 1992, led to a host of subsequent findings on how insulin exerts its effects on adipose and muscle cells. Like other RTKs, ligand binding does not perturb the electrostatic environment sufficiently to stabilize a conformation sufficiently different in its C-terminal cytoplasmic region to serve as a signal. Instead, RTKs form stable receptor dimers and oligomers. The close proximity of the two cytoplasmic kinase domains enables cross or autophosphorylation of tyrosine residues situated in the activation loop leading to full activation of the kinase. These actions lead to further phosphorylations on tyrosine residues outside the activation loop that exposes docking sites for cytoplasmic signaling molecules.

The insulin receptor subfamily operates slightly differently from the standard RTK. Each receptor consists of two subunits, an extracellular α subunit and a transmembrane plus cytoplasmic β subunit. The receptor forms dimers consisting of four chains – two α subunits and two β subunits linked together by disulfide bridges. Ligand binding triggers conformational changes leading to autophosphorylation of tyrosine residues required for full kinase activity. This step is accompanied by additional phosphorylations. Whereas the standard case is to use autophosphorylation to open up binding sites for downstream signaling proteins, the insulin receptor's kinase domain phosphorylates a scaffolding protein called the insulin receptor substrate (IRS). Phosphorylation of these proteins by the IR exposes binding sites for the downstream signaling elements.

The insulin receptor has two isoforms generated by alternative splicing. The IR-A isoform lacks exon 11 consisting of 12 amino acid residues. The second isoform, IR-B includes the exon 11 coding sequence. The presence or absence of exon 11 does not seem to affect insulin binding but does have an influence on IGF binding. The IR-A isoform is capable of binding IGF-II while IR-B cannot. The IR-B isoform is found in tissues involved in metabolic responses to insulin, namely, liver, muscle, and adipose tissue. IR-A, in contrast, is expressed preferentially in hematopoietic cells and in the CNS. Whereas IR-B is involved in regulating glucose uptake and metabolism, the IR-A form and IGF receptors (which also lack exon 11 like structural elements) have roles in growth control and cancer.

The insulin receptor substrate (IRS) proteins function as molecular scaffolds that provide binding sites for downstream signaling proteins following insulin binding. These proteins range in mass from 60 to 180 kDa. Each of these proteins contains a pleckstrin homology (PH) domain in its N-terminal followed by a phosphotyrosine binding (PTB) plus a long C-terminal tail that is not only subject to tyrosine phosphorylation by the insulin receptor but also to serine phosphorylation by downstream serine/threonine kinases.

Four insulin receptor substrate isoforms are expressed; these are designated as IRS-1 through IRS-4. Their basic structures are similar; each contains an N-terminal PH domain that facilitates binding to membrane phospholipids and a PTB domain that mediates binding to the insulin receptor. They also each possess a large number of tyrosine phosphorylation sites that once phosphorylated serve as docking locations for attachment of proteins possessing Src homology 2 (SH2) domains. The distribution of these docking sites differs among the IRS proteins as does their cellular distribution. In humans, IRS1 and IRS2 have emerged as the key members of this family with regard to diabetes. IRS1 is expressed in skeletal muscle, adipose tissue, and pancreatic β-cells while IRS2 is found in pancreatic β-cells, adipose tissue, liver, and brain.

The IRS proteins function as key signaling nodes implicated in the development of insulin resistance. The structure of the IRS-1 protein is shown in Fig. 3.3. Tyrosine phosphorylation by the insulin receptors establishes binding sites for phosphoinositide-3-OH kinase (PI3K), for the Grb2 adapter protein,

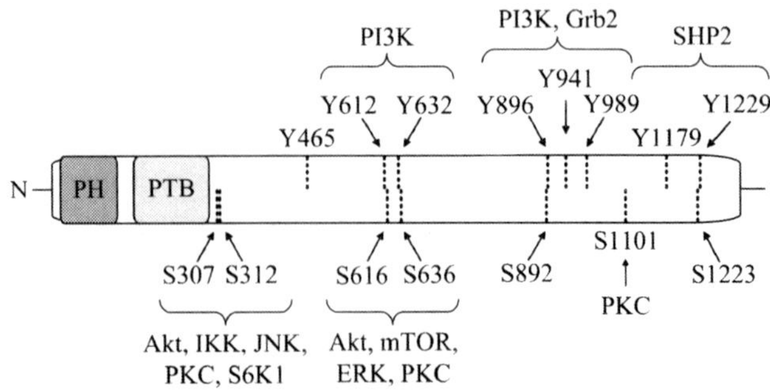

Fig. 3.3 Structure of the human insulin receptor substrate 1 (IRS-1) protein. *Dashed lines* indicate locations of some of the prominent phosphorylation sites. Upper set of *lines* indicate tyrosine (Y) phosphorylation sites and the lower set of *dashed lines* denote serine (S) phosphorylation sites. PH: Pleckstrin homology domain; PTB: phosphotyrosine binding domain. Proteins possessing SH2 domain found to bind tyrosine phosphorylated sites are listed as are serine/threonine kinases found to phosphorylate the serine sites either enhancing the feedforward signals or turning them off through feedback regulation

and for the protein tyrosine phosphatase SHP2. The latter along with the Grb2 adapter are essential, positive-acting elements of a number of signaling pathways most notably those leading to Ras/ERK activation. The events taking place downstream of insulin binding and IRS tyrosine phosphorylation will now be examined starting with the lipid kinase PI3K.

3.5 Phosphoinositide-3-OH Kinase (PI3K) and the PTEN Lipid Phosphatase

Phosphatidylinositol is a membrane phospholipid that serves as a source of lipid-signaling intermediates, or second messengers. The phosphorylated derivatives of phosphatidylinositol that function as signaling molecules are referred to as phosphoinositides. They along with a growing list of small molecules help organize and activate a number of signaling pathways involving initiating (first messenger) cytokines, neurotransmitters, and hormones. Among the small molecules intermediates are phosphoinositides, intracellular Ca^{2+} ($[Ca^{2+}]_i$), adenosine and guanine nucleotides, fatty acid metabolites, oxysterols, nitric oxide, and hydrogen peroxide.

The discovery that phospholipids are more then just passive components of cell membranes was made by Mabel and Lowell Hokin who in 1953 used P^{32} radioisotope labeling to observe the movement of these molecules in response to hormonal signals. Their discoveries led to a host of later breakthroughs concerning the actions of phosphoinositides. One of these, by Yatsutomi Nishizuka and coworkers in 1979, was that diacylglycerol (DAG) a product of phospholipase C (PLC)-catalyzed hydrolysis of PtdIns(4,5)P_2 functions as a signaling molecule that activates protein kinase C. A second product of PLC-catalyzed cleavage of PtdIns(4,5)P_2, namely Ins(1,4,5)P_3, was shown by Michael Berridge and colleagues to stimulate the release of $[Ca^{2+}]_i$ from intracellular stores thereby activating a number of calcium-dependent signaling responses. Thus, phosphatidylinositol 4,5 biphosphate (PIP_2) serves as the source of two lipid second messengers, diacylglycerol (DAG) and inositol 1,4,5-triphosphate, or Ins(1,4,5)P_3 or IP_3. The plasma membrane functions as a cellular repository for the PIP_2 and other phosphoinositides.

The hydrolysis of phosphoinositides is one of two ways of generating phospholipid-signaling intermediates. The second way is by covalent addition and removal of phosphoryl groups from the inositide head group by lipid kinases and phosphotases.

The inositol ring of the phosphatidylinositol molecule contains a phosphoryl group at position 1 that is tied to the glycerol backbone. All other OH groups of the inositol ring can be phosphorylated except those at positions 2 and 6. Phosphoinositide-3-OH kinase, or PI3K, is a lipid kinase that catalyzes the transfer of a phosphoryl group from an ATP molecule to the OH group at position 3 of the inositol ring of the lipid. The PI3Ks are composed of a catalytic subunit and a regulatory subunit. Like other signaling proteins, these heterodimers are highly modular in design and are classified according to the presence or absence of certain domains. The most important of these heterodimers with respect to insulin signaling are those composed of a p85α regulatory subunit plus a p110α catalytic subunit. The domain structure of these components is depicted in Fig. 3.4.

Class I PI3Ks are heterodimers composed of a 100-kDa catalytic subunit and an 85-kDa or 55-kDa joint regulatory/adaptor subunit. There are two kinds of Class I PI3Ks, determined by the presence or absence of an adaptor-binding (AB) domain in their N-terminal region and by the kind of receptor-binding events that activate them. Ligand-growth factor receptor binding stimulates the dissociation of the regulatory and catalytic subunits of PI3K from each other. The catalytic subunit phosphorylates the PIP_2 proteins at the 3′ position thereby making PIP_3, which then diffuses to and binds protein kinase B. Once bound by IP_3, PKB is able to be phosphorylated by PDK1 thereby completing the activation process.

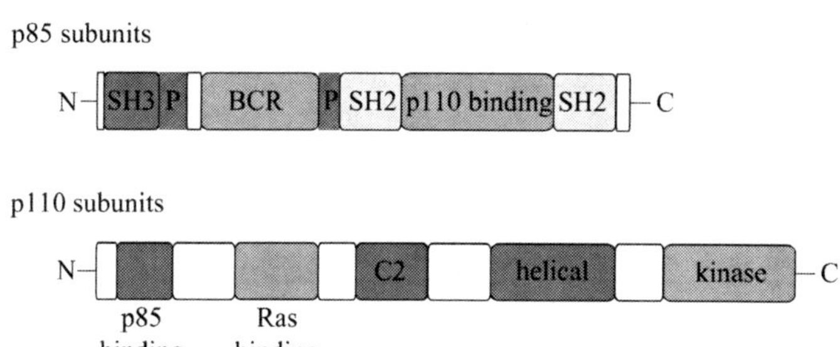

Fig. 3.4 Domain structures of the p85 and p110 subunits of PI3K Class I

Fig. 3.5 Opposing actions of PI3K and PTEN on phospholipids

$$\text{PtdIns(4,5)P}_2 \underset{\text{PTEN}}{\overset{\text{PI3K class I}}{\rightleftharpoons}} \text{PtdIns(3,4,5)P}_3$$

There is a corresponding set of lipid phosphatases that catalyzes the removal of phosphoryl groups from inositol rings. The most prominent of these is PTEN (phosphatase and tensin homolog deleted on chromosome ten). PTEN acts in opposition to PI3K and catalyzes the removal of phosphoryl groups from position 3 on inositol rings. It acts on PIP_3 to return it to a PIP_2 form, thereby reversing the catalytic effects of PI3K (Fig. 3.5).

One of the key signaling proteins recruited to the IRS scaffolding proteins is PI3K. Once the IRS proteins bind to phosphotyrosine residues in the kinase domain of the insulin receptor and have undergone tyrosine phosphorylation, they expose docking sites for recruitment and binding of the p85α subunit of the PI3K.

3.6 Activation of Protein Kinase B (PKB) and Protein Kinase C (PKC)

Protein kinase C like protein kinase B is a member of the AGC family of serine/threonine kinases. There are several subfamilies of protein kinase C, each characterized by a slightly different domain structure and associated coactivator requirements. As presented in Table 3.1, the different isozymes of protein kinase C belong to either the classical, novel, or atypical subfamilies. Each member of these families requires phosphorylation and coactivator binding, but the order of events and their consequences differ slightly from member to member. The various isozymes of protein kinase C are listed in Table 3.1 along with the associated coactivators.

The atypical protein kinases are involved in insulin signaling and will be the focus of the discussion. Like protein kinase B, full activation of atypical protein kinase C occurs when key residues located in the catalytic loop are phosphorylated by upstream kinases such as PDK1. However, that is not all that is required for activation. The phosphorylation step is preceded by interactions with the plasma membrane-associated phosphatidylserine (PtdSer) mediated by the C1 and C2 domains of the protein kinase. This step triggers conformational changes that relieve the inhibition imposed by the pseudosubstrate (PS) domain and attendant

Table 3.1 Protein kinase C families, isozymes, and activating second messengers

Protein kinase C	Abbreviation	Isozymes	Regulation
Classical	cPKC	α, βI, βII, γ	Ca^{2+}, DAG
Novel	nPKC	δ, ε, η, θ	DAG
Atypical	aPKC	ζ, ι/λ	

DAG, diacylglycerol.

prevention of autophosphorylation. And like PKB-required PIP_3 binding (discussed below), PtdSer and Ca^{2+}/DAG binding prepare the way for PDK1 phosphorylation of key residues in the activation loop of PKC.

There are three protein kinase B isoforms each coded by a separate gene. The genes are designated as PKBα, PKBβ, and PKBγ. The corresponding proteins produced are called Akt1, Akt2, and Akt3. The first of these, Akt1 (originally Akt), is broadly distributed among different cell types where it functions in pathways regulating growth and cell survival, the latter through its inhibition of apoptosis. The second isoform, Akt2, is highly expressed in insulin-responsive tissues such as liver, skeletal muscle, and adipose tissue. The third type, Akt3, functions as a growth regulator in the brain.

The Akts are organized into three domains. A PH domain occupies the N-terminal regions. This is followed by a kinase domain and a C-terminal regulatory domain. Full activation takes place in several spatiotemporal steps. These are illustrated in Fig, 3.6(a). The initial step is activation of a class I PI3K. This protein catalyzes the conversion of PIP_2 into PIP_3, which then serves to recruit the PKBs to the plasma membrane. Binding to the PH domain induces a conformational change that exposes the two phosphorylation sites, one in the activation loop and the other in the C-terminal hydrophobic motif (HM). PDK1 serves as an activator of AGC kinases. This serine/threonine kinase uses the HM as a docking site from which it is able to phosphorylate PKB in the activation loop. The kinase then dissociates from the PKB allowing a second kinase, initially designated as "PDK2" because of uncertainties in its identity (Fig. 3.6(a)). Once the PDK2 kinase is recruited, it phosphorylates PKB in its HM thereby completing the activating process.

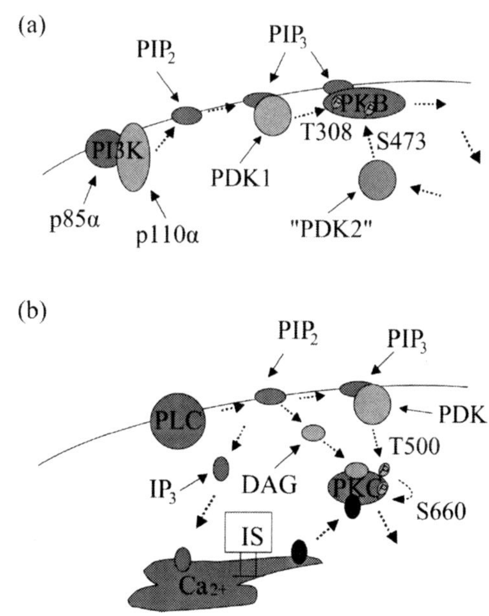

Fig. 3.6 Activation of protein kinase B/Akt (**a**) and protein kinase C (**b**) by phospholipids and the upstream kinase PDK1

Protein kinase C activation depicted untethered for simplicity and subsequent to PtdSer interactions is shown in Fig. 3.6(b). The starting step in this process is activation of phospholipase C, which catalyzes the hydrolysis of PIP_2 to generate DAG and IP3. In Fig. 3.6(b), the activation process for a classical protein kinase C subfamily member, in which both DAG and Ca^{2+} binding are required, is depicted. Activation of proteins belonging to the novel protein kinase C subfamily would need DAG but not Ca^{2+}, while atypical protein kinase C enzymes require neither DAG nor Ca^{2+}. The sequence of steps in the activation begins with phosphorylation in the activation loop by the AGC kinase PDK1. As shown in the figure, this step is followed by autophosphorylation of the second critical residue in the HM whereupon the protein is primed for activation by the second messenger(s). The atypical protein kinase C subfamily members do not possess the second critical serine/threonine residue in the HM. Rather there is a glutamic acid [Glu (E)] residue in its place. The steps leading to full activation of these enzymes are not yet fully known other than the requirement for PDK1 phosphorylation.

The Akts are key effectors of insulin's regulation of metabolism. Some of the activities regulated though Akt-mediated phosphorylation are depicted in Fig. 3.7. As can be seen, these activities include phosphorylation of phosphofructokinase 2 (PFK2), glycogen synthase kinase 3 (GSK3), Akt substrate of 160 kDa (AS160), and ACC. These targets, in turn, regulate the metabolic activities shown in the figure.

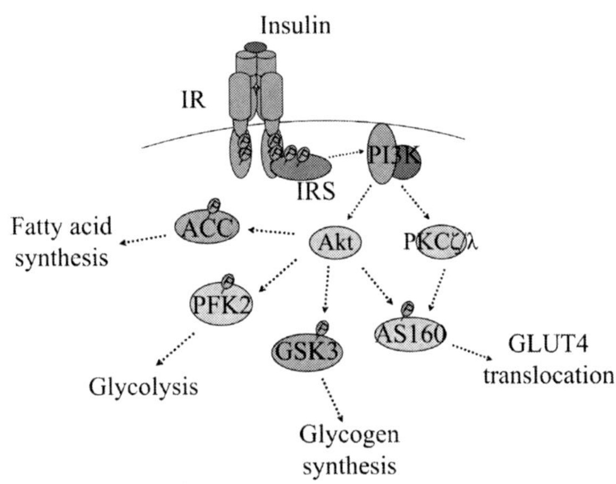

Fig. 3.7 Regulation of metabolism by insulin signaling via Akt

3.7 GLUT4 Transport Biomechanics and Regulation

T2D involves the failure to adequately transport glucose from the bloodstream into cells of tissues such as fat and muscle. Glucose import and metabolism in these tissues following a meal is critically dependent upon the presence or absence of an adequate number of glucose transporters in the plasma membrane of fat and muscle cells. The transporters responsible for the import, GLUT1-4, form pores in the plasma membrane that enable glucose molecules to enter cells by means of facilitated-diffusion down concentration gradients. This is the rate-limiting step in glucose metabolism and storage.

GLUT4 transporters continually cycle back and forth between intracellular and plasma membrane locations. In the absence of insulin signals or exercise-contraction, less than 10% of the GLUT4s are located at the plasma membrane while the bulk of the transporters, the other 90%, are sequestered in intracellular compartments. The transporters are rapidly removed from the cell surface and slowly recycled back to that location. They are packaged as cargo into transport vesicles and in that form move between the plasma membrane, trans-Golgi network, and endosomal compartments.

Insulin stimulation leads to a marked increase in the cell-surface transporter population. The sequence of steps through which this happens begins when insulin induces the activation of PKB and atypical protein kinase C (aPKC) concomitant with the cycling and membrane fusion of GLUT4-bearing vesicles. Several families of proteins link together the insulin-signaling pathways and vesicle biomechanics. Prominent among these are the SNARE proteins and several families of small GTPases.

Small GTPases belonging to the Ras superfamily function by passing through cycles of GDP/GTP binding and release. In this type of cycle, GTPases bind either GTP or GDP. They are converted from GDP-bound forms to GTP-bound forms by the catalytic actions of guanine nucleotide exchange factors (GEFs) while GTPase-activating proteins (GAPs) catalyze their subsequent hydrolysis. The GEFs carry out their function by kicking out the GDP molecule. In most situations, GTP is far more abundant than GDP and it readily binds the vacated pocket. In the reverse operation, the GAP catalyzes the hydrolysis of the GTP in which a phosphoryl group is removed leaving a GDP molecule bound to the GTPase.

The small GTPases function as molecular switches. They are turned ON when GTP is bound and turned OFF when GDP is bound. In the ON state, the GTPases are continually active and become inactive only when turned off by a GAP. In the active state, the GTPases bind downstream effectors thereby conveying a signal as depicted in Fig. 3.8.

The small GTPase superfamily can be divided into several subfamilies each distinguished by a characteristic set of cellular-signaling functions. The Rab family with more than 60 known members in humans is the largest group of small GTPases. These proteins coordinate the docking and fusion of cargo

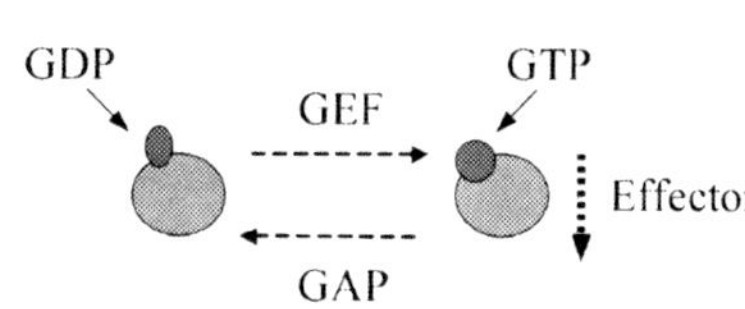

Fig. 3.8 The GTPase cycle is regulated by opposing actions of GAPs and GEFs

vesicles. Reciprocal pairs of SNARE proteins mediate the fusion of cargo vesicles. The binding of the SNAREs is preceded by tethering/protection protein-binding steps that ensure that only the appropriate fusion events take place. The Rab proteins are localized to the cytoplasmic face of organelles and vesicles and regulate the preliminary binding operations. The Rab protein cycle of GDP and GTP binding and release is synchronized with the movement and fusion of cargo vesicles. The process is a complex one and the GAPs and GEFs are assisted by two other sets of accessory proteins, GDP dissociation inhibitors, or GDIs, and the GDI displacement factors, or GDFs.

Rab GTPases link insulin-activated PKB signaling to GLUT4 vesicle transport. The Rab GTPases are cell specific. Three kinds of Rab GTPases are most likely involved in fat and muscle tissue – Rab 8A and Rab14 in muscle cells and Rab10 in adipocytes. The key signaling intermediate is a Rab GAP called AS160. The signaling route through which insulin-triggered PKB and aPKC signals influence GLUT4 vesicle cycling is shown in Fig. 3.9.

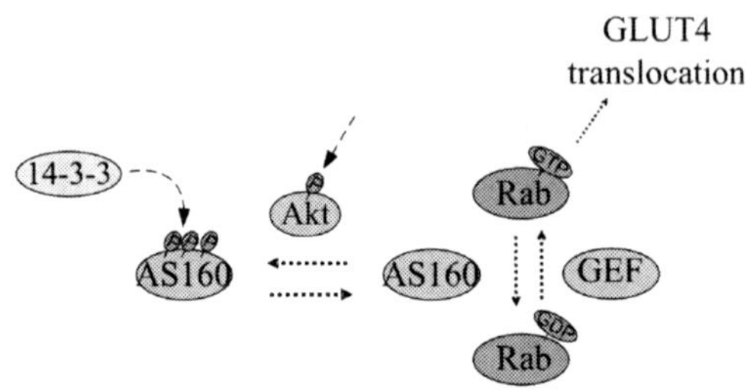

Fig. 3.9 GLUT4 translocation is regulated by Akt

3.8 The TOR Cassette Is the Downstream Target of Akt Signals

Tuberous sclerosis complex is the name given to a disorder characterized by benign tumors in multiple organ systems. The tumors arise from mutations in a pair of proteins named tuberous sclerosis 1, also known as hamartin, and tuberous sclerosis 2, alternatively called tuberin. These proteins serve as the input signaling unit for a protein complex consisting of hamartin (TSC1) – tuberin (TSC2), the GTPase Rheb, TORC1 and TORC2, two modules that contain the target of rapamycin (TOR) protein as their core element, and the S6K protein, with mTOR and S6K functioning as the output units (Fig. 3.10).

The TSC heterodimer functions as the input unit for the TOR cassette. A number of serine/threonine kinases phoshorylate residues in either TSC1 or TSC2 thereby conveying signals to the cassette. Some of the inputs to the TOR cassette are indicated in Fig. 3.11. In this figure, Akt along with the energy status inputs from AMPK and growth factor signaling pathways such as MAPKs impinge upon TSC2 while cell cycle kinases such as Cdc2 target TSC1. In addition to possessing a number of phosphorylation sites, TSC2 possesses a GAP motif, and the TSC complex functions as a GAP for the Rheb GTPase that connects the TSC heterodimer to the two TORCs.

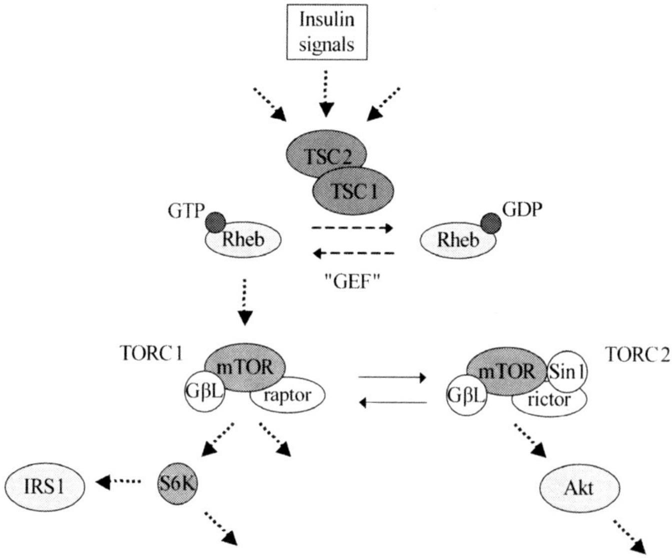

Fig. 3.10 The TOR cassette. Abbreviations: Raptor, regulatory associated protein of TOR; Rheb, Ras homologue enriched in brain; Rictor, rapamycin-insensitive companion of TOR; TOR, target of rapamycin; TORC, TOR complex; TSC, tuberous sclerosis complex

In the absence of insulin signals, the TSC is active and by functioning as a Rheb GAP inhibits Rheb signaling. Insulin signals sent into the TSC through Akt-catalyzed phosphorylation deactivate the TSC. In response, Rheb binds GTP and is turned on leading to its activation of the two TORCs. The key signaling element of the TOR complexes is the target of rapamycin kinase. This protein is a member of the PIKK family of serine/threonine kinases. Other

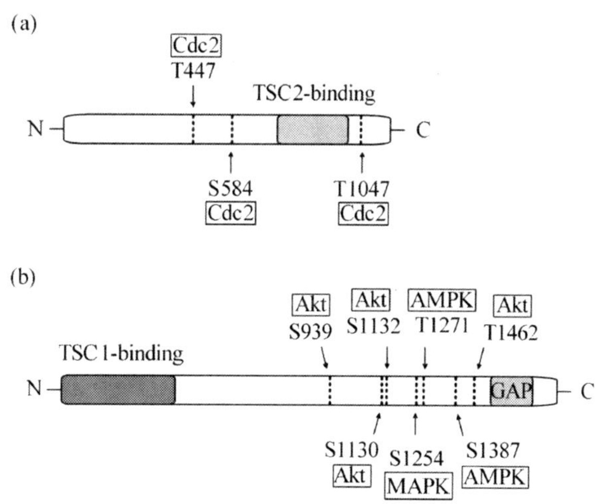

Fig. 3.11 The tuberous sclerosis complex. Shown are the structures of the (**a**) Hamartin (TSC1) and (**b**) Tuberin (TSC2) proteins. Shown in the figure are serine (s) and threonine (T) phosphorylation sites and the protein kinases that catalyze the attachment of the phosphoryl groups to the indicated serine (S) and threonine (T) residues

members of this family include the ATM/ATR and DNA-PKcs kinases that have major roles in DNA damage signaling and repair (discussed in Chapter 10). The two complexes are distinguished from one another by the presence of distinct scaffolding and stabilization proteins, the notable of these being raptor (TORC1) and rictor (TORC2).

3.9 Feedback Regulation of Akt by TORC2 and IRS by TORC1/S6K

The key structural element of the TOR cassette, of interest here, is the presence of two feedback loops, one leading to Akt and the other to IRS1. The insulin signals input to the cassette are generated when insulin binds the insulin receptor leading to tyrosine phosphorylation and activation of the IRS proteins which, in turn, activate PI3K and hence PDK1, which then phosphorylates Akt on Thr308/9 in the activation loop and activates and Akt1/2. The Akts once they are phosphorylated by PDK1 are able to phosphorylate Ser/Thr residues in downstream targets including the TSCs. Phosphorylation of the TSC proteins deactivates them thereby relieving their inhibition of Rheb. The Rheb GTPase is then able to activate the TORCs leading to TOR phosphorylation of Akt on Ser473/4 in the HM, thereby functioning as a PDK2. This additional phosphorylation event alters the mix of proteins that serve as substrates for Akt. One of the ones made available by the second phosphorylation is the FOXO proteins.

The second feedback loop has a different role. Whereas the PDK2 phosphorylation of Akt modifies its substrate specificity, phosphorylation of the IRS proteins by S6K operates as a termination signal that shuts down insulin signaling. This negative feedback loop serves a useful function under normal healthy conditions by preventing excessive signaling.

3.10 Insulin Resistance Develops from Inflammation and Metabolic Overload

Free fatty acids can influence both metabolism and inflammation through their ability to serve as ligands for Toll-like receptor 4 (TLR4), a member of the Toll/IL-1 family of cytokine receptors. This family of receptors has a prominent role in recognition of lipopolysaccharides (LPS) characteristic of bacterial pathogens and helps launch the body's innate immune response. The signaling pathway activated by LPS involves a number of adapter proteins among which are MyD88, TIRAP, and TRAF6 leading to activation of two major pathways – the NF-κB and JNK mitogen-activated protein (MAP) cascade leading to NF-κB and AP-1 gene expression, respectively. These pathways will be examined in greater detail in Chapter 7. The immediate effect

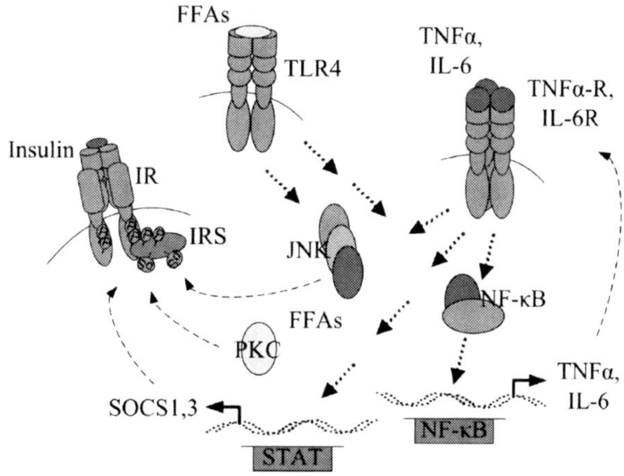

Fig. 3.12 Insulin resistance develops from the synergistic effects of inflammation and excessive amount of metabolites

of FFA binding to these receptors, as depicted in Fig. 3.12, is to reduce insulin transduction by phosphorylating the IRS1/2 insulin-signaling adapters on negative regulatory residues. These actions prevent tyrosine phosphorylation and signaling to downstream effectors. These actions prevent signaling through the insulin-signaling pathway to the Akt targets discussed earlier and, in particular, prevent an adequate uptake of glucose by GLUT4 as a result of the failure of Akt to become activated and phosphorylate the negative regulator AS160.

These actions are reinforced by inflammatory signaling through TNFα receptors. This signaling pathway activates JNK and NF-κB modules leading to a positive feedback loop in which NF-κB gene expression upregulates pro-inflammatory cytokines, which act through autocrine and/or paracrine routes to further stimulate inflammatory responses and inhibit insulin signaling. The IL-6 signaling pathway contributes to the burgeoning insulin resistance in yet another way. Cytokines such as IL-6 signal through Jak kinses and STAT transcription factors. One of the genes upregulated by these signals encodes a family of negative regulators, the suppressor of cytokine signaling (SOCS) proteins, which act back on the cytokine receptors to terminate signaling. Two of these negative regulators, SOCS1 and SOCS3, also target the IRS proteins. In particular, they promote ubiquitination and degradation of the IRS proteins thereby further short-circuiting insulin signaling.

Yet another pathway linking excessive FFAs and insulin resistance involves signaling through protein kinase C. Excess plasma fatty acid levels give rise to excessive amounts of long-chain fatty acyl-CoA molecules within hepatocytes and adipocytes. These molecules serve as precursors for diacylglycerol (DAG), which functions as a second messenger that activates a large number of protein kinase C (PKC) family members among which are PKCβ and PKCδ. Once activated PKCδ and others phosphorylate the IRS protein on Serine 307 as does JNK1.

Malonyl-CoA is a key regulator of fatty acid metabolism. Recall from Figs. 2.4 and 2.5 that malonyl-CoA inhibits the transport by CPT1 of long-chain fatty CoAs across the mitochondrial membranes into the matrix where they are oxidized. Instead, the LCACoAs are used for the synthesis of triglycerides and diacylglycerol. Under normal conditions, these actions are used to maintain energy balance with AMPK as the upstream sensor of energy status. During exercise, for example, falling energy levels in skeletal muscle activate AMPK leading to its inhibition of ACC and malonyl-CoA thereby allowing for transport of fatty acids into the matrix where they are oxidized. However, when there is a chronic oversupply of nutrients, this regulatory shift promotes activation of PKC and contributes to insulin resistance in muscle and liver. Along these same lines activation of malonyl-CoA decarboxylase, or MCD, catalyzes the conversion of malonyl-CoA back to acetyl-CoA, thereby acting like AMPK in opposition to ACC negative regulation of malonyl-CoA-mediated inhibition of fatty acid oxidation.

Reactive oxygen species (ROS) may be present and act in a second messenger capacity that strengthens the linkage between PKC and the IRS proteins. The central element in this linkage is the ability of PKC to activate superoxide-producing NADPH oxidase by phosphorylating p47phox, one of its regulatory subunits. The superoxide is rapidly converted to hydrogen peroxide, which acts through a positive feedback loop to further activate PKC. In addition, NADPH oxidase may not be the only source of ROS in the cells given the presence of stress-causing high levels of glucose and lipids.

Key steps linking excessive FFAs and JNK activation may involve endoplasmic reticulum (ER) stresses brought on by the FFAs. ER stresses are known to activate the JNK module. As indicated in Fig. 3.12, this serine/threonine kinase can then phosphorylate and inactivate the IRS1/2 proteins.

Insulin resistance in muscle may differ in its detailed causative chain of events from that in liver to which the above-mentioned chains may best apply. Studies in skeletal muscle have led to a second mitochondrial-based model. In this second model, β-oxidation is not reduced, but rather is accelerated leading to a mismatch between output from fat burning and the utilization of metabolic intermediates by the TCA cycle. This mismatch produces yet another set of metabolites that can impair insulin signal transduction and generate insulin resistance. Under this scenario, ROS produced by the mitochondria as a result of this mismatch would further contribute to the overall signaling dysregulation. These two mitochondrial models – one mostly liver-centric and the other muscle-centric – reflect the very different functionalities of hapatocytes and myocytes, and pose some challenges in devising treatments directed at the mitochondria.

Leptin resistance is another early event that feeds and reinforces obese and diabetic states. Recall from the last chapter that leptin along with insulin and ghrelin passes through the blood-brain barrier and binds to receptors expressed on neurons residing in the arcuate nucleus of the hypothalamus. These signals are part of a central nervous system regulatory network that maintains energy balance in the body. This system is inactivated in obese individuals; that is, leptin signals sent out by adipose tissue fail to reduce feeding and restore lean body weight in response to increasing adipose tissue mass. This failure is termed leptin resistance.

Leptin signals can be inactivated in one or more of several ways. One of these is through a failure in the transport across the blood-brain barrier. Another is through mismanaged receptor expression, and a third is through dysregulated leptin signal transduction. The latter is particularly interesting as it provides a potential link with inflammation.

The main components of the leptin-signal transduction pathway were presented in Fig. 2.11. As shown in this figure, one of the genes upregulated by the STAT3s encodes SOCS3. As is the case for leptin, once these proteins are upregulated through cytokine-signaling pathways they translocate to the plasma membrane where they terminate the inflammatory signaling. In the absence of inflammation, these proteins are expressed at low basal levels but are strongly upregulated by cytokines. Once upregulated, they bind and inhibit the Jaks thereby terminating the signaling.

These proteins, especially SOCS1 and SOCS3, contribute to both insulin resistance and leptin resistance. In the case of insulin resistance, these two SOCS family members target not only cytokine-associated Jaks proteins but also receptor tyrosine kinase signaling elements such as the IRS proteins. The SOCS proteins target the IRS protein for ubiquitin-mediated proteolysis, a lifetime-shortening process that will be discussed in Chapter 7. As a result, signaling downstream from the insulin receptor is blocked. The picture that emerges from these actions is one where pro-inflammatory cytokines in the form of TNFα and others trigger the upregulation of SOCS1 and SOCS3, which then target the IRS proteins and contribute to the obesity- and inflammation-induced insulin resistance. In the case of leptin resistance in the hypothalamus, chronic stimulation arising from the increased mass of adipose tissue eventually leads to a strong induction of SOCS3, which then shuts down the leptin signaling in the presence of elevated plasma leptin levels.

3.11 Glucose-Stimulated Hormone Release by Pancreatic Islet Cells

The buildup of glucose in the blood following feeding leads to glucose entry into pancreatic β-cells. This process is mediated by glucose transporters that convey the polar glucose molecules across the plasma membrane into the various cell types in the body. The tertiary structure of the transporters is that of twelve membrane-spanning helices with both the N- and C-terminals in the cytoplasm. These helices form water-filled channels that enable passive (energy-independent) glucose transport. The particular family member responsible for transport of glucose into pancreatic β-cells is the GLUT2 transporter. A summary of the four main facilitated glucose transporters was presented in Table 2.2.

The next step in the chain leading to insulin release is centered about the mitochondria, and involves the metabolism of glucose. Once in the cell, glucose is metabolized into pyruvate by means of glycolysis. The pyruvate is then metabolized, producing NADH and $FADH_2$ through the citric acid (Krebs or TCA) cycle. These reducing power agents are then utilized in the production of ATP by the mitochondrial electron transport chain that is responsible for oxidative phosphorylation. The net outcome of this series of metabolic steps, from a signaling standpoint, is that glucose entry is converted into an increase in the ATP/ADP ratio as depicted in Fig. 3.13.

Pancreatic Islet cells express a number of ion channels and this property plays a key role in the next step in the signal transduction cascade – entry of Ca^{2+} into the cell as a result of the increased ATP/ADP levels. Two classes of ion channels mediate this next step. The first type of ion channel is the ATP-dependent potassium channel (K_{ATP}). These channels are closed upon ATP binding; when open they enable the efflux of potassium from the cell and hyperpolarize the cell membrane. The second kind of ion channels is the voltage-gated calcium channel (L or N type). When open through depolarization of the membrane, these channels permit the entry of calcium into the cell. Thus, the increase in ATP/ADP ratio serves to close K_{ATP}; when this happens the membrane depolarizes, and the calcium channels open leading to an increase in cytosolic Ca^{2+} levels.

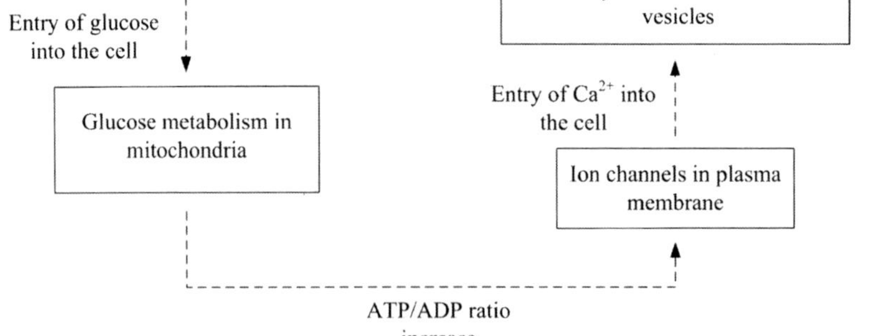

Fig. 3.13 Chain of events involved in glucose-induced insulin release from pancreatic islet β-cell

Like neurotransmitters, hormones are packaged into secretory vesicles. These are filled and transported to the plasma membrane. The vesicles are then docked and primed for release, awaiting triggering signals. Calcium binding serves as the trigger leading to vesicle fusion and release of the contents. Many of the same molecular mediators as seen in neurotransmitter release from presynaptic terminals are encountered in insulin exocytosis. Among these are the calcium channels, calcium binding proteins, vesicle regulators called SNARE proteins, and guanine nucleotides.

Glucagon release is tightly coordinated with insulin release so that their opposing actions operate properly over time. This is achieved by signaling between pancreatic β-cells, δ-cells, and α-cells. Pancreatic α-cells express many of the same ion channels as the do the Islet β-cells and δ-cells. Mitochondria do not seem to play a major role in the glucose-mediated exocytosis. Instead, cell-to-cell paracrine signals from β-cells and δ-cells regulate potassium channel opening and closure, thereby enabling opening of the calcium channels leading to Ca^{2+} entry and glucagon release at the appropriate times. Signals conveyed by insulin, zinc, and γ-aminobutyric acid (GABA) coordinate these responses. As a result, glucagon is released when glucose levels are low and inhibited when glucose levels are elevated.

3.12 K_{ATP} Channels and Their Regulation by Cellular Fuel Status

The ATP-sensitive potassium (K_{ATP}) channel provides a link between the firing properties of the neurons, that is, its electrophysiology, and its metabolic status – are its nutrients and energy supply plentiful or are they scarce. The channel consists of two subunits (Fig. 3.14). One of the subunits, called Kir6, forms a potassium-conducting pore that enables potassium ions to exit the neuron. The second component, the sulphonylurea receptor, or SUR, functions as a regulatory subunit. The functional conducting channel consists of four sets of Kir6-SUR subunits with the Kir6 subunits inside and the SUR subunits on the outside. In this configuration, two α helices contributed by each Kir6 subunit line the pore, and their P-loops act as a selectivity filter that permits passage of potassium ions but not others.

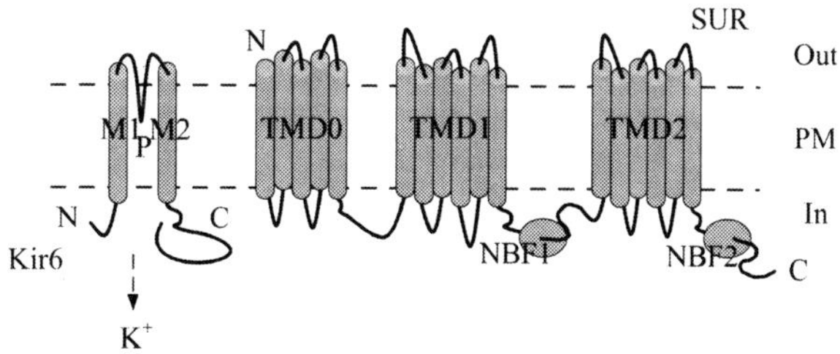

Fig. 3.14 The ATP-sensitive potassium channel. Abbreviations: TMD, Transmembrane domain; NBF, nucleotide binding fold

These channels are inward-rectifying (potassium ions pass more easily inward from the outside than in the other direction) and when open act as a brake on neural excitability. As their name indicates, they are regulated by the ATP/ADP ratio. When supplies are plentiful, ATP binds the large looped C-terminal domain of the Kir6 channel, which in response alters its orientation and blocks the channel. Low energy supplies are signaled by elevated ADP levels; the ADP molecules bind the regulatory subunit, and in the absence of ATP binding the C-terminal domain adopts an open configuration. The K_{ATP} channels open, potassium ions enter the cells and lower the resting potential thereby effectively damping down the neural excitability and shutting off the voltage-gated calcium channels.

The K_{ATP} channel is sensitive to a variety of cellular metabolites. First, it is responsive to its electrostatic milieu, particularly to phospholipids such as PIP_2 and PIP_3. These carry a net negative charge that influences channel opening and closing through interactions with the net-positively charged C-terminal Kir6 domain. When these phospholipids are situated adjacent to the C-terminal "lid" they will "flip" it to the open conformation. Conversely, fuel molecules such as ketones and long-chain fatty acids will, like the phospholipids, induce the channel to open. Thus, the K_{ATP} channel transduces cellular fuel status into membrane electrical responsiveness and activity.

3.13 Islet β-Cell Failure and Diabetic Complications

Hyperglycemia and aberrant metabolism leading to formation of sugar-derived metabolic intermediates and generation of ROS and apoptosis: under normal conditions glucose enters the glycolytic pathway and is metabolized in a series of steps leading from glycolysis to oxidative phosphorylation. This pathway becomes overwhelmed by the elevated glucose levels characteristic of hyperglycemia. Instead of being completely metabolized by the normal pathways, the excess glucose enters alternative pathways. These lead to the generation of DAG leading to protein kinase C activation, and to formation of sugar-derived metabolic intermediates, termed advanced glycation end products (AGEs). These products become appended to protein, lipids, and DNA, and alter their conformations and generate cross-linkages. This ensemble of metabolic alterations stimulates accumulation of reactive oxygen species, promotes apoptosis, and creates an environment conducive to amyloid buildup.

Amylin, or islet amyloid polypeptide (IAPP), is synthesized within the Golgi apparatus, and is packaged into secretory granules along with insulin by Islet β-cells. It is initially formed as an 89 amino acid residue prohormone that is subsequently processed into a 39 amino acid residue active form. Under hyperglycemic conditions, the alterations in the intracellular milieu discussed above generate ER stresses and result in misfolding of IAPP leading to a form that tends to aggregate and produce extracellular amyloid deposits. The aggregation-prone amylin proteins, like other amyloids, adopt an antiparallel, crossed β-pleated sheet structure. A key feature promoting adoption of this alternative secondary structure is the presence of a hydrophobic patch spanning residues 20–29. As will be seen for the amyloid deposits formed in connection with neurodegenerative disorders such as Alzheimer's disease, it seems to be a small oligomeric species formed by the misfolded proteins that are cytotoxic.

Endoplasmic reticulum stresses accumulate in the β-cells as they increase their output of insulin. They respond to these stresses by stimulating the upregulation of antioxidants and ER resident chaperones. The increased tendency for protein to misfold is further countered by a throttling back of protein synthesis. These changes are part of the cell's unfolded protein response to stresses in the ER. If the proteins cannot be restored to their functional form, the cell responds with an endoplasmic reticulum degradation of the misfolded proteins, and if conditions still worsen the cells trigger apoptosis. As the Islet β-cells increase their rate of apoptosis the Islet mass decreases. A similar course of action is taken by neurons, another class of cells that like the Islet β-cells secrete large amounts of proteins and are subjected to ER stresses, especially under unhealthy, neurodegeneration-promoting conditions. Protein misfolding and amyloid buildup are early-disease causing events in neurodegeneration; these subjects are treated in depth in Part III.

"Diabetic complications" is the term given to a number of debilitating disorders brought on by β-cell failure. These include blindness, kidney failure, nerve damage, and diabetes-associated metabolic disorders among which is atherosclerosis. One highly plausible model of how these complications arise is one in which the causative event linking one to the other is hyperglycemia-induced overproduction of reactive oxygen species, most notably superoxide. In this model, the metabolic pathway alterations discussed above lead to superoxide production and enable the generation of positive feedback that sustains the dysregulated metabolic and signaling pathways in these susceptible cell types. Other, complementary models emphasize the toxic effects of advanced glycations end products and ER stress and misfolded IAPP oligomers.

Further Reading

Grundy SM et al. 2004 Definition of metabolic syndrome. *Circulation* 109: 433–438.

Reaven GM 2005 Why syndrome X? From Harold Himsworth to the insulin resistance syndrome. *Cell Metab.* 1: 9–14.

Insulin Receptor and Substrates

Pirola L, Johnston AM and van Obberghen E 2004 Modulation of insulin action. *Diabetologia* 47: 170–184.

Taniguchi CM, Emanuelli B and Kahn CR 2006 Critical nodes in signaling pathways: insights into insulin action. *Nat. Rev. Mol. Cell Biol.* 7: 85–96.

White M 2002 IRS proteins and the common path to diabetes. *Am. J. Physiol. Endocrinol. Metab.* 283: E413–E422.
Zick Y 2004 Uncoupling insulin signalling by serine/threonine phosphorylation: a molecular basis for insulin resistance. *Biochem. Soc. Trans.* 32: 812–816.

PI3K and PTEN

Berridge MJ 1984 Inositol triphosphate and diacylglycerol as second messengers. *Biochem. J.* 220: 345–360.
Hawkins PT et al. 2006 Signalling through Class I PI3Ks in mammalian cells. *Biochem. Soc. Trans.* 34: 647–661.
Hokin MR and Hokin LE 1953 Enzyme secretion and the incorporation of P^{32} into phospholipids of pancreas slices. *J. Biol. Chem.* 203: 967–977.
Nishizuka Y 1995 Protein kinase C and lipid signaling for sustained cellular responses. *FASEB J.* 9: 484–496.
Toker A 2002 Phosphoinositides and signal transduction. *Cell. Mol. Life Sci.* 59: 761–779.

PKB and PKC

Brazil DP, Yang ZZ and Hemmings BA 2004 Advances in protein kinase B signaling: AKTion on multiple fronts. *Trends Biochem. Sci.* 29: 233–242.
Newton AC 2003 Regulation of the ABC kinases by phosphorylation: protein kinase C as a paradigm. *Biochem J.* 370: 361–370.
Yang J et al. 2002 Crystal structure of an activated Akt/protein kinase B terniary complex with GSK3-peptide and AMP-PNP. *Nat. Struct. Biol.* 9: 940–944.

GLUT4 Transport

Bonifacino JS and Glick BS 2004 The mechanisms of vesicle budding and fusion. *Cell* 116: 153–166.
Bryant NJ, Govers R and James DE 2002 Regulated transport of the glucose transporter GLUT4. *Nat. Rev. Mol. Cell Biol.* 3: 267–277.
Ishikura S, Bilan PJ and Klip A 2007 Rabs 8A and 14 are targets of the insulin-regulated Rab-GAP AS160 regulating GLUT4 traffic in muscle cells. *Biochem. Biophys. Res. Commun.* 353: 1074–1079.
Jahn R and Scheller RH 2006 SNARES – engines for membrane fusion. *Nat. Rev. Mol. Cell Biol.* 7: 631–643.
Sano H et al. 2006 Rab10, a target of the AS160 Rab GAP, is required for insulin-stimulated translocation of GLUT4 to the adipocyte plasma membrane. *Cell Metab.* 5: 293–303.
Shepherd PR and Kahn BB 1999 Glucose transporters and insulin action. *N. Engl. J. Med.* 341: 248–257.

TOR Cassette

Um SH et al. 2004 Absence of S6K1 protects against age- and diet-induced obesity while enhancing insulin sensitivity. *Nature* 431: 200–204.

Mitochondrial Overload, Inflammation, and Insulin Resistance

Boden G et al. 2005 Free fatty acids produce insulin resistance and activate the proinflammatory nuclear factor-κB pathway in rat liver. *Diabetes* 54: 3458–3465.
Bouzakri K et al. 2008 Malonyl coenzyme A decarboxylase regulates lipid and glucose metabolism in human skeletal muscle. *Diabetes* 57: 1508–1516.
Cai DS et al. 2005 Local and systemic insulin resistance resulting from hepatic activation of IKK-β and NF-κB. *Nat. Med.* 11: 183–190.
Choi CC et al. 2007 Continuous fat oxidation in acetyl-CoA carboxylase 2 knockout mice increases total energy expenditure, reduces fat mass, and improves insulin sensitivity. *Proc. Natl. Acad. Sci. USA* 104: 16480–16485.
Hancock CR et al. 2008 High fat diets cause insulin resistance despite an increase in muscle mitochondria. *Proc. Natl. Acad. Sci. USA* 105: 7815–7820.
Hirosumi J et al. 2002 A central role for JNK in obesity and insulin resistance. *Nature* 420: 333–336.
Itani SI et al. 2002 Lipid-induced insulin resistance in human muscle is associated with changes in diacylglycerol, protein kinase C, and IκB-α. *Diabetes* 51: 2005–2011.
Koves TR et al. 2008 Mitochondrial overload and incomplete fatty acid oxidation contribute to skeletal muscle insulin resistance. *Cell Metab.* 7: 45–56.
McGarry JD 2002 Dysregulation of fatty acid metabolism in the etiology of type 2 diabetes. *Diabetes* 51: 7–18.
Özcan U et al. 2004 Endoplasmic stress links obesity, insulin action, and type 2 diabetes. *Science* 306: 457–461.
Plomgaard P et al. 2005 Tumor necrosis factor-α induces skeletal muscle insulin resistance in healthy human subjects via inhibition of Akt substrate 160 phosphorylation. *Diabetes* 54: 2939–2945.
Song B et al. 2008 *Chop* deletion reduces oxidative stress, improves β cell function, and promotes cell survival in multiple mouse models of diabetes. *J. Clin. Invest.* 118: 3378–3389.

Insulin Resistance and SOCS Proteins

Rui L et al. 2002 SOCS-1 and SOCS-3 block insulin signaling by ubiquitin-mediated degradation of IRS1 and IRS2. *J. Biol. Chem.* 277: 42394–42398.
Ueki K et al. 2004 Suppressor of cytokine signaling 1 (SOCS-1) and SOCS-3 cause insulin resistance through inhibition of tyrosine phosphorylation of insulin receptor substrate proteins by discrete mechanisms. *Mol. Cell. Biol.* 24: 5434–5446.

K_{ATP} Channels and Glucose-Stimulated Insulin Release

Inagaki N et al. 1995 Reconstitution of I_{KATP}: an inward rectifier subunit plus the sulfonylurea receptor. *Science* 270: 1166–1170.
Lam TKT et al. 2005 Hypothalamic sensing of circulating fatty acids is required for glucose homeostasis. *Nat. Med.* 11: 320–327.
Ma WY, Berg J and Yellen G 2007 Ketogenic diet metabolites reduce firing in central neurons by opening K_{ATP} channels. *J. Neurosci.* 27: 3618–3625.
Nichols CG 2006 K_{ATP} channels as molecular sensors of cellular metabolism. *Nature* 440: 470–476.

Islet β-Cell Failure

Brownlee M 2005 The pathobiology of diabetic complications. *Diabetes* 54: 1615–1625.
Huang CJ et al. 2007 High expression rates of human Islet amyloid polypeptide induce endoplasmic reticulum stress-mediated β-cell apoptosis, a characteristic of humans with type 2 but not type 1 diabetes. *Diabetes* 56: 2016–2027.
Robertson RP 2004 Chronic oxidative stress as a central mechanism for glucose toxicity in pancreatic Islet beta cells in diabetes. *J. Biol. Chem.* 279: 42351–42354.

Chapter 4
Metabolic Program Execution and Switching

Many, if not most, of the rate-limiting enzymes involved in executing metabolic programs are regulated at the transcriptional levels. A small number of transcription factors and cofactors connect upstream hormonal and metabolic signals to the downstream metabolic enzymes. Chief among these are the nuclear receptors, PGC-1α cofactors, and FoxO transcription factors. These function as downstream targets of hormonal signals and as sensors and effectors of cellular nutrient and energy status signals. These will be discussed in this chapter along with another member of this elite group, the NAD^+/NADH-sensitive sirtuins. A fifth group of important factors, the SREBPs, will be explored in the next chapter where the focus will be on the management of cholesterol.

4.1 Nuclear Receptors Are Ligand-Activated Transcription Factors

Nuclear receptors are intracellular transcription factors (TFs) that regulate gene expression in a ligand-dependent way. There are two main groups of nuclear receptor ligands. One group consists of steroid hormones. Steroid hormones, unlike polypeptide growth factors and cytokines, are able to pass through the plasma membranes of target cells and bind to nuclear receptors. The second group of ligands is intracellular lipid metabolites. The nuclear receptors that bind these molecules were originally referred to as orphan receptors but are now recognized as functioning as lipid sensors and regulators of lipid (and carbohydrate) metabolism and thus of overall energy balance. The focus in this chapter will be on this second group of nuclear receptors.

Transcription coactivators and corepressors work in conjunction with this family of transcription factors to promote and repress gene transcription. These transcription cofactors carry out several functions essential for gene transcription. This includes chromatin remodeling, histone modification, DNA unwinding, and RNAP II recruitment. In the absence of activating signals, many if not most TFs associate with corepressors that prevent initiation of transcription. Initiating signals stimulate the decoupling of the corepressors from the TFs and promote recruitment and attachment of the coactivators in their place.

M. Beckerman, *Cellular Signaling in Health and Disease*,
Biological and Medical Physics, Biomedical Engineering,
DOI 10.1007/978-0-387-98173-4_4, © Springer Science+Business Media, LLC 2009

Coactivators often serve as signaling nodes for elements of upstream pathways. A key aim of this chapter is to examine the roles of the prototypic transcription coactivation PGC-1 as it operates in conjunction with the PPAR family members in energy-balance pathways.

The nuclear receptor family has 49 mammalian members. These receptors can be grouped according to their ligand-binding and dimerization properties as belonging to either Type 1 or Type 2. Members of the Type 1 group are not able to influence transcription in the absence of ligand binding. Rather, in the unbound state they associate with and are sequestered by heat-shock proteins. In response to ligand binding, they form homodimers and in that form they bind palidromic DNA repeats. Type 2 nuclear receptors can influence transcription, typically as inhibitors, in the absence of ligand binding. They form heterodimers with 9-cis retinoid receptors (RXR) and in that mode bind direct DNA repeats.

Representative examples of the two families of nuclear receptors are presented in Table 4.1. The first column lists Type 1 ligands and their nuclear receptors. This group is often referred to as the steroid hormone family because it contains the sex hormones (androgens, estrogens, and progestins), and adrenal hormones, mineralocorticoids such as aldosterone, and glucocorticoids such as cortisol. The type 2 family members have been subdivided into two subtypes. Type 2a has as its members extracellular ligands whereas the second subgroup, Type 2b, encompasses the intracellular lipids (sterol, fatty acid, bile acid, and xenobiotic) messengers. The constitutive androstane receptor and pregnane X receptor have a primary role in the detoxification of endobiotics and foreign chemicals (xenobiotics). The farnesoid X and liver X receptors regulate bile acids and cholesterol transport and absorption. There are three peroxisome proliferator-activated receptor (PPAR) isotypes. These are designated as PPARα, PPARγ, and PPARβ or alternatively PPARδ. The first of these, PPARα, acts predominately in the liver, heart, and skeletal muscle; the second, PPARγ, operates in adipose tissue, and the third, PPARβ/δ, is broadly distributed among many different tissue types. These receptors regulate lipid and glucose metabolism and homeostasis.

Table 4.1 Nuclear receptor groups and representative members

Type 1	Type 2a	Type 2b
Androgens (AR)	Ecdysone (EcR)	Constitutive Androstane Receptor (CAR)
Estrogens (ER)	All-trans Retinoids (RAR)	Farnesoid X, Bile Acids (FXR)
Glucocorticoids (GR)	9-cis Retinoids (RXR)	Liver X, Oxysterols (LXR)
Mineralocorticoids (MR)	Thyroid Hormone (TR)	Peroxisome Proliferator-Activated, Fatty Acids (PPAR)
Progestins (PR)	Vitamin D_3 (VDR)	Pregnane X/Steroid and Xenobiotic, Xenobiotics (PXR/SXR)

4.2 Nuclear Receptors Contain Five or Six Domains

Nuclear receptors have a modular architecture consisting of five or six domains labeled A through E or F. They each have an N-terminal ligand-independent activation function AF-1, or A/B domain, followed by a DNA binding C domain consisting of a pair of zinc fingers organized into a globular structure. The next region of the chain, the D domain, is a flexible hinge region that facilitates DNA binding and dimerization, and lastly the E domain is a large C-terminal domain required for ligand binding and dimerization. The ligand binding region may also contain a ligand-dependent activation function (AF-2) (Helix 12) domain, and in some family members there is an extreme C-terminus F domain (Fig. 4.1).

The length of the A/B domain varies from family member to family member and is subject to alternative splicing. The DNA binding domain is highly conserved and conveys sequence-specific DNA recognition. This region is characterized by a pair of zinc finger motifs. Other elements of the C domain are referred to as boxes (A-, D-, P-, and T-), and these DNA binding elements make contact with either the major or minor groove of the DNA.

Domain E is the key to the signaling actions of the nuclear receptor. As shown in Fig. 4.2 the LBDs of these receptors consist of 12 α-helices, 11 of which form a compact ligand-binding pocket while the 12th serves as a flexible lid. The lid is controlled by the structure of the ligand through allosteric mechanisms

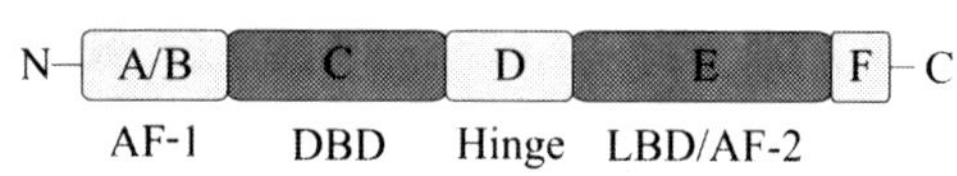

Fig. 4.1 Modular nuclear receptor structures

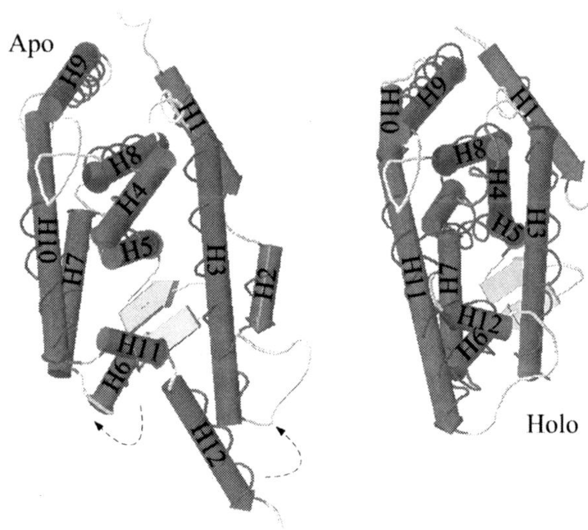

Fig. 4.2 Structure of the ligand-binding domain of the Liver X receptor in unliganded (apo) and liganded (holo) conformations determined through X-ray crystallography. The figure was prepared using Jmol with atomic coordinates deposited in the PDB under accession codes 1lbd (apo) and 1fby (holo)

and the action of this domain can be likened to a molecular mousetrap. In the absence of ligand binding, helix 12 is oriented in an open position away from the other helices. Ligand binding triggers the movements of several of the helices, the most important of which are helices 12 and 11. Helix 12, in particular, swings over the binding pocket and shuts the trap while helix 11 aligns itself with helix 10 to solidify the wall of the binding pocket. Other, somewhat smaller movements involve helices 2 and 3. Overall, the structural core defined by helices 1, 4, 5, 7, 8, 9, and 10 is rigid and the portion involving helices 2, 3, 11, and 12 is dynamic.

The binding of coactivators and corepressors is dependent upon many of the same structural features as are involved in ligand binding. Corepressors contain binding motifs called corepressor nuclear receptor boxes (CoNR boxes). These binding motifs are of the general form LXXXIXXL, which folds into a three-turn helix that binds a hydrophobic groove formed by helices 3 and 4 and stabilizes the NR in an inactive conformation. When a ligand binds to AF-2, helix 12 swings into a position in this region and disrupts corepressor binding. Coactivators typically possess LXXLL motifs. These shorter sequences form two-turn helices that are able to fit into the groove vacated by the corepressors, defined now by helices 3, 4, and 12, and stabilize the active conformation. The overall behavior is that of a molecular switch centered about the exchange of one type of coregulator for another.

4.3 The CAR Activates and Deactivates in a Manner Distinct from Other Nuclear Receptors

The constitutive androstane receptor (CAR) is active in the absence of ligand binding and deactivates when bound to its ligand. The receptor is abundantly expressed in liver and intestine, and controls expression of genes involved in clearing xenobiotics and endogenous toxins. In the absence of inducers, this receptor is sequestered in the cytoplasm where it forms a complex with the Hsp90 chaperone and a cochaperone unique to the CAR. Xenobiotics stimulate the translocation of the receptor to the nucleus where it binds its RXR coreceptor to initiate transcription of genes involved in their clearance – P450s, Phase II conjugating enzymes, and efflux transporters.

The structural features that endow the CAR with the ability to remain in an active conformation in the absence of ligand binding are depicted in Fig. 4.3. As can be seen by comparing this figure to the previous figure for RXR in its apo and holo conformations, the positioning of helix 12 is the key to the structure. In the CAR, helix 12 has a closed "mousetrap" orientation in the absence of ligand binding. The helix is held rigidly in place by a short helix (helix HX) whereas other NRs have a flexible loop in this locale. Secondly the CAR does not possess the usual C-terminal segment and as a result helix 12 is anchored at the other end thus further stabilizing the active state. This situation is disrupted

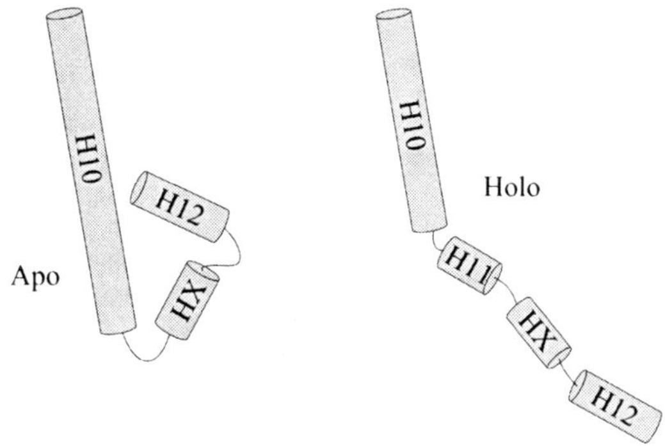

Fig. 4.3 Orientations of the mousetrap helices of the ligand-binding domain (LBD) of the constitutive androstane receptor (CAR) in activate (apo) and inactivate (holo) conformations

by the presence of a ligand which flips out helix 12 into an open mousetrap orientation resembling the RXR apo state.

4.4 Peroxisome Proliferator-Activated Receptors Are Lipid Sensors and Effectors

Peroxisome proliferator-activated receptors function as intracellular fatty acid sensors and regulators of energy balance. All three of the PPAR isoforms contribute to this key function as well as to inflammation control and insulin sensitivity (Fig. 4.4). Members of this family bind PPAR response elements (PPREs) as heterodimers with members of the RXR family of nuclear receptors providing feedback control through which lipid metabolites regulate the transcription of genes involved in lipid metabolism and homeostasis.

PPARα is expressed in tissues that are metabolically active. These include liver, kidney, heart, skeletal muscle, and brown fat. Other places where these receptors are active are macrophages, endothelial cells, and smooth muscle cells. In tissues such as liver and heart, it senses long-chain fatty acids and in response to elevated levels of this lipid stimulates energy production and maintains lipid homeostasis. PPARα upregulates genes involved in peroxisomal and mitochondrial β-oxidation and genes needed for lipoprotein synthesis and metabolism including those required for the cellular uptake and shuttling of fatty acids and for HDL assembly.

Ligand binding activates at least three families of genes – those encoding P450 enzymes, ABC transporters, and lipid shuttle proteins. The latter, termed liver fatty acid-binding proteins (LFABPs), shuttle lipid ligands to PPARα and PPARγ receptors in the nucleus. The P450 enzymes catalyze the conversion of the ligands to inactive forms while the ABC transporters facilitate the passage of the lipids from the cytosol into peroxisomes or other cellular compartments.

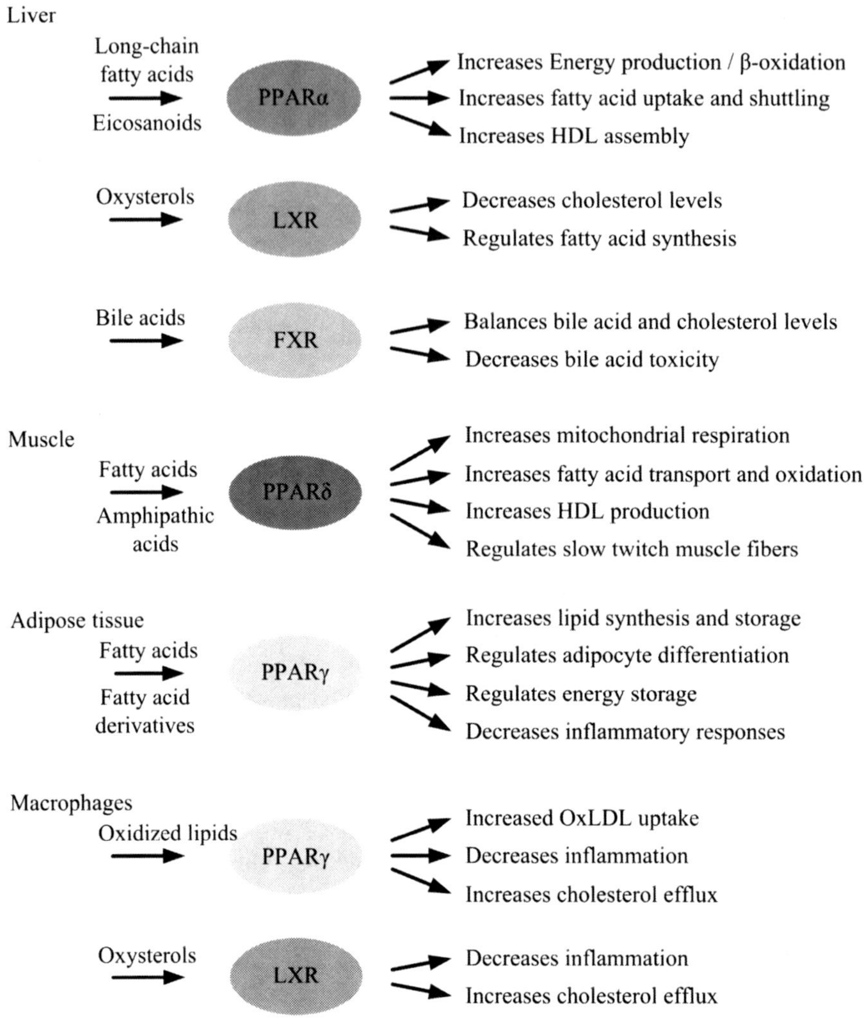

Fig. 4.4 Transcriptional actions of Type 2b nuclear receptors in different tissues and cell types

PPARδ is broadly expressed in many tissues. This receptor has a large binding pocket enabling it to bind to a variety of fatty acid and amphipathic acid ligands. In skeletal muscle, heart, and adipose tissue this member of the PPAR family targets genes involved in fatty acid transport and oxidation; it raises HDL cholesterol levels, and in skeletal muscle where it is more strongly expressed than the other isoforms it stimulates fatty acid oxidation and mitochondrial respiration, and regulates slow twitch muscle fibers.

PPARγ is expressed in adipose tissue, in epithelial and endothelial cells, and in macrophages, liver, and muscle but not as strongly in the latter. Ligands are

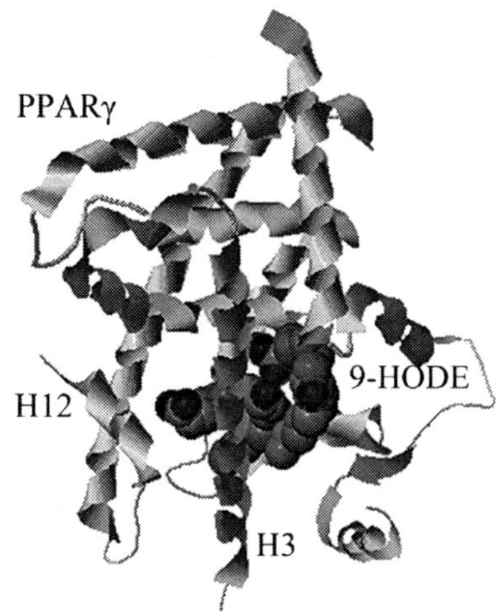

Fig. 4.5 Three-dimensional structure of the PPARγ ligand-binding domain in a complex with the oxidized fatty acid 9-HODE as determined through X-ray crystallography. This figure was prepared using Jmol with atomic coordinates deposited in the PDB under accession number 2vsr

fatty acids and fatty acid derivates. An example of FA binding is presented in Fig. 4.5. In this example, the ligand is the oxidized fatty acid 9-(s)-hydroxyoctadecadienoic acid (9-HODE). Nitrated fatty acids are another group of recently identified natural ligands for these nuclear receptors.

This nuclear receptor functions as a master regulator of adipocyte differentiation, energy storage, and glucose metabolism. It acts as an anti-inflammatory mediator through its negative regulation of pro-inflammatory cytokines in adipose tissue and macrophages. It also upregulates expression of the scavenger receptor CD36 used in the uptake of OxLDL by macrophages. This activity occurs along with this receptor's accompanying stimulation of lipid efflux from macrophages both impacting atherosclerosis.

4.5 Nuclear Receptors Require Coactivators and Corepressors

Transcription factors by themselves do not initiate transcription in eukaryotic cells. Rather, they require the assistance of cofactor complexes. These complexes contain (1) chromatin-modifying enzymes that remodel and modify chromatin structure to make the DNA more accessible to the transcription factors and the basal transcription machinery and (2) nonenzymatic cofactors that form bridges between transcription factors and the transcription machinery and serve as control points for integration of upstream signals.

There are two kinds of chromatin modifiers. There are (a) modifiers of chemical affinity between histones and DNA, and (b) hydrolyzing enzymes that break and reform bonds between histones and DNA. The first group consists of enzymes such as the histone acetyltransferases (HATs) and histone deacetylases (HDACs) that catalyze the attachment or detachment of acetyl groups to histone tails. These particular covalent modifications are usually made to amino groups situated at the end of lysine side chains. The covalent addition of the acetyl groups to the side chains reduces the net positive charge on the histone tails thereby weakening their attraction to the DNA strands. The addition of these groups counteracts the natural tendency for the chromatin fibers to fold into compact nucleosomal units making transcription difficult to impossible. Upon acetylation, the promoter sites become far more accessible to the transcription machinery. Other covalent additions catalyzed by chromatin modifiers are phosphorylation, methylation, ubiquitylation, SUMOylation, and ADP-ribosylation.

The second major group of chromatin modifiers consist of enzymes and enzyme-bearing complexes that use the energy of ATP hydrolysis to disrupt interactions between histones and the DNA. The switch/sucrose nonfermentable (SWI/SNF) complex is a prominent example of this type of chromatin modifier. A second prominent family of large complexes is the imitation SWI (ISWI) group and a third family is the Mi-2 group.

The nonenzymatic cofactor group consists of bridging cofactors that tend to form large or small complexes that coordinate transcription by serving as platforms for recruitment of other cofactors and as control points where upstream signals can be integrated together in a way that influences transcription. A prominent example of a large bridging cofactor complex is the Mediator complex.

Nuclear receptors are a prominent example of transcription factors that require the assistance of smaller cofactor complexes functioning either as coactivators or corepressors. Transcriptional coactivators are recruited by nuclear receptors to transcription sites and once there enhance transcription. As their name indicates, transcriptional repressors reduce transcriptional activity. These regulators typically bind unliganded nuclear receptors, and are displayed by the activators following nuclear receptor ligand binding. Transcriptional regulators are themselves regulated by upstream signaling events conveyed to them through phosphorylation, methylation, acetylation, and other posttranslational modifications. The first coactivators and corepressors to be discovered were SRC-1, N-CoR, and SMRT. Since their discovery in 1995, about 300 transcriptional regulators have been identified. Some of the most prominent members of these regulatory complexes involved in the maintenance of energy balance are listed in Table 4.2.

Protein arginine *N*-methyltransferases (PMRT) catalyze the transfer of methyl groups from S-adenosylmethionine to guanidine nitrogens on arginine residues. CARMI, a PMRT, methylates hostone H3, CBP/p300, and p/CIP, a HAT that functions as a bridging factor for assembly of other cofactors. p/CIP has several LXXLL motifs that mediate binding to liganded NRs. Its C-terminal region has a pair of transcription activation domains, designated as AD1 and AD2. AD1 binds to CBP, p300, and PCAF/GCN5; AD2 provides an interface to PRMTs such as CARMI. The two corepressors, NCoR and SMRT, associate with holocorepressor complexes. They are not enzymes but instead promote the recruitment of HDACs and other corepressors such as HDACs, and serve as a control point for upstream kinases such as Akt, MAPKs, and casein kinase 2 (CK2).

Table 4.2 Prominent transcriptional cofactors

Transcriptional regulator	Physiological role(s)
PGC-1	Scaffold for assembly of coactivators and corepressors; signaling node
CBP, p300	HAT
PCAF/GCN5	HAT
p160 family: SRC-1/NCoA1 TIF2/GRIP-1/SRC-2 p/CIP/ACTR/SRC-3	HAT
CARMI	Arginine methyltransferase
N-CoR	Recruits HDACs and DNA methylase factors
SMRT	Recruits HDACs

SRC-1: steroid receptor coactivator 1; CBP: CREB binding protein; p/CAF: p300/CBP associated factor; TIF2: transcriptional intermediary factor 2; pCIP: CBP-interacting protein; SMRT: silencing mediator for retinoid and thyroid receptors; N-CoR: nuclear receptor corepressor.

In the absence of ligand binding, nuclear receptors form complexes with corepressors that bind nuclear receptors through their corepressor interaction motifs (e.g., LXXXIXXL as discussed earlier). Other proteins recruited to the complex include the adapters TBL and TBLR1. These proteins recruit components of the ubiquitin-mediated proteolysis machinery needed to effect the displacement of the corepressors with coactivators. This replacement is initiated by ligand arrival. Coactivators possessing LXXLL motifs replace the inhibitory proteins resulting in activation of transcription. The overall process is a cyclical one with coactivator release following the activated transcription stage as pictured in Fig. 4.6.

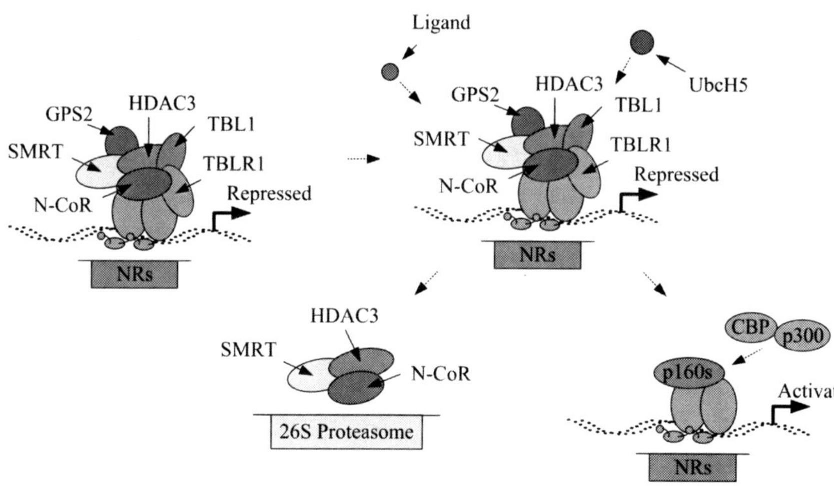

Fig. 4.6 Cycle of nuclear receptor repression and activation

4.6 PGC-1 Scaffold Protein in Regulation of Lipid Homeostasis

Members of the peroxisome proliferator-activated receptor γ coactivator 1 (PGC-1) family serve as platforms for recruitment and aggregation of factors needed for gene transcription. They are highly modular in design with domains specialized for binding nuclear receptors and other transcriptional activators, for binding corepressors, and for binding RNA. These coactivators are also highly regulated by upstream signals and as a result they function as key signaling nodes for the control of genes involved in energy balance.

There are three PGC-1 isoforms. The domain organization of PGC-1α, the most intensively studied of these, is depicted in Fig. 4.7. PGC-1β is similar in structure and function while the third isoform, PGC-related coactivator (PRC), is less homologous. As shown in Fig. 4.7 these proteins possess an N-terminal activation domain that binds nuclear receptors and transcription coactivators. The central region serves as a platform for assembly of corepressors, while the C-terminal domain mediates RNA binding. The N-terminal region designated in the figure as LRR encompasses several nuclear binding LXXLL and LLXXL sequences.

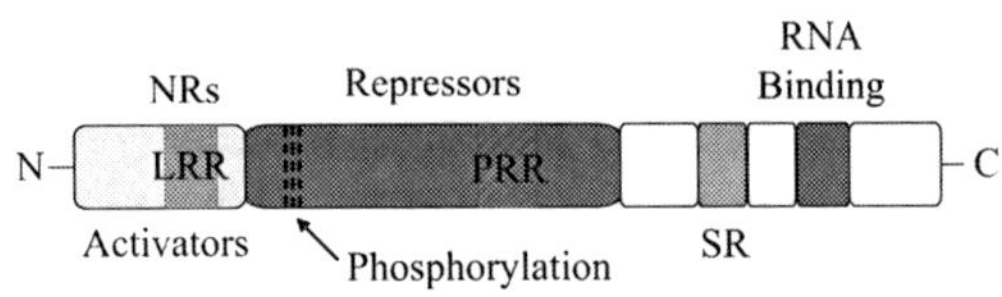

Fig. 4.7 PGCα structure. LRR: leucine-rich region; PRR: proline-rich region

4.7 FoxOs Mediate Survival, Metabolic, and Stress Responses

The insulin and insulin-like growth factor (IIS) pathway leading to PI3K and Akt activation promotes survival. The two most prominent of these targets of Akt are the pro-apoptotic Bcl2 family protein Bad and the forkhead transcription factor family (Fig. 4.8). The Bad protein is a member of the so-called BH3-only subgroup of Bcl2s. These regulators of apoptosis will be discussed in detail in Chapter 11. What is important here is to note that these proteins, when activated, translocate to the mitochondria where they promote apoptosis. This translocation is inhibited by Akt phosphorylation. This posttranslational

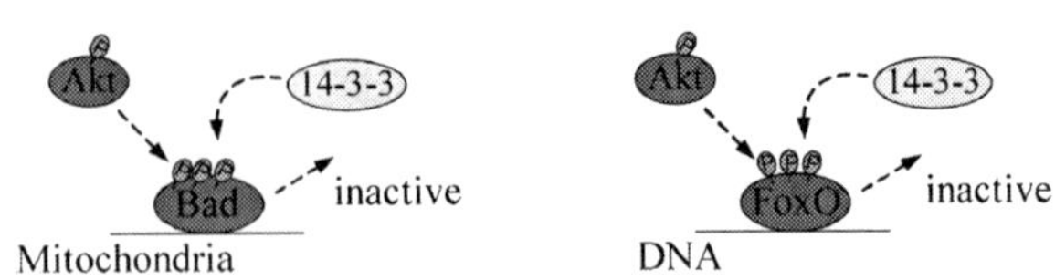

Fig. 4.8 Survival signaling by Akt in which phosphorylation of the pro-apoptotic factor Bad (*left panel*) and the FoxO transcription factors (*right panel*) immobilizes these proteins in the cytosol

modification creates binding sites for 14-3-3 proteins, which then bind and immobilize the Bad proteins in the cytosol away from the mitochondria.

A similar set of actions occurs with regard to the FoxOs. In the absence of Akt phosphorylation, the FoxOs localize to the nucleus where they stimulate transcription of pro-apoptotic genes. Phosphorylation by Akt at T24, S256, and S319 creates 14-3-3 binding sites and subsequent binding by these proteins antagonizes DNA localization and transcription. Instead, the 14-3-3 bound FoxOs shuttle out of the nucleus to the cytosol and become inactive. The appellation "survival" is used as shorthand for the antiapoptotic actions brought on by insulin and insulin-like growth factors signaling.

FoxOs are also regulated by acetylation. A number of transcription coactivators, most notably, p300 and CBP acetylate the FoxOs at multiple sites, as indicated in Fig. 4.9(c). These actions enhance the ability of transcribe genes with pro-apoptotic roles. The sirtuins, especially SIRT1 (and SIRT2), act as

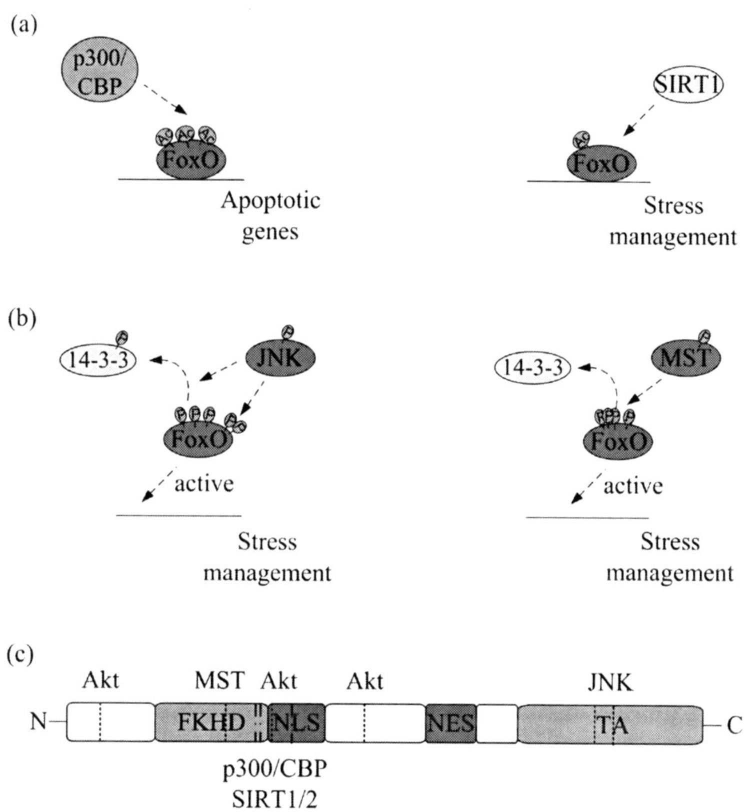

Fig. 4.9 Direction of FoxOs toward either apoptotic or stress management responses by acetylation and phosphorylation. (**a**) Effect of acetylation and (**b**) effects of phosphorylation. (**c**) Structure of the FOXO1 protein

FoxO deacetylases, and these removals promote a shift in promoter targets from those that support apoptosis to those that support stress management as depicted in Fig. 4.9(a). Under this label are genes involved in cell cycle arrest, allowing time for repair of DNA damage, and those that relieve oxidative stress, a major cause of DNA, protein, and lipid damage. A key contributor to these actions is the transcription coactivator PGC1. This factor, as well as appropriate mixes of additional cofactors, is recruited to the key promoter sites to stimulate transcription of metabolic, stress response, and/or apoptotic genes according to the integrated mix of upstream signals.

Phosphorylation has several different effects upon the actions by FoxOs. While phosphorylation by Akt disables FoxO transcription, other kinases that phosphorylate the FoxOs at sites different from those targeted by Akt promote stress responses. Two examples of this type of response are the phosphorylation of FoxO3 by MST in response to oxidative stresses and of FoxO4 by the JNK stress-responsive MAP kinase. These actions are illustrated in Fig. 4.9(b). In these instances, phosphorylation does not prevent transcription but rather changes which promoters and/or co-factors are bound.

The preceding set of exemplars brings into sharp focus a number of fundamental properties of the IIS signaling pathway(s). Increased insulin signaling counters insulin resistance and in doing so promotes health by reducing the chances that Type 2 diabetes will develop. However, it achieves this aim in part by reducing the likelihood that subject cells will undergo apoptosis in response to elevated stresses including those associated with DNA damage. As will be discussed in detail in Part II, apoptosis is one the body's chief anticancer barriers, and sustained strong IIS signaling increases the chances that cancer will develop later in life by making it harder to remove damaged cells.

4.8 14-3-3 Protein Function as Small, Mobile Phosphoprotein Binding Modules

14-3-3 proteins are small, independent phosphoserine and phosphothreonine binding modules. These mobile modules act as signaling chaperones that shuttle appropriately phosphorylated-signaling proteins out and away from sites of action to parking locations in the cytosol. Several examples of this type of regulation have been encountered already, the most recent being Bad and the FoxO proteins. These proteins are components, many if not most, of the signaling pathways implicated in type 2 diabetes and cancer.

Two consensus motifs are recognized – RSXpSXP and RXXXpSXP – but there are departures, small and large, and in some cases phosphorylation is not even required. In the case of the FoxOs, there is a high-affinity consensus site in the N-terminal region. Interactions take place between the 14-3-3s and this site along with a second site in the FKHD domain. The 14-3-3 proteins function as dimers and these may function through hierarchical binding to their ligands.

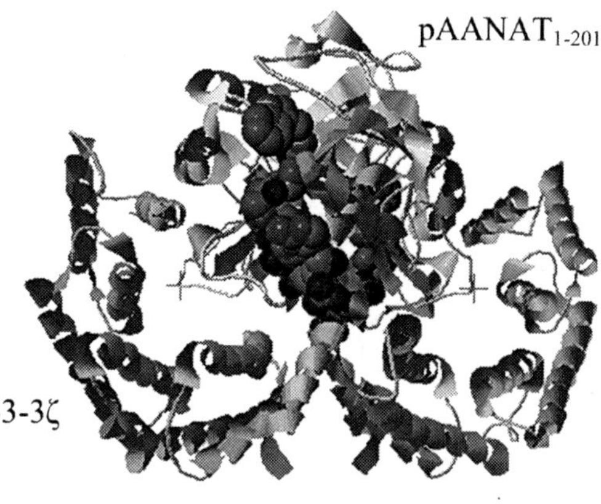

Fig. 4.10 Three-dimensional structure of the 14-3-3ζ dimer in a complex with a peptide fragment spanning amino acid residues 1-201 from serotonin *N*-acetyltransferase AANAT determined by means of X-ray crystallography. The figure was prepared using Jmol with atomic coordinates deposited in the PDB under accession number 1ib1

The emerging picture is one in which a high-affinity binding site serves as a gatekeeper. Binding to this site is followed by binding to a second, low-affinity site leading to formation of a 2:2 complex, or alternatively with a 2:1 stoichiometry in which two 14-3-3s bind one FoxO protein.

The 14-3-3 proteins have a high α-helical content and form rigid bowl-like scaffolds that can accommodate numerous structures. This biophysical property is illustrated in Fig. 4.10 for a 14-3-3 dimer bound to serotonin *N*-acetyltransferase (AANAT). The 14-3-3s differ from other phosphoprotein-binding modules in that they are capable of regulating ligand protein function by inducing conformational changes. They influence Daf-16 (FoxO) DNA binding and in the case of AANAT modify its enzymatic activity.

4.9 Gluconeogenesis in the Liver Is Stimulated by Glucagon and Repressed by Insulin

One of the many actions of insulin is to repress gluconeogenesis in the liver. The key steps in this process are sketched in Fig. 4.11. As was the case for many of the other actions of insulin, Akt is the key signal transducer. In response to insulin signals activated Akt phosphorylates SIK2 at serine 358. This kinase is a member of the salt-inducible kinase (SIK) family of serine/threonine kinases. Upon phosphorylation, SIK2 phosphorylates the transducer of regulated CREB activity 2 (TORC2) protein at serine 171. This posttranslational modification creates a binding site for 14-3-3 proteins, which bind to TORC2 and sequester it in the cytosol.

In the absence of Akt-mediated phosphorylation, FoxOs such as FoxO1 promote gluconeogenesis in the liver. In particular, this transcription factor stimulates expression of glucose 6 phosphatase (G6P) and the rate-limiting enzyme phosphoenol pyruvate carboxykinase (PEPCK). Insulin-promoted Akt signaling terminates these operations by sequestering TORC2 in the cytosol during feeding.

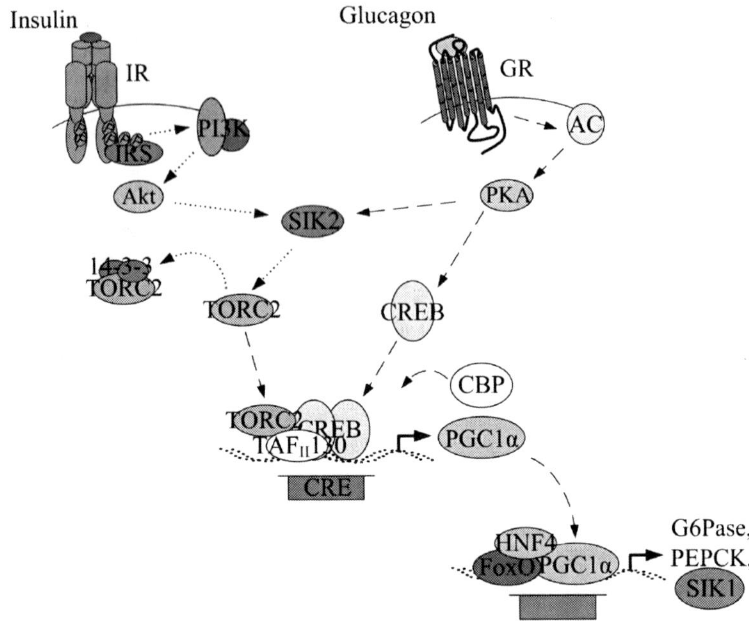

Fig. 4.11 Joint regulation by insulin and glucagon of TORC2-mediated gene transcription leading first to upregulation of PGC1α and second to upregulation of the enzymes G6Pase and PEPCK, and the end-point negative regulator SIK1

This action is countered by glucagon, which stimulates hepatic gluconeogenesis during fasting. This process is controlled by CREB, a transcription factor that is one of the primary downstream targets of protein kinase A signaling. A sketch of the glucagon signaling route that begins with glucagon and ends at CREB promoter sites is shown alongside the insulin pathway in Fig. 4.11. The glucagon receptor (GR) is a member of the G-protein-coupled receptor family and acts through heterotrimeric G-protein alpha subunits that activate adenylyl cyclase (AC), which produces cAMP second messengers. Increased production of this second messenger results in the activation of protein kinase A, which targets the cAMP-responsive element binding (CREB) protein. Once CREB is phosphorylated by PKA, it dimerizes and collects at cAMP responsive element (CRE) promoter sites where it recruits several cofactors to stimulate transcription.

In addition to stimulating CREB, glucagon-activated protein kinase A opposes the actions of Akt upon SIK2 thereby facilitating TORC2 relocation and retention in the nucleus. The assemblage of CREB, FoxOs, and TORC2 recruits several other essential cofactors such as CBP and TAF$_{II}$130 resulting in the transcription of the gene encoding PGC1α. Upregulation of PGC1α is a crucial step. This cofactor stimulates transcription of the rate-limiting enzyme PEPCK along with G6Pase to stimulate gluconeogenesis and also triggers transcription of SIK1, which acting as an end-point negative regulator terminates the process by phosphorylating TORC2.

4.10 Catecholamine Signaling Targets PGC1α to Promote Diet-Induced Thermogenesis in Brown Adipose Tissue

White adipose tissue is a repository of lipid droplets. These fat deposits, containing mostly triglycerides, are used as fuel when glucose supplies run low. Lipid droplets are not inert amalgams of lipids but instead are dynamic organelles. Their triglyceride contents are encased in a monolayer of phospholipids, cholesterol, and lipid droplet-associated proteins. Several kinds of phospholipids are found in the surface layer including phosphatidylcholine (PC), phosphatidylethanolamine (PE), and phosphatidylinositol (PI). Vesicle transport proteins are a prominent group of lipid droplet-associated proteins (LDAPs). In this group are small GTPases such as Rab18 and caveolin. The latter is an integral membrane protein found in plasma membrane microdomains called caveolae. A third abundant group of proteins are members of the perilipin-ADRP-TIP47-related protein domain (PAT) family. Perilipin assists in the storage of the lipids; it interacts with hormone-sensitive ligase (HSL), a key enzyme involved in lipolysis and mediates its activation.

The release of lipids is controlled by signaling sent by glucagon and the catecholamines, epinephrine and norepinephrine. The catecholamines bind to α2- and β-adrenergic receptors, members of the GPCR family, that activate cAMP second messengers through G-protein stimulation of adenylyl cyclase resulting in protein kinase A activity. This kinase phosphorylates HSL on a number of residues promoting its translocation to the lipid droplets and also phosphorylates perilipin. Lipolysis, the hydrolysis of the triglycerides by ligases (mainly HSL), then ensues.

Brown adipose tissue mediates an adaptive (nonshivering) thermogenesis response to cold exposure. It is also responsive to glucose deprivation that takes place during strenuous exercise, between meals, and at night during sleep. These cells, like the cells of white adipose tissue, contain lipid droplets and receive signals from the sympathetic nervous system. The drop in blood glucose triggers the release of catecholamines and their subsequent binding to β adrenergic receptors (βARs) stimulates thermogenesis. As was the case for WAT, the G-proteins associated with these receptors activate adenylyl cyclases, which generate cAMP and thus activate protein kinase A. In the case of catecholamine signaling, protein kinase A does not act through the CREB transcription factors but instead in the case of BAT activates Raf and the p38 mitogen-activated protein kinase cascade (Fig. 4.12). The p38 protein then activates the AFT-2 transcription factor to stimulate transcription of PGC1α. Once upregulated, the PGC1αs work along with PPAR-RXRs and ATF-2 to stimulate transcription of UCP1, which leads to the generation of heat.

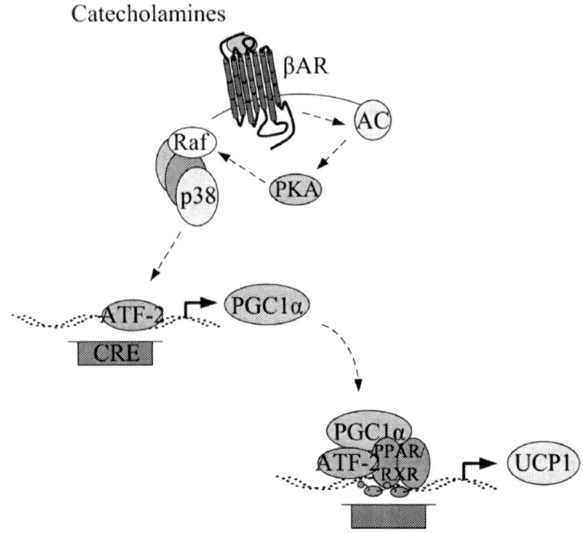

Fig. 4.12 Catecholamine binding to β adrenergic receptors stimulates thermogenesis

4.11 Caloric Restriction Extends Lifespan by Activating Protective Stress Responses

The term "caloric restriction" (CR) denotes a regimen in which the caloric input to a laboratory test subject is reduced 20–40%. This reduction in energy input dramatically extends lifespan in species ranging from yeasts to worms to flies to mice. It does not accomplish this reduction by slowing metabolism, a widely held, intuitively appealing but incorrect idea. Instead, CR seems to exert its effects by activating signaling pathways that turn on protective stress responses and initiate metabolic reprogramming. These actions delay the onset of age-related disorders – metabolic and growth related, immune system promoted, and neurodegenerative.

A highly conserved set of signaling agents and pathways underlie the longevity-promoting effects of CR. Foremost among these are the insulin/insulin-like growth factor pathway and a small set of transcription factors and cofactors, most notably the sirtuins, FoxOs, and PGC-1α. Hormonal signaling relayed through the insulin and insulin-like growth factor pathways regulates life span in mammals through alterations in gene expression. Insulin signaling has been discussed in depth in the previous chapter. Insulin-like growth factor signaling operates in a pathway parallel to the insulin-signaling pathway and merges with it to activate PKB and FoxO transcription factors. Mammals express four different FoxOs in place of one, Daf-16, expressed in yeasts and found to be prominently involved in lifetime extension under CR.

The sirtuins, the first of the above-mentioned families of transcription factors, are highly conserved from yeasts to worms to flies to mammals and

Table 4.3 Mammalian sirtuins, their cellular locations, and main enzymatic actions

Sirtuin	Site(s) of action	Catalytic function
SIRT1	Nucleus	Deacetylase
SIRT2	Nucleus, cytoplasm	Deacetylase
SIRT3	Mitochondria	Deacetylase
SIRT4	Mitochondria	ADP-ribosyltransferase
SIRT5	Mitochondria	Deacetylase
SIRT6	Nucleus	ADP-ribosyltransferase
SIRT7	Nucleolus	Deacetylase

seem to be centrally involved in the longevity extension properties of caloric restriction. The first identified member of this family of "longevity genes" is the Sir2 gene of the baker's yeast *Saccharomyces cerevisiae*. This gene product, along with the other members of this family, function as histone deacetylases, a post-translational modification that in general leads to gene silencing. (The acronym Sir2 stands for silent information regulator 2.) Unlike other histone deacetylases the sirtuins require the metabolite NAD^+ as a cofactor thus linking their action to cellular metabolism. There are seven mammalian sirtuins and these have been listed in Table 4.3. All require NAD^+ as a cofactor, and two of them, SIRT4 and SIRT6, function exclusively as ADP-ribosyltransferases.

4.12 SIRT1 Promotes Fatty Acid Oxidation in Liver and Skeletal Muscle

When glucose supplies run low as they do under caloric restriction, SIRT1 shifts the metabolic program in the liver and skeletal muscle toward fatty acid oxidation. It accomplishes this shift in two ways. At the transcriptional level it deacetylates the key coactivator PGC-1α (Fig. 4.13(b)). This posttranslational alteration leads to an increased transcription of genes encoding enzymes needed for fatty acid oxidation such as pyruvate dehydrogenase kinase 4 (PDK4). This enzyme phosphorylates pyruvate dehydrogenase (PDH) thereby deactivating an enzyme required for entry of pyruvate into the Krebs cycle. It also acts on AMPK, and by deacetylating LKB1 promotes the binding of LKB1 to the scaffold proteins STRAD, and this leads to activation of AMPK and increased fatty acid oxidation (Fig. 4.13(a)).

Resveratrol (RSV) is a polyphenol found in red wine. It has been found to counter age-related disorders ranging from atherosclerosis to cancers to neurodegeneration. In exploring the mechanisms underlying its disease-protective effects, it has been found to bind to SIRT1 and allosterically stimulate its enzymatic activity. As a result SIRT1 is more effective at binding and deacetylating PGC-1α, leading to PGC-1α activation and gene transcription of

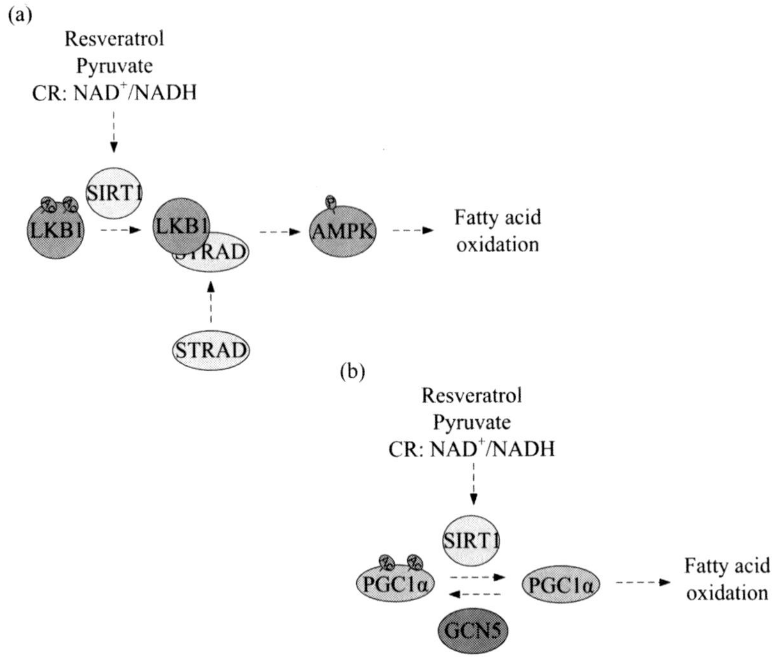

Fig. 4.13 Promotion of fatty acid oxidation by SIRT1. (**a**) Regulation of LKB1-AMPK activity by SIRT1 and (**b**) regulation of PGC1α activity by SIRT1

stress-reducing and metabolic genes. Under fasting conditions, SIRT1 is induced in WAT and skeletal muscle by the elevation in NAD + /NADH ratio.

Metabolic reprogramming by SIRT1 and PGC-1α in energy-responsive tissues may well underlie the positive effects of caloric restriction. Among the alterations brought on are changes in adaptive thermogenesis and mitochondrial gene expression in brown adipose tissue and muscle. The lowering of glucose levels through dietary means stimulates metabolic reprogramming in muscle resulting in an increased utilization of fatty acid oxidation and oxidative phosphorylation, and increased mitochondrial biogenesis and oxygen consumption. As a result, laboratory subjects are better able to resist diet-induced obesity; there is an increased responsiveness to insulin, and a decrease in IGF-1 levels.

Further Reading

Nuclear Receptors and Their Ligand-binding Domain

Gronemeyer H, Gustaffson JA and Laudet V 2004 Principles for modulation of the nuclear receptor superfamily. *Nat. Rev. Drug Disc.* 3: 950–964.

Nagy L and Schwabe JWR 2004 Mechanism of nuclear receptor molecular switch. *Trends Biochem. Sci.* 29: 317–324.

Wurtz JM et al. 1996 A canonical structure for the ligand-binding domain of nuclear receptors. *Nat. Struct. Biol.* 3: 87–94.

PGC-1

Finck BN and Kelly DP 2006 PGC-1 coactivators: inducible regulators of energy metabolism in health and disease. *J. Clin. Invest.* 116: 615–622.
Kelly DP and Scarpulla RC 2004 Transcriptional regulatory circuits controlling mitochondrial biogenesis and function. *Genes Dev.* 18: 357–368.
Lin J, Handschin C and Spiegelman BM 2005 Metabolic control through the PGC-1 family of transcription coactivators. *Cell Metab.* 1: 361–370.
Spiegelman BM and Heinrich 2004 Biological control through regulated transcriptional coactivators. *Cell* 119: 157–167.

Peroxisome Proliferator-Activated Receptors (PPARs)

Barish GD, Narkar VA and Evans RM 2006 PPARδ: a dagger in the heart of the metabolic syndrome. *J. Clin. Invest.* 116: 690–597.
Brown JD and Plutzky J 2007 Peroxisome proliferator-activated receptors as transcriptional nodal points and therapeutic targets. *Circulation* 115: 518–533.
Dean M, Hamon Y and Chimini G 2001 The human ATP-binding cassette (ABC) transporter superfamily. *J. Lipid Res.* 42: 1007–1017.
Itoh T et al. 2008 Structural basis for the activation of PPARγ by oxidized fatty acids. *Nat. Struct. Mol. Biol.* 15: 924–931.
Lee CH, Olson P and Evans RM 2003 Lipid metabolism, metabolic diseases, and peroxisome proliferator-activated receptors. *Endocrinology* 144: 2201–2207.
Li Y et al. 2008 Molecular recognition of nitrated fatty acids by PPARγ. *Nat. Struct. Mol. Biol.* 15: 865–867.
Semple, RK, Chatterjee VKK and O'Rahilly S 2006 PPARγ and human metabolic disease. *J. Clin. Invest.* 116: 581–589.
Waxman DJ 1999 P450 gene induction by diverse xenochemicals: central role of nuclear receptors CAR, PXR and PPAR. *Arch. Biochem. Biophys.* 369: 11–23.
Wolfrum C, Bormann CM, Börchers T and Spencer F 2001 Fatty acids and hypolipidemic drugs regulate peroxisome proliferator-activated receptors α- and α-mediated gene expression via liver fatty acid binding protein: a signaling path to the nucleus. *Proc. Natl. Acad. Sci. USA* 98: 2323–2328.

FoxOs

Greer EL and Brunet A 2005 FoxO transcription factors at the interface between longevity and tumor suppression. *Oncogene* 24: 7410–7425.
Lehtinen MK et al. 2006 A conserved MST-FoxO signaling pathway mediates oxidative-stress responses and extends life span. *Cell* 125: 987–1001.
Van der Horst A and Burgering BMT 2007 Stressing the role of FoxO proteins in lifespan and disease. *Nat. Rev. Mol. Cell Biol.* 8: 440–450.
Wang MC et al. 2005 JNK extends life span and limits growth by antagonizing cellular and organism-wide responses to insulin signaling. *Cell* 121: 115–125.

14-3-3 Proteins

Obsil T et al. 2001 Crystal structure of the 14-3-3ζ: Serotonin *N*-acetyltransferase complex: a role for scaffolding in enzyme regulation. *Cell* 105: 257–267.

Yaffe MB 2002 How do 14-3-3 proteins work? Gatekeeper phosphorylation and the molecular anvil hypothesis. *FEBS Lett*. 513: 53–57.

Hepatic Gluconeogenesis

Dentin R et al. 2007 Insulin modulates gluconeogenesis by inhibition of the coactivator TORC2. *Nature* 449: 366–369.

Koo SH et al. 2005 The CREB coactivator TORC2 is a key regulator of fasting glucose metabolism. *Nature* 437: 1109–1114.

Puigserver P et al. 2003 Insulin-regulated hepatic gluconeogenesis through FoxO1–PGC-1α interaction. *Nature* 423: 550–555.

Sirtuins

Cao W et al. 2004 p38 Mitogen-activated protein kinase is the central regulator of cyclic AMP-dependent transcription of the brown fat uncoupling protein 1 gene. *Mol. Cell. Biol.* 24: 3057–3067.

Gerhart-Hines Z et al. 2007 Metabolic control of muscle mitochondrial function and fatty acid oxidation through SIRT1/PGC1α. *EMBO J*. 26: 1913–1923.

Hou XY et al. 2008 SIRT1 regulates hepatocyte lipid metabolism through activating AMP-activated protein kinase. *J. Biol. Chem*. 283: 20015–20026.

Lan F et al. 2008 SIRT1 modulation of the acetylation state, cytosolic localization, and activity of LKB1. *J. Biol. Chem*. 283: 27628–27635.

Rogers JT et al. 2005 Nutrient control of glucose homeostasis through a complex of PGC1α and SIRT1. *Nature* 434: 113–118.

SIRT1 and Resveratrol

Baur JA et al. 2006 Resveratrol improves health and survival of mice on a high-calorie diet. *Nature* 444: 337–342.

Lagouge M et al. 2006 Resveratrol improves mitochondrial function and protects against metabolic disease by activating SIRT1 and PGC-1α. *Cell* 127: 1109–1122.

Chapter 5
Cholesterol

Cholesterol is essential for eukaryotic life. Cholesterol comprises about 20–25% of the plasma membrane lipids; it is almost entirely absent from the mitochondria and endoplasmic reticulum, but is increasingly present in the endocytic and Golgi compartments. The importance of cholesterol is a consequence of its striking biophysical ability to create a unique state called the liquid-ordered phase. In creating this state, cholesterol increases the cohesion and packing of neighboring lipids. It generates microdomains that differ in their lipid content and are demarcated by semipermeable boundaries, and regulates eukaryotic signaling.

The chapter will begin with an examination of how cholesterol influences the biophysical properties of lipid membrane. Cholesterol is an essential component of cellular membranes. As a result cholesterol homeostasis is maintained at all times in a healthy cell through multiple routes involving SREBPs, Scaps, Insigs, LXRs, and PPARγ. These regulatory pathways will be examined following the exploration of cholesterol's unique ability to generate the liquid-ordered state.

Cholesterol cannot freely move about the body but instead is conveyed by lipoprotein particles. There are five different kinds of lipoproteins each with their mix of lipids and apolipoproteins. These transport systems will be explored next along with discussions of the different kinds of apolipoproteins, cholesterol esterification, and how cholesterol enters and leaves the cells. These last-named processes are mediated by receptors for import and transporters for export, and by enzymes that catalyze the conversions back and forth between free cholesterol and cholesterol esters.

5.1 Membrane Lipids Form Gels and Liquid States

Membrane lipids form gels and liquid states through the cooperative effects of multiple weak noncovalent interactions such as van der Waals forces and hydrogen bonds. As a result there is considerable fluidity of movement within the membrane, and the constituents are free to diffuse laterally and rotationally.

M. Beckerman, *Cellular Signaling in Health and Disease*,
Biological and Medical Physics, Biomedical Engineering,
DOI 10.1007/978-0-387-98173-4_5,

The overall structure is that of a fluid of lipids and membrane-associated proteins undergoing Brownian motion. The motions of the molecules are not completely free, but instead are limited to specific regions of the membrane, giving rise to a mosaic of membrane compartments.

Membrane lipids consist of a head group, a backbone, and a tail region, as depicted in Fig. 5.1. Phosphoglycerides contain a glycerol backbone that is linked to a phosphoryl group bonded to a phosphorylated alcohol group. A different backbone component, sphingosine, is used in the sphingolipids. Lipids found in biological membranes vary in chain (tail region) length and degree of saturation. Chains vary in length, possessing even numbers of carbons typically between 14 and 24 with 16, 18, and 20. Chains with one or more double bonds are unsaturated. These bonds are rigid and introduce kinks in the chain. In a fully saturated acyl chain, the carbon–carbon atoms are covalently linked by single bonds. The carbon atoms in this type of chain can each establish a maximum possible number of bonds with hydrogen atoms, hence the term saturated. Such chains are free to rotate about their carbon–carbon bonds and can be packed tightly. In contrast, the kinks present in an array of unsaturated lipid cause irregularities or voids to appear in the array and these molecules cannot be packed as tightly.

The degree of saturation of the acyl chains and the cholesterol content influence the melting point and fluidity of the lipids in the membrane. Cholesterol plays an important role in determining the fluidity of the membrane compartments. It is smaller than the phospho- and sphingolipids as depicted in Fig. 5.1(a) and is distributed between both leaflets. As the concentration of cholesterol increases, the lipid membrane becomes less disordered gel like and more like an ordered liquid in which the lipids are more tightly packed together especially when saturated sphingolipids are present A depiction of the differences between loose and tight packing is presented in Fig. 5.1(b).

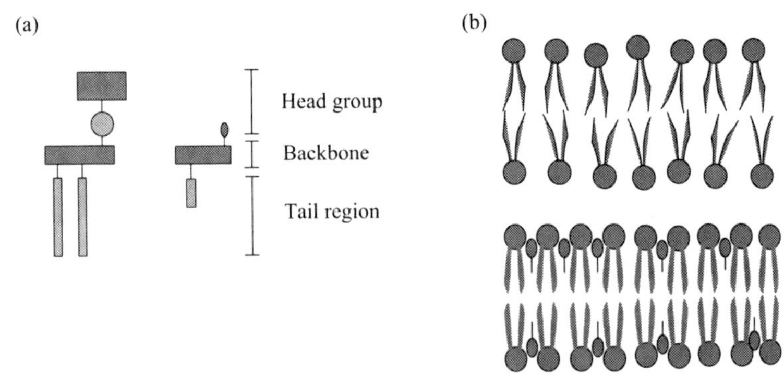

Fig. 5.1 Membrane lipids. (**a**) Structure of a membrane lipid (*left*) and cholesterol (*right*) showing their organization into head group, backbone, and tail region. (**b**) Lipid bilayers composed of unsaturated lipids (*upper*) and saturated lipids plus cholesterol (*lower*)

In the upper panel, unsaturated lipids, for example, phosphatidylcholine (PC), are loosely and irregularly packed. This may be contrasted with the bottom panel in which a mixture of cholesterol and a more ordered saturated lipid such as sphingomyelin are more tightly and regularly arranged.

The plasma membranes of eukaryotes are not uniform, but rather contain several kinds of lipid domains, each varying somewhat in its lipid composition. Compartments enriched in cholesterol and sphingolipids contain high concentrations of signaling molecules – GPI-anchored proteins in their exoplasmic leaflet and a variety of anchored proteins in its cytoplasmic leaflet. The best characterized of these signaling compartments are the caveolae (little caves). These are tiny flask-shaped invaginations in the outer leaflet of the plasma membrane. They play an important role in signaling as well as in transport. They are detergent insoluble and are enriched in coatlike materials, caveolins, which bind to cholesterol.

Another kind of lipid domain is the lipid raft. These structures differ from caveolae in that they do not contain caveolins. They are formally defined as small (10–200 nm), heterogeneous, highly dynamic sterol- and sphingolipid-enriched domains that compartmentalize cellular processes. Because of their small, dynamic nature, lipid rafts have proven to be far more elusive than caveolae to detect and characterize with respect to health and disease.

Cholesterol- and sphingolipid-enriched microdomains can float within the more diffuse lipid bilayer. The fluid and detergent insoluble properties arise from the tight packing of the long and straight acyl chains of the sphingolipids and from the high cholesterol content. The cholesterol molecules not only rigidify the compartment but also facilitate the formation of signaling complexes and the initiation of signaling by them.

The phase states that can be formed by a particular mixture of lipids can best be visualized through phase diagrams, an example of which is shown in Fig. 5.2. In this diagram, a mixture of cholesterol and two other lipids is represented. One of the lipids, Y (e.g., PC), is unsaturated and thus more fluid (l_d) like; the other, Z (e.g., sphingomyelin), is saturated and forms a gel (s_o) phase. As the

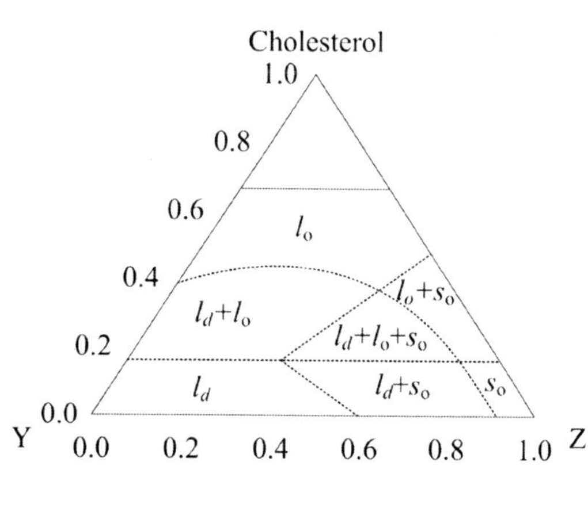

Fig. 5.2 Three-component lipid phase diagram. In this diagram, the various phases for a hypothetical mixture of cholesterol, an unsaturated lipid Y, and a saturated lipid Z are illustrated for a temperature at or near and below the physiological temperature of 37°C

Table 5.1 Biophysical properties of the common lipid bilayer phases*

Phase	Symbol	Chain segment order	Diffusion constant D ($\mu m^2/s^{-1}$)
Solid gel	s_o	High	10^{-3}
Liquid-crystalline	l_d	Low	1
Liquid-ordered	l_o	High	1

*Adapted from van Meer, Voelker, and Feigenson 2007 *Nat. Rev. Mol. Cell Biol.* 9: 112.

cholesterol content increases, the cholesterol-associated liquid-ordered phase (l_o) increasingly dominates. In between the pure phases, various combinations of two and three phases coexist. A summary of the main biophysical characteristics of the three phases is presented in Table 5.1.

5.2 Feedback Regulation of Cholesterol Synthesis by Insigs

Cholesterol is not only derived from foodstuffs but is also synthesized de novo in the cytoplasm and microsomes of cells. In the cholesterol biosynthetic pathway, twenty or more enzymes acting sequentially generate cholesterol from acetyl CoA. The five major stages in this process are sketched in Fig. 5.3. In this pathway, five products are sequentially synthesized starting with acetyl-CoA and passing through mevalonate, farnesyl diphosphate/farnesyl pyrophosphate (farnesyl PP), squalene, and lanosterol and ending with cholesterol.

Cellular cholesterol levels are regulated in several ways, one of which is by endpoint feedback inhibition. The key enzyme targeted for inhibition is 3-hydroxy-3-methylglutaryl coenzyme A reductase, or HMG CoA reductase (HMGR-CoA). This enzyme catalyzes the key reaction in step (1), and is the overall rate-limiting enzyme in cholesterol synthesis, steps (1) through (5). In more detail, in step (1) acetyl-CoA is converted to acetoacetyl-CoA by the enzyme acetoacetyl CoA thiolase, and this first intermediate product is then

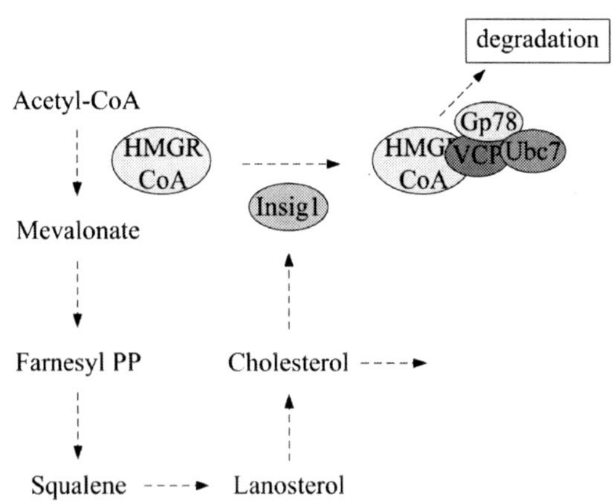

Fig. 5.3 Synthesis of cholesterol and its feedback regulation through HMG-CoA: Insig1-mediated degradation

converted to HMG-CoA by HMG-CoA synthase. HMG-CoA reductase catalyzes the third intermediate step, the conversion of HMG-CoA to mevalonate.

Figure 5.3 illustrates how Insig proteins regulate cholesterol synthesis through its end-point regulation of HMGR-CoA lifetime. As shown in this figure, the buildup in the concentration of sterols activates Insig1. This protein functions as a scaffold that brings together HMGR-CoA and the proteolytic machinery consisting of the E3 ubiquitin ligase Gp78, its E2 Ubc7, and an adapter VCP. (Ubiquitin-mediated proteolysis is discussed in detail in Chapter 7). These actions greatly reduce the lifetime of the HMGR-CoAs and terminate cholesterol synthesis.

HMGR-CoA is the target of the cholesterol-lowering drugs known as statins. These drugs bind to the active site of the enzyme and block access to that site by the substrate. The three-dimensional structure of the catalytic part of HMGR-CoA, as revealed by means of X-ray crystallography, is presented in Fig. 5.4. The enzyme functions as a tetramer as shown in Fig. 5.4(a); and in this arrangement utilizes pairs of subunits, as shown in Fig. 5.4(b) to bring together oxidized $NADP^+$ and CoA in the vicinity of the active site.

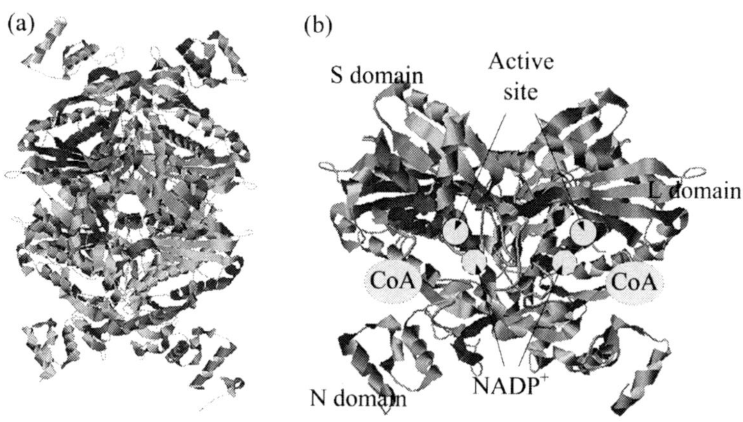

Fig. 5.4 Three-dimensional structure of the catalytic portion, amino acid residues 426–888, of the rate-limiting enzyme HMGR-CoA determined by means of x-ray crystallography. (**a**) Arrangement of the subunits to form a tetramer and (**b**) organization of the dimer. The N-terminal (N), large (L), and small (S) domains for one of the subunits have been labeled. The figure was prepared using Jmol with atomic coordinates deposited in the PBD under accession code 1dq8

5.3 Feedback Regulation of Cholesterol Synthesis by SREBPs

Sterol regulatory element binding proteins (SREBPs) are transcription factors that provide feedback regulation of cholesterol synthesis. These proteins are anchored to membranes of the endoplasmic reticulum and nuclear envelop. In the absence of cholesterol, the SREBPs leave their location and translocate to the nucleus where they stimulate transcription of genes encoding proteins involved in uptake and synthesis of cholesterol. When cholesterol is present, the SREBPs remain sequestered in these membranes. This negative regulatory loop provides end-point feedback control of cholesterol homeostasis in the cell.

The SREBP polypeptide chains pass through the ER membrane twice, with the N- and C-terminal ends residing in the cytosol. Two other proteins associate with the SREBPs. One of these is the SREBP-cleavage-activating protein, or Scap, which acts as a chaperone that escorts the SREBPs to the Golgi. The second partner is Insig1. This protein functions as the sterol sensor. Insigs (Insig-1 and -2) are 6 pass intrinsic membrane proteins embedded in the ER (Fig. 5.5). When adequate supplies of cholesterol are present, the Insig1 protein binds to the SREBP protein and prevents translocation of the SREBPs to the Golgi. This association and inhibition does not occur in the absence of cholesterol. Under low sterol (cholesterol and oxysterol) conditions, Scap associates with COPII proteins leading to vesicle-mediated transport to the Golgi.

In the Golgi, a pair of proteases cleaves the SREBPs chopping off the N-terminal bHLH-ZIP domain from the C-terminal regulatory domain. The N-terminal domain subsequently translocates to the nucleus where it binds SRE in enhancer and promoter regions of target genes. The process whereby a membrane-bound transcription factor is proteolytically cleaved at multiple sites to free up the physiologically active segment is known as regulated intramembrane proteolysis (RIP). This process is not unique to the SREBPs (SREBP-1a, SREBP-1c, and SREBP-2), but rather is encountered in the activation of the Notch protein, a major transcriptional regulator of embryonic development. It is also utilized in

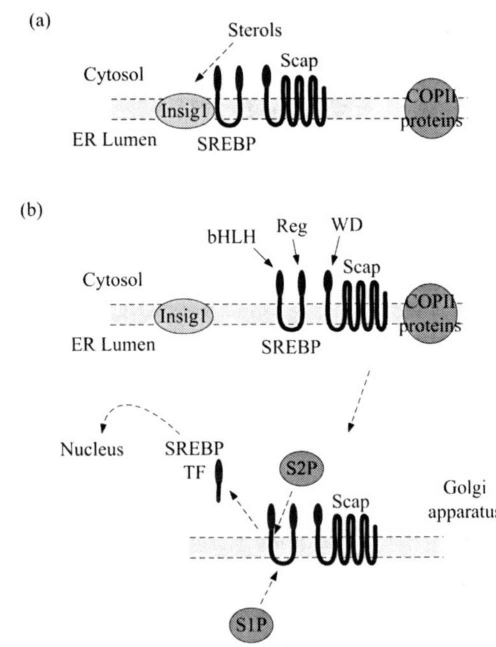

Fig. 5.5 SREBP processing and activation. (**a**) Inhibition by sterols and (**b**) activation in the absence of sterols through regulated intramembrane proteolysis (RIP)

activating key ER stress-response mediators, and in the formation and activation of the amyloid β protein, a major factor, if not the key causative agent, in Alzheimer's disease.

5.4 SREBPs, Liver X Receptors, and Farnesoid X Receptors Regulate Transcription

The SREBP proteins stimulate transcription of enzymes critical for cholesterol and fatty acid biosynthesis, metabolism, and uptake. Among the genes found to be regulated by the SREBPs are the ones that encode

1. HMG-CoA synthase,
2. HMG-CoA reductase,
3. Acetyl CoA carboxylase (ACC),
4. Farnesy diphosphate synthase,
5. Fatty acid synthase (FAS), and
6. Low-density lipoprotein (LDL) receptor.

In general, SREBP-2 manages cholesterol while the SREBP-1s treat both cholesterol and fatty acids with SREBP-1c predominant in the liver. The SREBPs do not work alone in their management of cholesterol, but rather operate together with the liver X receptors (LXRs) and farnesoid X receptor (FXR) to handle sterol storage and transport, movement between peripheral tissue, and hepatic excretion.

Liver X receptors are activated by oxysterol ligand binding. Oxysterols are endogenous oxidative metabolites of cholesterol, and thus LXR serves as a cholesterol sensor that together with the sterol regulatory element-binding protein 1c (SREBP-1c) regulates fatty acid synthesis. It does so in large part through upregulation of CYP7A1, the rate-limiting enzyme in conversion of cholesterol into bile acids. Two closely related isoforms are LXRα and LXRβ. They form dimers with RXRs and when activated by ligand binding through the replacement of corepressors with coactivators stimulate the transcription of LXRE genes involved in cholesterol metabolism, most notably in the (reverse) transport of excess cholesterol to the liver for removal from the body.

Cholesterol is removed from the body through a route involving conversion to bile acids in the liver. This allows for the solubilization of the cholesterol and excretion into bile. Farnesoid X receptors sense the presence of bile acids and respond by maintaining bile acid homeostasis in which the balance between bile and cholesterol is maintained. Through a set of feedback loops, this nuclear receptor regulates bile acid synthesis, detoxification of bile acids and their efflux, these last two functions help prevent excessive toxicity in the liver. The rate-limiting enzyme for synthesis of bile acids is the CYP7A1 members of the P450 family of phase I enzymes. The LXR receptor exerts its regulatory actions through repression of CYP7A1 gene expression while as discussed above LXR exerts positive regulation on this liver-specific gene.

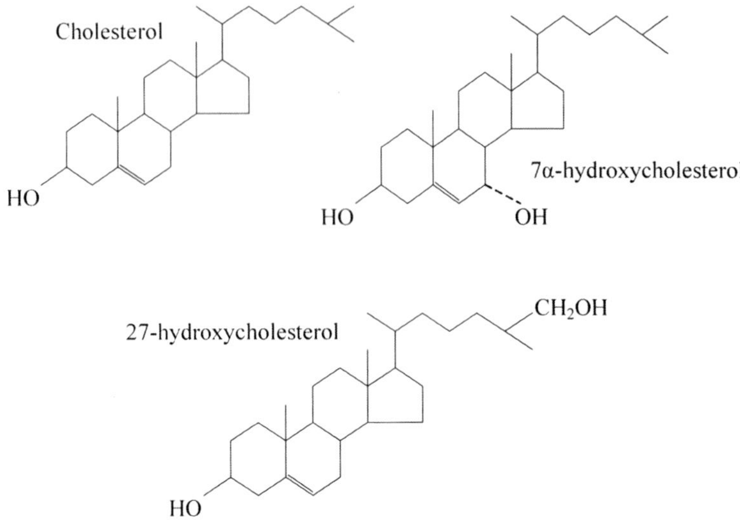

Fig. 5.6 Chemical structures of cholesterol and the two most common oxysterols, 7α-hydroxycholesterol and 27-hydroxycholesterol

Oxysterols are oxygenated derivatives from, and precursors to, cholesterol. They may be produced by enzymes such as members of the cytochrome P450 family or alternatively by reactive oxygen species. As discussed earlier in this chapter, oxysterols are generated as intermediates in the cholesterol excretion pathway that generates bile acid end products from cholesterol. They are also a component of OxLDL particles associated with atheromas. The two most prominent oxysterols occurring in OxLDLs are depicted in Fig. 5.6. As illustrated in the figure, the cholesterol molecules have been modified at the 7 and 27 positions in the cases of 7α-hydroxycholesterol and 27-hydroxycholesterol, respectively.

5.5 Lipoproteins Are Carriers of Cholesterol and Triglycerides

Lipids in the form of triglycerides, phospholipids, and sterols such as cholesterol first enter the body through foodstuffs. In the stomach, triglycerides are partially hydrolyzed and large fat globules consisting of triglycerides, their hydrolyzed products, phospholipids, and sterols are formed. The next stage in the processing takes place in the intestinal lumen. Bile salts are mixed in along with pancreatic lipases. The lipases act on the trigylcerides breaking them down into monoacylglycerols and free fatty acids with the aid of the bile salts that help emulsify the fat globules into small droplets that can be acted upon easily by the lipase. The emulsified and hydrolyzed products then enter intestinal absorptive cells called enterocytes through fatty acid transporters.

Once inside the enterocytes, free fatty acids and monoacylglycerides are converted back to triglycerides and, together with the uptaken cholesterol and apolipoprotein B, are packaged into cholesterol-rich lipoprotein particles called chylomicrons. These particles are secreted into the extracellular space and enter the circulation via the lymphatics.

Cholesterol cannot circulate by itself but rather requires a carrier, the lipoproteins. There are five kinds of lipoproteins, each differing in size, density, and lipid and protein composition. The physical and chemical properties of these cholesterol transport molecules are summarized in Table 5.2. These entries have been ordered from the largest and least dense chylomicrons to the smallest and most dense HDLs. The alterations in size and density reflect differences in composition. For example, low-density lipopoteins (LDLs) contain more lipids and less proteins than the high-density lipoproteins (HDLs).

Table 5.2 Lipoproteins

Lipoprotein class	Size (nm)	Density (g/ml)	Chol. (%)	Tri. (%)	Prot. (%)
Chylomicron	1000	< 0.96	4	88	1
VLDL	25–90	0.96–1.006	23	56	8
IDL	40	1.006–1.019	43	29	11
LDL	26	1.019–1.063	58	13	21
HDL	6–13	1.063–1.210	38	14	45

The LDLs transport lipids from the liver and intestines to other organs in the body. HLDs, in contrast, remove cholesterol returning it to the liver for eventual removal from the body. In the liver, it is catabolized to bile acids and eliminated.

Chylomicrons are produced in the small intestine by absorptive cells (enterocytes). They transport lipids (cholesterol and triglycerides) derived from foodstuffs to the liver, adipose tissue, cardiac, and skeletal tissues. They enter the bloodstream via the lymphatic system, and acquire ApoC-II and ApoE from HDLs.

Very low-density lipoproteins (VLDLs) are synthesized in the liver. They are the main vehicle for transport of triglycerides from the liver to adipocytes and muscle tissue for energy storage and production of energy through oxidation. Fatty acid part of the VLDLs are released to the muscle cells and adipocytes; this loss together with the loss of the ApoCs leaves a VLDL remnant, termed an intermediate-density lipoprotein, or IDL. Some IDLs are taken up by the liver while others become low-density lipoproteins, or LDLs.

5.6 Apolipoproteins are Amphipathic, Lipid-Binding Constituents of the Lipoproteins

There are two different kinds of apolipoproteins. Some are exchangeable while others are not. In the group of exchangeable apolipoproteins, one finds the ApoAs, APoCs, and ApoE; these proteins are able to move between lipoproteins whereas nonexchangeable ones, the ApoBs, are firmly anchored and cannot move from one class of lipoprotein to another.

The dominant feature of the exchangeable apolipoproteins is their secondary structure. These proteins contain 11- and 22-mer repeats of amino acid sequences that permit their folding into amphipathic α-helices. Amphipathic α helices have the important biophysical property of having opposing polar and nonpolar faces oriented along the long axis of the helix. This organizational structure permits associations with both lipids and the aqueous blood plasma through which it shuttles the lipids and is essential for their ability to move from between the different kinds of lipoproteins.

The predominant apolipoprotein in HDLs is ApoA-I accounting for about 70% of the protein content. This 240–243 amino acid residue protein not only shuttles cholesterol from tissues to the liver, but also serves as a cofactor for the enzyme that converts free cholesterol to cholesterol esters in the plasma (LCAT, to be discussed shortly) and as a ligand for cell surface receptors. Other apolipoproteins present in HLDs are listed in Table 5.3.

ApoC-II plays an analogous role in chilomicrons and VLDLs. Chilomicrons and VLDLs are rich in triglycerides. Like ApoA-I, ApoC-II is an essential cofactor for a crucial enzyme; in this situation, the enzyme is lipoprotein lipase, which is responsible for hydrolyzing the triglycerides. This enzyme is encountered in small capillaries of adipose and muscle tissues, where it breaks down triglycerides into free fatty acids and glycerol. The chilomicrons begin as nascent chilomicrons possessing ApoB48, and ApoA-I, II, and IV but not ApoC-II or ApoE. These are acquired in circulation from HDLs and at the same time the HDLs acquire phospholipids from the chilomicrons. Once the chilomicrons have acquired their ApoC-IIs, they are able to hydrolyze the TGs; the resulting fatty acids are imported into adipose tissue, converted back to TGs, and stored, while in muscle tissue the FAs are oxidized to generate energy. The series of steps followed by the VLDLs is similar with ApoB-100 in place of ApoB-48. Again ApoC-II and ApoE are acquired from HDLs in circulation. As the VLDLs lose their contents they, like the chilomicrons, transfer their phospholipids to the HDLs and also transfer back the acquired apolipoproteins. As the VLDLs lose their content, they become IDLs and eventually LDLs containing ApoB-100 as their sole apolipoprotein (Table 5.3).

Table 5.3 Apolipoproteins, their sources, and their distribution among the lipoproteins

Lipoprotein class	Apolipoproteins	Source tissue
Chylomicron	ApoA-1, ApoA-II, ApoA-IV, ApoB-48, ApoC-I, ApoC-II, ApoC-III, ApoE	Intestine
VLDL	ApoB-100, ApoC-I, ApoC-II, ApoC-III, ApoE	Liver
IDL	ApoB-100, ApoC-I, ApoC-II, ApoC-III, ApoE	VLDL Catabolism
LDL	ApoB-100	IDL Catabolism
HDL	ApoA-1, ApoA-II, ApoA-IV, ApoC-I, ApoC-II, ApoC-III, ApoE	Liver, Intestine

High-density lipoproteins are first formed as two-dimensional disciodal (disk-like) structures bearing phospholipids and free cholesterol. This form of nascent HDL is then converted into a mature three-dimensional HDL particle containing a nonpolar core consisting of cholesterol esters and triglycerides in addition to the phospholipids and free cholesterol localized to its periphery. One of the keys to the HDL metabolism is the ability of ApoA-I to alter its shape according to its lipid associations. ApoA-I has one shape under lipid-poor or free-solution conditions, another shape when bound to nascent discoidal HDLs, and a third shape when occupying the surface of mature spherical HDLs.

In the absence of lipids, amphipathic α helical proteins such as ApoA-I, ApoA-II, ApoC-II, and ApoE may partially unfold and assume alternate conformations leading to disease-associated amyloid formation. Amyloids such as these consist of long, semiregular fibers exhibiting a twisted ribbon or rod-like shape. The underlying biophysical feature of the folding intermediates is the presence of β-sheets in place of the repeating α helices. Apolipoproteins are not the only proteins possessing amphipathic α helical secondary structures with a propensity in the absence of lipids to form amyloid fibrils. Serum amyloid A (SAA) and α-synuclein are two other prominent examples. The SAAs comprise a family of apolipoproteins that associate with HDL under acute inflammatory conditions. The latter are present in Lewy bodies formed in Parkinson's disease, while SAAs and Apo amyloids are found in atherosclerotic plaques. Amyloids are a common feature of several neurodegenerative diseases and will be examined in detail in Part III of the text.

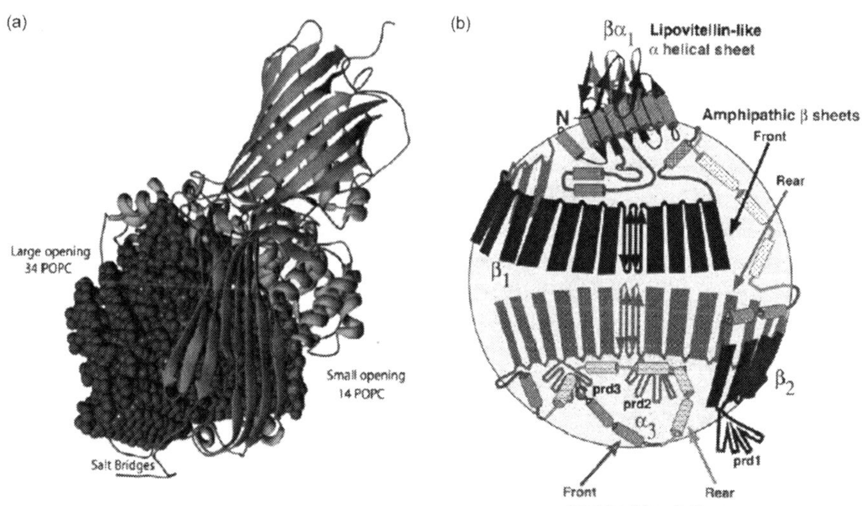

Fig. 5.7 Lipoprotein particles. (**a**) N-terminal 1000 residues of ApoB-100 and (**b**) LDL particle showing the arrangement of full-length ApoB-100 about its surface. Reprinted by permission from Elsevier from Richardson PE et al. 2005 *Biophys. J.* 88: 2789, copyright 2005 by the Biophysical Society

Atherosclerosis involves the inappropriate accumulation of ApoB-100-containing lipoproteins in the intima. ApoB-100 proteins are synthesized in the liver. They are exceptionally large monomeric proteins, the primary sequence consisting of 4536 amino acid residues (approximately 550 kDa in mass). They are highly insoluble molecules consisting of three amphipathic α-helical domains and two amphipathic β-stranded domains. A detailed three-dimensional model of the first 1000 residues of ApoB-100 is presented in Fig. 5.7(a). In this model, there is a highly visible lipid-binding pocket that facilitates formation of the lipoprotein particle. A more complete but less detailed representation of how the remaining portions of ApoB-100 organize the LDL particle is presented in Fig. 5.7(b).

5.7 Cholesterol Comes in Two Forms – As a Sterol, i.e., as a Free Cholesterol (FC) Molecule, and as a Cholesterol Ester (CE)

Cholesterol exists in two forms. One of these, free cholesterol, is the form depicted in Fig. 5.6. In its alternative, esterified form cholesterol molecule has an acyl "tail" that shields the OH group. The conversion from FC to CE is catalyzed in the plasma by an enzyme known as lecithin:cholesterol acyltransferase, or LCAT. In its CE form, the cholesterol molecules are able to reside in the interior of the lipoprotein whereas in the FC form the molecules are constrained by the OH group to lie on the outer surface. The esterification of cholesterol is crucial for proper loading of cholesterols into the hydrophobic interior of the HDLs. When this happens, nascent small HDLs become large mature HDLs carrying appreciable quantities of cholesterol molecules to the liver for disposal.

Cholesterol esterification occurs in the forward cholesterol transport pathway too. This process is catalyzed by acyl coenzyme A: cholesterol acyltransferases 1 and 2 (ACAT1 and ACAT2). These enzymes catalyze the conversion of free cholesterol to cholesterol esters intracellularly. The ACATs are integral membrane proteins localized to the rough endoplasmic reticulum. These molecules use acyl-coenzyme A molecules as acyl donors. ACAT2 transfers these chains to ApoB-containing lipoproteins in enterocytes (intestines) and hepatocytes (liver), while ACAT1 is active in macrophages where the CEs are packaged into lipid droplets. The conversion is not one-way only. Other enzymes, CEases, can convert CE back to FC to facilitate movement in and out of cells.

The cholesterol ester transfer protein (CETP) is another crucial participant in cholesterol management. This protein is secreted by the liver and circulates in the bloodstream bound to HDL. This enzyme mediates the distribution of cholesterol esters and triglycerides among the various types of lipoproteins. As discussed earlier cholesterol esters enter the bloodstream in HDLs in reactions catalyzed by LCAT. Triglycerides enter the bloodstream and are transported primarily by chylomicrons secreted from the intestines and by

VLDLs produced by the liver. The CETPs act in a bidirectional manner to balance the distribution of cholesterol and triglycerides among the five populations of lipoproteins. If the distributions of cholesterol and triglycerides are in balance, CETP does very little. If there is too much HDL cholesterol, the CETPs shift some of the transport burden to the other lipoproteins. The cholesterols are subsequently delivered by the LDLs either to peripheral tissues or to the liver through LDL receptor-mediated uptake and by the HDLs to the liver by scavenger receptor-mediated uptake.

5.8 ABC Transporters Export Cholesterol from Macrophages

As will be discussed in the next chapter, atherosclerosis is characterized by the unrestricted buildup of lipid droplets due to excessive amounts of cholesterol. Eventually, this buildup leads to apoptosis and plaque rupture resulting in heart attack and stroke. A key element in preventing this from happening is the efficient efflux of cholesterol from macrophages and foam cells. The pathways for this involve several members of the ABC family of transporters and the nuclear receptor family.

Two members of the ATP binding cassette (ABC) transporter family are responsible for exporting cholesterol into HDL particles thereby reducing cholesterol buildup in macrophages and foam cells. One is ABCA1 and the other is ABCG1 (and/or ABCG4). These appear to operate in tandem. ABCA1 mediates the transport of cholesterol from cells to lipid-poor Apo-A while ABCG1 mediates the transport of cholesterol from cells to lipid-rich lipoproteins. The sequential operation – first ABCA1 and then operating on both Apo-A and lipid-rich products – are converted by ABCG1 (or ABCG4) to a single population of cholesterol-rich particles.

The membrane topology of the ABCA1 transporter is depicted in Fig. 5.8. As can be seen, ABCA1 possesses 12 TM helices; the N- and C-termini are cytoplasmic and the first and last six helices are separated by an extensive cytolasmic loop, and there are a pair of large extracellular loops, one between helices 1 and 2 and the second between helices 7 and 8. The second transporter is similar except that the two sets of six helices are separated and reversed in the sense that the NH2 termini are at the "ends" of helices 6 and 12. These proteins

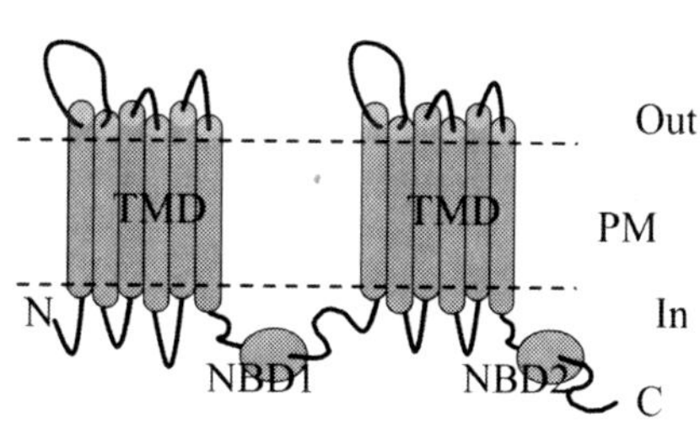

Fig. 5.8 Membrane topology of the ABCA1 cholesterol transporter. Abbreviations: TMD, transmembrane domain; NBD, nucleotide binding domain

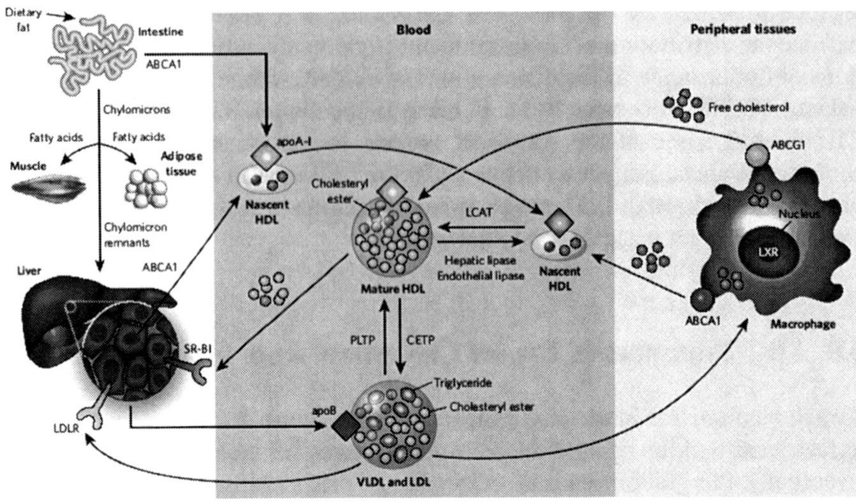

Fig. 5.9 HDL trafficking between liver and peripheral tissues. Reprinted by permission from Macmillan Publishers Ltd.: [Nature] Rader DJ and Daugherty A 2008 *Nature* 451: 904

appear to operate by forming a channel in the membrane and flopping the lipids from inner to outer leaflets in an ATP-dependent manner.

Liver X receptors regulate cholesterol efflux from the body. They respond to excessive cholesterol levels by upregulating the expression of the ABCA1 cholesterol transporter. This action enhances the formation, loading, and cycling of HDL particles between the liver and macrophages. The constant cycling of HDLs and LDLs between liver and peripheral tissues is depicted in Fig. 5.9.

Further Reading

Cholesterol and Lipid Membrane Biophysics

De Almeida RFM, Fedorov A and Prieto M 2003 Sphingomyelin /phosphatidylcholine/ cholesterol phase diagram: boundaries and composition of lipid rafts. *Biophys. J.* 85: 2406–2416.

Huang J and Feigenson GW 1999 A microscopic interaction model of maximum solubility of cholesterol in lipid bilayers. *Biophys. J.* 76: 2142–2157.

Kasumi A et al. 2005 Paradigm shift of the plasma membrane concept from the two-dimensional continuum fluid to the partitioned fluid. *Annu. Rev. Biophys. Biomol. Struct.* 34: 351–378.

Simons K and Toomre D 2000 Lipid rafts and signal transduction. *Nat. Rev. Mol. Cell Biol.* 1: 21–39.

Regulation of Cholesterol Homeostasis

Brown MS et al. 2000 Regulated intramembrane proteolysis: a control mechanism conserved from bacteria to humans. *Cell* 100: 391–398.

Goldstein JL, DeBose-Boyd RA and Brown MS 2006 Protein sensors for membrane sterols. *Cell* 124: 35–46.

Horton JD, Goldstein JL and Brown MS 2002 SREBPs: activators of the complete program of cholesterol and fatty acid synthesis by the liver. *J. Clin. Invest.* 109: 1125–1131.

Istvan ES et al. 2000 Crystal structure of the catalytic portion of human HMG-CoA reductase: insights into regulation of activity and catalysis. *EMBO J.* 19: 819–830.

Istvan ES and Deisenhofer J 2001 Structural mechanism for statin inhibition of HMG-coA reductase. *Science* 292: 1160–1164.

Radhakrishnan A et al. 2007 Sterol-regulated transport of SREBPs from endoplasmic reticulum to Golgi: oxysterols block transport by binding to Insig. *Proc. Natl. Acad. Sci. USA* 104: 6511–6518.

Rawson RB 2003 The SREBP pathway – insights from Insigs and insects. *Nat. Rev. Mol. Cell Biol.* 4: 631–640.

Shi YG and Burn P 2004 Lipid metabolic enzymes: emerging drug targets for the treatment of obesity. *Nat. Rev. Drug Discov.* 3: 695–710.

Sun LP et al. 2007 Sterol-regulated transport of SREBPs from endoplasmic reticulum to Golgi: Insig renders sorting signal in Scap inaccessible to COPII proteins. *Proc. Natl. Acad. Sci. USA* 104: 6519–6526.

Apolipoproteins

Brouillette CG et al. 2001 Structural models of human apolipoprotein A-I: a critical analysis and review. *Biochim. Biophys. Acta* 1531: 4–46.

Hatters DM and Howlett GJ 2002 The structural basis for amyloid formation by plasma apolipoproteins: a review. *Eur. J. Biophys.* 31: 2–8.

Richardson PE et al. 2005 Assembly of lipoprotein particles containing apolipoprotein-B: structural model for the nascent lipoprotein particle. *Biophys. J.* 88: 2789–2800.

Segrest JP et al. 2001 Structure of apolipoprotein B-100 in low density lipoproteins. *J. Lipid Res.* 42: 1346–1367.

Cholesterol Esterification

Jonas A 2000 Lecithin cholesterol acyltransferase. *Biochim. Biophys. Acta* 1529: 245–256.

Reverse Cholesterol Transport

Barter PJ et al. 2003 Cholesterol ester transfer protein: a novel target for raising HDL and inhibiting atherosclerosis. *Arterioscler. Thromb. Vasc. Biol.* 23: 160–167.

Lewis GF and Rader DJ 2005 New insights into the regulation of HDL metabolism and reverse cholesterol transport. *Circ. Res.* 96: 1221–1232.

Linsel-Nitschke P and Tall AR 2005 HDL as a target in the treatment of atherosclerotic cardiovascular disease. *Nat. Rev. Drug Dis.* 4: 193–205.

Vaughan AM and Oram JF 2006 ABCA1 and ABCG1 or ABCG4 act sequentially to remove cellular cholesterol and generate cholesterol-rich HDL. *J. Lipid Res.* 47: 2433–2443.

Chapter 6
Atherosclerosis

Atherosclerosis, the main cause of heart attacks, stroke, and gangrene of the extremities, is a disease of the medium and large arteries. It often begins early in life and progresses in a series of stages throughout adulthood. This progression, as shown in Fig. 6.1, begins with some form of injury or disturbance to the normal flow of blood near the arterial wall, and with an accumulation of cholesterol in that region. These two abnormalities combine to produce inflammation.

Macrophages, T-cells, and free and macrophage-containing lipid droplets accumulate in the intima and form fatty streaks. As would be the case for an injury, repair cells, most notably macrophages and smooth muscle cells, migrate to the site of the disturbance. LDL particles are oxidized by various arterial wall factors and the oxLDL particles are taken up by the macrophages, which when loaded with these deposits become foam cells. The smooth muscle cells that migrate in to help affect the repair secrete extracellular matrix remodeling factors and proteins. Eventually, a cap forms over the growing lesion. The buildup does not terminate but instead continues and further impedes blood flow; the cap weakens over time, and upon rupturing and loosening, a clot generates a heart attack or stroke.

6.1 The Arterial Wall Consists of Three Layers

The arterial wall contains several types of cells organized into three layers. The layer closest to the endothelium is called the intima. It consists of extracellular matrix molecules such as collagens and proteoglycans. This layer is bounded by an internal elastic lamina composed as its name suggests of elastic fibers. The middle layer of the arterial wall, termed the media, is composed of smooth muscle cells (SMCs) and lastly, the outermost layer, the adventitia, contains connective tissue, fibroblasts, and SMCs.

A thin monolayer of endothelial cells sits in between the cells of tissues and organs and the circulating blood. Individual endothelial cells are fairly thin,

M. Beckerman, *Cellular Signaling in Health and Disease*, Biological and Medical Physics, Biomedical Engineering,
DOI 10.1007/978-0-387-98173-4_6, © Springer Science+Business Media, LLC 2009

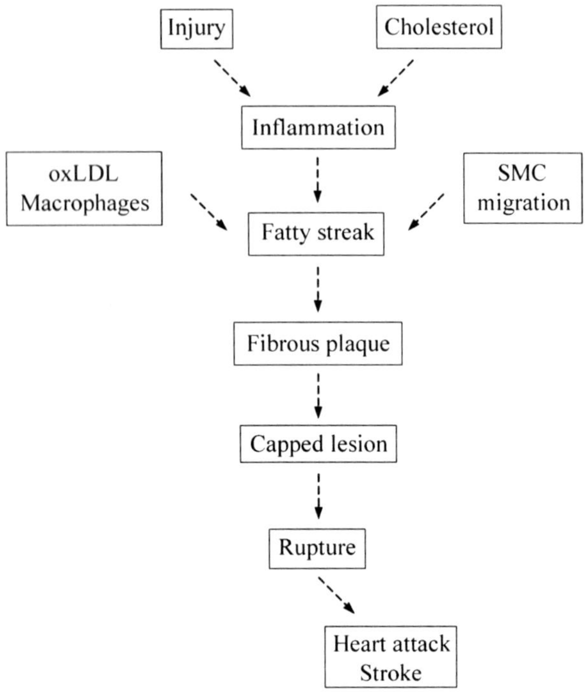

Fig. 6.1 Sequences of stages in the development of atherosclerosis

measuring from 0.2 to 4 μm from the apical to the basolateral surfaces, and elongated, ranging from 10 to 50 μm in length and width (Fig. 6.2). The total endothelial lining consists of more than one trillion endothelial cells, covering several tens of thousands of km of blood vessels and occupying a surface area of hundreds of m^2.

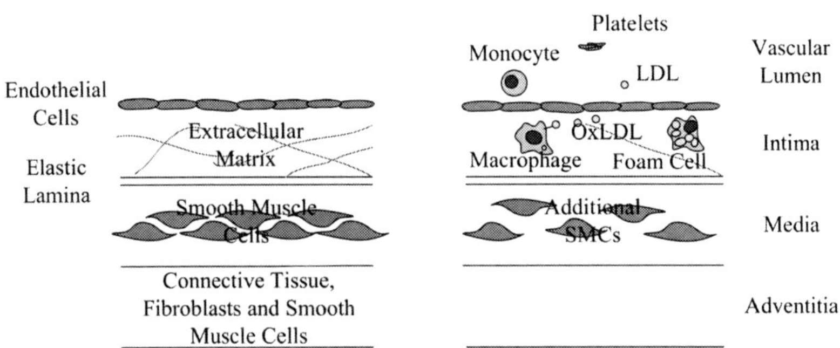

Fig. 6.2 Arterial walls consisting of three layers. *Left panel*: Normal composition of the wall; *Right panel*: Additional cellular elements recruited and activated by injury, cholesterol buildup, and inflammation

6.2 Cells Are Continually Subjected to Forces

Cells are exposed to gravity, fluid and air pressure, and tensions associated with the intracellular cytoskeleton, the extracellular matrix, and neighboring cells. Endothelial cells form a layer between the vascular wall and the flowing blood. As a result they are continually exposed to biomechanical forces resulting from blood flow. One kind of force is cyclic strain. This type of force is due to the pulsatile nature of the blood flow; that is, from systolic/diastolic pressure changes. These forces are compressive; they act perpendicular to epithelial cell surface, either pushing down or pulling up. The second type of force is wall shear stress. Blood flows faster in the center of a blood vessel and slower near the vessel walls. This difference between flow in the center and flow near walls is due to frictional forces within the blood and between the blood and the walls. These frictional forces are encapsulated by the notion of fluid viscosity – the greater the viscosity the larger the frictional forces. These stress forces act tangentially to the cell surfaces lining the blood vessel and are oriented in the direction of the flow.

The rate at which blood flows through an artery depends on the geometry of the blood vessel, the pressure gradient, and the viscosity of the fluid. The relationship of these quantities to one another is given by Poiseuille's law:

$$Q = \frac{\pi(\Delta P)R^4}{8\eta L} \tag{6.1}$$

In this expression, Q is fluid flow rate; ΔP is the pressure gradient; R denotes the radius of the artery and L its length, and η represents the viscosity of the blood.

One of the main features of blood flow through arteries is as noted above that it flows faster in the center of the artery and slower near the walls. This variation in flow rate from one side to the other is illustrated in Fig. 6.3. As can be seen, the velocity profile is approximately parabolic, and this kind of movement, typical of a healthy artery, is termed laminar flow. Parabolic flow can be represented by a velocity equation of the form

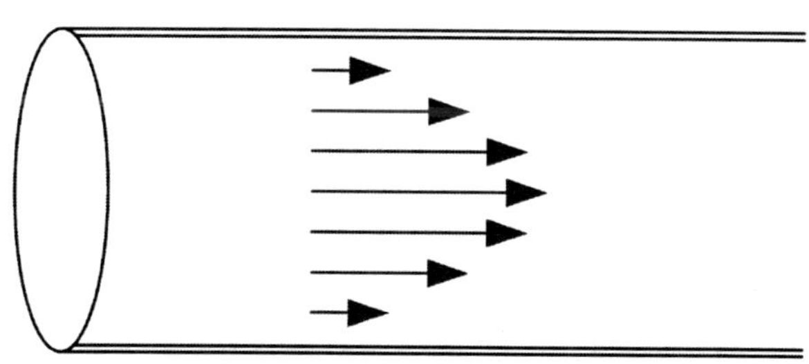

Fig. 6.3 Parabolic flow through arteries. *Arrow* lengths denote relative speed of the flowing blood

$$v = \frac{(\Delta P)(R^2 - r^2)}{4\eta L} \tag{6.2}$$

where v is the velocity, and r is the radial distance measured from the center to the point where the velocity is determined. As can be seen from this equation, the velocity is maximal at the center where $r = 0$ and minimal (zero) at the blood vessel wall, $r = R$. The flow Q and velocity v are related to one another through the relation that defines Q as the cross-sectional area, πR^2, times the mean velocity.

The shear rate, γ, and shear stress, τ, can now be introduced. The shear rate is by definition the derivative of the velocity with respect to radial distance; that is,

$$\gamma = \frac{dv}{dr} = \frac{(\Delta P)r}{2\eta L} \tag{6.3}$$

The shear stress is then given by the product of the shear rate and the viscosity. The value of prime interest is the magnitude of the shear stress at the blood vessel wall. This quantity can be fixed by equating the flow force $(\Delta P) \cdot \pi R^2$ to the frictional force, which is the wall shear stress τ_R multiplied by the lateral surface area $2\pi R \cdot L$:

$$\tau_R = \frac{(\Delta P)R}{2L} = \frac{4Q\eta}{\pi R^3} \tag{6.4}$$

where in the last expression Eq. (6.1) has been used to eliminate pressure gradient (ΔP) in favor of the more easily measured flow rate Q.

6.3 Atherosclerotic Lesions Occur Preferentially in Regions of Disturbed Blood Flow

Atherosclerosis occurs preferentially in regions of low and disturbed shear stress. Examples of where this occurs are at vessel branch points, where bifurcations occur, and at places where there is narrowing and bending of the arteries as depicted in Fig. 6.4. Several mechanisms have been identified that might explain why the disease develops in these regions. One of the earliest observations is that in these regions the dwell time for LDL as well as for leukocytes is increased relative to regions of constant flow. A more recent set of observations tie disturbed blood flow to altered signaling leading to endothelial dysfunction.

The core concept is that endothelial cells are normally exposed to a blood flow that is uniform in terms of its average speed and its direction. In response, the cells orient their cytoskeleton stress-bearing components and elongate their shape along the direction of the flow. Once these adjustments are complete, the signaling events and pathways responsible for managing these changes are turned off. In situations where the cells encounter disturbed blood flow, the signaling pathways responsible for the cellular adaptations do not shut down in a timely manner. Instead, the signaling pathways remain active continually trying to successfully trigger an appropriate set of cellular adjustments.

Table 6.1 Influences of high and low shear stress on cell physiology and function

Property	High shear stress	Low shear stress
Regime	> 14 dyn/cm^2	< 4 dyn/cm^2
State	Healthy	Unhealthy
Cell shape	Elongated	Compact
Nitric oxide production	Stimulated	Repressed
Platelet coagulation	Discouraged	Encouraged
Monocyte recruitment	Discouraged	Encouraged
Cell growth	Discouraged	Encouraged

Table 6.1 includes an estimate of the shear stresses in units of force per unit area associated with normal high shear stress and abnormal low shear stress regimes. As indicated in the table, low shear stresses give rise to the unhealthy situation of sustained signaling. Nitric oxide production (to be discussed in the next section) is atheroprotective. It is repressed when shear stresses are low. The remaining three entries in the table are highly suggestive that the endothelial cells are responding as through there is an injury.

At its most fundamental level, atherosclerosis involves the biophysics of disturbed blood flow, cellular signaling and adhesion, and the migration and actions of cells that mediate tissue repair and inflammation. It starts with an accumulation of cholesterol in a small arterial region, and with an inflammatory response triggered by an aberration in the normal flow of blood. The sequences of events that follow are summarized in a high-level fashion in Fig. 6.1. In the remainder of this chapter, these events will be examined in mechanistic and molecular detail.

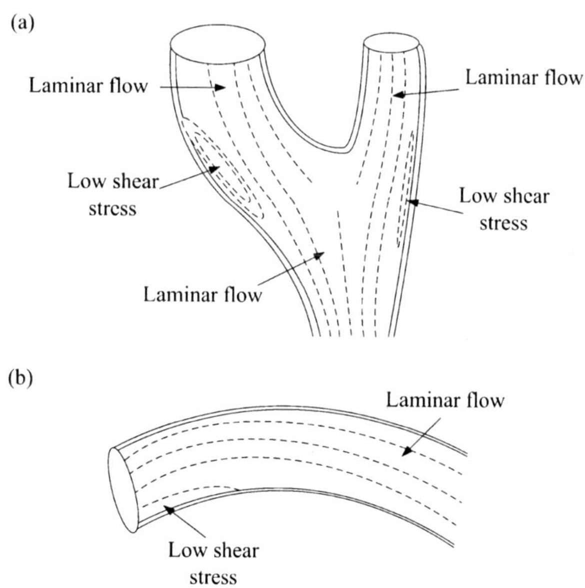

Fig. 6.4 Blood vessel architecture generates regions of undisturbed and disturbed blood flow. Abnormally low shear stresses occur (**a**) near vessel branching and (**b**) in regions that are curved

6.4 Cells Utilize Multiple Mechanotransduction Pathways That Convey Information About Blood Flow

One of the key systems involved in mechanotransduction is the cytoskeleton. This system functions as a distributed network that relays shear stress information from the apical surface to focal adhesion complexes at the basal surface of the epithelial cells and to cell-cell junctions connecting adjacent cells. Other modules involved in mechanotransduction are stress-responsive ion channels, receptor tyrosine kinases, G-protein-coupled receptors, and integrins. These work together with the cytoskeleton and with signaling elements situated within caveolae to activate intracellular signaling pathways leading to the expression of stress-responsive genes at shear-stress-responsive element (SSRE) promoters.

The endothelial lining is not a passive medium but rather an active one that is responsive to changes in its local microenvironment. These structures, critically situated in between tissue and blood, are responsive to both. The endothelial cells present a large surface area for the interchange of materials between themselves and blood. The apical surface of the cells forms the interface between the solid tissue and flowing blood. It contains of two layers of macromolecules that extend out from the cells into the blood stream. The inner layer, called the glycocalyx, is composed of macromolecules that remain attached to the plasma membrane. This thin layer is about 70 nm thick and consists of glycoproteins and proteoglycans. A second thicker layer commonly referred to as the endothelial surface layer or ESL is approximately 500 nm in thickness. This layer extends out from the glycocalyx and makes contact with the circulating blood. It is composed of solubilized glycosaminoglycans and adsorbed plasma proteins.

The glycocalyx may serve as the outermost elements of a cellular mechanotransduction network that distributes stress information to the actin cytoskeleton and various control points. This layer of macromolecules operates jointly with other stress-responsive elements embedded in the plasma membrane such as ion channels and receptors, with focal adhesions, and with signaling proteins strategically situated in caveolae. In response, the cells adjust their morphology, structure, and functions. Key components that are capable of relaying stress information are heparan sulfate proteoglycans (HSPGs), most notably the syndecans. These molecules possess single-pass transmembrane segments that make contact with the cytoskeleton. Another group of HSPGs are the glypicans; these attach to the outer cell surface of the endothelial cells by means of a GPI anchor.

6.5 Mechanotransduction Pathways Relay Information About Blood Flow to Endothelial Caveolae and Nitric Oxide Synthase

Blood vessels are able to respond and adjust to changes in blood flow, which as shown by Eqs. (6.1) and (6.3) is a function of several quantities among which are blood viscosity and vessel diameter. Changes in blood flow are sensed by the endothelium and, in response, the vessel diameter is either increased (vasodilation) or decreased (vasoconstriction) in order to restore proper flow. The key step in this process is the mechanotransduction of wall shear stresses into cellular responses. Several pathways jointly operate as mechanotransducers and convert the conditions of low endothelial shear stresses into changes in gene expression. Under normal healthy conditions, these alterations are protective and restore hemodynamic stability. When local conditions do not return to normal but instead the aberrant flow conditions persist, actions that would otherwise mediate protective tissue remodeling create a milieu conducive to atherosclerosis.

One of the chief mechanotransduction pathways involves the caveolae introduced in the last chapter. Within caveolae, glycosphingolipids influences integrin signaling and clustering; cholesterol molecules rigidify the compartment and facilitate the formation of signaling complexes, and caveolin-1 regulates endothelial nitric oxide synthase (eNOS) localization and activation. The eNOS enzyme is responsible for synthesis of nitric oxide (NO), a powerful vasodilator. Nitric oxide's actions are atheroprotective and anti-inflammatory. They include, in addition to regulating vascular tone, reduced expression of pro-inflammatory genes, preventing leukocytes migration into the vascular wall, inhibition of platelet aggregation, and reduced apoptosis through post-translational modification of caspase activity.

Caveolins bind and regulate the activity of endothelial nitrogen oxide synthase. Caveolins are a family of three proteins – Caveolin-1, -2 and -3. Their NH3 and COOH terminals lie in the cytosol while their central, transmembrane (TM) region is bent into a hairpin shape. This segment extends into but does not entirely pass through both leaflets of the plasma membrane. Caveolins oligomerize, forming large clusters. The scaffolding domain situated just N-terminal to the TM segment is involved in membrane attachment and oligomerization. The corresponding region C-terminal to the TM region provides sites for palmitoylation. The interactions of calveolins with eNOS involve the scaffolding domain of caveolins and the Ca^{2+}/CaM binding site of the eNOS. When bound by caveolins activation of eNOS by calcium is inhibited and thus caveolins negatively regulate its activity, thereby assisting in the maintenance of NO homeostasis in the healthy cell.

6.6 oxLDL Is Atherogenic and Acts in Opposition to eNOS and NO

Caveolae serve as gateways for the efflux of cholesterol. Caveolins bind and transport cholesterol. They bind cholesterol in the endoplasmic reticulum (ER) and transport it to the caveolae. Upon arrival cholesterol either leaves the cell, undergoing transport by HDL to the liver, or becomes part of the plasma membrane. One of the ways that the cholesterol content of caveolae can be depleted is through the actions of oxLDL. This form of LDL, but not the normal one, attracts cholesterol, leeching it out of the caveolae. In response, the caveolins detach from the caveolae and are internalized along with the attached eNOS enzymes. In this unhealthy situation, the buildup of oxLDL downregulates the atheroprotective activity of eNOS. This sets up a deleterious positive feedback loop centered about too much oxLDL and too little NO. This activity is opposed by the HDL, which when present will bind the SR-BI scavenger receptor and stimulate eNOS activity.

The transition from a cholesterol-associated injury occurring in regions of reduced shear stresses and turbulent blood flow to early-stage atherosclerosis occurs through a buildup in oxidized low-density lipoprotein (oxLDL). This begins to happen when LDLs become trapped in the regions of disturbed blood flow. Once the LDLs become trapped in the walls they become susceptible to oxidation by a number of mechanisms. Initial mild oxidation is followed by monocyte recruitment leading to further oxidation by the monocytes resulting in modification of the LDL structure. The modified oxLDLs cannot be taken up by the LDL receptor. Instead, they are recognized by scavenger receptors. This alternative pathway to cholesterol uptake bypasses the normal LDL-receptor-regulated negative feedback, which would halt any further uptake when certain cholesterol levels were reached. Endothelial cells, smooth muscle cells, and macrophages are all capable of oxidatively modifying the LDLs to a form recognized by the scavenger receptors.

When there is too much LDL over time, cholesterol-rich oxidized low-density lipoproteins (oxLDL) accumulate in the intima of medium and large arteries. In response, endothelial cells express adhesive molecules and send out chemical messages that attract monocytes and T-cell lymphocytes to the intima and trigger an inflammatory response. The monocytes mature into cells that take up and remove the oxLDL molecules, and smooth muscle cells migrate into the area of injury and multiply as part of the tissue repair process. If the injury process, in the form of an oxLDL buildup, persists then a chronic inflammatory response ensues leading to atherosclerosis.

Leukocytes migrate into the site of oxLDL injury from the bloodstream in order to initiate an inflammatory response. They move out of the rapidly moving center of the bloodstream to the arterial walls and establish an initial tethering contact with the luminal surface of the endothelial cells. They begin to roll in the direction of the blood flow until they reach the vicinity of the infection. At this point, their forward motion is arrested. The leukocytes then crawl out of the bloodstream, squeezing their way through the EC layer to reach the site of the injury/oxLDL infection in the intima.

This process is beautifully choreographed by the cellular residents of the arterial wall, and by the leukocytes and platelets situated in the bloodstream. They carry out this dance by expressing and releasing an ensemble of cell adhesion molecules and chemokines, one set for rolling contact, another for arrest, and a third distinct set for diapedesis, the transendothelial migration of leukocytes from the bloodstream to the intima. Before discussing the choreography further it is worthwhile to briefly introduce the dancers – the cell adhesion molecules and chemokines.

6.7 Cell Adhesion Molecules and Chemokines Mediate Leukocyte Migration into Sites of Inflammation

Four types of cell adhesion molecules participate in leukocyte homing from the bloodstream to the injury/infection site. These are

- Selectins
- Integrins
- IgCAMs
- Cadherins

Selectins are expressed on the surface of leukocytes, the endothelium lining the blood vessels, and platelets that form adhesive plugs, or clots, at sites of wounds. There are three kinds of selectins. L-selectins are expressed on leukocytes, E-selectins are expressed on the endothelial cells, and P-selectins are expressed on platelets and the endothelium. These adhesion molecules mediate the capture of the circulating leukocytes by the walls of the blood vessels and make possible their subsequent rolling (Table 6.2).

Integrins are bi-directional signaling receptors. Integrins are membrane-spanning glycoproteins composed of noncovalently attached α and β subunits. In vertebrates, 18 distinct α subunits and 8 different β subunits have been identified so far. Not all of the $18 \times 8 = 144$ combinations of α and β subunits can be formed. Instead, a far smaller number, namely 24, αβ heterodimers can occur. Each integrin subunit contains a large extracellular domain, a single transmembrane segment, and a short cytoplasmic domain. There is one exception to this rule. The β_4 subunit has a large cytoplasmic domain. The α subunits are larger than the β subunits. The α subunits vary in size from 120 to 170 kDa (up to 1114 amino acid residues) while β subunits range in size from 90 to 100 kDa (up to 678 residues).

Integrins and cadherins differ from most transmembrane signal transducers that transmit signals in one direction, from outside the cell inward. Integrin and cadherin receptors transmit signals in both directions – from outside the cell inward (outside-in) and from inside the cell outward (inside-out). In outside-in signaling, binding of the integrins to the ECM triggers changes in the pattern of

Table 6.2 Selectins, cells and structures that express them, their functions, and ligands

Selectin	Expression	Functions	Ligands
L-selectin	Leukocytes	Leukocyte trafficking, rolling adhesion	PSGL-1, GlyCAM-1, MAdCAM-1, CD34
P-selectin	Platelets, endothelium	Rolling adhesion	PSGL-1
E-selectin	Endothelium	Rolling adhesion induced by inflammation	PSGL-1

PSGL-1, P-selectin glycoprotein ligand-1; MAdCAM-1, mucosal-addressing cellular adhesion molecule-1; GlyCAM-1, glycosylation-dependent cell adhesion molecule-1.

gene expression. Inside-outside signals produce changes in the integrin conformation resulting in changes in adhesiveness. Integrins tie the ECM to the cellular cytoskeleton and anchor cells in a fixed position, while cadherins are a primary element of cell-to-cell junctions.

Cell adhesion molecules (CAMs) of the immunoglobulin superfamily (Ig-SF) mediate cell–cell and cell-ECM adhesion. Cell-surface receptors belonging to the immunoglobulin superfamily, the IgCAMs, are characterized by the presence of one or more immunoglobulin-like domains in their extracellular region. These adhesion molecules mediate cell-to-cell contact by binding to cell surface counterreceptors, and help establish and maintain contact with the extracellular matrix by binding to ECM constituents (Table 6.3).

Cadherins (calcium-dependent adherins) are present on most cells of the body. There are four classical cadherins – epithelial (E), neural (N), retinal (R), and vascular endothelial (VE). Their extracellular domains consist of five repeats of a 100 amino acid EC domain and bind in a homophilic manner to cadherins attached on the opposing surface. The intracellular segment of the cadherin attaches to the actin cytoskeleton by means of linker proteins called catenins. There are at least four types of catenin proteins. Of these the best characterized are α-catenin and β-catenin. In the simplest model of cytoskeleton attachment, the β-catenin binds to the catenin domain of the cadherin and to the α-catenin. The α-catenin, in turn, connects to the actin cytoskeleton. In recent more detailed models, β-catenin along with plakoglobin and p120 interacts with cadherins while α-catenin functions as an allosteric regulator of actin filament assembly and organization at the cadherin-mediated cell-to-cell junctions.

Chemokines are a family of small, 7–15 kDa protein chemoattractants. These molecules are secreted by several different kinds of cells at an infection

Table 6.3 Members of the IgCAM group of cell adhesion receptors

Ig-SF CAM	CD designation	Ligand	Distribution/role
ICAM-1	CD54	LFA-1; Mac-1; fibrinogen	Lymphocytes, endothelial cells when inflammation is present
ICAM-2	CD102	LFA-1, Mac-1	Recirculating leukocytes
ICAM-3	CD50	LFA-1	Leukocytes
LFA-2	CD2	LFA-3	Lymphocytes
LFA-3	CD58	LFA-2	Broad distribution
NCAM	CD56	NCAM, collagen, heparin	Neural tissue
PECAM-1	CD31	PECAM-1	Platelets, leukocytes
VCAM-1	CD106	$\alpha_4\beta_1$ and $\alpha_4\beta_7$ integrins	Lymphocytes, endothelial cells when inflammation is present

ICAM, Intercellular cell adhesion molecule; LFA, lymphocyte function-associated antigen; NCAM, neural cell adhesion molecule; PECAM, platelet endothelial cell adhesion molecule; VCAM, vascular cell adhesion molecule; CD, cluster of differentiation.

site in order to recruit leukocytes from blood to sites of infection in tissue. Neutrophils, monocytes, fibroblasts, endothelial cells, SMCs, and epithelial cells secrete chemokines. Chemokines have a positive charge and bind to heparins and heparin sulfate, negatively charged polysaccharides found within the extracellular matrix, and on the surface of endothelial cells. The heparin and heparin sulfate molecule act as receptors for the chemokines. The chemokines become immobilized when bound to the polysaccharide-covered substrates and form stable gradients of chemokine concentration. The migratory leukocytes navigate up the gradients to the target sites.

6.8 Leukocyte Migration Occurs Through a Multistep Adhesion Cascade

Monocytes and other leukocytes migrate into sites of inflammation through a series of stages, each of one of which is mediated by a distinct set of cell adhesion molecules. The first stage is capture, or tethering. In this stage, the leukocytes that have moved out of the center of the blood vessel to the wall establish initial contacts with the surface of the wall. This stage is mediated by the three kinds of selectins and their PSGL-1 ligands. One they have made contact with the wall, they begin to loosely roll along the surface. In these two stages, slip and catch bonds established by the selectins and their ligands play essential roles as will be discussed shortly. In the next stages, integrins and their cell adhesion molecule counter-receptors assume progressively greater roles. Chemokines mediate activation of the integrins leading to a halting of forward movement near the target inflammation sites. The monocytes then migrate either through or between the endothelial cells into the intima. This stage, diapedesis, is mediated by yet another set of adhesion molecules as indicated in Fig. 6.5. These stages will now be examined in more detail beginning with capture.

Recall that platelets are released from the bone marrow and circulate in the bloodstream. Their role in the inflammatory response is to form clots or plugs that block blood flow at sites of injury. Adhesion and aggregation are central to platelet function, and they express integrins, selectins, and IgCAM receptors. Microvilli are adhesion molecule-rich extensions of the cell surface. They are

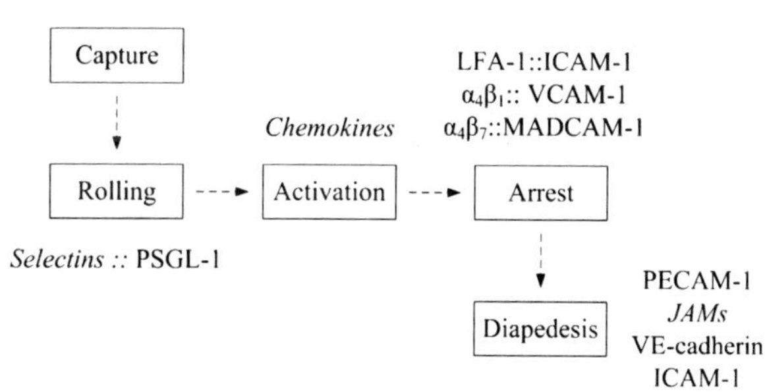

Fig. 6.5 The multistep, leukocyte adhesion cascade (see text for details)

formed at the outside facing, or apical, surface in a variety of different cell types in order to increase the effective surface area. The primary platelet selectin ligand is the P-selectin glycoprotein-1 (PSG1). It is constitutively expressed on the tips of microvilli of leukocytes such as neutrophils and monocytes.

The surface of ECs as well as migrating leukocytes is studded with microvilli. On the EC surface the microvilli are populated by L-selectins, P-selectin glycoprotein ligand-1 (PSGL-1), ICAM-1, and integrins $\alpha_4\beta_7$ and $\alpha_4\beta_1$. These cell-surface molecules along with the chemokines mediate initial tethering and rolling contact.

This stage is followed by one where the $\alpha_L\beta_2$ (LFA-1) integrins, ICAM-1, and VCAM-1 situated on the smooth surface of the ECs take over to mediate hard contact and stopping. The LFA-1 integrins are heterodimers maintained in an inactive (bent) conformation on the circulating leukocytes. Chemokines signaling through their GPCRs on the leukocyte surface active the integrins by stimulating a conformational change to a more extended conformation having a higher affinity for ligands. It is then primed to bind its ICAM-1 ligand and mediate firm adhesion. (In its closed conformation the integrin heterodimer spans just 5 nm of cell surface compared to 25 nm in its extended conformation).

6.9 Selectins Are Key Mediators of Leukocyte Tethering and Rolling

Blood flow in small vessels such as the postcapillary venules takes place under low flow rates, reasonably high viscosities, and small vessel diameter. Under these conditions, the velocity of the blood flow is greatest in the center of the vessel and decreases to zero as the vessel walls are approached as described in the last chapter. During inflammation, the blood vessels dilate and the overall flow rate is slowed. Red blood cells aggregate into large assemblies called rouleaux that collect into the center of the vessels displacing the leukocytes, which migrate from the center of the blood vessel to the vessel walls. Cell-surface receptors expressed on the plasma membranes of the endothelial cells and leukocytes can then engage one another to promote the stages of motion leading to firm adhesion and extravasation.

Selectins are key mediators of leukocyte rolling. These are mosaic proteins with a common structural organization. They each have an NH_2 terminal lectin domain followed by an epidermal growth factor (EGF) domain followed by a number of consensus repeats (CRs) of a complement-like binding sequence, a single transmembrane segment, and a short cytoplasmic tail. The lectin domain is a carbohydrate-binding domain that enables the selectin to bind to carbohydrates structures on its ligand. The CRs are thought to function as spacers that extend the molecule a distance that supports optimal rolling. The three selectins vary in the number of CRs. Human P-selectin has nine CRs, while E-selectin has six, and L-selectin just two.

The key feature of the selectin structure and its regulation by force is the presence of a hinge between the lectin and EGF domains (see Fig. 6.6).

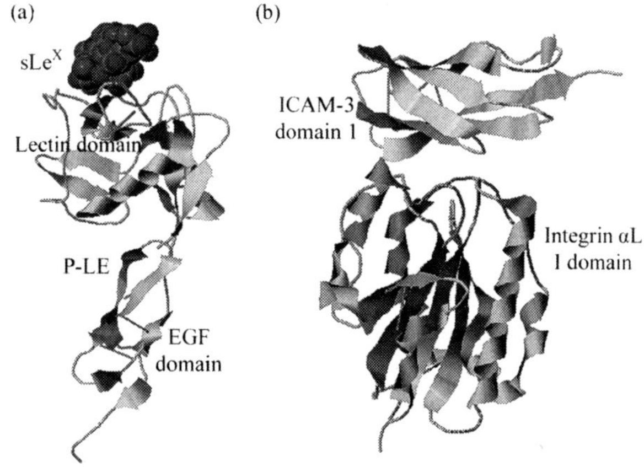

Fig. 6.6 (**a**) Selectin in a complex with its sLeX ligand. Shown is the extracellular carbohydrate-binding lectin domain and the EGF domain. PDB code 1g1r. (**b**) Integrin αL I domain in a complex with its ICAM-3 ligand. PDB code 1t0p

The angle made by this hinge is regulated by force and changes in relative orientation of the two domains are converted allosterically into changes in ligand-binding affinity. There is, in fact, a threshold shear below which leukocyte capture does not occur.

Bonds form and break during leukocyte rolling. Bond lifetimes must be long enough to permit formation of multiple bonds extending from the leukocyte's leading edge to its trailing edge. If the off (bond dissociation) rates are too high, multiple bonds cannot form. One bond will dissociate before another can be formed. At the other extreme, too tight a bonding at the trailing edge will either immobilize the cell or will lead to situations where large forces can pull a receptor molecule out of the membrane. The association and dissociation rates should be rapid, but not too rapid. Bond lifetimes for selectins vary from seconds to milliseconds.

6.10 Slip and Catch Bonds Play Important Roles in Selectin-Mediated Rolling

The mechanisms by which adhesion molecules such as selectins form and then release their attachments are revealing. Chemical bonds are not as is usually through static entities. Rather, they are dynamic structures whose strength and lifetimes can vary according to the forces they experience and the manner in which these forces are applied. In leukocyte rolling, the bonds that tether leukocytes to the endothelial wall are subjected to stretching, or tensile, forces as a result of the shear stresses. The usual situation is one in which applied forces weaken bonds reducing their lifetimes. These bonds are referred to as "slip" bonds, since they allow leukocytes experiencing shear forces to slip away and not become immobilized prematurely. A second kind of bond is termed a "catch" bond. This allows for the capture of the leukocytes and the start of rolling. The enhancements in the off rates can be appreciable – the off rate rises exponentially with increasing force. The enhanced dissociation rates arising from the tensile forces shift the bond lifetimes into the range needed to ensure proper rolling.

The off rate amplification is a consequence of the lowering of the energy barriers by the applied forces. The dependence of bond lifetimes on the presence and strength of applied forces is not specific to rolling or leukocytes. Instead, these considerations apply equally well to other bonds between receptors and ligands. The acceleration in dissociation rates due to the presence of applied forces provides a general mechanism for transducing mechanical stresses into signaling responses. Membrane, cytoskeletal, and signaling elements coming together at control points will sense and respond to mechanical stresses by dissociating far more rapidly than would be the case in the absence of stresses. The selectins that mediate leukocyte rolling exhibit both kinds of bonds. The catch bonds are formed first. They enable the leukocytes to be caught by the surface. Once the leukocytes are caught and begin rolling, a transition takes place from a catch to a slip bond regime so that the latter can mediate proper detachment from the surface.

This dependence of the off rate on external forces is captured by the expression

$$k_{off}\,(F) = k^0_{off}\,\exp[-E(F)/k_BT] \tag{6.5}$$

In this equation, the usual dependence of the rate on a constant barrier height E has been replaced by a force-dependent barrier $E(F)$. The quantity F represents the applied force on the bond formed during rolling. The applied force amplifies the off (dissociation) rate for the receptor-ligand complex by reducing the height of the transition barrier. This force-driven reduction in barrier heights by an external force is depicted in Fig. 6.7. The amplification in dissociation rate due to the presence of an applied force is found by inserting the simplified expression for $E(F)$ into the off rate equation. This gives

$$k_{off}(F) = k_{off}(0)\exp[Fx_{av}/k_BT] \tag{6.6}$$

where $k_{off}(0)$ is the unstressed rate, that is,

$$k_{off}(0) = k^0_{off}\,\exp[-E/k_BT] \tag{6.7}$$

is the conventional value for the off rate in the absence of an applied force. In the above, the angle-dependent expression has been replaced by the quantity x_{av} representing a thermally averaged distance between receptor and ligand interfaces over which the bond weakens but does not yet break. The expression for the amplification in the off rate is known as Bell's equation. According to Bell's equation, the off rate rises exponentially with increasing force. In situations

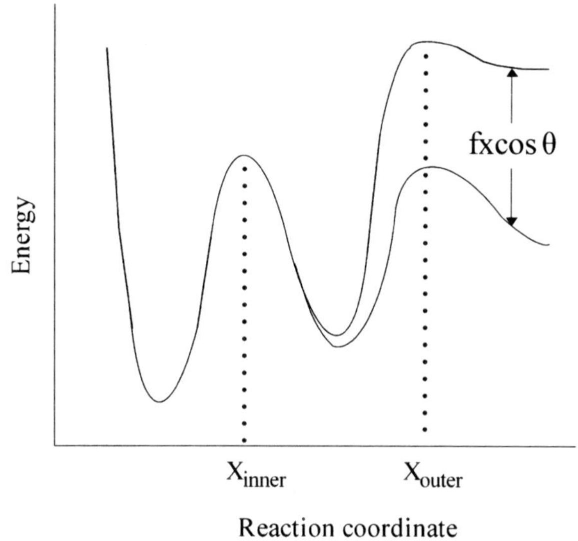

Fig. 6.7 Lowering of the transition barriers by applied forces

where the barriers are sharp, the principal consequence is the linear lowering of the barrier heights as a function of distance, with little change in shape or location of the peaks and valleys. The influences of an applied force on the energy landscapes are illustrated in Fig. 6.7 for the cases where there are two barriers. Dissociation rates are influenced not only by applied forces but also by how rapidly they are applied. A bond that will appear strong if the forces are rapidly ramped up to a maximum value will appear far weaker and more easily broken if the same external forces are applied slowly.

6.11 Leukocyte Arrest Through the Joint Actions of Chemokines and Integrins

Integrins are bidirectional signaling, cell adhesion molecules. These are membrane-spanning glycoproteins composed, as mentioned in the last chapter, of noncovalently attached α and β chains. Each chain has a large extracellular domain, a single membrane-spanning domain, and a short cytoplasmic domain. The extracellular portions are each composed of five or more domains and many of these contain multiple subdomains.

Representative arrangements of the major extracellular domains of the integrin α and β subunits are presented in Fig. 6.8. In this depiction, the outermost unit is the inserted (I) domain, the subunit shown in Fig. 6.8(b). Not all integrins contain I or I-like domains. Conformations also vary somewhat from integrin to integrin. In Fig. 6.8, three conformations are depicted. The bent conformation is a low-affinity conformation. This orientation of subunits is in

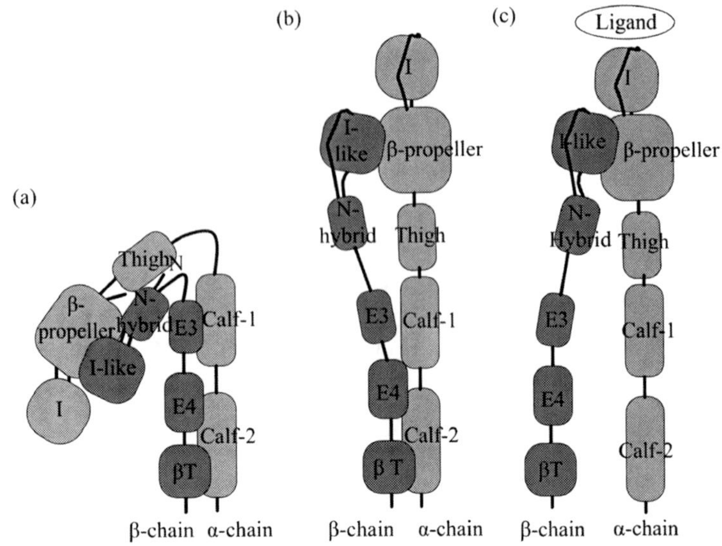

Fig. 6.8 Integrin extracellular domains illustrating the three main conformations for activation and signaling. (**a**) Bent, inactive, (**b**) extended, inactive, and (**c**) extended, active, ligand-bound conformation

sharp contrast to the two other conformations, an intermediate affinity open or extended conformation and a high-affinity ligand-binding conformation. A crucial difference between the two extended conformations is the movement of the lower portions of the chains, the tails, away from one another. These movements are reflected in corresponding movements of the transmembrane and cytoplasmic portions of the chains away from one another. Those movements along with allosteric changes at the upper end domain that either promote or inhibit ligand binding convey the bidirectional, inside-out and outside-in signals.

Integrins link the extracellular matrix to the cytoskeleton and transduce mechanical signals from one to the other. The cytoskeleton linker protein talin is the key element connecting integrins to the cytoskeleton. Binding of extracellular ligands such as the ICAMs initiates outside-in signaling, whereas talin functions as the cytoplasmic ligand that triggers inside-out signaling. Talin is a rod-shaped protein composed of bundles of α helices and contains an N-terminal head region, referred to as a FERM domain. This head domain encompasses binding sites that mediates attachment of talin to the cytoplasmic portion of the integrin β chain. Talin also binds to F-actin and to actin-binding proteins thereby connecting the integrins to the cytoskeleton. Mechanical signals can convey information about events taking place on one side of the cell through the cytoskeleton to signaling nodes residing on other sides and to organelles such as the nucleus. Outside-in signaling serves several functions one of which is to provide a survival signal indicative that the cell remains in proper contact with its surrounding. These signals are relayed via the Src/focal adhesion kinase (FAK) complex to downstream targets such as Akt. This role of integrin signaling, in cancer, will be discussed in Part II of the text.

Integrins are responsible for the arrest phase of leukocyte migration from the circulation into sites of injury and infection. Although selectins are able to promote tethering and rolling, they cannot arrest a leukocyte. Instead, integrins working together with chemokines abruptly stop rolling leukocytes at the right place. Rolling occurs until the leukocytes encounter sufficient numbers of arrest chemokines juxtapositioned next to integrin ICAM ligands on endothelial cells. Chemokine binding to their G-protein-coupled receptors triggers an intracellular signaling cascade leading to talin activation and changes in integrin conformations from the bent inactive conformation to the extended intermediate conformation primed for ICAM binding, full activation, and arrest of rolling. One of the events contributing to arrest is the increased mechanical stability of the integrin-cytoskeleton system brought on by chemokine binding. These adjustments strengthen the integrin-ICAM bonds.

6.12 Epithelial Cell-to-Cell Adhesions Are Maintained by Junctional Complexes

Endothelial cell-to-cell contacts are established by members of the above-mentioned families of cell adhesion molecules. These are organized into two main classes of complexes that mediate both adhesion and intracellular signaling. These are adherens junctions (AJs) and tight junctions (TJs). The core molecular constituents of these complexes are summarized in Table 6.4. The cadherin-catenin complexes discussed in the preceding section are the core members of adherens junctions.

Tight junctions contain a different set of cell adhesion molecules. Claudins are a family of more than twenty proteins. They are four-pass transmembrane proteins mediating cell-to-cell adhesion. Their N- and C-termini are located in the cytosol.Occludin is another four-pass transmembrane protein with terminals in the cytosol. These cell adhesion molecules are found in both endothelial and epithelial cells. Highest expression levels are in brain endothelial tissue. The third group of transmembrane tight junction proteins is the junction adhesion molecules, or JAMs. These are members of the immunoglobulin superfamily. Three JAMs have been found to date – JAM-A, JAM-B, and JAM-C. Two other loosely related CAMs have been identified as well. These are known as endothelial cell-selective adhesion molecule (ESAM) and Coxsackie- and adeno-virus receptor (CAR).

Zonula occludens (ZO) proteins are members of the membrane-associated guanylate kinase (MAGUK) family. These proteins have prominent roles in organizing the tight junctions. In endothelial cells, the ZO proteins are loosely speaking the TJ counterparts of the catenins. These are not the only

Table 6.4 Junctional complexes that regulate cell-to-cell adhesion and permeability

Complex	Function
Adherens junctions	
- VE-cadherin	Transmembrane cadherin
- Catenins	Cytosolic
α-catenin	
β-catenin	
γ-catenin (plakoglobin)	
p120-catenin (p120)	
Tight junctions	
- Claudins	Transmembrane CAM
- Occludin	Transmembrane CAM
- JAMs	Transmembrane IgCAM
- ZO proteins	Cytosolic
ZO-1	
ZO-2	
ZO-3	
PECAM-1 complexes	
- PECAM-1	Transmembrane IgCAM
Nectin-afadin system	
- Nectin	Transmembrane IgCAM
- Afadin	Cytosolic

submembrane proteins resident in TJs. The region around tight junctions is rich in signaling proteins, so much so that they are commonly referred to as forming a cytoplasmic plaque. The ZO proteins function as molecular scaffolds that help organize these signaling meshes.

Two other mediators of cell-to-cell adhesion are included in Table 6.4. Platelet endothelial cell adhesion molecule 1 (PECAM-1) is a member of the immunoglobulin superfamily. It associates with VE-cadherin and for that reason may be considered to be a component of the adherens junctions. It binds to partner PECAM-1s on opposing cell surfaces. The last set of entries in Table 6.4 is the nectin-afadin system. The nectins are another family of immunoglobulin-like cell adhesion molecules. There are four members of this family and they form homo- and heterophilic cell–cell contacts. They associate with the cytosolic protein afadin, which connects the nectins to the actin cytoskeleton. This system organizes both adherens and tight junctions by activating and recruiting required signaling elements to the junction sites. Prominent among the signaling molecules recruited and activated are the Cdc42 and Rac small GTPases. These GTPases are members of the Rho family, whose members play key roles in organizing the actin cytoskeleton.

There are differences between the two kinds of junctions. In the case of tight junctions, the outer leaflets of the plasma membranes on the opposing cells come into close contact, at the so-called kissing points. In adherens junctions, the two surfaces are separated from one another by distances of 15–20 nm. The latter links adjacent cell surfaces while the former act as barriers to unwanted diffusion.

6.13 Leukocytes Enter the Intima by Passing In-Between Epithelial Cells and by Passing Through Them

As noted earlier, the movement of leukocytes into sites of injury and infection occurs is stages – leukocyte capture and tethering is followed by rolling, firm arrest, and, lastly, diapedesis. Both kinds of cell – leukocytes and endothelial cells (ECs) – cooperate with one another to promote effective migration. Microvilli are adhesion molecule-rich extensions of the cell surface. They are formed at the outside facing, or apical, surface in a variety of different cell types in order to increase the effective surface area. The surface of ECs as well as migrating leukocytes is studded with microvilli rich in adhesion molecules and chemokines. In these structures, reciprocal pairs of ligands and their receptors situated on opposing surfaces promote the various stages of migration and attachment. The first stages of the dance are dominated first by selectins and their ligands, and then by integrin receptors and their counterreceptors.

Monocytes have two ways of getting through the epithelial cell layer. One way is to pass through the cells and the other is to squeeze between a pair of adjacent cells. The former is known as the transcellular pathway and the latter as the paracellular pathway. It seems that both pathways are utilized by migrating leukocytes. Of the two, the paracellular pathway is best understood and is regarded as being the dominant route. In its early locomotion-to-diapedesis stages, Rho GTPases are activated in endothelial cells. These cytosolic proteins along with other focal adhesion associated proteins such as FAK are relocated and activated by ICAM-1 signaling leading to a loosening of mechanical cell-to-cell contacts. Other early acting adhesion molecules during diapedesis include junctional proteins prominent among which are PECAM-1, vascular endothelial cadherin (VE-cadherin), possibly the junctional surface protein designated as CD99, and the JAMs. These proteins form homophilic contacts with partner proteins on the opposing cell surface.

In traversing the endothelium by means of the transcellular pathway, the leukocytes must avoid damaging the nucleus and other cellular structures. Both the endothelial cells being traversed and the invading leukocytes send out microvilli-like projections. Those sent by the endothelial cells form a transmigration-promoting cup-like structure about the leukocytes that guides the leukocytes. The leukocytes, in turn, send out podosomes, actin- and adhesion-molecule-rich invasive organelles in which integrins drive actin polymerization that probe the endothelial cell for a permissive region through which to pass. Sites of passage are associated with endothelial cell caveolae and caveolins, and ICAM-1, which relocates, forms clusters within the caveolae and interacts with the leukocyte integrins.

Endothelial cells are especially equipped to detect and respond to abnormal, low wall shear stresses. Changes in shear stresses can be detected by PECAM-1 and by integrins, and by stretch-sensitive ion channels. In particular, shear stresses induce the phosphorylation and activation of PECAM-1 and promote the association of VE-cadherin with vascular endothelial growth factor (VEGFR) 2 receptor. In these activities, VEGFR2, VE-cadherin, and PECAM-1 operate as a shear stress mechanotransduction complex that stimulates changes in cell–cell adhesion through direct actions and through gene expression.

6.14 Rupture of the Fibrous Cap and Not the Lesion Itself Causes Thrombosis

One of, if not the, earliest consequence of low shear stress is the triggering of an inflammatory response that marks the site of potential damage and recruits the monocytes. The early developmental stages of the disease as depicted in Fig. 6.9 provide a more detailed summary of the earliest stages of atherosclerosis than shown at a high level in Fig. 6.1. In Fig. 6.9, the initiating events are low shear stress and cholesterol in excessive amounts that leads to a buildup of oxLDL within macrophages. The low shear stresses trigger an inflammatory response that together with the oxLDL trigger a cascade of events starting with the initial recruitment by the endothelial cells of the monocytes. The subsequent inflammatory cascade generated within the local microenvironment by endothelial cells, macrophages, and smooth muscle cells creates a progressively more serious lesion over time as the inflammation becomes chronic and self-sustaining.

Over time, a thin fibrous cap develops that sits on top of a large lipid core. The latter, often referred to as the necrotic core contains large numbers of macrophages along with both extracellular and intracellular deposits of lipids. By themselves these structures do not cause thrombosis but their rupture and subsequent lodgings do. Rupture is promoted by the inflammatory and apoptotic activities of the macrophages.

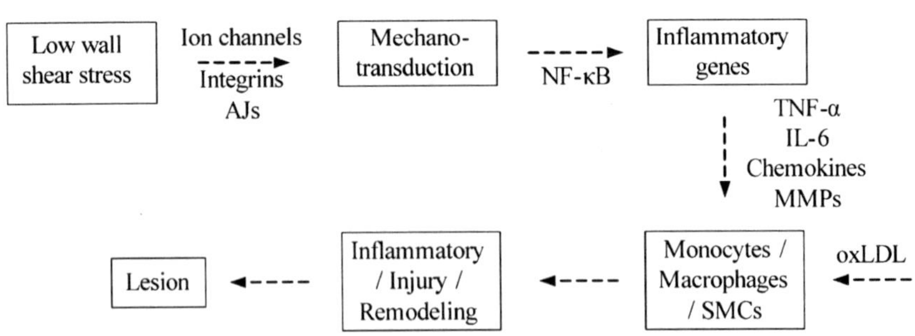

Fig. 6.9 Development of an atherosclerotic lesion

The stimulation of gene expression by the shear stresses leads to upregulation of pro-inflammatory cytokines such as TNF-α and IL-6, chemokines, extracellular matrix remodeling factors, most notably matrix metalloproteinases (MMPs) such as MMP2 and MMP9, and tissue factor (TF), a leadoff signaling molecule in the coagulation cascade leading to thrombin formation and blood coagulation. In triggering these responses, the signals generated by the various mechanotransducers converge on the NF-κB signaling node which, upon activation, stimulates transcription from NF response elements in promoters. These actions continue throughout the evolution of the disease. Of particular importance in the rupture phase of the disease is the inflammatory responses brought on by oxLDL-laden foam cells within the growing lesion.

The extracellular matrix (ECM) is not a passive medium for attachment but rather a dynamic one that influences cellular behavior in many ways. The ECM provides sites for attachment of signaling molecules, sequesters them, supplies ligands for cell surface adhesion molecules, and regulates cell survival through interactions with the cell surface and cytoskeleton. The extracellular matrix is remodeled continually. It undergoes remodeling during development and wound healing, and ECM remodeling participates in the transition from the normal to the disease states characteristic of atherosclerosis and cancer.

Matrix metalloproteinases (MMPs) are a family of more than two dozen secreted proteolytic enzymes that cleave components of the ECM and signaling proteins associated with the ECM and cell surface resulting in their activation and solubilization. These enzymes are normally synthesized in small quantities but production is increased in response to cytokines and growth factors associated with mechanical injury and inflammation (Table 6.5).

Growth factor and cytokines secreted by cells in the microenvironment stimulate a progressively broader spectrum of MMP to be secreted. Vascular smooth muscle cells constitutively express MMP-2. Stretch/injury responses

Table 6.5 Metalloproteinase classification

Group	MMP	Common name
Collagenases	MMP-1	Collagenase-1 (Interstitial collagenase)
	MMP-8	Collagenase-2 (Neutrophil collagenase)
	MMP-13	Collagenase-3
Gelatinases	MMP-2	Gelatinase-A
	MMP-9	Gelatinase-B
Stromelysins	MMP-3	Stromelysin-1
	MMP-10	Stromelysin-2
	MMP-11	Stromelysin-3
Matrilysins	MMP-7	Matrilysin
Membrane type MMPs	MMP-14	MT1-MMP
	MMP-15	MT2-MMP
	MMP-16	MT3-MMP
	MMP-17	MT4-MMP
Others	MMP-12	Macrophage metalloelastase

lead to further secretion of MMP-2 and MMP-9. Monocytes recruited to ECs secrete MMP-9, MMP-14, and MMP-16. During that same period the SMCs respond to inflammatory cytokine IL-1 and growth factor PDGF signals by secreting additional MMPs, namely, MMP-1, MMP-3, and MMP-9. Macrophages take up oxLDL, become foam cells, and secrete a still broader spectrum of MMPs, specifically, MMP-1, MMP-3, MMP-8, MMP-11, MMP-12, and MMP13. These agents weaken the fibrous cap and by that means promote plaque rupture.

An emerging goal in the prevention of thrombosis is to find ways to strengthen the cap since it seems that the mechanical stability of the cap and not its mere size that is the determining factor in rupture. Several targets are available starting with the MMPs and including upstream elements of pro-apoptotic and pro-inflammatory signaling pathways activated by macrophages that promote apoptosis of vascular smooth muscle cells. Increasing the presence of vascular smooth muscle cells can strengthen the overall structure, and thus finding ways of reducing the various negative influences of the macrophages are being sought. These therapeutic approaches are accompanied by others, including agonists and inhibitors of selected elements of LDL and HDL metabolism, of biologically active lipids, and of components of the multistep adhesive cascade.

Further Reading

Hemodynamic Forces and Mechanotransduction

Davies PF 1995 Flow-mediated endothelial mechanotransduction. *Physiol. Rev.* 75: 519–560.

Malek AM, Alper SL and Izume S 1999 Hemodynamic shear stress and its role in atherosclerosis. *JAMA* 282: 2035–2042.

Resnick N et al. 2003 Fluid shear stress and the vascular endothelium: for better and for worse. *Prog. Biophys. Mol. Biol.* 81: 177–199.

Traub O and Berk BC 1998 Laminar shear stress: mechanisms by which endothelial cells transduce an atheroprotective force. *Arterioscler. Thromb. Vasc. Biol.* 18: 677–685.

Mechanotransduction and the Glycocalyx

Florian JA et al. 2003 Heparan sulfate proteoglycan is a mechanosensor on endothelial cells. *Circ. Res.* 93: e136–e142.

Weinbaum S et al. 2003 Mechanotransduction and flow across the endothelial glycocalyx. *Proc. Natl. Acad. Sci. USA* 100: 7988–7995.

Mechanotransduction: The Cytoskeleton and the Matrix

Discher DE, Jammey P and Wang YL 2005 Tissue cells feel and respond to the stiffness of their substrate. *Science* 310: 1139–1143.

Ingber DE 2006 Cellular mechanotransduction: putting all the prices together again. *FASEB J.* 20: 811–827.
Vogel V and Sheetz M 2006 Local force and geometry sensing regulate cell functions. *Nat. Rev. Mol. Cell Biol.* 7: 265–275.

Fluid Shear Stress, Nitric Oxide, and Caveolae

Tzima E et al. 2005 A mechanosensory complex that mediates the endothelial cell response to fluid shear stress. *Nature* 437: 426–431.
Yu J et al. 2006 Direct evidence for the role of caveolin-1 and caveolae in mechanotransduction and remodeling of blood vessels. *J. Clin. Invest.* 116: 1284–1291.
Landmesser U, Hornig B and Drexler H 2004 Endothelial function: a critical determinant in atherosclerosis? *Circulation* 109 (Suppl II): 27–33.
Szmitko PE et al. 2003 New markers of inflammation and endothelial cell activation: Part I. *Circulation* 108: 1917–1923.

Multistep Adhesion Cascade

Ley K et al. 2007 Getting to the site of inflammation: the leukocyte adhesion cascade updated. *Nat. Rev. Immunol.* 7: 678–689.
Luster AD, Alon R and von Adrain UH 2005 Immune cell migration in inflammation: present and future therapeutic targets. *Nat. Immunol.* 12: 1182–1190.

Mechanisms of Cell Adhesion: Catch and Slip Bonds

Bell GI 1978 Models for specific adhesion of cells to cells. *Science* 200: 618–627.
Chen S and Springer TA 2001 Selectin receptor-ligand bonds: formation limited by shear rate and dissociation governed by the Bell model. *Proc. Natl. Acad. Sci. USA* 98: 950–955.
Dembo M et al. 1988 The reaction-limited kinetics of membrane-to-surface adhesion and detachment. *Proc. R. Soc. Lond. B* 234: 55–83.
Marshall BT et al. 2003 Direct observation of catch bonds involving cell-adhesion molecules. *Nature* 423: 190–193.
Merkel R et al. 1999 Energy landscapes of receptor-ligand bonds explored with dynamic force spectroscopy. *Nature* 397: 50–53.
Somers WS et al. 2000 Insights into the molecular basis of leukocyte tethering and rolling revealed by structures of P- and E-selectin bound to sLe^X and PSGL-1. *Cell* 103: 467–479.
Zhu C et al. 2008 Mechanisms for flow-enhanced cell adhesion. *Ann. Biomed. Eng.* 36: 604–621.

Integrins, Conformations and Signaling

Kim M, Carman CV and Springer CV 2003 Bidirectional transmembrane signaling by cytoplasmic domain separation in integrins. *Science* 301: 1729–1725.
Tadokoro S et al. 2003 Talin binding to integrin β tails: a final common step in integrin activation. *Science* 302: 103–106.

Takagi J et al. 2002 Global conformational rearrangements in integrin extracellular domains in outside-in and inside-out signaling. *Cell* 110: 599–611.
Wegener KL et al. 2007 Structural basis of integrin activation by talin. *Cell* 127: 171–182.

Adhesion, Integrins, and Cell Arrest

Alon R and Ley K 2008 Cells on the run: shear-regulated integrin activation in leukocyte rolling and arrest on endothelial cells. *Curr. Opin. Cell Biol.* 20: 625–632.
Schmitz J, Benoit M and Gottschalk KE 2008 The viscoelasticity of membrane tethers and its importance for cell adhesion. *Biophys. J.* 95: 1448–1459.
Shamri R et al. 2005 Lymphocyte arrest requires instantaneous induction of an extended LFA-1 conformation mediated by endothelial-bound cytokines. *Nat. Immunol.* 6: 497–506.

Diapedesis

Carman CV and Springer TA 2004 A transmigratory cup in leukocyte diapedesis both through individual vascular and endothelial cells and between them. *J. Cell Biol.* 167: 377–388.
Dejana E 2004 Endothelial cell–cell junctions: happy together. *Nat. Rev. Mol. Cell Biol.* 5: 261–270.
Millán J et al. 2006 Lymphocyte transcellular migration occurs through recruitment of endothelial ICAM-1 to caveola- and F-actin-rich domains. *Nat. Cell Biol.* 8: 113–123.
Nieminen M et al. 2006 Vimentin function in lymphocyte adhesion and transcellular migration. *Nat. Cell Biol.* 8: 156–162.
Ostermann G et al. 2002 JAM-1 is a ligand of the β2 integrin LFA-1 involved in transendothelial migration of leukocytes. *Nat. Immunol.* 3: 151–158.
Shay-Salit A et al. 2002 VEGF receptor 2 and the adherens junction as a mechanical transducer in vascular endothelial cells. *Proc. Natl. Acad. Sci. USA* 99: 9462–9467.
Schenkel AR, Mamdouh Z and Muller WA 2004 Locomotion of monocytes on endothelium is a critical step during extravasation. *Nat. Immunol.* 5: 393–400.

Plaque Rupture

Boyle JJ 2005 Macrophage activation in atherosclerosis: pathogenesis and pharmacology of plaque rupture. *Curr. Vasc. Pharmacol.* 3: 63–68.
Geng YJ and Libby P 2002 Progression of atheroma: a struggle between death and procreation. *Arterioscler. Thromb. Vasc. Biol.* 22: 1370–1380.
Newby AC 2008 Metalloproteinase expression in monocytes and macrophages and its relationship to atherosclerotic plaque instability. *Arterioscler. Thromb. Vasc. Biol.* 28: 2108–2114.
Shah PK 2003 Mechanisms of plaque vulnerability and rupture. *J. Am. Coll. Cardiol.* 41: 15S–22S.
Tabas I 2005 Consequences and therapeutic implications of macrophage apoptosis in atherosclerosis. *Arterioscler. Thromb. Vasc. Biol.* 25: 2255–2264.

Chapter 7
Chronic Inflammation

The emergence of inflammation as a major ingredient in the development of atherosclerosis has parallels in other disease states and classes. It is a contributing factor to cancer, especially metastasis, and can be either beneficial or harmful in neurodegeneration. Like atherosclerotic tissues, the microenvironments of other diseased tissues contain leukocytes, pro-inflammatory cytokines, and additional potentially harmful agents such as reactive oxygen species. Multiple initiating factors can trigger the loss of tissue homeostasis including injury, metabolic syndrome-associated loss of glucose and lipid homeostasis, malfunctions in growth and cell death regulation, and aggregation-prone misfolded proteins. These factors stimulate and interact with inflammation and its attendant effects lead over time to self-sustaining disease states associated with atherosclerosis, cancers, and neurodegenerative disorders.

The goal of this chapter is to examine the intracellular signaling routes that generate inflammatory responses. The first of these is the tumor necrosis factor (TNF) pathway activated by TNFα. Tumor necrosis factor α is a major pro-inflammatory cytokine implicated in the metabolic syndromes and a variety of autoimmune disorders. The initial discovery of TNFα involved its ability to kill tumor cells once its expression was induced by the immune system. Subsequent discoveries led to a far more general picture in which these proteins form a large superfamily of cytokines that bind cognate TNF receptors and trigger inflammatory, immune, developmental, and apoptotic responses.

A second major pathway involved in inflammatory signaling is the interleukin 1 (IL-1)/Toll pathway. This pathway is a central component of the innate immune response to bacterial lipopolysaccharides (LPS), a main component of the outer membrane of Gram-negative bacteria such as *Escherichia coli*. Examples of genes upregulated through activation of this pathway include IL-1, IL-6, TNFα, the chemokine IL-8, and the cell adhesion molecules ICAM-1 and E-selectin. This cytokine, like TNFα, functions as a major "we need help" signal that sounds an alarm to muster leukocytes to the site of an injury or infection. When cells are continually stimulated to secrete these molecules the result is an excessive one that may lead to disease rather than health.

M. Beckerman, *Cellular Signaling in Health and Disease*,
Biological and Medical Physics, Biomedical Engineering,
DOI 10.1007/978-0-387-98173-4_7,

These two signaling pathways lead through the NF-κB signaling node resulting in the upregulation of pro-inflammatory genes. The chapter will begin with an exploration of the NF-κB signaling node. These discussions will be followed by examinations of the upstream TNFα and IL-1/Toll signaling routes. As might be imagined, turning off inflammatory signaling in a timely manner is important. Several families of nuclear receptors do just that, thereby establishing a connection at the transcriptional level between metabolism and inflammation. That topic will be explored in the last part of the chapter. Ubiquitination and SUMOylation have important roles in activation, regulation, and termination of inflammatory signaling. The agents of these posttranslational modifications will be introduced in conjunction with the pathways where they are utilized.

7.1 The NF-κB Signaling Node Consists of IKKs, IκBs, and NF-κBs

Nuclear factor-κB (NF-κB) proteins are transcription factors that respond to stresses, inflammation, infection, and injury. They respond to shear stresses by stimulating the expression of genes for chemokines and cell adhesion molecules, and respond to inflammation, injury, and infection by regulating genes controlling inflammation, stress responses, tissue remodeling, and cell death. As shown in Fig. 7.1, the NF-κB family has five members. They each possess an N-terminal Rel homology domain responsible for DNA binding, dimerization, and nuclear localization. The Rel proteins (Rel-A or p65, Rel-B, and c-Rel) contain a C-terminal transactivation domain (TAD) while the other two

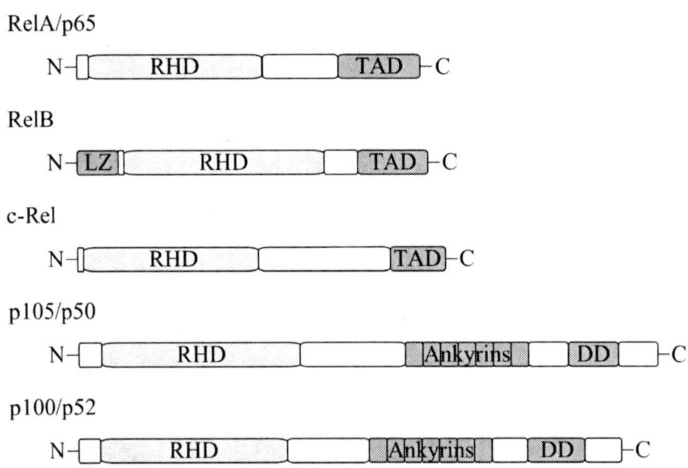

Fig. 7.1 Nuclear factor-κB domain structures. RHD: Rel-homology domain; LZ: Leucine-zipper-like motif; TAD: Transactivation domain; DD: Region homologous to death domain; Ankyrins: Ankyrin repeats

proteins exist in either long precursor (p105, p100) or short postprocessed (p50, p52) forms, and each possesses a death-like domain (DD) and a set of ankyrin repeats. These proteins also referred to as NF-κB1 and NF-κB2 are formed and sequestered in the cytosol as p105 and p100 precursors, respectively. They are proteolytically processed to make the shorter p50 and p52 forms. The C-terminus portion that is removed from p105 and p100 contains a series of ankyrin repeats that characterizes the IκB proteins and mediates their ability to sequester NF-κBs.

In the absence of upstream activating signals, NF-κB transcription factors are sequestered in the cytosol and also in the nucleus in inactive forms by IκB inhibitory proteins. The characteristic structural feature of these proteins (Fig. 7.2) is the presence of the aforementioned string of ankyrin repeats. Ankyrin repeats are protein–protein interaction modules. Six or seven ankyrin repeats are present in the IκBs and this family may be regarded as including the long form NF-κBs (p105 and p100). The secondary structure of the Bcl3 ankyin repeats is depicted in Fig. 7.2. As shown in this figure, each repeat consists of a pair of α-helices connected by a loop, each repeat containing 33 amino acid residues. The helices are arranged in a way that forms an extended, slightly curved surface whose binding specificity is determined by the surface amino acid residues.

The NF-κB transcription factors, IκB inhibitory proteins and members of a third family of proteins – activating kinases called IKKs – form a core NF-κB signaling module (Table 7.1). The IKK proteins consist of two catalytic subunits, IKKα and IKKβ, and a regulatory subunit, IKKγ, or NF-κB essential modulator (NEMO). Both of the IKK catalytic subunits are competent to

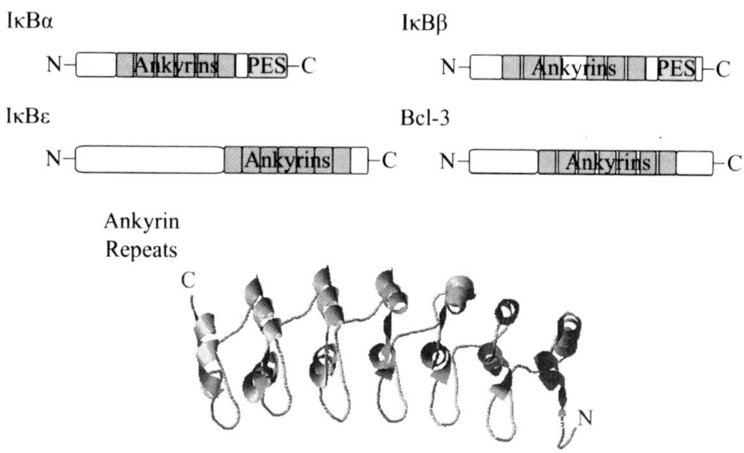

Fig. 7.2 *Upper panel*: IκB domain structure. PES: Proline (P), Glutamate (E), serine (S), threonine (T)-rich (PEST) domain. *Lower panel*: Ankyrin repeats from the Bcl3 protein. The figure was prepared using Jmol with atomic coordinates deposited in the PDB under accession code 1k1a

Table 7.1 Members of the NF-κB signaling node

NF-κB family	IκB family	IKK family
RelA/p65	IκBα	NEMO (IKKγ)
RelB	IκBβ	IKKα
c-Rel	IκBε	IKKβ
p105/p50	Bcl3	
p100/p52		

phosphorylate IκB proteins and function upstream of the IκBs. A simplified picture (Fig. 7.3) of how the NF-κB system operates is as follows. The IKKs are the key point of convergence for regulatory signals. Once their kinase activity is turned on through upstream signaling, they phosphorylate the IκBs. This step relives the inhibition placed upon the NF-κB by the IκBs. The tagged IκB proteins are then ubiquitinated and directed to the 26S proteasone for destruction or alternatively serve as shuttling agents that help convey the NF-κBs to the nucleus. The NF-κBs form homo- and hetero-dimers, the most common combination being p65/p50, and once translocated to the nucleus stimulate transcription of a suite of inflammatory genes.

Ubiquitination and proteolysis play important roles in NF-κB signaling. These processes will be reviewed in the next section and then a more complete picture of NF-κB inflammatory signaling will be presented. Events connecting the NF-κB node to signaling through the upstream receptor systems will be examined as will events occurring in the nucleus involving coactivation and corepression complexes.

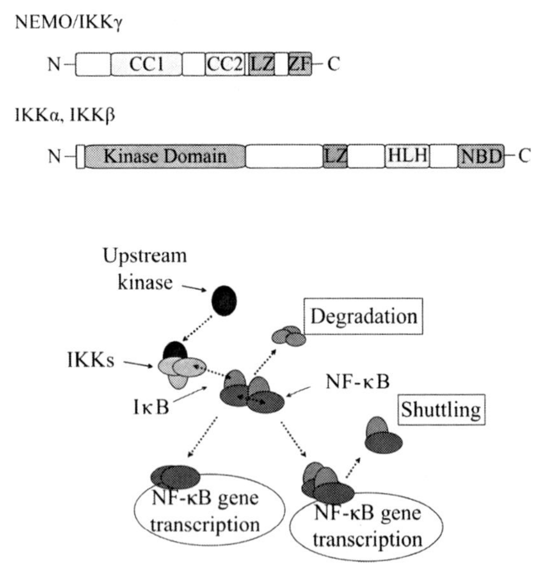

Fig. 7.3 *Upper panel*: IKK domain structure. CC: coiled-coil motif; ZF: zinc-finger domain; HLH: helix-loop-helix domain; NBD: nucleotide-binding domain. *Lower panel*: Canonical pathway for activating NF-κB signaling illustrating the overall arrangement of the IKK, IκB, and NF-κB proteins

7.2 Protein Ubiquitination Plays a Central Role in Cellular Signaling

Ubiquitin is a small protein, consisting of 76 amino acid residues with a mass of 8.5 kDa. It gets its name because of its ubiquitous presence in the cytosol and nucleus of bacteria, yeast, higher plants, and animal cells. It was discovered in 1975 by Goldstein and coworkers and given the provisional name UBIP (ubiquitous immunopoietic polypeptide). The reversible attachment of this molecule to signaling proteins has emerged as a major theme in cellular signaling since its role in degradation of short-lived species was first explored in the early 1980s by Ciechanover, Hershko, and Rose. The importance of this covalent modification was lent further support by discoveries in just the last few years of deubiquitinating enzymes.

Ubiquitin-dependent proteolysis is carried out in two stages. In the fist step, the protein to be degraded is tagged by covalent attachment of a string of ubiquitin molecules; in the second, the protein so-tagged is degraded within a large chambered multisubunit 26S proteasome. Three types of enzymes operating in a sequential fashion are responsible for attachment of the ubiquitin molecules to the target protein. The first enzyme to act is the ubiquitin-activating, or E1, enzyme. It is followed by an ubiquitin-conjugating, or E2, enzyme and finally by a ubiquitin protein ligase, or E3 enzyme. To begin the process, the E1 catalyzes the attachment of an ubiquitin molecule to the sulfhydryl group on a catalytic cysteine residue within the E1. A high-energy thioester bond is formed between the sulfhydryl group and the carboxyl group of a ubiquitin C-terminus glycine residue. This is done in an ATP-dependent way as depicted in Fig. 7.4. Two other steps follow – the ubiquitin protein is transferred from the E1 to an E2 catalytic cysteine residue to which it is again linked by means of a

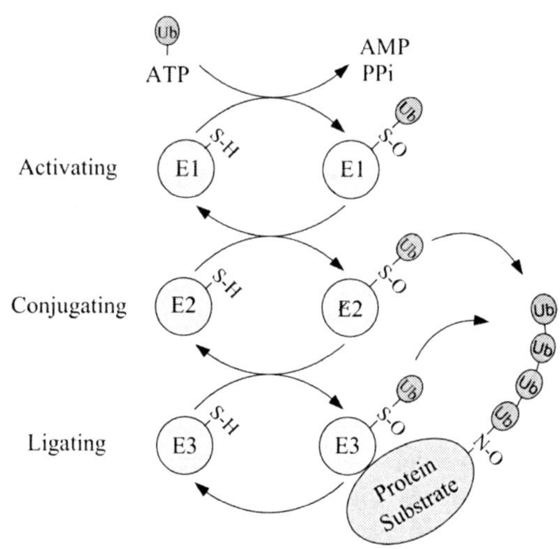

Fig. 7.4 Protein ubiquitination. Illustrated in this figure is the attachment of ubiquitin molecules to an SH group on cysteine residues of E1, E2, and E3 enzymes. Two routes for the final step are illustrated – direct transfer from the E2 to the substrate mediated by the E3, and intermediate transfer of the ubiquitin to the E3 prior to transfer to the substrate

thioester bond. The E3 then catalyzes the transfer of the ubiquitin either to an amino group of a side chain substrate lysine residue or by addition to multiubiquitin chains already being formed. Transfer can occur either using the E3 as a transfer intermediate or directly from the E2.

Ubiquitin possesses 7 lysine residues and can be conjugated on any of a number of lysine residues among which are K48 and K63. These two alternative sites for ubiquitination lead to different outcomes. In the case of K48, ubiquitination of the chain so produced leads to degradation of the tagged proteins by the 26S proteasome. In situations where K63 is the site for ubiquitination, the chain serves not as a tag for degradation but rather as a platform for assembly of a complex that activates downstream signaling. In either situation, the added ubiquitin is linked to either a substrate lysine or to the preceding ubiquitin's K48 or K63 through its C-terminal carbonyl group. The selection of the substrate lysine is dictated by that lysine's ability to attack the thioster bond holding the ubiquitin to the E2/E3. Inflammatory signaling pathways involving K48 ubiquitination are referred to in the literature as the canonical pathway, and signaling routes involving K63 lysines are termed the alternative pathway.

Ubiquitin-handling enzymes are organized in a hierarchical way. A small number of E1s interact with a larger number of E2s and these interact with a yet larger number of E3s. The E3s confer substrate specificity on the degradation process acting either by themselves or together with the E2s. The E3 ligases may be divided into two groups according to whether they possess a RING (really interesting new gene) finger domain or a HECT (homologous to E6-AP carbonyl terminus) domain.

The second step in ubiquitin-mediated proteolysis is the chopping up of ubiquitin-tagged proteins by the 26S proteasome. These proteins are constructed from multiple subunits arranged in a symmetric fashion in several layers to form a hollow chamber with a lid. The 26S proteasome has a 20S barrel-shaped core and a 19S capping structure at one or both ends. Ubiquitin-tagged proteins are passed in one end and the proteolytic products are released from the other end along with the recyclable ubiquitin molecules.

The proteasome core consists of four rings of subunits, stacked on top of one another. Each ring consists of an oligomeric arrangement of seven subunits (see Fig. 19.6). The core contains the catalytic part of the structure. They chop up the substrate proteins in a nonspecific manner. In order to accomplish this task, the proteins must be inserted in an unfolded form in order to enter the barrel-shaped core. The catalytic sites are located in the central-most two rings, while positioning and control over entry and exit are maintained by the outer two rings. Substrate proteins are unfolded and inserted into the core by the capping structure. The capping structure consists of a lid and a base, each composed of multiple regulatory subunits. It is responsible for recognizing polyubiquitinated protein substrates.

7.3 TNFα Signaling Occurs Through Complex I and Complex II

Ligands for this signaling system are synthesized as trimers. They are initially formed as transmembrane proteins that are proteolytically cleaved by the TNFα converting enzyme (TACE) to form a soluble ligand. This arrangement gives rise to trimeric assemblages of ligands, receptors, and cytoplasmic adapters that transduce messages into the cellular interior. Several different classes of adapters associate with members of the TNF family. One group of adapter is characterized by the presence of a death domain (DD). Prominent in the group are the Fas-associated death domain (FADD) and TNF-R-associated death domain (TRADD) proteins. Another group of adapters is the tumor necrosis factor (TNF) receptor-associated factors, or TRAFs. Tumor necrosis factor signaling through these adapters is depicted in Fig. 7.5.

The TRAF adaptors act as key intermediaries between the receptors and downstream signaling elements. The N-terminal domains of the TRAFs contain zinc and RING fingers motifs. These motifs consisting of arrangements of four cysteine and/or histidine residues that bind zinc ions facilitating the formation of a compact domain that, along with a C-terminus coiled-coil domain, mediates downstream signaling. In particular, the RING finger motif binds E2 ubiquitin-conjugating enzymes such as Ubc13 and the TRAFs can function as E3s to ubiquitinate either nearby proteins such as RIP1 or themselves.

There are six mammalian TRAFs, and they are named TRAF1 through TRAF6. The TRAF2 and TRAF6 proteins are crucially involved in TNF and IL-1 signaling, where they function as adaptors that link the receptors to downstream NF-κBs and MAP kinase cascades leading to AP-1 gene expression. The MAP kinase signaling routes activated by TNF, IL-1, and TCR receptors will be examined in Part II of this text.

TRAF6 differs from other TRAFs, which only mediate TNF signaling. In contrast, TRAF6 also participates in IL-1 signal transduction and also assists in T-cell receptor (TCR) signaling. This difference is due to the presence

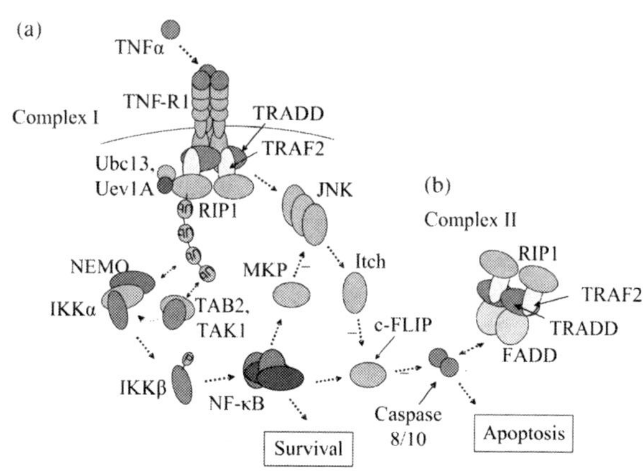

Fig. 7.5 TNF signaling. (**a**) Pathway leading to activation of NF-κB and JNK signaling modules in Complex I. (**b**) Formation of the death-inducing (apoptosis) signaling by Complex II

of a TRAF-C domain and recognition by TRAF6 of binding sites on receptors that differ from those recognized by other TRAFs. In the case of IL-1 signal transduction, motifs in the IRAKs are recognized by TRAF6.

The steps involved in conveying TNF signals to the NF-κBs are as follows. Ligand binding stimulates the recruitment of the TRADD protein to the receptor's cytoplasmic death domain. Prior to ligand binding, this domain was masked by resident silencer of death domain (SODD) protein binding. The TRAFFs serve as scaffolds for assembly of a signaling complex containing TRAF2s and other adapter proteins called RIP1. These two proteins – TRAF2 and RIP1 – work on concert to recruit and activate the IKKs. This activation process is facilitated by members of the TAK1 complex, consisting of TAK1 and TAB2. Steps in the activation of the IKKs include phosphorylation and K63-mediated attachment of polyubquitin chains that serve as platforms for further assembly of proteins in a close vicinity where they may interact with one another.

TNFα signaling can not only lead to cell survival through the upregulation of pro-inflammatory genes and other survival factors, but can also lead to the opposite outcome – cell death. Failure to activate the IKKs leads to the activation of an alternative signaling route centered about the death-inducing signaling complex, or DISC. Key adapters in this complex are the TRADD and FADD. These lead to recruitment and activation of the apoptosis-initiating enzymes caspases 8 and 10. The assemblage of adaptors and other intermediaries leading to survival signaling is known as Complex I while the signal proteins that assemble into the DISC are referred to as Complex II.

The decision between survival and apoptosis is determined in large measure by the balance between NF-κB and c-Jun NH_2-terminal kinase (JNK) signaling. When there is strong IKK activation and NF-κB signaling, the JNK branch of the TNFα signaling pathway is shut down. This occurs when NF-κB activates MAP kinase phosphatases (MKPs), which then inhibit JNK signals, and when NF-κB activates the caspase inhibitor c-FLIP (cellular FLICE-like inhibitory protein) which prevents activation of caspase 8 signaling. If the opposite situation occurs in which IKK and NF-κB are not strongly induced, then JNK signals activate the E3 ubiquitin ligase Itch, which triggers the proteolytic degradation of c-FLIP resulting in Complex II assembly and apoptotic signaling.

7.4 Reactive Oxygen Species (ROS) Influences the Choice Between Survival and Death

Reactive oxygen species in excess of normal amounts function as second messengers that shift the balance away from survival toward apoptosis. It accomplishes this by strengthening signaling through the JNK pathway; in particular, it stimulates ASK1, the furthest upstream kinase in the JNK pathway. At the same time, it inhibits the MKPs and thus prevents their inhibition of JNK signaling. In those situations where NF-κB gene expression occurs, antioxidants are upregulated as part of the program.

In more detail, the cell death route requires sustained activation of JNK signaling. This is promoted by ROS such as hydrogen peroxide, which reacts with the MPKs and prevents their ability to dephosphorylate and turn off JNK signaling. The mechanisms whereby hydrogen peroxide inhibits PTPs such as the MPKs will be examined in the next chapter. In brief, the cell survival branch of this pathway leads through NF-κB-mediated gene transcription. Among the genes upregulated by the route are ferritin heavy chain (FHC) and superoxide dismutases that oppose ROS activity. Ferritin heavy chain sequesters iron and prevents the buildup of highly reactive hydroxyl radicals through Fenton reactions.

The RIP1 protein sites are at the juncture of the survival and apoptotic pathways. Different upstream signals result in formation of adapter complexes that reflect the character of the upstream signals and in response activate specific subsets of the downstream signal transducers. In the case of necrotic cell death induced by TNF, a complex is formed containing RIP1, TRADD, and the membrane-bound NADPH oxidase, which generates ROS. The latter then supports sustained JNK signaling leading to necrosis.

7.5 Toll-like Receptor 4 Responds to Bacterial Lipopolysaccharides and Mammalian Lipids

Members of the interleukin-1 (IL-1) family of cytokines stimulate the expression of genes involved in inflammation and autoimmune diseases. Like TNF these molecules was first cloned in 1984. Prior to that time, a host of physiological responses associated with diseases and stresses had been attributed to this protein without certainty as to whether one or more different agents were in fact responsible for the strong responses evoked by receptor binding. Since that time other members of this family have been found. Prominent among these are the Toll-like receptors (TLRs) of which there are 10 mammalian members.

The IL-1 family plays a key role in the innate immune response to pathogens and works together with TNFα to mount appropriate cellular responses to a particular type of pathogen. The receptors of the IL-1 family function as pattern recognition receptors (PRRs) that bind certain distinct pathogen-associated molecules patterns (PAMPs). In response to ligand binding, a set of adaptors are assembled that mediate downstream signaling to transcription factors that upregulate elements of the inflammatory response such as interferons (in the case of virus attacks) and acute phase proteins. The IL-1 family is an ancient one. It has been identified in plants, in insects (*Drosophila*) where it is known as the Toll/Dorsal pathway and in vertebrates where it is referred to as the IL-1R/NF-κB pathway.

The main bacterial lipopolysaccharides (LPSs) detector is the Toll-like receptor 4 (TLR4). Significantly, this receptor also binds mammalian fatty acids (FAs) and actions of this kind link together inflammation and excess

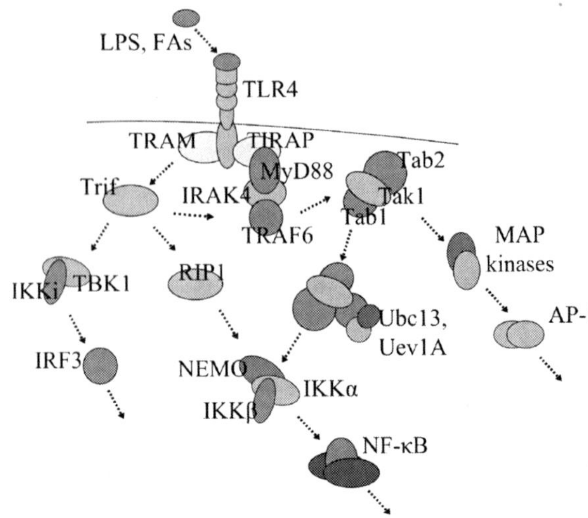

Fig. 7.6 TLR4 signaling showing its complex branching into IRF3, NF-κB, and AP-1 gene expression

lipids. The TLR4 signaling pathway contains two main branches as shown in Fig. 7.6, one that passes through the adapters Toll/IL-1 receptor domain-containing adapter protein (TIRAP) and myeloid differentiation factor 88 (MyD88) and the other that utilizes Trif-related adapter molecule (TRAM) and TIR domain-containing adapter inducing IFNβ (Trif), where TIR stands for Toll/Interleukin-1 receptor homology. The TIRAP/MyD88 pathway is common to almost all TLRs; it is the one utilized by TLR4 in response to fatty acids, and leads to activation of NF-κB and AP-1 gene expression.

The main branch of the TLR4 signaling pathway includes a number of adapters among which are, besides TIRAP and MyD88, members of the IR-1R-associated kinase (IRAK) and TRAF families. As depicted in Fig. 7.6, IRAK4 is recruited along with TRAF6. The TRAF6 protein dissociates from the receptor complex and forms a new complex with Tak1, Tab1, and Tab2 at the plasma membrane. These proteins associate with the E2 ubiquitin-conjugating enzyme to form a complex that recruits and activates the IKKs, which then activate the NK-κBs. In addition, Tak1 functions as the upstream-most member of the MAP kinase cascade. Activation of the MAP kinase cascades such as JNK and p38 lead to activating protein-1 (AP-1) gene expression.

As noted in Chapter 3, fatty acid binding to TLR4 leads to NF-κB signals, gene expression, JNK-mediated phosphorylation and inactivation of the IRS proteins, and insulin resistance. The two pathways – TNFα and TLR4 – act synergistically in reaction to cytokines and fatty acids to establish positive feedback actions that exacerbate the situation and generate a strong, sustained inflammatory response. Macrophages and adipocytes in adipose tissue, and macrophages, endothelial cells and vascular smooth muscle cells in the arterial wall generate inflammatory cytokines and ROS that contribute strongly to T2D and atherosclerosis.

7.6 Downstream and into the Nucleus with NF-κBs

Events taking place downstream, in the nucleus, result in either repression or activation of transcription as was the case with nuclear receptors. Again, several factors interact with chromatin to promote or inhibit the transcription of genes. In the absence of activating signals from the cell surface, p50 or p52 homodimers reside at NF-κB transcription sites in the nucleus and interact with corepressors such as SMRT and N-CoR to repress transcription. Upstream activating signals relayed by IKKα set off a chain of events leading to the replacement of the corepressor with coactivators and replacement of p50 (or p52) with p65.

The steps leading to transcription activation are illustrated in Fig. 7.7. The initial step is IKKα phosphorylation of SMRT on chromatin. As shown in the figure, corepressors SMRT and N-CoR form complexes with the histone deacetylase HDAC3. The arrival of IKKα results in the recruitment of two histone-binding proteins, transducin β-like protein 1 (TBL1) and the closely related protein TBLR. These last-named proteins have roles both in repressing transcription and promoting it. They join with SMRT and N-CoR in forming transcription-repression complexes, and alternatively by recruiting the ubiquitin-mediated proteolysis machinery (represented in this figure by Ubc5) promote transcription activation in a ligand-dependent manner. Recruitment of the aforementioned factors leads to dissociation of the repressor complex from

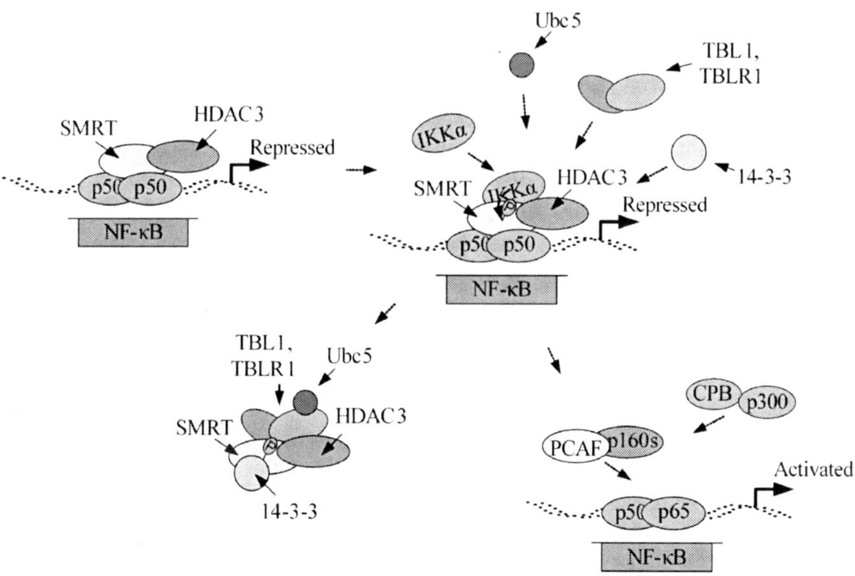

Fig. 7.7 Regulation of NF-κB gene expression through subunit replacements and formation of repressor and activator complex at promoter sites (see text for details)

the NF-κB promoter. This step enables coactivators to bind to the NF-κB promoter to which p65 has now been recruited thereby forming a transcription-activating complex.

7.7 Glucocorticoids Terminate Inflammatory Responses and Restore Homeostasis

It is important for there to be mechanisms for turning off inflammatory responses once they have completed their protective roles in dealing with injury and infection. Key elements in terminating inflammatory and adaptive and innate immune responses are glucocorticoids and the hypothalamic-pituitary-adrenal (HPA) axis responsible for its secretion. These homeostatic actions are triggered by cytokines that signal to the paraventricular nucleus (PVN) of the hypothalamus, which then secretes corticotropin releasing hormone (CRH), which upon reaching the pituitary stimulates the secretion of adrenocorticotropic hormone (ACTH). This hormone, in turn, stimulates the secretion of glucocorticoids by the adrenal grand. The glucocorticoids circulate throughout the body to throttle back inflammatory responses in the thymus, bone marrow, lymph nodes, and by resident cells of the immune system.

Inflammatory disorders include allergies, asthma, and autoimmune disorders among which are multiple sclerosis, rheumatoid arthritis, and type 1 diabetes (Table 7.2). Glucocorticoids are prescribed for treatments of many of these inflammatory disorders, but because of their multifaceted ability to inhibit inflammatory responses they have long-term deleterious effects as well as beneficial ones.

The glucocorticoid receptor (GR) is a member of the nuclear receptor family. It is a steroid hormone (Type 1) nuclear receptor that along with the mineralocorticoid (aldosterone) receptor binds glucocorticoids such as cortisol. Like other nuclear receptors, GRs possess A/B, C, D, E, and F domains as depicted in Fig. 4.1. These receptors form complexes with members of the Hsp70 and Hsp90 families of molecular chaperones, which facilitate ligand binding and trafficking. (Chaperone systems are discussed in detail in Chapter 18.) Ligand binding stimulates further conformational changes leading to their translocation to the nucleus from the cytosol. The GRs repress inflammation in two distinct ways. First, they form homodimers and in that form promote transcription of

Table 7.2 Autoimmune disorders

Addison's disease	Pernicious anemia
APECED	Rheumatoid arthritis
Graves' disease	Sjögren's syndrome
Hashimoto's thyroiditis	Systemic lupus erythematosus
Multiple sclerosis	Type 1 diabetes
Myasthenia gravis	

anti-inflammatory genes. Second, they directly interact with and repress the NF-κB and AP-1 gene expression apparatus. The latter process is referred to as transrepression.

7.8 LXRs and PPARγ in Transrepression of Inflammation Through SUMOylation

Glucocorticoid receptors are not the only members of the nuclear receptor family that limit inflammation. Liver X receptors and PPARγ do so as well, and in the process function, especially in macrophages, as integrators of metabolic and inflammatory signals. Lipids originating in the walls of bacterial pathogens, oxysterols, and cholesterol are each capable of triggering inflammatory responses. In these responses, macrophages converge on the site of lipid buildup and internalize them. Internalization and the subsequent enzymatic actions within the macrophages produce metabolites that activate nuclear receptors, most notably LXRs and PPARγ. These, in turn, downregulate genes that mediate inflammatory responses through transrepression. In this process, inflammatory gene-expression programs that proceed through the NF-κB pathway are blocked by the nuclear receptors. There is some reciprocity, and the metabolic and inflammatory programs reciprocally regulate one another. Key targets of repression are inducible nitric oxide synthase (iNOS), tumor necrosis factor α (TNF-α), cycloxygenase (COX) 2, chemokines such as monocyte chemotactic protein (MCP) 1, and pro-inflammatory interleukins such as IL-6 and IL-1β.

An example where all three of these receptor types are active in transrepression is in the transrepression of TLR signaling in macrophages. Acting downstream from the TRL4/MyD88 complex, the glucocorticoid receptor binds and disrupts p65 – interferon regulatory factor (IRF) complexes, while LRXs and PPARγ repress TLR signaling in macrophages by means of SUMOylation. The three nuclear receptor types repress overlapping yet distinct sets of genes in a signal- and gene-specific manner to limit inflammatory response to bacterial LPS, oxysterols, and cholesterol. This process whereby LXRs and PPARγ limit inflammation will be examined, next, following a short introduction to SUMOylation.

Like ubiquitin, the small ubiquitin-related modifier (SUMO) proteins (SUMO-1, SUMO-2, and SUMO-3) are covalently attached to target proteins through the sequential actions of E1, E2 SUMO conjugating, and E3 SUMO ligase enzymes. In humans, SAE1/SAE2 functions as the E1 and Ubc9 as the E2, while PIAS (protein inhibitor of activated STAT) proteins frequently operate as E3s. The 76-residue ubiquitin and 100-residue SUMO proteins are structurally similar as shown in Fig. 7.8. Both ubiquitin and SUMO proteins are attached to lysine side chains on target proteins, and both have key but different roles in the nucleus.

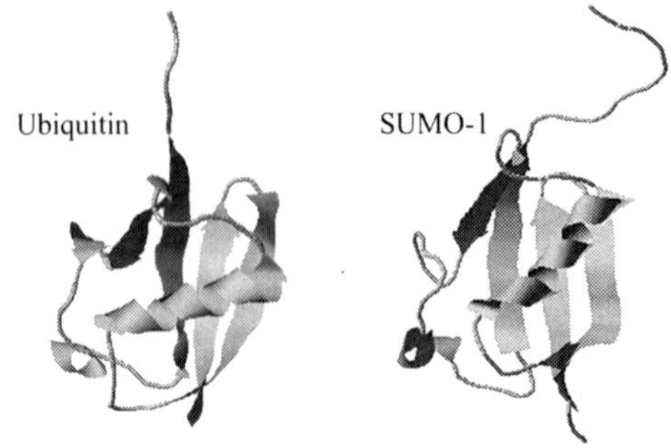

Fig. 7.8 Comparison of the three-dimensional structure of ubiquitin and SUMO (with the N-terminal-most residues omitted to better facilitate comparison). The N-terminals are at the *bottom* and the C-terminals are at the *top* in this depiction. The figure was prepared using Jmol with atomic coordinates deposited in the PDB under accession numbers 1ubq (ubiquitin) and 1a5r (SUMO)

The downstream transrepression of signals initiated when ligands such as LPS bind TLR4 receptors is depicted in Fig. 7.9. The pathway from ligand-receptor binding leads through adapters such as MyD88 and others to the NF-κB module. Translocation of an activating p50/p65 dimer to the nucleus leads to the displacement and subsequent proteolytic destruction of the NF-κB corepressor complex containing N-CoR and HDAC3 resulting in the assembly of a transcription-activation complex in its place.

This series of steps is prevented from occurring by ligand-activated nuclear receptors such as the GR, LXRs, and PPARγ. In the case of the LXRs and PPARγ, SUMOylation leads to conformational changes that enable the nuclear receptors to interact with the corepressors and prevent their dislodging from the promoter site. The corepressors are not degraded and the upregulation of

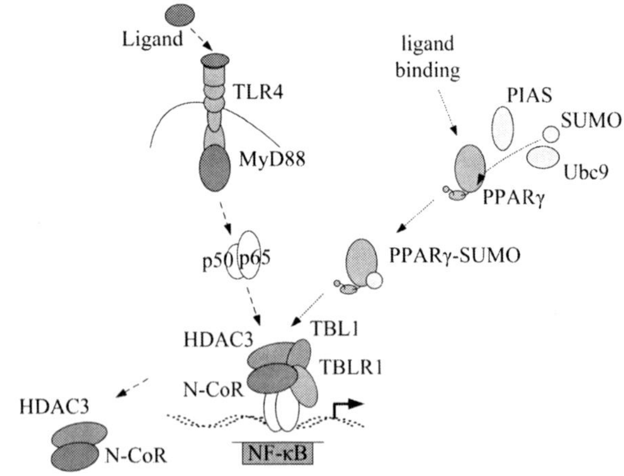

Fig. 7.9 PPARγ-SUMO-mediated transrepression of TLR4 signaling

inflammatory mediators such as iNOS is prevented. The LXRs and PPARγ pathways are similar but distinct. SUMO-1, which participates in PPARγ transrepression is replaced by SUMO-2/3 and HDAC4 E3 ligase contributes to the SUMOylation of the LXRs, and the actions at the promoter site are slightly different.

7.9 The Local Microenvironment Is a Key Organizational Unit in Health and Disease

The tissue is a basic functional unit in a multicellular organism lying just above the cellular level of organization. Using this observation as a starting point, the local microenvironment may be regarded as a small piece of tissue that encompasses the defining cell type of the tissue, connective tissue/extracellular matrix, and supporting cells such as fibroblasts, macrophages, and endothelial cells. The microenvironment (in the small) and the tissue (in the large) are the basic level of organization to which a state of health or disease, injury or repair, may be assigned.

Table 7.3 lists the main cellular residents of several key tissue microenvironments. In each of these disease situations, there is a disease trigger that sets off a series of alterations in the local microenvironment of the affected cells. The triggers responsible for onset of T2D and atherosclerosis have been discussed. The key triggering event in cancer is, of course, onset of tumor growth; in the case of the neurodegenerative disorders, it aggregates of misfolded proteins, amyloid beta (Aβ) in Alzheimer's disease and α-synuclein in Parkinson's disease.

The presence of chronic inflammation in a tissue is a sure sign of disease. An important indicator of a tissue's inflammatory status is not only the presence of macrophages but also the nature of their activities. Some tissues have their own resident macrophages; these tissues and others acquire specific kinds of macrophages in response to disease and injury. Another key determinant of tissue health is the presence or absence of angiogenesis, the process whereby new

Table 7.3 Tissue disease-associated microenvironments

Disease	Trigger	Microenvironment
MDs		
- T2D	Lipid buildup	Adipocytes, macrophages, stromal vascular fraction (SVF)
- Athero.	Cholesterol buildup, hemodynamic stress	Endothelial cells, platelets, smooth muscle cells, macrophages
Cancers	Tumor growth	Epithelial cells, endothelial cells, fibroblasts, stem cells, adipocytes, macrophages
Neurodeg.		
- AD, PD	Misfolded Aβ, α-synuclein	Neurons, astrocytes, microglia

blood vessels are generated from pre-existing ones. In adipose tissue, this last-mentioned element appears in Table 7.3 as the stromal vascular fraction (SVF). Metastasis is by its very nature microenvironment-centric, and tumors have been likened to wounds that do not heal. The tumor microenvironment (stroma) contains a rich assortment of cell types including, besides the tumor (epithelial) cells, endothelial cells, stem cells, fibroblasts, macrophages, neutrophils, and lymphocytes. Inflammatory and angiogenic responses are central to the pathogenesis of cancer with macrophages and fibroblasts playing particularly important roles as will be discussed in Chapter 15.

Central nervous system microenvironments contain not only neurons but also astrocytes (macroglia) and microglia. In the presence of disease and/or injury, reactive astrocytes function as CNS counterparts to fibroblasts, while microglia function as the central nervous system's resident macrophages. The chief triggers for activation of these repair cells are dead and dying cells, and as mentioned above disease-associated accumulations of misfolded Aβ (Alzheimer's disease) and α-synuclein (Parkinson's disease). These events stimulate changes in the astrocytes and microglia from quiescent and patrolling states to more activated ones as is the case for their counterparts elsewhere in the body. These signaling processes will be explored in Part III.

The term 'stromal vascular fraction' denotes a heterogeneous mix of cell types present in adipose tissue that serves as a reservoir of adipose tissue precursors and supporting angiogenic cells and signaling molecules. This structure supports the plastic properties of adipose tissue required to meet ever-changing fat storage needs. The SVF contains blood cells, endothelial cells, and adipocyte precursors. Adipose-derived stem cells can be found in the SVF and these can be made to differentiate into several types of cells. As expected from a structure that supports increases in cell mass a variety of cytokines are secreted that promote angiogenesis, adipocyte differentiation, and cell survival along with a variety of proteolytic enzymes that facilitate matrix remodeling. The most prominent of these with regard to angiogenesis is VEGF; others found to have a role in angiogenesis, adipogenesis, and endothelial cell survival are hepatocyte growth factor (HGF), fibroblast growth factor (FGF), transforming growth factor β (TGFβ), and granulocyte-macrophage colony stimulating factor (GM-CSF).

7.10 The Inflammatory Response Is a Biphasic One with Distinct Clear Up and Reconstruction Phases

The first phase of an inflammatory response is aimed at destroying pathogens, removing dead and dying cells, damaged ECM material, and cellular debris. This phase is followed by a recovery phase in which the tissue is restored to a healthy, fully functional condition. Different ensembles of signaling molecules are utilized during each of these phases. These signals (1) guide the recruitment into the tissue of cells needed to effect the removal and repair phases and (2) within the microenvironment instruct the cells which of several states of differentiation are the appropriate ones to assume at that particular time.

In the initial stage of the response, chemokines and other recruitment signals are sent out by the tissue cells and other permanent cellular residents of the microenvironment to guide leukocytes, fibroblasts, and, if needed, stem cells to sites of infection and injury. Other agents modify the vasculature in order to facilitate recruitment of leukocytes. These actions are the ones typically associated with inflammation of the tissue. Blood monocytes that are recruited in this way enter the tissue and differentiate into macrophages. T cells, dendritic cells, and neutrophils enter, as well. The main inflammatory mediators that regulate this phase are interferon γ (IFNγ), interleukin-6 (IL-6), TNFα, transforming growth factor β (TGFβ), reactive oxygen and nitrogen species, the aforementioned chemokines, and enzymes that degrade the ECM such as the matrix metalloproteinases that help create channels and clear away damaged material.

Recall that interferon and hematopoietin (interleukin) signals are transduced into the cell via plasma membrane receptors that recruit and activate members of the Jak (Janus kinases) family of tyrosine kinases and through them to members of the signal transducer and activator of transcription (STAT) family of transcription factors. Interferon γ signals through STAT1 and, together with TLR4 and TNFα induced NF-κB activation, support phase 1 pro-inflammatory signaling. Other interleukins, most notably IL-4 and IL-13, signal through STAT6, and this transcription factor promotes a phase 2 pro-recovery program of gene expression. Under normal conditions, phase 1 is followed by phase 2, which is terminated in a timely manner at the completion of the repair. Yet another interleukin that helps orchestrate this orderly program of events is IL-10. This cytokine utilizes a STAT3 pathway to mediate a changeover from innate to adaptive immunity effectors, from neutrophil to T-cell-centric processes, during phase 1 of the inflammatory response.

7.11 Macrophages Are Inflammatory Cells with Key Roles in the Body's Response to Infection and Injury

Macrophages function as phagocytes and as antigen-presenting cells with key roles in both innate and adaptive immunity. They are derived from myeloid stem cells that differentiate into circulating blood monocytes, which circulate in blood and migrate into tissue in response to infection and injury. In some special tissues and organs, the migratory macrophages join tissue-specific resident macrophages such as Kupffer cells (liver), osteoclasts (bone), alveolar macrophages (lung), and microglia (brain).

Macrophages are highly plastic and, in response to environmental cues, may carry out either an inflammatory response program or a repair program. In the simplest model, derived from studies of murine macrophages, they are regarded as in either an M1 state or an M2 state. The M1 (classical) state functions as a pro-inflammatory responder. This state is induced by pro-inflammatory cytokines such as IFNγ secreted by T cells and others in the microenvironment and, in response, the M1 macrophages secrete pro-inflammatory mediators TNFα and IL-6 along with reactive oxygen/nitrogen species such as nitric oxide. The M2 (alternative) state responds to IL-4 and IL-13 signals from nearby cells. Macrophages in this state facilitate the repair of the damaged tissue. In this role, they secrete repair mediators such as VEGF that stimulates angiogenesis and cytokines such as IL-10 that oppose inflammation and upregulate factors such as arginase that reduce NO production. These states are not irrevocably fixed in time; rather, in response to changing signals, the macrophages can reversibly switch from one phenotype to the other to facilitate the current needs whether they are inflammatory or repair in nature.

The process whereby blood monocytes move from circulation into tissue, and differentiate into macrophages, was examined in Chapter 6, dealing with atherosclerosis. The process involves loose initial attachment to the blood vessel walls away from the center of the rapidly moving blood. Rolling is followed by firm attachment, stoppage, and then extravasation into tissue, a process that is choreographed by chemokines and cell adhesion molecules. The two-state model described in the preceding paragraph can be applied to blood monocytes, which undergo patrolling behavior and can be grouped into inflammatory and repair/tissue residential classes. The M1 monocytes respond to MCP-1/CCR2 chemokines while the M2 group is more strongly attuned to CX_3CR1 chemokines. Trafficking in atherosclerotic and other tissues is guided by the expression patterns of these and other chemokines and by cell adhesion molecules such as L-selectins and LFA-1 integrins.

The liver protects the host from gut-derived pathogens and to carry out this task it possesses a population of resident macrophages, the Kupffer cells. Under quiescent conditions these macrophages as well as resident adipose tissue macrophages adopt the M2 form. In more detail, both adipocytes and hepatocytes are able to generate Th1 (e.g., IFNγ) and Th2 (e.g., IL-13) cytokines, the former favoring the M1 polarized state and the latter the M2 state. The relative amount of these two classes of cytokines determines the state adopted by the macrophages. The switch in nuclear programs to support the M2 state is regulated in Kupffer cells by PPARγ and PPARδ, and in adipose tissue resident macrophages by PPARδ, both acting downstream of the hepatocyte- and adipocyte-released cytokines. In adipose tissue, like other tissues migratory macrophages respond to inflammatory signals by adopting an M1 phenotype and, when the inflammatory conditions are dealt with, switch to an M2 form to resolve the inflammation and reset the tissue microenvironment to a quiescent state. Continued overfeeding creates a condition where the abnormal conditions are not resolved leading to continued increase in adipose tissue mass, sustained M1 (plus M2) macrophage presence, and release of FFAs.

7.12 Fibroblasts Are Connective Tissue Cells

Fibroblasts maintain the health of the extracellular matrix. They do so by synthesizing ECM fibrous material, especially collagens and ground substance, a term that encompasses glycosaminoglycans such as hyaluronan along with glycoproteins and proteoglycans. Fibroblasts have a prominent role during wound healing and tissue repair following infection and injury. Like macrophages these versatile cells differentiate into different functional forms. A key event during tissue repair is the differentiation of fibroblasts into myofibroblasts. Unlike fibroblasts, myofibroblasts contain a contractile apparatus that uses α-smooth muscle (α-SM) actin along with its attendant motor proteins. Myofibroblast stress fibers generate traction forces that enable these cells to mechanically pull together ends and close up wounds. Once the tissue has been repaired the myofibroblasts revert back to the fibroblast form. In atherosclerotic lesions, migrating smooth muscle cells dedifferentiate into myofibroblasts in response to TGFβ signals and, together with migrating adventitial (myo)fibroblasts, secrete collagens that serve major components of the protective fibrous cap over the atherosclerotic lesion.

Transforming growth factor β (TGFβ) is not only an important regulator of cellular function and differentiation but also a key regulator of the extracellular matrix. This protein and the signaling pathway it activates have key roles in both tumor growth and metastasis, and these aspects will be discussed in detail in Chapter 15. Transforming growth factor β is synthesized and secreted by many if not most cells in the body. Once secreted it resides as part of a complex in an inactivated, latent form in the extracellular matrix. It may be thought of as a sentinel, awaiting activation by perturbations to the ECM that disrupt its immobilization within the complex. Examples of activating signals are reactive oxygen species, matrix metalloproteinases (MMPs), and mechanical forces. The latter can be readily supplied by the myofibroblasts through integrins that link TGFβ complexes to internal stress fibers.

Excessive myofibroblast activity promotes fibrotic diseases. In the heart, one of the key signaling targets of TGFβ is the fibroblasts. In response to transforming growth factor β and mechanical tension, fibroblasts migrate into damaged tissue, secrete collagens in order to generate new connective tissue, and differentiate into myofibroblasts. These steps are part of the normal wound repair response to, for example, myocardial infarction and ischemia. These cells then affect the phase II repair process. Failure to terminate these remodeling processes in a timely fashion leads to fibrosis, the excessive accumulation of fibrous materials in a target tissue or organ. Hepatic stellate cells (HSCs) and TGFβ are key mediators of liver fibrosis. In the absence of inflammation, the HSCs serve as hepatic storage depots for vitamin A (retinoids). Persistent inflammation stimulates the transdifferentiation of these cells into α-SM actin-bearing myofibroblasts. These cells then function in a wound healing remodeling mode in which they secrete cytokines and MMPs as well as matrix components.

7.13 Mesenchymal Stem Cells Are Located Throughout the Body

Another class of organ and tissue repair cells is the mesenchymal stem cell. These are bone marrow–derived cells capable of differentiating into non-hematopoietic cells of connective tissue (mesenchyme) such as adipocytes (adipose tissue), chondrocytes (cartilage), and osteoclasts (bone). These cells are alternatively called myeloid-derived stromal cells. They have been found within connective tissue of most organs in the body. Not all of these cells are the same as one another, but rather the term encompasses heterogeneous mixture of similar cell types. Surface markers that characterize these cells include CD29 (integrin β1), CD44 (hyaluronan receptor), and CD49a-f, along with their lack of hematopoietic cell surface markers. These cells reside in bone marrow and in the circulation, and when needed home to sites of injured and inflamed tissue.

These cells have engendered great interest because of their potential in tissue engineering and repair in response to disease and injury. They can participate in the repair process in several ways. First, they may differentiate into tissue resident cells to replace those that have been lost. These cells, for example, can transdifferentiate into myocytes and vascular endothelial cells and have been found to be of benefit in recovery from scarring induced by myocardial infarction. Second, they may serve as precursors of fibroblasts. Third, they may release soluble factors that recruit hematopoietic cells and other cell types needed for repair. Fourth, they may secrete soluble factors that support survival and proliferation of the resident cells, most prominently through promotion of angiogenesis and suppression of inflammatory and immune responses. With regard to their role as immunosuppressors, they have been studied for their potential role in ameliorating autoimmune disease responses. Some caution in their use in tissue repair has been noted; in some situations, they have been found to promote cancer and metastasis. The latter is due at least in part to their ability to secrete chemokines that act on nearby tumor cells to enhance motility and invasiveness.

Further Reading

Protein Ubiquitination

Goldstein G et al. 1975 Isolation of a polypeptide that has lymphocyte-differentiating properties and is probably represented universally in living cells. *Proc. Natl. Acad. Sci. USA* 72: 11–15.

Hicke L, Schubert HL and Hill CP 2005 Ubiquitin-binding domains. *Nat. Rev. Mol. Cell Biol.* 6: 610–621.

Hochstrasser M 2006 Lingering mysteries of ubiquitin-chain assembly. *Cell* 124: 27–34.

Petroski MD and Deshaies RJ 2005 Mechanism of lysine48-linked ubiquitin-chain synthesis by the cullin-RING ubiquitin-ligase complex SCF-Cdc34. *Cell* 123: 1107–1120.

NF-κB Node

Mosavi LK et al. 2004 The ankyrin repeat as molecular architecture for protein recognition. *Prot. Sci.* 13: 1435–1448.
PerkinsND 2007 Integrating cell-signalling pathways with NF-κB and IKK function. *Nat. Rev. Mol. Cell Biol.* 8: 49–62.
Scheidereit C 2006 IκB kinase complexes: gateways to NF-κB activation and transcription. *Oncogene* 25: 6685–6705.

TNFα Signaling

Chang L et al. 2006 The E3 ubiquitin ligase Itch couples JNK activation to TNFα-induced cell death by inducing c-FLIP$_L$ turnover. *Cell* 124: 601–613.
Ea CK et al. 2006 Activation of IKK by TNFα requires site-specific ubiquitination of RIP1 and polyubiquitin binding by NEMO. *Mol. Cell* 22: 245–257.
Kamata H et al. 2005 Reactive oxygen species promote TNFα-induced death and sustained JNK activation by inhibiting MAP kinase phosphatases. *Cell* 120: 649–661.
Kim YS et al. 2007 TNF-induced activation of the Nox1 NADPH oxidase and its role in the induction of necrotic cell death. *Mol. Cell* 26: 675–687.
Papa S 2006 The NF-κB-mediated control of the JNK cascade in the antagonism of programmed cell death in health and disease. *Cell Death Diff.* 13: 712–729.
Pham CG et al. 2004 Ferritin heavy chain upregulation by NF-κB inhibits TNFα-induced apoptosis by suppressing reactive oxygen species. *Cell* 119: 529–542.
Wajant H, Pfizenmaier K and Scheurich P 2003 Tumor necrosis factor signaling. *Cell Death Diff.* 10: 45–65.
Wu H and Arron JR 2003 TRAF6, a molecular bridge spanning adaptive immunity, innate immunity and osteoimmunity. *BioEssays* 25: 1096–1105.

TLR4 Signaling

Dinarello CA 1994 The interleukin-1 family: 10 years of discovery. *FASEB J.* 8: 1314–1325.
Kawai T and Akira S 2006 TLR signaling. *Cell Death Diff.* 13: 816–825.
Michelsen KS et al. 2004 Lack of Toll-like receptor 4 or myeloid differentiation factor 88 reduces atherosclerosis and alters plaque phenotype in mice deficient in apolipoprotein E. *Proc. Natl. Acad. Sci. USA* 101: 10679–10684.
Shi HS et al. 2006 TLR4 links innate immunity and fatty acid-induced insulin resistance. *J. Clin. Invest.* 116: 3015–3025.

NF-κB in the Nucleus

Anest V et al. 2003 A nucleosomal function for IκB kinase-α in NF-κB-dependent gene expression. *Nature* 423: 659–663.
Baek SH et al. 2004 Exchange of N-CoR corepressor and Tip60 coactivator complexes links gene expression by NF-κB and β-amyloid precursor protein. *Cell* 110: 55–67.
Hoberg JE et al. 2004 SMRT derepression by the IκB kinase α: a prerequisite to NF-κB transcription and survival. *Mol. Cell* 16: 245–255.

Perissi V et al. 2004 A corepressor/coactivator exchange complex required for transcriptional activation by nuclear receptors and other regulated transcription factors. *Cell* 116: 511–526.
Yamamoto Y et al. 2003 Histone H3 phosphorylation by IKK-α is critical for cytokine induced gene expression. *Nature* 423: 655–659.
Yoon HG et al. 2003 Purification and functional characterization of the human N-CoR complex: the roles of HDAC3, TBL1 and TBLR1. *EMBO J.* 22: 1336–1346.

Anti-Inflammatory Actions of Glucocorticoids

Pratt WB and Toft DO 2003 Regulation of signaling protein function and trafficking by the Hsp90/Hsp70-based chaperone machinery. *Exp. Biol. Med.* 228: 111–133.
Rhen T and Cidlowski JA 2005 Antiinflammatory actions of glucocorticoids – new mechanisms for old drugs. *N. Engl. J. Med.* 353: 1711–1723.

LXR- and PPARγ-Mediated Transrepression and SUMOylation

Bensinger SJ et al. 2008 LXR signaling couples sterol metabolism to proliferation in the acquired immune response. *Cell* 134: 97–111.
Castrillo A and Tontonoz P 2004 Nuclear receptors in macrophage biology: at the crossroads of lipid metabolism and inflammation. *Annu. Rev. Cell Dev. Biol.* 20: 455–480.
Ghisletti S et al. 2007 Parallel SUMOylation-dependent pathways mediate gene- and signal-specific transrepression by LXRs and PPARγ. *Mol. Cell* 25: 57–70.
Gill G 2004 SUMO and ubiquitin in the nucleus: different functions, similar mechanisms? *Genes Dev.* 18: 2046–2059.
Hay RT 2005 SUMO: a history of modification. *Mol. Cell* 18: 1–12.
Joseph SB et al. 2003 Reciprocal regulation of inflammation and lipid metabolism by liver X receptors. *Nat. Med.* 9: 213–219.
Ogawa S et al. 2005 Molecular determinants of crosstalk between nuclear receptors and Toll-like receptors. *Cell* 122: 707–721.
Pascual G et al. 2005 A SUMOylation-dependent pathway mediates transrepression of inflammatory response genes by PPARγ. *Nature* 437: 759–763.
Repa JJ and Mangelsdorf DJ 2002 The liver X receptor gene team: potential new players in atherosclerosis. *Nature Med.* 8: 1243–1248.
Ricote M, Valledor AF and Glass CK 2004 Decoding transcriptional programs regulated by PPArs and LXRs in the macrophage: effects on lipid homeostasis, inflammation, and atherosclerosis. *Arterioscler. Thromb. Vasc. Biol.* 24: 230–239.
Zelcer N and Tontonoz P 2006 Liver X receptors as integrators of metabolic and inflammatory signaling. *J. Clin. Invest.* 116: 607–614.

Macrophages

Auffray C, et al. 2007 Monitoring of blood vessels and tissues by a population of monocytes with patrolling behavior. *Science* 317: 666–669.
Bouhlel MA et al. 2007 PPARγ activation primes human monocytes into alternative M2 macrophages with anti-inflammatory properties. *Cell Metab.* 6: 137–143.

Geissmann F, Jung S and Littman DR 2003 Blood monocytes consist of two principle subsets with distinct migratory properties. *Immunity* 19: 71–82.
Kang K et al. 2008 Adipocyte-derived Th2 cytokines and myeloid PPARδ regulate macrophage polarization and insulin sensitivity. *Cell Metab.* 7: 485–495.
Lumeng CN et al. 2007 Obesity induces a phenotype switch in adipose tissue macrophage polarization. *J. Clin. Invest.* 117: 175–184.
Odegaard JI et al. 2008 Alternative M2 activation of Kupffer cells by PPARδ ameliorates obesity-induced insulin resistance. *Cell Metab.* 7: 496–507.
Tacke F et al. 2007 Monocyte subsets differentially employ CCR2, CCR5, and CX3CR1 to accumulate within atherosclerotic plaques. *J. Clin. Invest.* 117: 185–194.

Fibroblasts

Gressner AM and Weiskirchen R 2006 Modern pathogenic concepts of liver fibrosis suggest stellate cells and TGF-β as major players and therapeutic targets. *J. Cell. Mol. Med.*10: 76–99.
Hinz B 2007 Formation and function of the myofibroblast during tissue repair. *J. Invest. Dermatol.* 127: 526–537.
Tomasek JJ et al. 2002 Myofibroblasts and mechanoregulation of connective tissue remodeling. *Nat. Rev. Mol. Cell Biol.*3: 349–363.
Wipff PJ et al. 2007 Myofibroblast contraction activates latent TGF-β1 from the extracellular matrix. *J. Cell Biol.* 179: 1311–1323.

Mesenchymal stem cells

Chamberlain G et al. 2007 Mesenchymal stem cells: their phenotype, differentiation capacity, immunological features, and potential for homing. *Stem Cells* 25: 2739–2749.
De Silva Meirelles L, Chagastelles PC and Nardi NB 2006 Mesenchymal stem cells reside in virtually all post-natal organs and tissues. *J. Cell Sci.* 119: 2204–2213.
Miyahara Y et al. 2006 Monolayered mesenchymal stem cells repair scarred myocardium after myocardial infarction. *Nat. Med.* 12: 459–465.

Chapter 8
Redox Signaling

The realization that potentially damaging free radicals are present in cells began with observations by Rebeca Gerschman in 1954 that free radicals produced as by-products of oxidative metabolism cause damage to cells in a way similar to the actions of ionizing radiation such as X-rays. In the same year Commoner, Townsend, and Pake using an electron-spin resonance spectrometer, which can detect the presence of unpaired electrons, found that ROS were present in biological materials. The importance of these observations was reinforced by the discovery that cells express enzymes that modify free radicals in order to render them less harmful. Superoxide dismutase was discovered by McCord and Fridovich in 1969, and this was just the first of a growing list of enzymes that target free radicals. That free radicals can damage lipids, DNA, and proteins led to the publication of a highly influential paper by Denham Harman in 1956 postulating that free radical-induced cellular damage is a key contributor to aging.

The concept that certain reactive oxygen and nitrogen species are helpful rather than serving just as damaging metabolic byproducts, and actually participate in cellular signaling, has taken longer to become accepted. Two species – hydrogen peroxide and nitric oxide (NO) – are now recognized as having positively acting cellular roles. An important step leading to this realization was the discovery that nitric oxide generated by endothelial cells relaxes vascular smooth muscle cells. That discovery took place in several phases. Robert Furchgott first made the discovery in 1980 that some sort of signal was being sent from vascular endothelial cells to smooth muscle cells instructing them to relax. Then Ferid Murad established in 1977 that nitric oxide can signal smooth muscle cells to relax, and finally Louis Ignarro and Salvador Moncada (working independently) reported in 1987 that the signaling molecule identified by Furchgott and called by him endothelium-derived relaxing factor, or EDRF, was in fact nitric oxide.

The understanding that hydrogen peroxide carries out substantive signaling functions has gained credence as its role in signaling was gradually uncovered starting in the mid-1990s. During that time it was found that hydrogen peroxide was used by cells to coordinate the temporal sequences of events associated with

M. Beckerman, *Cellular Signaling in Health and Disease*,
Biological and Medical Physics, Biomedical Engineering,
DOI 10.1007/978-0-387-98173-4_8,

phosphorylation by protein kinases and dephosphorylation by protein phosphatases. Growth factor signal transduction associated with the coordinated actions by receptor tyrosine kinases and cytosolic protein phosphatases was the first of these processes to be uncovered. This was followed by the recognition of the role of hydrogen peroxide in activating kinases central to a number of signal transduction pathways. The finding of new roles has continued to the present for both nitric oxide and hydrogen peroxide.

8.1 Hydrogen Peroxide and Nitric Oxide Are Signaling Molecules

Hydrogen peroxide and nitric oxide have all four properties essential for signaling. First, these particular radicals are capable of diffusing an appreciable distance. Second, enzymes have been identified that generate them and others have been found that degrade them. Consequently, hydrogen peroxide and nitric oxide can be regulated both temporally and spatially. Third, these radicals bind to sulfhydryl (SH) groups on cysteine residues. The modifications are reversible as are other forms of posttranslational modification such as protein phosphorylation. Fourth, these actions can be triggered by physiological stimuli, most notably, stresses.

In contrast to hydrogen peroxide, reactive oxygen species such as the hydroxyl radical are poorly suited for signaling. They are simply too reactive. They interact readily with almost any molecule in their immediate vicinity, and cannot freely diffuse over a meaningful distance. In contrast to the hydroxyl radical, hydrogen peroxide and nitric oxide are fairly stable. They can diffuse over relatively large distances and readily pass through membranes.

The third criterion implies that enzymes initiate and also terminate the signaling. Otherwise it would take too long to launch the signaling and to end it. The fourth entry in this list, useful, means that the enzymes that launch the signaling do so in response to stresses and other physiological stimuli and that the signals themselves in the form of posttranslational modification do something meaningful such as turning on or off, activating or inhibiting, its downstream signaling targets.

Some of the most prominent examples of how hydrogen peroxide is generated and destroyed are presented in Fig. 8.1. First superoxide is produced in several ways. One of these, as shown in Fig. 8.1(a), is by NAD(P)H oxidase, the enzyme used by phagocytes to kill pathogens. This enzyme is also utilized by nonphagocytic cells. Superoxide dismutase converts superoxide to hydrogen peroxide. A landmark in understanding the properties and importance of reactive oxygen species was the discovery of superoxide dismutase. As indicated in Fig. 8.1(b), this enzyme catalyzes the conversion of superoxide to oxygen and hydrogen peroxide. There are three isoforms of superoxide dismutase (SOD). These are sometimes referred to as manganese superoxide dismutase (MnSOD) found in mitochondria, copper/zinc superoxide dismutase (Cu/ZnSOD) localized

(a)

$$O_2 + e^- \xrightarrow[\text{oxidase}]{\text{NAD(P)H}} O_2^{\bullet -}$$

(b)

$$2H^+ + 2O_2^{\bullet -} \xrightarrow{\text{Cu/Zn Superoxide dismutase}} H_2O_2 + O_2$$

(c)

$$H_2O_2 + e^- \xrightarrow{Fe^{2+}} {}^{\bullet}OH + OH^-$$

(d)

$$2H_2O_2 \xrightarrow{\text{Catalase}} O_2 + 2H_2O$$

Fig. 8.1 Generation and destruction of hydrogen peroxide (**a**) Generation of superoxide by the NAD(P)H oxidase complex; (**b**) conversion of superoxide to hydrogen peroxide by superoxide dismutases (SODs) such as Cu/ZnSOD; (**c**) generation of the dangerous hydroxyl radical through metal-catalyzed (Fenton) reactions; and (**d**) elimination of hydrogen peroxide by enzymes such as catalase

in the cytosol, and extracellular superoxide dismutase (EC-SOD) found outside the cell and which also utilizes copper-zinc. All three are metalloenzymes – they rely on prosthetic groups containing transition metals to facilitate their catalytic actions. Extracellular superoxide dismutase also utilizes copper-zinc prosthetic groups. Nowadays these enzymes are designated as SOD1 (cytosolic), SOD2 (mitochondrial), and SOD3 (extracellular).

The hydroxyl radical causes a far greater degree of damage than hydrogen peroxide. As shown in Fig. 8.1(c), its production can be catalyzed by iron. Enzymatic processes of this general type are called Fenton reactions. Detoxification enzymes such as catalase and glutathione peroxidase prevent this reaction from happening by catalyzing the conversion of hydrogen peroxide to molecular oxygen and water as depicted in Fig. 8.1(d).

Superoxide and nitric oxide are at worst only weakly damaging, but when they come together they form peroxynitrite ($ONOO^-$), a highly toxic free radical. Nitric oxide has a biological half life of about 1 s. It easily diffuses through membranes and passes out of the cell into circulation where red blood cells convert it to nitrate through interactions with oxyhemoglobin. Superoxide is also removed rapidly from the cellular environment through actions of superoxide dismutases, and this reaction is rapid, limited only by the diffusion rate, and prevents superoxide from combining with nitric oxide, another diffusion-limited process, to make peroxynitrite. However, if the sites of superoxide and NO generation are near one another then the two species will combine to generate peroxynitrite.

8.2 Nox Enzymes

There is perhaps no better starting point for exploring the multiple roles of redox signaling in health and disease than the Nox enzymes. The founding member of this family is the phagocyte NADPH oxidase. This enzymatic complex is used by phagocytes to kill invading pathogens. It does so by generating a burst of superoxide radicals. (This release is known as the oxidative burst.) The deliberate generation of superoxide is not limiting to phagocytes and killing. Rather, Nox enzymes are utilized for hydrogen peroxide signaling throughout the body. There are at least eight members of the Nox enzyme family. In addition to Nox2, the phagocyte enzymes, Nox1 and Nox4, are utilized by endothelial cells and vascular smooth muscle cells to generate hydrogen peroxide. This signaling molecule is generated from superoxide produced by the NADPH oxidases through the actions of resident SOD enzymes. Nox enzymes are used by cells not only in the vascular wall but also elsewhere in the body. As prime generator of ROS in the vascular wall, these enzymes contribute to atherosclerosis and hypertension.

In phagocytes, the NADPH oxidase complex catalyzes the generation of superoxide anions (O_2^-) from molecular oxygen (O_2) using NADPH as the electron donor. The superoxide anions, along with more potent reactive oxygen species derived from them, are used to kill the bacterial and fungal targets of the phagocytes. The active NAPDH oxidase is a multiprotein enzyme assembled from six subunits. Two of the subunits are responsible for catalysis while the other four subunits are regulatory in character. The catalytic subunits, gp91^{phox} (Nox2) and p22^{phox}, are membrane bound and remain inactive in the absence of the regulatory subunits. The Nox2 chain topology and its electron transferring heme and flavin domains are presented in Fig. 8.2. As shown in this figure the electron donor, NADPH, binds to the cytoplasmic NAD domain. Electrons are transferred to the heme domains resulting in generation and release of superoxide, which is converted by superoxide dismutase to hydrogen peroxide. The two transmembrane subunits, gp91^{phox} and p22^{phox}, are collectively referred to as cytochrome b_{558}. In its inactive state, some of the cytochrome b_{558} heterodimers are embedded in the plasma membrane while others (most) are sequestered within the cell awaiting activating signals.

Activation and inactivation of NADPH are controlled by its regulatory subunits. In the inactive (resting) state, the four regulatory subunits – p40^{phox}, p47^{phox}, p67^{phox}, and the small GTPase Rac – are sequestered in the cytosol

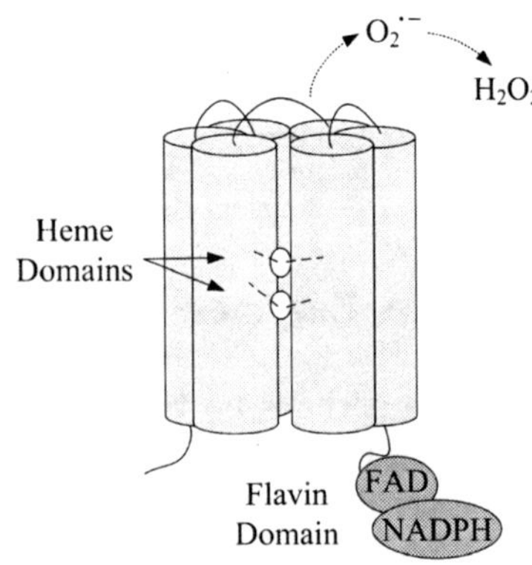

Fig. 8.2 Organization of the gp91^{phox} (Nox2) subunit of phagocytic NADPH oxidase enzyme complex

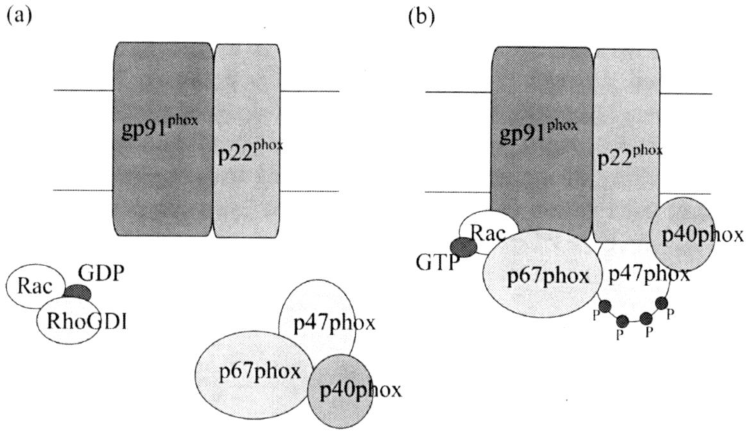

Fig. 8.3 The NADPH oxidase complex. (**a**) Inactive state and (**b**) active state. Seven Nox enzymes have been identified to date. These are gp91^{phox}, or Nox2, in phagocytes, and Nox1, 3, 4, 5, and DUOX1, 2 in nonphagocytic cells. Abbreviations: NADPH oxidase, Nox; Dual oxidase, DUOX; phagocyte NADPH oxidase, phox

away from the catalytic subunits. Activation follows assembly of the complex at the membrane in response to the appropriate stimuli (microorganisms, inflammatory agents). These two alternative arrangements are illustrated in Fig. 8.3. Phosphorylation of p47phox by upstream kinases is one of the main ways of activating this complex. The Rac GTPase provides another means of activation.

8.3 Oxidation of Sulfhydryls and Hydrogen Peroxide Signaling

The signature motif of protein tyrosine phosphatases consists of the sequence His-Cys-X-X-Gly-X-X-Arg-[Ser/Thr]. This sequence contains a conserved cysteine that functions as a nucleophile during catalysis. The presence of the Arg residue gives rise to the Cys with a low pKa, where Ka is the acid dissociation constant. As a result the Cys is particularly susceptible to posttranslational modification by oxidation. When oxidized the cysteine residue can no longer function as a nucleophile and the PTP loses its catalytic ability. This process is reversible and the reversible oxidation of PTPs is an example of redox signaling. Key elements in this form of signaling are the hydrogen peroxide (H_2O_2) molecules acting as intracellular messengers, the NADPH oxidase (Nox) complex that generates these messengers, and the peroxiredoxins that negatively regulate them.

Signals through growth factor receptors lead to Akt activation and downstream from there as discussed, for example, in the chapter on insulin signaling and diabetes, and as will be discussed again with regard to cancer. It was briefly noted that the tyrosine phosphatases PTEN acts as a negative regulator in that pathway. PTEN is constitutively active and in order to fully activate signaling through the growth factor receptor PTEN has to be inactivated for a while. This is accomplished through H_2O_2 signaling. In this pathway, H_2O_2 binds and transiently inactivates peroxiredoxis (Prx) II allowing for a buildup of intracellular H_2O_2. The hydrogen peroxide then reversibly binds and inactivates PTEN thereby permitting strong growth factor signaling to take place via Akt. Once the transient inactivation of Prx II is relieved, the hydrogen peroxide levels are reduced, the block on PTEN's catalysis is relieved, and growth factor signaling terminates. The transient inactivation of PTPs by hydrogen peroxide is depicted in Fig. 8.4.

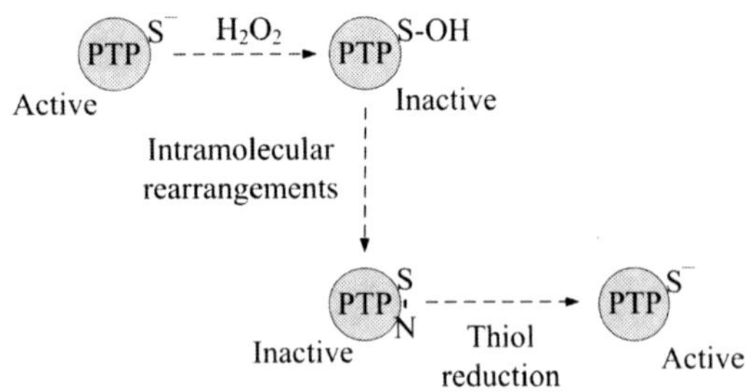

Fig. 8.4 Inactivation of protein tyrosine phosphatases (PTPs) such as PTEN through hydrogen peroxide mediated oxidation

Hydrogen peroxide is a mild oxidant that targets sulfhydryl (SH) groups located at the termini of the side chains on cysteine residues. These side chains of cysteine residues consist of CH_2–SH groups, or thiols. Most of the cysteine residues found in proteins are not susceptible to oxidation by H_2O_2. However, there are a select number of proteins possessing low pKa cysteines and these serve as the oxidation targets of hydrogen peroxide. These oxidative modifications like other posttranslational modifications that have a signaling role are fully reversible. The hydrogen peroxide-derived additions can be removed by the small molecule GSH, and by the enzymes glutaredoxin and thioredoxin.

The lipid phosphatase PTEN and the cell cycle regulator Cdc25 are also targets of hydrogen peroxide oxidation. In these proteins, a second cysteine (Cys^{71}) participates in the inactivation process. Oxidation of the active site cysteine (Cys^{124}) triggers the formation of a disulfide bond between the two cysteines that prevents further oxidations and ensures reversibility of the oxidative modification.

The tight connection between redox balance and redox signaling is illustrated in Fig. 8.5, which illustrates the actions of periredoxins in their regulation of hydrogen peroxide levels. At low hydrogen peroxide concentrations, thioredoxin peroxidase (Prx) behaves as a redox enzyme that, like catalase, converts hydrogen peroxide to water thereby preventing its conversion to dangerous hydroxyl radicals by Fenton reactions, and maintaining hydrogen peroxide concentrations in the absence of signaling at low baseline levels. A key step in this protective action is the transfer of an oxygen to the thiol present on a critical cysteine. This catalytic process is temporarily halted at high hydrogen

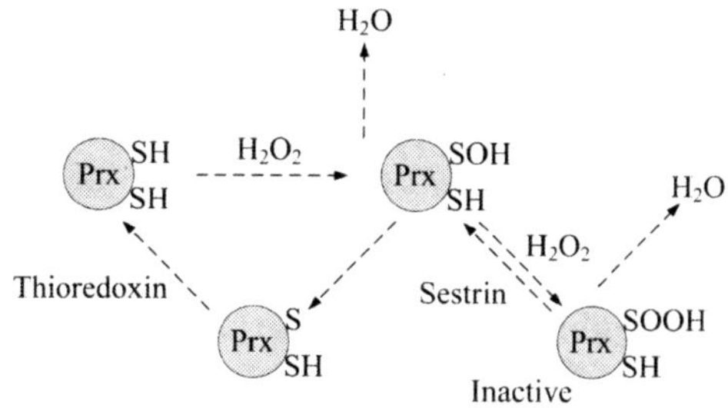

Fig. 8.5 Hydrogen peroxide detoxification and signaling in the thioredoxin system

peroxide concentrations such as those generated by NADPH oxidases. In these situations, the Prx SOH (sulfenic acid) group gains an additional oxygen to become a SO_2H (sulfinic acid) group, whose presence inactivates the catalytic ability of the Prx enzyme. This action is revered by Sulfiredoxin/Sestrin, thereby ending the inactivation and terminating the hydrogen peroxide signal in accordance with the criteria laid down earlier in this chapter for signaling.

There are two systems – thioredoxin (Trx) and the glutaredoxin (Grx) – that utilize thiol-based mechanisms for regulating redox balance and signaling. These systems maintain a correct balance between oxidizing and reducing conditions within the cell. The targets of both of these systems are side chains belonging to critical cysteine residues situated in catalytic active sites or protein-protein interfaces. The thioredoxin system contains at its core thioredoxin and thioredoxin reductase. The balance between reduced and oxidized forms of thioredoxin (and glutathione) reflects the oxidative state of the cell or compartment. Thioredoxin in its reduced form appears in the peroxiredoxin cycle depicted in Fig. 8.5. The return from oxidized to reduced form is catalyzed by thioredoxin reductase using NADPH as the electron donor. The glutaredoxin system operates in a similar manner. The core units are glutaredoxin and glutaredoxin reductase, with glutathione peroxidase (Gpx) appearing in place of thioredoxin peroxidase (Prx). Glutathione reductase then catalyzes the return conversion of oxidized glutathione (GSSG) to reduced glutathione (GSH) using NADPH.

The transient inhibition of protein tyrosine phosphatase actions by H_2O_2-mediated oxidation is an important step in insulin signaling. Several PTPs actively suppress insulin signaling in the absence of insulin-ligand binding. Prominent among these are PTEN, which counters the actions of PI3Ks, and PTP1B, which targets the insulin receptor and IRS-1 proteins and renders them inactive. One of the cellular responses to insulin binding is activation of plasma membrane-associated Nox enzymes such as Nox4, triggered by phosphorylation of their regulatory subunits. Once activated these enzymes generate superoxide, which is converted hydrogen peroxide by SOD enzymes. The hydrogen peroxide molecules then temporarily inactivate the PTPs thereby enabling insulin signals to be relayed into the cell and reach their downstream targets.

8.4 Nitric Oxide Synthases and Nitric Oxide Signaling

Nitric oxide is synthesized in the cell by a trio of enzymes, the NO synthases (NOSs), each constitutively active in a particular type of tissue – neuronal (nNOS), endothelial (eNOS) and inducible, or immune and cardiovascular specific, (iNOS). These enzymes catalyze the conversion of L-arginine into L-citrulline and nitric oxide (NO). As shown in Fig. 8.6, these enzymes combine in a single chain both an oxygenase domain and a reductase domain. Several cofactors are required for activation among which are the Ca^{2+} binding molecule calmodulin (CaM), flavins, namely, flavin mononucleotide (FMN) and flavin adenine dinucleotide (FAD), and NADPH. These latter cofactors bind to elements contained within the reductase domain. Recall that calmodulin is a small calcium binding molecule. Binding to Ca^{2+}/CaM displaces an adjacent autoinhibitory loop in the enzyme. In most endothelial cells, eNOS is constitutively expressed. The enzyme is sensitive to intracellular calcium levels, humoral factors (hormones and growth factors), and shear stress.

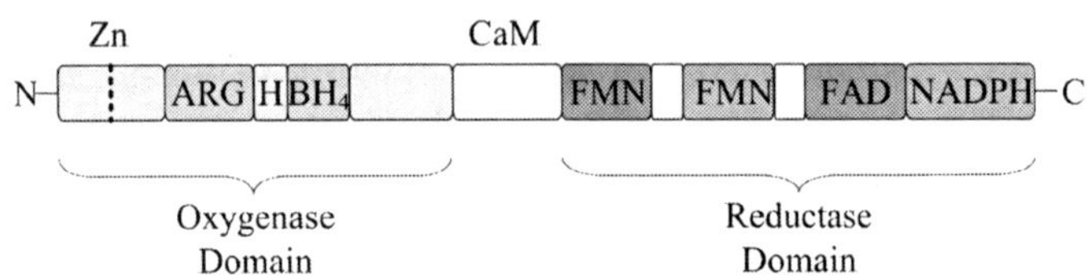

Fig. 8.6 Nitric oxide synthase domain arrangements. Shown is the organization of eNOS and iNOS, nNOS possesses an N-terminal PDZ domain in addition to the structural elements depicted above

Recall from the last chapter that nitric oxide is one of the agents released by M1 macrophages in response to inflammatory stimuli. This ability is closely related to the arginine-metabolizing capabilities of inducible nitric oxide (NOS2) synthase. In those situations where arginine is largely absent, the electron flow through the NOS enzymes is diverted towards producing superoxide instead of nitric oxide. In the case of nNOS (NOS1), the oxidase domain is crucial whereas the reductase domain functions as the superoxide generator in iNOS. These differences reflect the different functional roles of the various NOS enzymes. In the case of nNOS, the primary role is to generate NO signals, and generation of superoxide is minimized. In the immune system, the roles are broader and include pathogen killing. In macrophages, iNOS enzymes work together with another arginine-utilizing enzyme, arginase 1 (ARG1). The latter hydrolyzes arginine to ornithine and urea; and when this catalytic operation dominates, the supply of arginine available for NO production is curtailed. Joint utilization of iNOS and ARG1 results in generation of both nitric oxide and superoxide leading to the formation of the far more potent free radical peroxynitrite.

Endothelial nitric oxide synthase (NOS3) is responsible for most of the nitric oxide production in the vasculature. Like the other NOS enzymes eNOS can undergo decoupling, producing superoxide in place of nitric oxide. One of the essential cofactors for eNOS is tetrahydrobiopterin, or BH_4. Oxidative stress brought on by a mix of factors including those associated with the metabolic syndrome (e.g., high blood pressure, elevated glucose levels) by cigarette smoke associated free radicals, and stepped-up generation of superoxide by NADPH enzymes, impairs BH_4 and triggers eNOS uncoupling leading to peroxynitrite production, further uncoupling, and a complete switch over time to superoxide production.

Nitric oxide carries out its signaling tasks in one of two ways. In the first way, NO acts through soluble guanylyl cyclase (sGC) and particulate guanylyl cyclase (pGC) to produce cGMP from GTP. These messengers bind and modulate the response properties of ion channels, regulate phosphodiesterase actions, and participate in pathways that regulate gene expression. The second mode of NO signaling is by *S*-nitrosylation. *S*-nitrosylation is the name given to the covalent attachment of nitrogen monoxide groups to thiol (SH) side chains on cysteine residues. This type of posttranslational modification can modulate signals by attaching to cysteines involved in the regulatory and catalytic functions of the affected proteins.

Nitric oxide is responsive to hemodynamic stimuli and is atheroprotective. Under normal healthy conditions, fluid shear stresses ensure the constant production of NO by eNOS. The enzyme localizes to the plasma membrane. The regulatory pathway involves lipid kinase phosphoinositide-3-OH kinase (PI3K) production in response to the fluid shear stresses. This production leads to activation of several protein kinases, among which are protein kinase A and protein kinase B (Akt). Protein kinase A and Akt phosphorylation of eNOS on Ser1177 is activating. eNOS is phosphorylated constitutively by protein kinase C on Thr495, which is a negative phosphorylation site. Dephosphorylation by protein phosphatases in response to upstream signaling events triggers a strong activation of eNOS when accompanied by Ser1177 phosphorylation.

The continued production of nitric oxide by eNOS helps protect the endothelial cells against the onset of apoptosis, and the recruitment of platelets and monocytes. It exerts these atheroprotective effects by acting as a signaling molecule that inhibits gene transcription of cytokines and cell adhesion molecules and by functioning as a reactive oxygen/nitrogen species to stimulate inhibitory *S*-nitrosylation of factors that promote apoptosis.

Nitric oxide is a key regulator of vasodilation. Before this aspect is examined it is worthwhile to review the basic pumping actions that take place in the heart starting with the Frank-Starlings law and excitation–contraction coupling, then energy supply to contractile apparatus, and regulation at the transcriptional levels by nuclear receptors.

8.5 The Frank-Starlings Law and Excitation–Contraction Coupling

The Frank-Starling law states that the greater the volume of blood entering the heart during diastole, the greater of the volume of blood ejected during systolic contraction (the stroke volume). This pacing happens because the increase in volume entering the left ventricle stretches the wall of the chamber. This stretching, in turn, puts an increased load on the muscle fibers, which respond by contracting more forcefully. That is, the force generated by an individual muscle fiber (sarcomere) is proportional to its length, and stretch (length) is determined by the end-systolic volume. By this means, the heart maintains proper blood circulation over a broad range or work demands. The work cycle is depicted in Fig. 8.7.

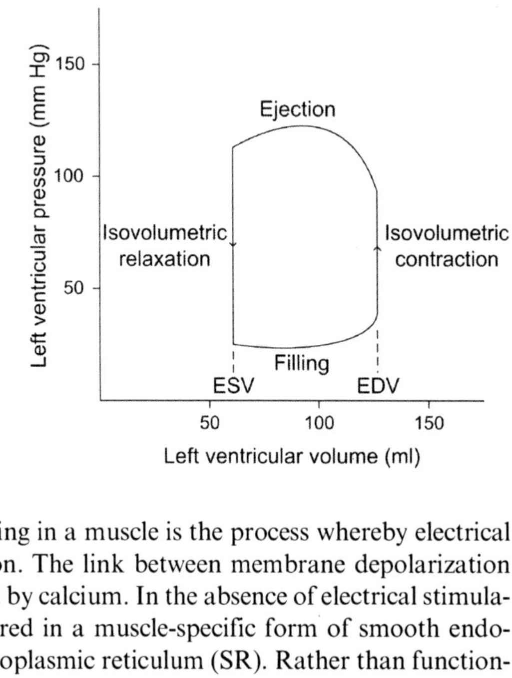

Fig. 8.7 Left ventricular pressure-volume work loop. The difference between the end-diastolic volume (EDV) and end-systolic volume (ESV) is the stroke volume

Excitation–contraction coupling in a muscle is the process whereby electrical excitation produces a contraction. The link between membrane depolarization and fiber contraction is provided by calcium. In the absence of electrical stimulation, intracellular calcium is stored in a muscle-specific form of smooth endoplasmic reticulum called the sarcoplasmic reticulum (SR). Rather than functioning as a site of protein synthesis, the SR stores and releases calcium upon demand. In particular, it releases calcium in response to electrical stimulation. The released calcium acts as a second messenger that arrives at the myofibers, binds, and activates (derepresses) the myofilaments. Calcium levels in the cytosol are increased from two sources – from extracellular calcium that enters upon depolarization of voltage-gated calcium channels, and from the SR when triggered by just entered Ca^{2+}, a release step termed Ca^{2+}-induced calcium release (CICR). During the relaxation phase of the cycle, cytosolic levels of calcium are returned to their base level through the actions of various calcium pumps.

In pumping, blood heart muscle cells, myocytes, convert energy stored in high-energy chemical bonds to mechanical work. The myocytes contain long fiber bundles, the myofibrils, and these are built from still smaller units, the sarcomeres which contain the contractile apparatus. Key components of the sarcomeres are myosin, actin, tropomyosin, and troponin. These are organized into two kinds of filaments, thick and thin. The myosin motor proteins reside in the thick filaments, while the actin contractile fibers, tropomyosin proteins, and troponins are arranged in the thin filaments. In the absence of calcium, tropomyosin blocks the myosin-binding sites on the actin fibers. Troponin contains Ca^{2+} binding sites and functions as the regulatory unit in the assemblage. Calcium binding to troponin (troponinC) induces troponin-tropomyosin interactions and these alter the tropomysin conformation resulting in an opening of the binding sites. Myosin then binds to actin and exerts a contractile force.

The application of contractile forces by myosin is accomplished through the utilization of a high-energy chemical bond. Myosin remains inactive and bound to ATP in a low-energy rest state. It can spontaneously hydrolyze ATP to ADP and P_i and store the energy in a populates a high-energy bond. As indicated in Fig. 8.8 it remains in this state in the absence of calcium-troponin binding. The arrival of calcium sets off a series of conformational changes; myosin engages actin and performs work, and ADP and Pi are released. Myosin remains bound to actin until another ATP molecule arrives. Binding and release follows, and myosin returns to its rest state.

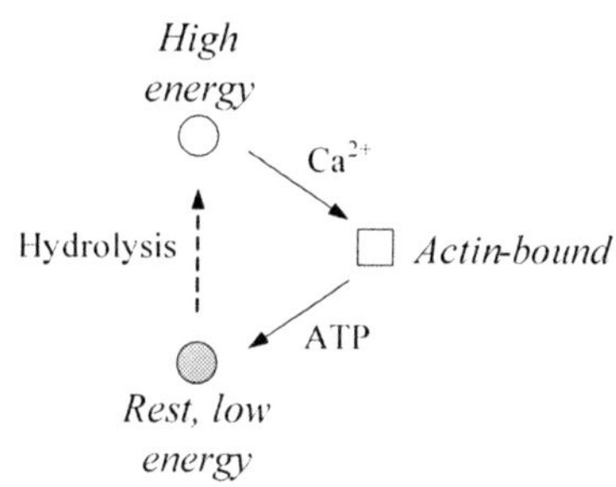

Fig. 8.8 Simplified state diagram for myosin and its interactions with actin

A key component of the overall system is the presence of an extensive network that shuttles ATP from the mitochondria to the myofibrils. Several kinases, chief among which is creatine kinase, function as shuttle mediators. Creatine kinase promotes the transfer of high-energy phosphate bonds in ATP to creatine thereby forming phosphocreatine. In that form, energy is transmitted from the electron transport chain in the mitochondria to the myofibrils. In that locale, another enzyme, myofibrillar creatine kinase, recovers ATP from phosphocreatine. The ATP is then available for use while creatine returns by means of passive diffusion to the mitochondria. The responses of the three major shuttle mediators to energy demand are illustrated in Fig. 8.9. As can be seen there is a linear variation in phosphotranfer flux with increasing heart

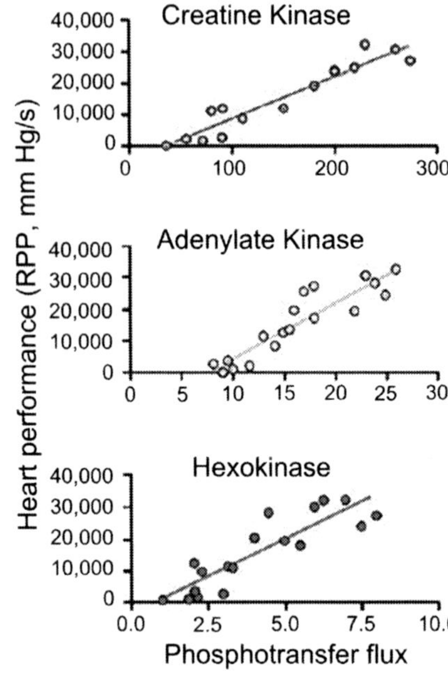

Fig. 8.9 Phosphotransfer network transfers ATP from mitochondria to myofibrils and heart performance varies linearly with phosphotransfer flux. Reprinted with permission from the American Physiological Society from Saks V et al. 2006 *J. Physiol.* 571: 253

demand and performance for each of the shuttle networks. Together, the creatine kinase, adenylate kinase, and hexokinase networks provide some of the needed underpinnings for excitation–contraction coupling and the Frank-Starling law.

8.6 Transcriptional Regulation of the Metabolic Programs

A balance between lipid and glucose burning is essential for proper heart function. Departures in one direction or the other are associated with different forms of heart disease. In the normal heart, fatty acid oxidation is the main energy source, but in the myocardial diseases, such as congestive heart failure, fatty acid oxidation is impaired. As a result, there is a shift toward a greater than normal utilization of glucose metabolism. In contrast, in diabetic and insulin-resistant situations a form of lipid-toxicity arises. Glucose uptake is impaired and in response FA metabolism is increased. Over time there is too much reliance on FA metabolism resulting in mitochondrial abnormalities and ROS accumulation.

Fatty acid utilization is promoted by the transcription factors PPARα and PPARδ and by the transcriptional coactivator PGC-1α. These transcription factors are critical regulators fatty acid uptake, fatty acid oxidation, and oxidative phosphorylation. Downregulation of PPARα and PGC-1α is observed in situations of underutilization of fatty acids, while all the three factors are involved in the adjustments to diabetic and insulin-resistant situations. One of the activities promoted by these factors is an increase in mitochondrial biogenesis. Over time this too can promote cardiac dysfunction. Yet another adjustment to diabetes is an increased use of uncoupling proteins. Recall that UCP1 is involved in uncoupling in brown adipose tissue. In contrast, UCP2 and UCP3 function in the heart to protect against the ROS buildup mentioned above, but at the expense of reduced efficiency in generating ATP.

Skeletal muscle contains a mix of myocyte fibers. Slow twitch (Type 1) fibers rely on oxidative-processes while fast twitch (Type 2) fibers are glycolytic. The latter are able to generate contractions more speedily than the aerobic slow twitch types. There are several intermediate fiber types as well. Striate muscle is highly adaptive and it can change the mix of fiber types to match demands – endurance (slow twitch) versus short and rapid bursts of activity such as in sprints and weight lifting. PPARδ supports endurance by promoting expression of genes needed for oxidative, slow twitch activity, and is the dominant PPAR isoform in muscle. Genes controlled by PPARδ in striate muscle include those required for FA oxidation, oxidative phosphorylation, and the contractile apparatus. PGC-1α has an important role in skeletal muscle plasticity, as well. It promotes the formation of slow twitch fibers, while its sibling, PGC-1β, drives the formation of Type 2X fibers, an intermediate form that combines fast twitch biophysics with oxidative metabolism.

Cardiac myocytes express both nNOS and eNOS isoforms. These are each located in different subcompartments of the cell. NOS1 is localized in the SR in close proximity to ryanodine receptors (RyRs) and positively regulates its function. NOS3 is situated at the plasma membrane where is localizes within caveolae and negatively regulates L-type Ca^{2+} channels. Another resident of the SR that colocalizes with NOS1 is the enzyme xanthine oxidoreductase.

Xanthine oxidoreductase (XOR) is a metabolic enzyme, best known for its ability to catalyze the last steps in the degradation of purines. In two half reactions, it converts hypoxanthine to xanthine and then transform xanthine to uric acid. In the process, it generates superoxide anions and hydrogen peroxide. It occurs in two forms – xanthine oxidase (XO) and xanthine dehydrogenase (XDH); these transfer electrons from xanthine to oxygen and NAD^+, respectively. Its functions are not restricted to purine degradation, however, and in the cardiovascular system it has been identified as a causal agent of ischemia-reperfusion injury.

Superoxide generated by XOR and NO produced by NOS enzymes have opposite actions on ECC. The superoxide produced by XOR damps down myocyte responsiveness to calcium signals and thus negatively impacts excitation–contraction coupling. Increased activity of XOR has been associated with heart disease and as mentioned above with ischemia-reperfusion injury. Nitric oxide produced by the two isoforms positively influences excitation–contraction coupling by interacting with the RyR receptor, most likely through *S*-nitrosylation of one or more cysteine residues leading to an increase in receptor channel opening probability. It further improves ECC by negatively regulating superoxide availability through formation of peroxynitrite.

Nitric oxide and calcium reciprocally control vasodilation and vasoconstriction by smooth muscle cells. The signaling pathways as shown in Fig. 8.10 intersect at myosin, the ATP-hydrolyzing motor protein that supplies the contractile forces. Recall that myosin contains both a heavy chain and a light chain, with the latter providing a regulatory role. The light chain is regulated by means of phosphorylation. It is phosphorylated by myosin light chain kinase (MCLK) and dephosphorylated by myosin light chain phosphatase (MLCP). The former, in turn, is activated by calcium and the latter by NO mediated by its second messenger cyclic guanosine monophosphate (cGMP).

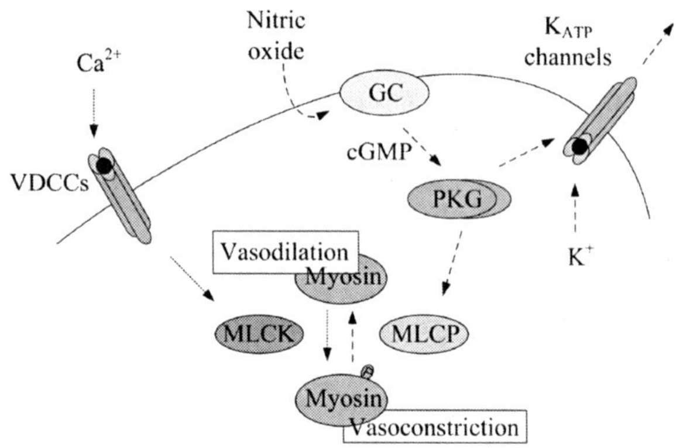

Fig. 8.10 Reciprocal regulation of vasoconstriction and vasodilation by calcium and nitric oxide

The vascular tone pathway functions in the following manner. In the absence of NO signals, the ATP-activated potassium channel (K_{ATP}) is closed. Calcium ions that enter the cell through voltage-gated calcium channels (VGCCs) form complexes with calmodulin. The calcium/calmodulin complexes activate the MCLKs leading to myosin phosphorylation and vascular constriction. When NO is present, it strongly enhances the production of the intracellular messenger cGMP by guanylyl cyclases (GCs). This messenger activates protein kinase G, which activates MLCP and this phosphatase, in response, stimulates the dephosphorylation of the myosin light chains leading to an increase in vasodilation. The cGMPs also act through kinases that phosphorylate K_{ATP} and promote its opening. The ensuing hyperpolarization of the plasma membrane closes the VGCCs thereby shutting off entry of calcium into the cell.

8.7 Inappropriate *S*-Nitrosylation Contributes to Neurodegenerative Disorders

At some point redox imbalances and impaired H_2O_2 and NO signaling are implicated in all of the diseases covered in this textbook. They are involved not only in T2D, hypertension, heart disease, and atherosclerosis, but also in cancers and neurodegenerative disorders of all types. The following examples will illustrate some of N-nitrosylation signaling activities, positive and negative, occurring in the brain, first (negative) between redox biochemistry and protein folding biophysics, and second (positive) on the protective activities of eNOS against excessive excitation and oxidative damage in neurons.

Members of the protein disulfide isomerase (Pdi) family participate in the unfolded protein response. This pathway lies at the center of the ER machinery that ensures that nascent proteins are properly folded (a topic that will be examined in Chapter 18). The Pdis catalyze the formation of disulfide bonds in a process referred to as oxidative protein folding. These proteins greatly resemble thioredoxins. They possess four domains that are homologous to thioredoxin including two domains containing redox active CXXCs. This protein acts as an ER resident molecular chaperone that recovers misfolded proteins. However, if they are *S*-nitrosylated, their catalytic ability becomes impaired. This SNO posttranslational modification leads to increases in misfolded and polyubiquinated proteins that contribute to Alzheimer's disease and Parkinson's disease.

Excitotoxicity is the name given to the process whereby excitatory amino acid receptors in neurons such as the NMDA receptor become overstimulated. When this happens, excessive amounts of calcium enter the cell and calcium homeostasis is disrupted, large numbers of free radicals are produced, oxidative phosphorylation is impaired, and excessive glutatate uptake from the synaptic cleft is blocked. To prevent this happening, neural nitric oxide synathase molecules are tethered in the vicinity of the NMDA receptors, and NO production is stimulated when calcium enters the cell and binds to calmodulin. This event stimulates nNOS to produce NO and citrulline from arginine. *S*-nitrosylation of critical cysteine residues located in the regulatory subunits of the receptors then throttles back the calcium influx through the receptors. Under hypoxic conditions this nNOS-mediated regulation of NMDA receptor activity is increased. This is accomplished by making additional cysteine sites on the NMDA receptors available for modification. This increased sensitivity is made possible by a pair of disulfide-linked thiols that function as an oxygen-sensing motif. Under hypoxic conditions, these disulfide-liked thiols are reduced to free thiols and become available for *S*-nitrosylation.

NMDA receptors are involved in redox balance and signaling in yet another way. These receptors stimulate antioxidant production by promoting transcription of sestrin ATP-dependent reductases and by inactivating the thioredoxin inhibitor Txnip. The latter is inhibited through NMDA receptor activation of the PI3K-Akt signaling pathway that inhibits Foxo stimulation of Txnip. Synaptic activity thus behaves in a protective manner to guard against oxidative imbalances and damage.

Yet another target of *S*-nitrosylation in the brain is parkin. This protein functions as an E3 ubiquitin ligase and when subjected to certain mutations is one of the factors that can generate familial Parkinson's disease. Nitrosylation of this protein has been linked to sporadic Parkinson's disease.

Lastly, *S*-nitrosylation is used for feedback regulation of NOS activity during inflammatory responses. Nitrogen oxide synthases are one of the main transcriptional targets of TNFα-signaled inflammatory responses. Recall that in this response, TNFα signals to the NF-κB module, which upon activation upregulates a host of inflammatory mediators including the NOS enzymes. Several components of the NF-κB are subsequently *S*-nitrosylated including both p50 and p65 subunits of NF-κB, and upstream acting IKKβ. *S*-nitrosylation of the p50 and p65 subunits reduces their affinity for DNA while *S*-nitrosylation of IKKβ impedes its ability to phosphorylate the IκBs. One of the steps in activating the TNFα/NF-κB signaling pathway is denitrosylation of IKKβ. Examples of enzymes that can denitrosylate substrate proteins are thioredoxin acting in the cytosol and thioredoxin reducatase in mitochondria.

8.8 The Electron Transport Chain Can Generate Reactive Oxygen Species

In oxidative phosphorylation, electrons from NADH or $FADH_2$ are transferred in a series of steps to molecular oxygen. The NADH and $FADH_2$ molecules serving as the electron donors are synthesized in the mitochondrial matrix in the citric acid cycle. The electrons from the NADH and $FADH_2$ donors flow through a series of four enzyme complexes located in the inner mitochondrial membrane. In the electron transfer process, protons are pumped out of the mitochondrial matrix to the intermembrane space thereby creating a proton motive force (PMF) consisting of the sum of the resulting proton concentration gradient and transmembrane potential, the latter arising from the charge difference built up across the IMM. Energy released when protons flow back to the matrix through the enzyme complexes is then used in a fifth complex to synthesize ATP.

The arrangement of the five respiratory complexes in the mitochondrial inner membrane is depicted in Fig. 8.11. As shown in the figure, three of the

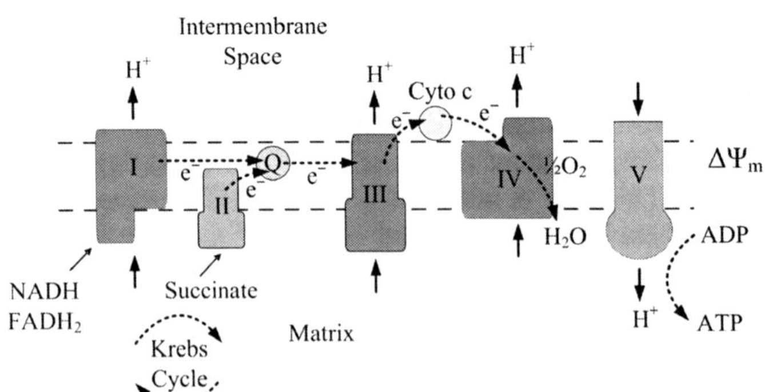

Fig. 8.11 The five respiratory complexes comprising the electron transport chain

respiratory complexes (I, III, and IV) pump protons (H^+) from the mitochondria matrix out to the intramembrane spaces (IMS). Protons move down the concentration gradient through chain (V) to return to the matrix. Complex II is an electron relay that funnels electrons from succinate to ubiquinone. The respiratory chain includes a pair of electron carriers. CoQ transfers electrons from complexes I and II to complex III and cytochrome c transports electrons from complex III to complex IV.

The respiratory chain operates in the following manner. When respiratory demand is low, few electrons flow through the respiratory chain. Input to the system consists of NADH at complex I, succinate at complex II, O_2 at complex IV, and ADP and inorganic phosphate (P_i) at complex V. Output (besides water) is ATP along with a small quantity of ROS. The flow of electrons depends on the redox state of the NADH input and by the proton motive force, which is high under resting conditions. The amount of ADP available for conversion to ATP is the key regulator of the system. If the supply is low there is little conversion to ATP and the flow of proteins back to the matrix is small. ADP is supplied to complex V through the adenine nucleotide translocator (ANT). The ANT translocates ADP from the IMS to the matrix and at the same time translocates newly formed ATP to the IMS from the matrix. When there is an increase in energy demand, the ANT opens permitting a greater influx of ADP. In response, there is an increase production of ATP. Protons are pumped back to the matrix reducing the PMF, which, in turn, permits an increased number of electrons to flow through the chain. When the demand is met the ADP supply is reduced, the PMF rises, and the flow of electrons is throttled back.

In the classical model of Chance and Williams, the amount of ADP available as a substrate for the ATPase controls the rate of oxidative phosphorylation. Control is exerted at the end of the process, at complex V, aided by the ANT which exchanges ATP for ADP, and which operates through feedback inhibition, keeping the membrane potential, $\Delta\Psi_m$, at a high setting, to throttle back the movement of protons and electrons. The amount of ATP available at complex V is not the only factor that can influence the operation of the electron transport chain. It depends on there being an adequate input supply of NADH to complex I. And this quantity can, in turn, be influenced by Ca^{2+}, which can exert its effects at several control points, that is, at Ca^{2+}-sensitive dehydrogenases, within the Krebs (TCA) cycle. The output of ATP from the electron transport chain can also be influenced by nitric oxide and peroxynitrite.

Nitric oxide functions as a regulator of oxidative phosphorylation. The key locale of interaction is complex IV. Nitric oxide reversibly inhibits the cytochrome c oxidase (COX) by competing with molecular oxygen. When it binds COX in place of molecular oxygen the ATP output is reduced. At high concentrations, it also interferes with ubiquinone–cytochrome reductase (complex III) and increases the rate of superoxide production by the ETC. This leads to peroxynitrate production, which inhibits NADH-ubquinone reducatase (complex I) and COX. This tuning down of the electron transport chain generates changes in cellular signaling. By inhibiting ATP production the AMP/ATP ratio increases leading to upregulation of AMPK signaling. It promotes continued oxygen consumption at reduced oxygen concentrations and this leads to an increased superoxide production that serves as a signal to increase NF-κB signaling and gene expression.

The electron transport chain is believed to be the main cellular source of reactive oxygen species. The main locus of ROS production is thought to be complex III, with a smaller contribution from complex I. Originally it was thought that some 2–5% of the electron flux was lost from ATP generation and produced instead superoxide. That number is now believed to be far smaller perhaps one tenth of the original estimates. In addition, mitochondria have effective means of limiting the buildup of these products. Since ROS and RNS serve a useful purpose as signaling intermediaries, it may be more fruitful to regard deleterious effects of ROS and RNS as arising from inappropriate and/or excessive signaling rather than as damaging agents per se with a large caveat that emphasizes the damaging nature of peroxynitrite. A second major caveat must be added that notes that under unhealthy metabolic conditions, such as the one where depleted NADH induces a stalling of the ETC at complex III, increases in ROS may be expected.

Further Reading

Free Radicals in Cells

Commoner B, Townsend J and Pake GE 1954 Free radicals in biological materials. *Nature* 174: 689–691.

Gerschman R et al. 1954 Oxygen poisoning and X-irradation: a mechanism in common. *Science* 119: 623–626.

Harman D 1956 Aging – a theory based on free radical and radiation chemistry. *J. Gerontol.* 11: 298–300.

McCord JM and Fridovich I 1969 Superoxide dismutase: an enzymic function for Erythrocuprein (Hemocuprein). *J. Biol. Chem.* 244: 6049–6055.

Szabó C, Ischiropoulos H and Radi R 2007 Peroxynitrite: biochemistry, pathophysiology and development of therapeutics. *Nat. Rev. Drug Disc.* 6: 662–680.

Nox Enzymes

Brandes RP and Kreuzer J 2005 Vascular NADPH oxidases: molecular mechanisms of activation. *Cardiovasc. Res.* 65: 16–27.

Groemping Y and Rittinger K 2005 Activation and assembly of the NAPDH oxidase: a structural perspective. *Biochem. J.* 386: 401–416.

Lambeth JD 2004 Nox enzymes and the biology of reactive oxygen. *Nat. Rev. Immunol.* 4: 181–189.

Hydrogen Peroxide and Thiol-based Redox Signaling

Adachi T et al. 2004 *S*-glutathiolation of Ras mediates redox-sensitive signaling by angiotensin II in vascular smooth muscle cells. *J. Biol. Chem.* 279: 29857–29862.

Berndt C, Lillig CH and Holmgren A 2007 Thiol-based mechanisms of the thioredoxin and glutaredoxin systems: implications for diseases in the cardiovascular systems. *Am. J. Physiol. Heart Circ. Physiol.* 292: H1227–H1236.

Budamov AV et al. 2004 Regeneration of peroxiredoxins by p53-regulated sestrins, homologs of bacterial AhpD. *Science* 304: 596–600.

Clempus RE and Griendling KK 2006 Reactive oxygen species signaling in smooth muscle cells. *Cardiovasc. Res.* 71: 216–225.

Rhee SG et al. 2005 Intracellular messenger function of hydrogen peroxide and its regulation by peroxiredoxins. *Curr. Opin. Cell Biol.* 17: 183–189.

Tonks NK 2005 Redox redux: revisiting PTPs and the control of cell signaling. *Cell* 121: 667–670.

Veal EA, Day AM and Morgan BA 2007 Hydrogen peroxide sensing and signaling. *Mol. Cell* 26: 1–14.

Woo HA et al. 2003 Reversing the inactivation of peroxiredoxins caused by cysteine sulfinic acid formation. *Science* 300: 653–656.

Woo HA et al. 2005 Reduction of cysteine sulfinic acid by sulfiredoxin is specific to 2-Cys peroxiredoxins. *J. Biol. Chem.* 280: 3125–3128.

Wood ZA et al. 2003 Peroxiredoxin evolution and the regulation of hydrogen peroxide signaling. *Science* 300: 650–653.

Nitric Oxide Signaling

Alderton WK, Cooper CE and Knowles RG 2001 Nitric oxide synthases: structure, function and inhibition. *Biochem. J.* 357: 593–615.

Boo YC and Jo HJ 2003 Flow-dependent regulation of endothelial nitric oxide synthase: role of protein kinases. *Am. J. Physiol. Cell Physiol.* 285: C499–C508.

Fleming I and Busse R 2003 Molecular mechanisms involved in the regulation of the endothelial nitric oxide synthase. *Am. J. Physiol. Regul. Integr Comp. Physiol.* 284: R1–R12.

Forman HJ et al. 2004 Redox signaling; thiol chemistry defines which reactive oxygen and nitrogen species can act as second messengers. *Am. J. Physiol.* 287: C246–C256.

Förstermann U and Münzel T 2006 Endothelial nitric oxide synthase in vascular disease: from marvel to menace. *Circ.* 113: 1708–1714.

Hess DT, et al. 2005 Protein *S*-nitrosylation: purview and parameters. *Nat. Rev. Mol. Cell Biol.* 6: 150–166.

Stamler JS, Lamas S and Feng FC 2001 Nitrosylation: the prototypic redox-based signaling mechanism. *Cell* 106: 675–683.

Stuehr D, Pau S and Rosen GM 2001 Oxygen reduction by nitric-oxide synthases. *J. Biol. Chem.* 276: 14533–14536.

Cardiac Bioenergetics

Bers DM 2002 Cardiac excitation–contraction coupling. *Nature* 415: 198–205.

Kaasik A et al. 2001 Energetic crosstalk between organelles – architectural integration of energy production and utilization. *Circ. Res.* 89: 153–159.

Saks V et al. 2006 Cardiac system bioenergetics and the Frank-Starling law. *J. Physiol.* 571: 253–273.

Transcriptional Regulation

Arany Z et al. 2007 The transcriptional coactivator PGC-1β drives the formation of oxidative Type 2X fibers in skeletal muscle. *Cell Metab.* 5: 35–46.

Berry CE and Hare JM 2004 Xanthine oxidoreductase and cardiovascular disease: molecular mechanisms and pathophysiological implications. *J. Physiol.* 555: 589–606.

Boudina S et al. 2007 Mitochondrial energetics in the heart in obesity-related diabetics – direct evidence for increased uncoupled respiration and activation of uncoupling proteins. *Diabetes* 56: 2457–2466.

Burkart EM et al. 2007 Nuclear receptors PPARβ/δ and PPARα direct distinct metabolic regulatory programs in the mouse heart. *J. Clin. Invest.* 117: 3930–3939.

Duncan JG et al. 2007 Insulin-resistant heart exhibits a mitochondrial biogenic response driven by the Peroxisome Proliferator-Activated Receptor-α/PGC-1α gene regulatory pathway. *Circulation* 115: 909–917.

Khan SA et al. 2004 Neuronal nitric oxide synthase negatively regulates xanthine oxidoreductase inhibition of cardiac excitation–contraction coupling. *Proc. Natl. Acad. Sci. USA* 101: 15944–15948.

Wang YX et al. 2004 Regulation of muscle fiber type and running endurance by PPARδ. *PLoS Biol.* 2: e294.

S-*Nitrosylation and Neurodegeneration*

Benhar M et al. 2008 Regulated protein denitrosylation by cytosolic and mitochondrial thioredoxins. *Science* 320: 1050–1054.

Hashemy SI and Holmgren A 2008 Regulation of the catalytic activity and structure of human thioredoxin 1 via oxidation and *S*-nitrosylation of cysteine residues. *J. Biol. Chem.* 283: 21890–21898.

Hess DT et al. 2005 Protein *S*-nitrosylation: purview and parameters. *Nat. Rev. Mol. Cell Biol.* 6: 150–166.

Papadia S et al. 2008 Synaptic NMDA receptor activity boosts intrinsic antioxidant defenses. *Nat. Neurosci.* 11: 476–487.

Takahashi H et al. 2007 A novel thiol oxygen sensor: Hypoxia enhances *S*-nitrosylation-mediated inhibition of NMDA receptors activity. *Neuron* 53: 53–64.

Uehara T et al. 2006 *S*-nitrosylated protein-disulfide isomerase links protein misfolding to neurodegeneration. *Nature* 441: 513–517.

Yao DD et al. 2004 Nitrosative stress linked to sporadic Parkinson's disease: *S*-nitrosylation of parkin regulates its E3 ubiquitin ligase activity. *Proc. Natl. Acad. Sci. USA* 101: 10810–10814.

Oxidative Phosphorylation and the Electron Transport Chain

Brookes PS et al. 2004 calcium, ATP, and ROS: a mitochondrial love-hate triangle. Am. J. Physiol. *Cell Physiol.* 287: C817–C833.

Chance B and Williams GR 1956 The respiratory chain and oxidative phosphorylation. *Adv. Enzymol. Relat. Areas Mol. Biol.*17: 65–134.

Cortessa S et al. 2003 An integrated model of cardiac mitochondrial energy metabolism and calcium dynamics. *Biophys. J.* 84: 2734–2755.

Moncada S and Bolaños JP 2006 Nitric oxide, cell bioenergetics and neurodegeneration. *J. Neurochem.* 97: 1676–1689.

Palacios-Callender M et al. 2004 Endogenous NO regulates superoxide production at low oxygen concentrations by modifying the redox state of cytochrome c oxidase. *Proc. Natl. Acad. Sci. USA* 101: 7630–7635.

Pfluger PT et al. 2008 SIRT1 protects against high-fat diet-induced metabolic damage. *Proc. Natl. Acad. Sci. USA* 105: 9793–9798.

St-Pierre J et al. 2006 Suppression of reactive oxygen species and neurodegeneration by the PGC-1 transcriptional coactivators. *Cell* 127: 397–408.

Szabó C, Ischiropoulos H and Radi R 2007 Peroxynitrite: biochemistry, pathophysiology and development of therapeutics. *Nat. Rev. Drug Disc.* 6: 662–680.

Turrens JF 2003 Mitochondrial formation of reactive oxygen species. *J. Physiol.* 552: 335–344.

Part II
Cancer

Chapter 9
The Cell Cycle

Cells are continually exposed to a host of agents that can damage DNA and cause cancer. In response, cells have evolved a trio of interlocking cancer barrier systems. The first-line barrier system is an integrated multicomponent DNA damage detection, cell cycle arrest, and repair system. This system halts the orderly progression through the cell cycle once damage is detected to allow time for repair, and when the repairs are completed the cell cycle is allowed to proceed. If the cell is subjected to excessive DNA damage and stresses, and inappropriate growth may occur, the cell is not allowed to proceed through the cell cycle. Rather, it either undergoes apoptosis (the second barrier) to remove a possibly dangerous cell, or the cell is put into a state of permanent growth arrest, or senescence (the third barrier).

Mechanistically, unscheduled growth stimuli and/or agents such as radiation and reactive oxygen species give rise to DNA damage and stalled replication forks. These stresses activate checkpoint pathways that halt the cell cycle to allow for time to repair the damage and restart DNA replication. If the damage is not repaired in a timely fashion, the checkpoint signaling system routes the cells away from growth toward apoptosis or senescence.

A prominent example of how this type of response may be viewed as a barrier is the cellular responses to oncogenic forms of the Ras GTPase. This protein acts as a signal transducer in the growth pathways. It is overexpressed or mutated to a constitutively active form in many tumors. However, if the same overexpressed or mutated Ras proteins are introduced into healthy cells, the result is senescence rather than growth/proliferation. It appears that the healthy cells are able to erect a barrier against overstimulation by Ras. A crucial difference between the healthy and disease states of these cells is the presence of a faulty barrier system in the latter. Checkpoint proteins such as p53 are mutated and ineffective in the cancerous cells but are vigorously active in the normal cells into which oncogenic Ras has been introduced.

Another example is presented by the cellular responses to c-Myc, a transcription factor that is overexpressed in many cancers. This protein contributes to the disease state by integrating together growth signals, promoting proliferation, and disrupting cellular homeostatic regulation. In healthy cells, overexpressed levels

M. Beckerman, *Cellular Signaling in Health and Disease*,
Biological and Medical Physics, Biomedical Engineering,
DOI 10.1007/978-0-387-98173-4_9,

of c-Myc generate DNA damage leading to activation of the barrier pathways. As was the case for overexpressed Ras, oncogenic c-Myc does not, by itself, trigger tumor growth. Instead, it must be accompanied by a lowering of the anticancer barriers.

A schematic representation of tumor development and metastasis is presented in Fig. 9.1. Two sets of events have to occur at the outset. First, a "driver" or throttle must be opened. That is, persistent and excessive growth signaling must be present to drive the development of a tumor. This may occur in ways that depend upon the specific tissue and cancer. Two widely studied examples of this type of event are mutations in Ras and c-Myc. The second requirement is the lowering of the anticancer barriers as discussed at the beginning of this section. The two most frequently mutated proteins in human cancers – p53 and pRb – are the central elements of the oncogenic barrier pathways.

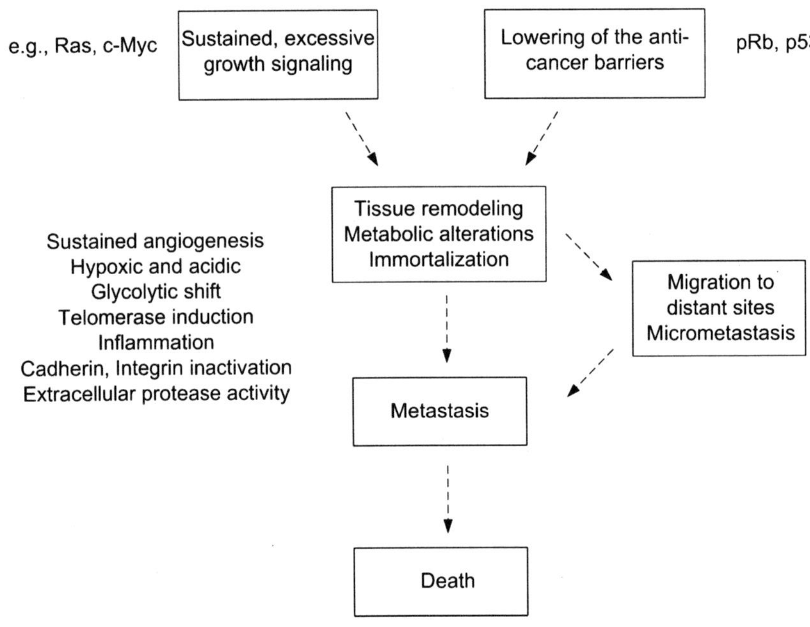

Fig. 9.1 Cancer progressions begin with oncogene-promoted excessive growth signals (such as mutated Ras and c-Myc) and the lowering and disabling of the tumor-suppressor-mediated antigrowth barriers (through mutations to, for example, p53 and pRb). Additional hallmarks of cancer are listed to the left in the middle of the figure

These events – persistent growth signaling and the lowering of antiproliferation barriers – are the first of several hallmarks of cancer, a term made popular through an influential paper published in 2000 by Douglas Hanahan and Robert Weinberg. Once excessive growth begins, several additional events must take place for the growth to continue. The first of these is access to blood supply. The growing and proliferating cells must stimulate angiogenesis. This requirement was pointed out by a landmark paper by Judah Folkman and colleagues in the early 1970s. More generally, interactions between cells and microenvironment generate extensive tissue remodeling. The tumor stroma that develops is characteristically hypoxic and acidic, and these events are coupled to a metabolic shift toward aerobic glycolysis, a striking metabolic property first pointed out by Otto Warburg in 1924 and known since then as the Warburg effect.

Most cancer deaths, 90% or so, arise from tumors that have metastasized. Metastasis has been regarded as a late event but this view has undergone considerable revision recently. This change in attitude is driven by clinical observations indicating this picture in many, if not most, cancers may be wrong, and by accumulating evidence from research studies showing early migration of cells to secondary sites. Early migrating cells are extremely rare events and are a challenge to detect. Two types of cells are now understood to become migratory and develop metastasis. These are differentiated cells that undergo changes in morphology enabling them to detach from their surrounding and migrate to distant sites and a special population of resident stem cells referred to as cancer stem cells.

Several other hallmarks of cancer are denoted in Fig. 9.1. In order for metastasis to occur, cell adhesion molecules such as cadherins and integrins must be inactivated and protein components of the ECM have to be degraded by proteases such as matrix metalloproteinases (MMPs). Cancers have similarities to wounds that do not heal and inflammation, another hallmark, is an essential ingredient of these reengineering processes. Finally, internal changes in the DNA organization must occur to sustain the continual growth and proliferation; the two most notable of these changes in avoidance of shutdown of DNA replication are telomere shortening and epigenetic silencing of certain regions of DNA.

A highly stylized depiction of the two interlocking sets of signaling and regulatory networks that maintain cellular health and guard against diseases such as cancer is presented in Fig. 9.2. The innermost of the two concentric circles in this figure represents the presence of a complex set of internal-signaling networks that maintain cellular health and control when and how a cell grows and proliferates. The outer circle denotes the tissue surrounding the cell. This includes the extracellular matrix components, the vasculature, and other cell populations resident nearby that comprise the tumor stroma and in the case of migratory (metastasizing) cells the potential new host tissue. The two curved arrows represent two-way communication between the subject cell and its surroundings. The presence of these elements is intended to emphasize that

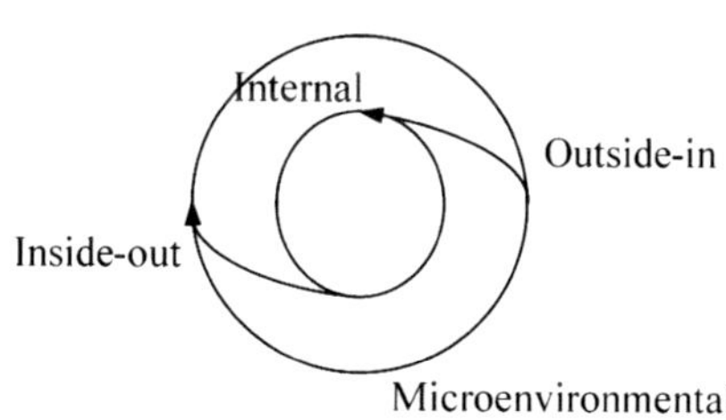

Fig. 9.2 The cell and its microenvironment mutually influence one another through inside-out and outside-in signaling

a cancerous cell profoundly influences its microenvironment, and vice versa, and these interactions as emphasized in Fig. 9.2 are a major factor in cancer development.

Referring back to Fig. 9.1, growth factors and their signal transduction pathways are, of course, prime examples of outside-in signals that, when dysregulated through mutations and overexpression influence cancer development. Ras and c-Myc act downstream and within these signaling pathways with receptor tyrosine kinase, nonreceptor tyrosine kinases, and mitogen-activated protein (MAP) kinases functioning as key intermediaries. Another set of key outside-in signaling proteins involved in this case in maintaining tissue homeostasis and opposing unrestrained growth are the members of the transforming growth factor-β (TGF-β) family. All of these signaling pathways will be discussed in the chapters that follow.

The list of hallmarks presented in Fig. 9.1 contains an impressive number of inside-out signaling events and microenvironmental alterations encompassed by the terms tissue remodeling and metastasis. These include angiogenesis, degradation of the ECM by external proteases, the acidic and hypoxic milieu, and inflammation. These events enable the cancerous cells to thrive, and a tumor to grow, in an otherwise hostile environment. With regard to metastasis, a host of signaling events between the metastasizing cells and their surroundings must occur; these are coordinated with and accompanied by major activity in multiple internal signaling pathways. These activities enable the migrating cells to successfully transit through the stages of metastasis – intravasation, transport in the bloodstream, preliminary colonization by precursors, homing, and extravasation.

The plan for Part II is, as follows. The remainder of this chapter will be devoted to a review of the cell cycle and its main regulators. This will be followed in the next chapter with an exploration of the first of the anticancer barriers – the DNA damage and cell cycle checkpoint response. The other two anticancer barriers, apoptosis and senescence, will be examined in the chapter immediately following Chapter 10. In addition to genetic changes and dysregulated cellular signaling, there is another dimension to the disease – its altered epigenetics. That layer of cellular signaling and regulation will be examined in Chapter 12. The last three chapters of Part II will be devoted to exploring the internal, inside-out, and outside-in signaling pathways that lie at the heart of all cancers and how they give rise to the hallmarks presented in Fig. 9.1.

9.1 The Cell Cycle Has Four Phases

The cell cycle encompasses two key events: replication of the genome and segregation of the replicated chromosomes away from one another into daughter cells at division. In eukaryotes, these events are separated in time. The genome is replicated early in the cell cycle in S-phase while segregation occurs later during M-phase. The replication phase is preceded by a gap, or G1, phase during which cell growth takes place and S-phase is separated from mitosis by a second gap, or G2, phase. The progression of a cell through these four phases is depicted in Fig. 9.3.

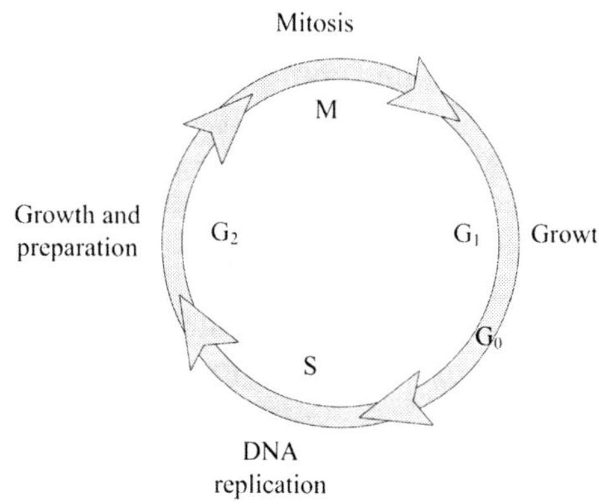

Fig. 9.3 The four phases of the cell cycle. Cells not committed to undergo the transition to S phase, and from there to mitosis, can arrest in a special state called G_0. The three not-mitosis phases are collectively referred to as interphase

The temporal separation of DNA replication and genome segregation is enforced by a control system termed the cell cycle engine containing at its core three classes of proteins – (i) cell cycle effectors, (ii) proteolytic enzymes, and (iii) inhibitors and activators. The most notable of the upstream cell cycle effectors are the serine/threonine kinases called cyclin-dependent kinases (Cdks) that have as their regulatory subunits proteins called cyclins. Cyclin-dependent kinases are present throughout the cell cycle but must bind to their regulatory cyclin to become active. There are several different kinds of cyclin-dependent kinases, each kind is associated with a particular cyclin subunit. The concentrations of the different cyclins oscillate with the cell cycle, and different Cdk-cyclin pairs are active at different times in the cell cycle. These associations are summarized in Table 9.1.

Table 9.1 Cyclin/cyclin-dependent kinases and associated phases of the cell cycle in which they are active

Regulatory subunit	Catalytic subunit	Cell cycle phase
Cyclin D	Cdk4, Cdk6	G1 phase
Cyclin E	Cdk2	G1/S transition
Cyclin A	Cdk2	S-phase
Cyclin A	Cdk1 (Cdc2)	S-phase, G2-phase
Cyclin B	Cdk1 (Cdc2)	G2-phase, M-phase

9.2 Ubiquitin-Mediated Proteolysis Is a Key Part of the Cell Cycle Machinery

Cyclins and cyclin-dependent kinase inhibitors are degraded in a precisely timed fashion by ubiquitin-mediated proteolysis. As a result of these actions, the concentrations of the different active Cyclin-Cdk complexes rise and then fall in a sequential manner. The variations in Cdk activity drive the cell from one phase to the next, in its cycle of growth and division, through timed phosphorylation of substrate proteins. A combination of cyclins and cyclin-dependent kinases, and their inhibitors and regulators, is responsible for this orderly progression.

The ubiquitin-dependent ligases that work together with protein phosphorylation and dephosphorylation to regulate the cell cycle are members of Cullin-RING superfamily of single- and multisubunit enzymes. Two families of E3s are involved in cell cycle regulation. These are members of the anaphase promoting complex/cyclosome (APC/C or APC) family and members of the Skp1-Cullin-F-box (SCF) family. Both families are composed of large multi-subunit complexes. A typical APC/C is composed of at least 13 subunits. SCF complexes do not have as many subunits as the APC/Cs but they are still quite large with at least four subunits each. These families, their substrate adapters (These will be discussed in Section 9.6), and associated phases of the cell cycle are listed in Table 9.2.

Table 9.2 E3 ubiquitin ligase complexes and associated cell cycle phases

E3 ubiquitin ligase	Substrate adaptor	Cell cycle phase
SCF	Skp2	G1/S Transition, S, G2 Phases
SCF	β-TrCP	S-phase, G2-phase, M-phase, and exit
SCF	Fbw7	G2-phase, M-phase
APC/C	Cdc20	Metaphase/anaphase transition
APC/C	Cdh1	Anaphase to G1/S transition

β-TrCP, β-transducin repeat-containing protein; Cdc20, cell division cycle 20.

9.3 Several Families of Activators and Inhibitors Are Part of the Cell Cycle Engine

The third component of the cell cycle engine is the cell cycle regulatory protein group. These proteins are listed in Table 9.3. The first entries in the table are members of the family known as Inhibitors of Cdk4, or Ink4s. These G-phase proteins repress Cdks associated with D-type cyclins, Cdk4 and Cdk6. Members of the CIP/KIP family, like the Ink4s, operate as cyclin-dependent kinase inhibitors, or CKIs. These proteins preferentially inhibit Cdk2 complexes but will repress Cdk4 complexes and others, too. The next two entries in the table inactivate Cdk1s by phosphorylating them on Thr14 and Tyr15. This action is opposed by members of the Cdc25 family of protein phosphatases that dephosphorylate the Cdk1s on Thr14 and Tyr15 thereby activating them. Then two final entries in the table, hEmi1 and Plk1, target the APC/C directly and indirectly rather than the Cyclin/Cdk complexes.

Table 9.3 Cell cycle inhibitors and activators

Protein family	Members	Description
Ink4	p16Ink4a p15Ink4b p18Ink4c p19Ink4d	Cyclin-dependent kinase inhibitors; repress Cyclin D/Cdk4/6 complexes
CIP/KIP	p21Cip1 p27Kip1 p57Kip2	Cyclin-dependent kinase inhibitors
Wee1	Wee1 Myt1	Wee1 phosphorylates (and inhibits) Cdk1 on Tyr15; Myt1 phosphorylates (and inhibits) Cdk1 on Thr14 and Tyr15
Cdc25	Cdc25A Cdc25B Cdc25C	Phosphatase that depohosphorylates (and activates) Cdk1 on Thr14 and Tyr15
hEmi1	hEmi1	Pseudosubstrate interphase APC/C inhibitor. Represses APC^{Cdh1}; G1/S transition. Represses APC^{Cdc20}; G2/M transition
Plk1	Plk1	Kinase that phosphorylates hEmi1 early in mitosis, tagging it for destruction by the SCF and derepressing the APC/C

Ink4, Inhibitors of cyclin-dependent kinase 4; CIP/KIP, Cdk interacting protein/kinase inhibitory protein; Myt1, Membrane-associated, tyrosine- and threonine-specific, Cdc2 inhibitory kinase 1; Cdc25, Cell division cycle 25; Emi1, Human early mitotic inhibitor 1; Plk1, Polo-like kinase 1.

9.4 The Retinoblastoma Proteins and E2F Transcription Factors Are Downstream Cell Cycle Effectors at the G1/S Transition

The main downstream targets of activated cyclin-dependent kinases are the retinoblastoma protein (pRb) and its two siblings, the pocket proteins p107 and p130 (Pocket proteins will be discussed shortly). These proteins bind to members of the E2F family of transcription factors. Eight members of the E2F family of proteins have been discovered to date. These members differ from one another in how they affect transcription and on their choice of pocket proteins. Most E2Fs form heterodimers with one of two members of the DP family of proteins – one E2F subunit and one DP subunit. The most recently identified members of this family, E2F7 and E2F8, are exceptions to this rule; they do not require a DP protein for DNA binding. As shown in Table 9.4 E2F1–E2F3 act primarily as transcriptional activators and bind to pRb. The next two E2Fs, E2F4 and E2F5, bind to p107 and p130 and function as repressors. The E2Fs that act as transcriptional activators are required for entry into S-phase. Genes upregulated by these E2Fs include cell cycle regulators such as Cyclin E and Cyclin A, Cdc25A, hEmi1, and Cdc2 along with essential components of the DNA-replication machinery such as Cdc6, Orc1, and the MCM proteins.

During the G_1 stages of the cell cycle, the E2Fs are maintained in an inactive state by their pRb binding. During this time, pRb is underphosphorylated (hypophosphorylated), and is able to form stable complexes with these

Table 9.4 The E2F family transcription factors grouped according to their structure and function

E2F subfamily	Role	Description
E2f1, E2F2, E2f3	Activator	Growth-regulated control of cell proliferation and cell cycle exit; main pRb targets
E2F4, E2F5	Repressor	Active throughout the cell cycle; regulators of differentiation; associate with p107 and p130
E2F6	Repressor	pRb-independent actions; associates with polycomb group proteins or Mga and Max proteins
E2F7, E2F8	Repressor	Bind promoters without requiring DP1 cofactors; negative regulator of E2F1 and p53-directed apoptosis during embryonic development

transcription factors. Its phosphorylation state changes near the G_1/S boundary when the protein becomes hyperphosphorylated. Cyclins and their associated cyclin-dependent kinases play a key role in inactivating pRb. Phosphorylation of pRb by these cell-cycle regulators and by the growth factor-activated kinases frees the E2F from their pRb-inhibition and they can then carry out their cell-cycle progression-driving transcriptional activities. Two noncontiguous domains, referred to as the A and B domains, form a binding locus known as an A/B pocket that binds many proteins to pRb. For that reason, pRb and other members of its family, p107 and p130, are known as A/B pocket proteins. Cyclins D and E recruit their partner CDKs to the A/B pocket and to a docking site located in the C terminus.

These proteins are arranged and operated in the circuit presented in Fig. 9.4. Growth signals impinge on G1-phase cyclins such as Cyclin D1 and on the CIP/KIP family member p27. In response to the mitogenic signals, Cyclin D1 and Cdk4/6 form active complexes. The activated kinases phosphorylate pRb on several of its regulatory sites. The pRb proteins are partially inhibited from inhibiting the E2F transcription factors. In response, the E2Fs begin to upregulate transcription of the Cyclin E gene. Cyclin E initially remains inactive due

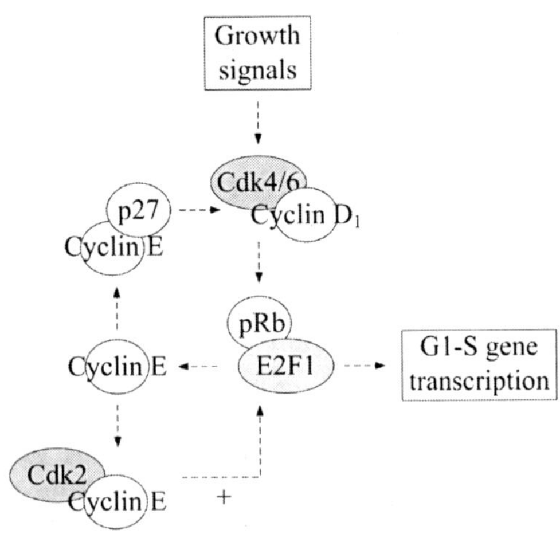

Fig. 9.4 The central circuit responsible for the transition from G_1-phase to S-phase. Included in the figure are some of the feedback loops that sculpt the transition. These and others are discussed in the text

to p27 binding. However, the Cyclin D1/Cdk4,6 complexes sequester some of the p27 molecules, and as Cyclin E/Cdk2 complexes slowly become active they phosphorylate pRB thereby further activating them leading to an increasing transcription of the Cyclin E gene. The Cyclin E/Cdk2s begin to phosphorylate their p27 inhibitors thereby triggering their ubiquitination by the SCF complex leading to their degradation.

This activation process is enhanced through a positive feedback loop involving Cdc25A and cyclin E/Cdk2. The phosphatase stimulates complex formation and the active Cyclin E/Cdk2 complexes activate the Cdc25As. The result of this regulatory process is a carefully sculpted CyclinE/Cdk2 buildup that starts out slowly due to the stimulation of both positive (Cyclin D1/Cdk4,6) and negative (p27) factors by mitogens, but then increases rapidly at the G1/S transition through the actions of the positive feedback loops involving Cyclin E.

The pocket proteins along with the E2Fs transcriptional factors are key regulators of the transition from G1 to S phase. The pRB protein interacts with not only with the E2Fs but also with chromatin-remodeling complexes. As a consequence, it is able to inhibit the transcriptional activity of the E2Fs by (1) direct binding in which the activation domain of the E2F is blocked and by (2) recruiting HDACs, components of the SWI/SNF complex, and other factors that interact with and modify chromatin.

9.5 Cell Cycle Effectors at the G2/M Transition

The key cell cycle effector pair that drives the transition from G2 phase to mitosis is Cyclin B/Cdk1. This complex is activated at the G2/M transition by the Cdc25s, which dephosphorylate Cdk1 on Thr14 and Tyr15 thereby overriding the inhibitory effects of the cell cycle inhibitors Wee1 and Myt1 (Table 9.3). Once they are fully activated, the Cyclin B/Cdk1s drive the transition to, and progression through, mitosis by shutting down energy-consuming anabolic processes and triggering the structural reorganization central to mitosis. Listed below are some of the activities carried out in these arenas by Cyclin B/Cdk1:

- Cessation of Anabolic Processes
    - Ribosome biogenesis
    - Transcription machinery
    - Translation machinery
- Structural Reorganization
    - Nuclear envelope breakdown
    - Centrosome separation and spindle assembly
    - Chromosome condensation

Ribosome biogenesis is an extremely energy-intensive process. It is halted along with mRNA processing and other RNA polymerase I, II, and III transcriptional activities not totally halted by chromosome condensation. Cyclin B/Cdk1 complexes achieve these results by phosphorylating components of the basal transcription machinery such as SL1, TFIID, TFIIB, RNA polymerase II, and ribosomal S6 kinase 1 (S6K1). Proteins targeted for phosphorylation by Cyclin B/Cdk1 include lamins, inner nuclear membrane proteins, and nucleoporins. Phosphorylation of these supporting components of the nuclear envelope loosens their associations and leads to membrane swelling and dissolution. Microtubule-associated proteins and kinesin-related motor proteins are targeted as are regulators of chromosome condensation such as condensins.

9.6 The SCF and APC/C Are Large Multisubunit Complexes

As noted earlier in Section 9.2, the anaphase-promoting complex and the Skp1-Cdc53-F-box complex are intimately involved in the progression through the cell cycle. As its name suggests, the anaphase-promoting complex is a crucial regulator of a cell's entry into anaphase, and, more generally, progression through the cell cycle. The SCF complex recognizes phosphorylated proteins and as a result plays a key role in numerous signaling pathways besides the cell cycle engine.

The SCF and APC complexes form rigid structures that select, bind, and position substrates for transfer of ubiquitins from the E2s. The crystal structure of the core components of the SCFSkp2 complex is depicted in Fig. 9.5. In this complex, the Cullins are seen to form a supporting base. These subunits are so named for their role in culling substrates. The NTD is composed of three repeats of a five-helix bundle. The Skp1 protein functions as an adaptor that connects the F-box protein Skp2 to the NTD. The CTD is linked to another adaptor, the Rbx1 protein. This RING finger protein provides an attachment site for the E2 ubiquitin-conjugating enzyme. When the substrates bind the

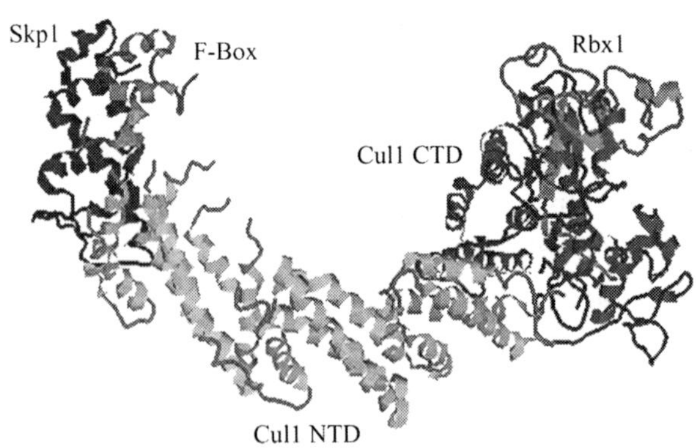

Fig. 9.5 The crystal structure of the core components of the SCFSkp2 complex determined by means of X-ray diffraction. The figure was prepared using Protein Explorer with atomic coordinates deposited in the PDB under accession number 1LDK

F-box protein Skp2, they are well positioned to receive ubiquitin molecules from the E2.

The core subunits of the SCF are arranged in a way that forms a platform for binding substrate and E2, positioning them in a manner that promotes transfer of the ubiqutin from E2 to the substrate. As can be seen in Fig. 9.5, the N-terminal domain of a cullin protein, in this case Cul1, forms a scaffold, a broad base for attachment of the other components of the ligase. The name Cullin is derived from the role of these proteins in culling substrates. A RING finger subunit, Rbx1, is situated at one end next to the Cul1 C-terminal domain. This subunit is responsible for binding the E2 ubiquitin-conjugating enzyme. Another adapter, Skp1, connects the substrate recognition F-box of the Skp2 protein to the rest of the complex.

F-box proteins function as adapters that link the core subunits of the SCF complex to the substrates. The defining feature of these proteins is the presence of an F-box motif, about 50 amino acids in size, in the amino terminal that serves as an interface for protein-protein interactions. Other common motifs are the WD (WD40) repeats and leucine-rich repeats (LRRs). These motifs are among the most common motifs encoded in the human genome. They are usually found in the carbonyl terminal of the F-box proteins and function as phosphorylated protein-recognition motifs while the amino terminal F-box motif links the F-box to the core components of the SCF E3 ligase.

The anaphase-promoting complexes are organized in a similar fashion to the SCF. These E3s have a cullin-containing complex analogous to the cullin base of the SCF, an adaptor complex that like Skp1 positions and binds the substrate adaptor, and a RING finger-adaptor complex that, like Rbx1, is for E2 attachment. In place of the Skp2 and other F-box proteins, the APC/C possesses two substrate-specifying adaptors, Cdc20 and Cdh1. These subunits each bind multiple substrates that possess motifs termed D boxes and KEN boxes. The APC activates when these regulator subunits, already bound to the substrates, recruit the core APC module. Some of the prominent SCF and APC/C adaptors are listed in Table 9.2. As is the case for the Cyclins and CDKs, different adaptors are active at different times in the cell cycle.

9.7 Mathematical Modeling Is an Essential Tool in Understanding Signaling Pathways and Networks

As has become increasingly clear over time, the cellular signaling and regulatory networks that maintain the body's health are complex. They incorporate multiple interacting partners, posttranslational modifications, proteolytic destruction, and positive and negative feedback loops. These operations all sit on top of the cell's normal turnover and recycling of proteins. Systems biology/bioengineering in the form of mathematical modeling provides the means of integrating this large body of information together into a coherent picture of the underlying mechanisms and processes. It can reveal the precise functions of the components, why certain feature may be present, and what happens when these components and features are altered by disease and therapeutic intervention.

The cell cycle engine is one, if not the, best studied of the cellular signaling networks in eukaryotes. It is fairly simple compared to some of the other major networks yet exhibits a full spectrum of insightful dynamical properties. The starting point for these studies as well as most others is the law of mass action and the Michaelis-Menton (MM) equation. In the remainder of this section, the MM equation will be derived. This equation will be incorporated into the expressions for the time rate of changes of the various enzymatic components of the cell cycle in the next section. When the set of ordinary differential equations is solved one sees how a set of nonoscillatory components can undergo perfectly regular oscillations simply as a result of the way the network is wired.

The law of mass action provides the basis for building kinetic model of the cell cycle process. This rule states that the rate of a chemical reaction is proportional to the product of the activities of the reactants. For a simple reaction of the form

$$\mathrm{A + B} \xrightarrow{\mathrm{k}} \mathrm{C}$$

The law of mass action states that:

$$\frac{d[C]}{dt} = k[A][B] \tag{9.1}$$

where k is the reaction rate. A typical enzyme-catalyzed reaction is more complex and assumes the form

$$\mathrm{E + S} \underset{\mathrm{k_{-1}}}{\overset{\mathrm{k_1}}{\rightleftharpoons}} \mathrm{ES} \xrightarrow{\mathrm{k_2}} \mathrm{E + P}$$

In the above, E denotes the enzyme; S is the substrate that is transformed into the product P by the catalytic actions of the enzyme. The reaction proceeds in two steps. First the enzyme and substrate come into close proximity to form an intermediate, ES. This intermediate is short lived and gives rise to the product. It is further assumed that the second step proceeds in the forward direction only. Applying the law of mass action to this situation yields four ordinary differential equations:

$$\begin{aligned}\frac{d[E]}{dt} &= -k_1[E][S] + (k_{-1} + k_2)[ES] \\ \frac{d[S]}{dt} &= -k_1[E][S] + k_{-1}[ES] \\ \frac{d[ES]}{dt} &= k_1[E][S] - (k_{-1} + k_2)[ES] \\ \frac{d[P]}{dt} &= k_2[ES]\end{aligned} \tag{9.2}$$

In many situations, the rate of buildup and breakdown of enzyme-substrate complexes are in balance and the concentration of enzyme-substrate complexes remains almost constant. That is, the time rate of change of $[ES]$ is near zero. Under these steady state conditions, one finds that the buildup in the reaction product depends upon the substrate in the following way:

$$V_0 \equiv \frac{d[P]}{dt} = k_2[ES] = \frac{V_{\max}[S]}{K_M + [S]} \tag{9.3}$$

In the above,

$$V_{\max} = k_2([E] + [ES]), K_M = (k_1 + k_{-1})/k_2$$

This expression, known as the Michaelis-Menton equation, serves to highlight that the simple linear dependence of a reaction product upon the reactant concentrations, implied by the simple formula Equation (9.1) does not hold for most enzyme-driven processes. The difference between these two types of kinetics is illustrated in Fig. 9.6. The first part of the Fig. 9.6(a), is the linear

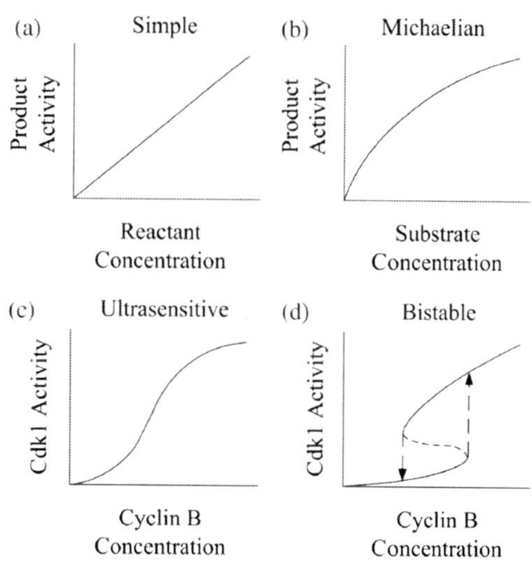

Fig. 9.6 Cell cycle dynamics: (**a**) Simple linear behavior implied by Equation (9.1); (**b**) Hyperbolic behavior according to the Michaelis-Menton enzyme kinetics; (**c**) Ultrasensitive behavior arising as a consequence of negative feedback, and (**d**) Bistability and hysteresis arising through the combined actions of positive and negative feedback

dependence expected of the simple reaction. The second part of Fig. 9.6(b), depicts a curve that is hyperbolic in shape in accordance with the Michaelis-Menton expression for V_0. At low substrate concentrations, the dependence of V_0 on $[S]$ is linear. However, as the substrate concentration increases the curve flattens out, becoming independent of $[S]$ and approaching the value $V_{\max}$ at high $[S]$ values. The latter regime is referred to as the zeroth order process and the low $[S]$ linear regime as the first order process.

9.8 The Goldbeter Model of Entry and Exit from Mitosis

The Goldbeter model, introduced in 1991, describes how cyclic behavior regulating mitosis can arise through negative feedback in a remarkably simple model. In this model, there are just three independent variables each representing a macroscopic quantity, namely, the amount or concentrations of the chemical species. The first variable, C, represents the amount of Cyclin. When the buildup in concentration of the Cyclin reaches a certain level it activates a protein called, at that time, the maturation promoting factor (MPF), which triggers entry into mitosis. The MPF is now known to be Cyclin B/Cdk1 complexes. The MPF when activated stimulates the degradation of the cyclin through activation of a protease. The third variable, X, represents the protease, which once activated degrades the cyclin thereby provides negative feedback, the key network property of the model.

The Goldbeter model is depicted schematically in Fig. 9.7. In this figure, the MPF is designated as M. It exists in two states, an inactive, phosphorylated form and an active, dephosphorylated one. Similarly, the protease X is either inactive (phosphorylated) or active (dephosphorylated). The kinases and phosphatases responsible for these actions are not explicitly modeled but their behavior is captured implicitly through the rates and associated kinetics, assumed to obey the Michaelis-Menton equation. The

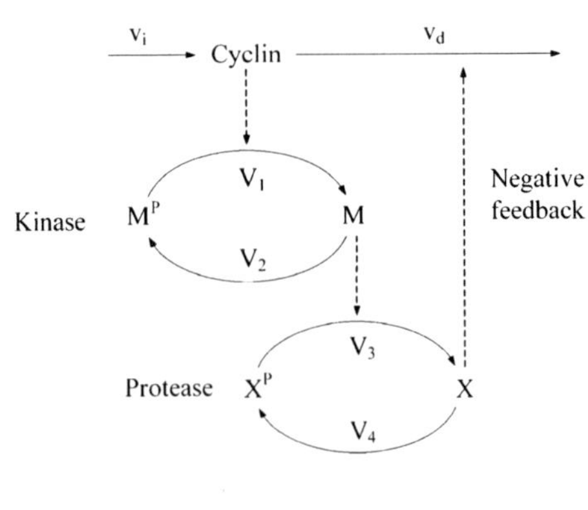

Fig. 9.7 The Goldbeter model. M: MFP (Cyclin E/Cdk1); X: protease

first equation in this model describes the time rate of change in the cyclin concentration, C:

$$\overbrace{\frac{dC}{dt}}^{\text{Rate of change of Cyclin}} = \underbrace{v_i}_{\text{Synthesis}} - \overbrace{v_d X \frac{C}{K_d + C}}^{\text{Protease-driven degradation}} - \underbrace{k_d C}_{\text{Decay}} \tag{9.4}$$

There are three terms in this formula. The first and third terms describe the synthesis and spontaneous degradation, respectively, of the cyclin. The second term describes the effect of the negative feedback on the cyclin. This is modeled as a linear dependence of the cyclin concentration on the protease variable, X. The next equation

$$\overbrace{\frac{dM}{dt}}^{\text{Rate of change of MFP}} = \overbrace{V_1 \frac{(1-M)}{K_1 + (1-M)}}^{\text{Activation (dephosphorylation)}} - \overbrace{V_2 \frac{M}{K_2 + M}}^{\text{Deactivation (phosphorylation)}} \tag{9.5}$$

describes the time rate of change in the MFP concentration. In deriving this equation, the total amount of MFP that is available is assumed to be constant. M denotes the fraction of phosphorylated MFP and $1-M$ is the corresponding fraction of unphosphorylated MFP. Similarly, the total amount of protease is taken as constant. X is the fraction of phosphorylated protease and $1-X$ is the corresponding fraction of unphosphorylated protease. As indicated in the figure, activation of the kinase is stimulated by the cyclin. This is taken into account through a cyclin-dependent V_{max} for MFP activation:

$$V_1 = \frac{C}{K_c + C} V_{M1} \tag{9.6}$$

The final formula of the model describes the time rate of change of the protease. Like the MFP, the protease is inactive when phosphorylated and active when dephosphorylated.

$$\overbrace{\frac{dX}{dt}}^{\text{Rate of change of Protease}} = \overbrace{V_3 \frac{(1-X)}{K_3 + (1-X)}}^{\text{Activation (dephosphorylation)}} - \overbrace{V_4 \frac{X}{K_4 + X}}^{\text{Deactivation (phosphorylation)}} \tag{9.7}$$

Activation of the protease is dependent upon active MFP. This interaction is taken into account through the use of an active MFP-dependent V_{max} for protease activation:

$$V_3 = MV_{M3} \tag{9.8}$$

The response of the Cdk1 activity to changes in Cyclin B concentration is shown in Fig. 9.6(c). In place of the hyperbolic dependence of the product activity upon the substrate concentration seen in Fig. 9.6(b), the Cdk1 activity has a sigmoidal dependence on the Cyclin B concentration – At low Cyclin B concentrations, the Cdk1 activity remains flat and is insensitive to the Cyclin B concentration. This response behavior changes in the middle portion of the response curve. In this regime, the Cdk1 activity increases rapidly as the Cyclin B concentration grows, and the Cdk1 activity becomes ultrasensitive to small changes in Cyclin B concentration. This type of curve has two important features, a threshold and a slope. The flat (insensitive) portion of the plot that occurs before the ultrasensitive rise serves as a threshold for the response, and the slope of the curve in the ultrasensitive regime describes the sharpness of the response.

In the Cyclin B – Cdk1 plot all other variables in the model are kept constant, and the cyclin is neither synthesized nor degraded. This is not the only kind of Cyclin B – Cdk1 plot that can be generated. A second kind of plot can be generated by turning on the entire dynamical systems and then plotting the Cdk1 activity that results from the dynamically varying Cyclin B concentration. When this is done for the Goldbeter model a closed loop trajectory is produced similar to the one shown in Fig. 9.8(a). This type of curve is referred to as a limit cycle because of the closed and repetitive (cyclical) nature of the trajectory. The Cdk1 waveform, that is, the variation in Cdk1 activity over time, is presented in Fig. 9.8(b). The oscillatory character of the activity in the Cyclin B – Cdk1 complex is clearly visible.

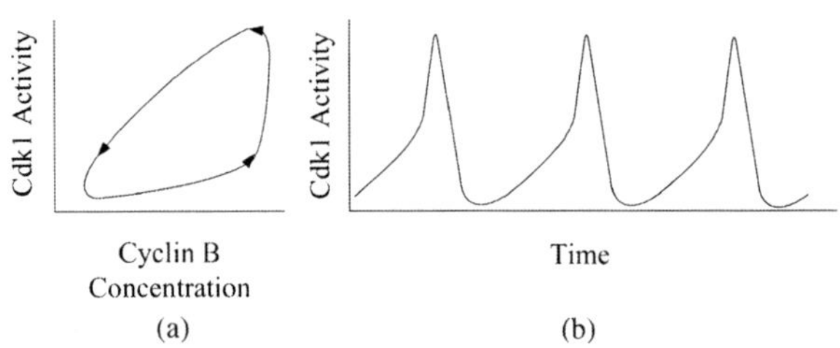

Fig. 9.8 Dynamic behavior in the Goldbeter model: (**a**) limit cycle oscillations and (**b**) corresponding waveform

9.9 Multiple Positive and Negative Feedback Regulate the Progression Through the Cell Cycle

The single negative feedback loop from the protease to the cyclin is not the only feedback loop present. Rather, multiple positive and negative feedback loops operating jointly regulate the cell cycle progression. The presence of some of the

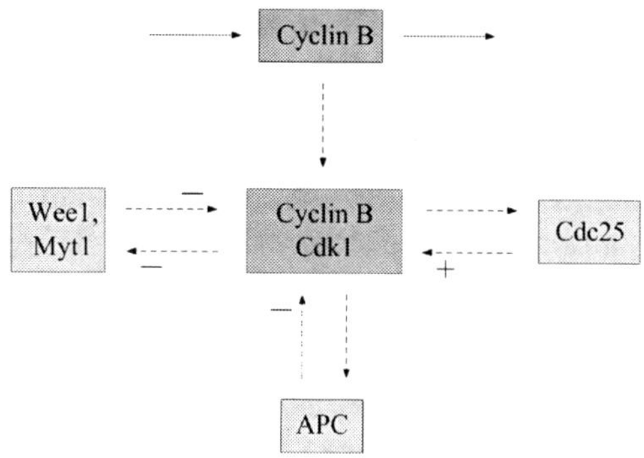

Fig. 9.9 Multiple positive and negative feedback loops ensure the correct sequences of events and timing of transitions

additional loops in the regulation of Cyclin B – Cdk1 activity is depicted in Fig. 9.9. As was the case in Fig. 9.7, Cyclin (Cyclin B) feeds the formation of MPF (Cyclin B – Cdk1 complexes) and the MPF activates the protease (APC) that provides the negative feedback. The full activation of the MPF is mediated by the Cdc25 phosphatase, which operates through a positive feedback loop. In addition, the Wee1 and Myt1 kinases that maintain the MPF in an inactive state function through a double negative feedback loop.

One of the consequences of adding the positive feedback contribution to the Cyclin B – Cdk1 regulatory system is illustrated in Fig. 9.6(d). The sigmoidal, ultrasensitive curve shown in Fig. 9.6(c) has changed into a hysteresis, bistable curve. There are two stable states, one of low Cdk1 activity and one representing a high level of Cdk1 activity. As the Cyclin B concentration increases, a point is reached where the Cdk1 activity jumps from the lower curve to the higher one. When the Cyclin B concentration starts decreasing, the Cdk1 activity does not simply retrace its previous behavior when the concentration was increasing. Instead, it remains in the high activity state until another point is reached where it jumps back to the lower curve. The delay in returning to the earlier levels is what is mean by the term hysteresis.

The limit cycle illustrated in Fig. 9.8(a) is representative of trajectory in which positive feedback makes a contribution. As positive feedback becomes more pronounced, the waveform becomes more spike-like. Sharp, regularly spaced spikes in Cdk1 activity are punctuated by long period in between where there is little change in activity. When positive feedback is decreased or is absent the waveform is more sinusoidal. The waveform presented in Fig. 9.8(b) is typical of situations in which there is a modest amount of positive feedback and hysteresis. Variations of this type are referred to relaxation oscillations.

The positive and double-negative feedback loops create a sharp threshold for committing the cell to mitosis. Either the cell commits to mitosis, or it doesn't. In either case, once the decision is made it is irrevocable. That is, the cell it does not switch back and forth, nor does it dwell somewhere in between.

9.10 Multisite Phosphorylation Helps Ensure the Correct Ordering of Events

Transitions from one phase to the next must be timed in a way that ensures that events occur in proper order. One way this is accomplished is through multisite phosphorylation. Many of the cell cycle regulators that have roles in the G1/S and G2/M transitions are phosphorylated on a series of distinct sites, by one or more kinases. Prominent examples include Cdc25, Weel, Cyclin E, and pRb. An example of this kind of regulation is the timing of the G1 to S transition in budding yeast by the SCF^{Cdc4} complex. This SCF targets a cyclin-dependent kinase inhibitor called Sic1 for proteolysis. This event only happens when Sic is phosphorylated on six or more of the nine possible phosphorylation sites. Once the Sic1s have passed their multisite phosphorylation threshold, they are able to bind the requisite Cdc4 subunit of the SCF complex. This delay allows sufficient time for proper entry into S-phase.

A similar strategy is encountered in the requirement for multiple ubiquitination events producing chains of ubiquitins. In this case, the timing requirement extends to the proteolytic processes whereby substrates are degraded in a strict orderly fashion some before others. The order at which substrates are degraded depends on how they are ubiquitinated by the APC (and SCF). Substrates are tagged for ubiquitination by the covalent addition of an entire chain of ubiquitin molecules to the lysines. If the APC is able to catalyze the addition of the necessary chain in one binding event that substrate will be ready for ubiquitination before a substrate that can only have its chain built up one step at a time through multiple rounds of binding and release. The former, all at once, way of adding a ubiquitin chain is referred to as processive while the far slower one at a time way is termed distributive. Thus, early degraded substrates are ubiquitinated in a processive manner and late ones in a distributive way.

9.11 Traversing the Cell Cycle with the APC and SCF

Phosphorylation and proteolysis work hand in hand. Recall that a cell passing through mitosis begins with prophase and then passes through prometaphase followed by metaphase, anaphase, telophase, and cytokinesis ending with a pair of separated interphase daughters ready to grow. As indicated in Table 9.1, Cyclin B activates cdk1, which along with several other kinases drives the onset of early mitosis. The activation of cdk1 by binding to cyclin in prophase is followed in anaphase and telophase by its deactivation through proteolysis of Cyclin B by the APC. This deactivation step is needed both for exit from mitosis and for DNA replication by the daughters later in the next cell cycle.

The APC activates when its regulator subunits recruit the core APC module. The Cdc20 regulatory subunit binds first and this step is necessary for passage from metaphase to anaphase. The second regulatory unit, Cdh1, takes over during anaphase and functions into the succeeding G1 phase. Cyclin B is bound by Cdc20 and ubiquitinated by the APC. Cdc20 binding occurs when the buildup of Cdk1, which as mentioned above occurs in prophase, exceeds a threshold level and phosphorylates the APC at multiple sites. These events increase Cdc20 binding while at the same time Cdh1 is phosphorylated (by Cyclin A Cdk2) and as a consequence does not bind the APC. As mitosis approaches its end, Cdc20 is ubiquitinated by the APC and degraded; Cdh1 is dephosphorylated and binds the APC. Cdh1 remains bound to the APC until the G1-S transition where it is phosphorylated once again and dissociates from the APC preparing the way for the next round of Cdc20 binding.

Cyclin B is one of the two major substrates of the APC. The second key substrate is cohesin, a complex that holds sister chromatids together and prevents their separation. In order for sister chromatid separation to take place followed by their movement toward opposite spindle poles in anaphase, this complex must be disrupted. The APC accomplishes this by ubiquitinating an enzyme called securin. This protein binds and inhibits a protease called separase, which when released from its inhibition by securin cleaves the Scc1 subunit of the cohesin complex, making possible sister chromatid separation.

The exit from mitosis to G1 phase is controlled in large measure by the APC/C and Cyclin A. The APC/C degrades Cyclin A during M phase but not during G1 phase when Cyclin A begins to accumulate once again. The APC/C is active during this time and degrades other substrates but not Cyclin A. This selectivity is achieved through a carefully timed switch in ubiquitin-conjugating E2 enzymes. In place of UBCH10 active during mitosis, UBCH5 is utilized during G1 phase. Most interestingly, UBCH10 remains active until it finishes ubiquitinating its various substrates, then ubiquitinates itself paving the way for UBCH5 binding. Timing once again takes the stage.

A second mechanism further promotes S phase entry. Stimulation of E2F transcription by relieving pRb inhibition leads to expression of hEmi1. This protein accumulates at the G1-S transition and inactivates APC^{Cdh1} leading to further accumulation of Cyclins A and E. To complete the cycle and permit APC^{Cdc20} activity when it is needed in mitosis, hEmi1 is phosphorylated at that time by the Polo-like protein 1 (Plk1). This event tags the protein for ubiquitination by $SCF^{\beta TrCP}$.

Further Reading

Bartkova J et al. 2005 DNA damage response as a candidate anti-cancer barrier in early human tunorigenesis. *Nature* 434: 864–870.

Braig M et al. 2005 Oncogene-induced senescence as an initial barrier in lymphoma development. *Nature* 436: 660–665.

Folkman J, Merler E, Abernathy C and Williams G 1971 Isolation of a tumor factor responsible for angiogenesis. *J. Exp. Med.* 133: 275–288.

Gimbrone MA Jr, Leapman SB, Cotran RS and Folkman J 1972 Tumor dormancy in vivo by prevention of neurovascularization. *J. Exp. Med.* 136: 261–276.

Gougoulis VG et al. 2005 Activation of the DNA damage checkpoint and genomic instability in human precancerous lesions. *Nature* 434: 907–913.
Hanahan D and Weinberg RA 2000 The hallmarks of cancer. *Cell* 100: 57–70.
Michaloglou C et al. 2005 BRAFE600-associated senescence-like cell cycle arrest of human naevi. *Nature* 436: 720–724.
Vaux DL, Cory S and Adams JM 1988 *Bcl2* gene promotes haemopoietic cell survival and cooperates with *c-myc* to immortalize pre-B cells. *Nature* 335: 440–442.

Replication of the Genome

Blow JJ and Dutta A 2005 Preventing re-replication of chromosomal DNA. *Nat. Rev. Mol. Cell Biol.* 6: 476–486.
Forsburg SL 2004 Eukaryotic MCM proteins: beyond replication initiation. *Microbiol. Mol. Biol. Rev.* 68: 109–131.
Machida YJ, Hamlin JL and Dutta A 2005 Right place, right time, and only once: replication initiation in metazoans. *Cell* 123: 13–24.
Mailand N and Diffley JFX 2005 CDKs promote DNA replication origin licensing in human cells by protecting Cdc6 from APC/C-dependent proteolysis. *Cell* 122: 915–926.
Tercero JA, Longhese MP and Diffley JFX 2003 A central role for DNA replication forks in checkpoint activation and response. *Mol. Cell* 11: 1323–1336.

Cell Cycle Engine

Hansen DV et al. 2004 Plk1 regulates activation of the anaphase promoting complex by phosphorylation and triggering $SCF^{\beta TrCP}$-dependent destruction of the APC inhibitor Emi1. *Mol. Biol. Cell* 15: 5623–5634.
Hsu JY et al. 2002 E2F-dependent accumulation of hEmi1 regulates S phase entry by inhibiting APC^{Cdh1}. *Nat. Cell Biol.* 4: 358–366.
Kellogg DR 2003 Wee1-dependent mechanisms required for coordination of cell growth and cell division. *J. Cell Sci.* 116: 4883–4890.
Miller JJ et al. 2006 Emi1 stably binds and inhibits the anaphase-promoting complex/cyclosome as a pseudosubstrate inhibitor. *Genes Dev.* 20: 2410–2420.
Moshe Y et al. 2004 Role of Polo-like kinase in the degradation of early mitotic inhibitor 1, a regulator of the anaphase promoting complex/cyclosome. *Proc. Natl. Acad. Sci. USA* 101: 7937–7942.
Sherr CJ and Roberts JM 1999 CDK inhibitors: positive and negative regulators of G1-phase progression. *Genes Dev.* 13: 1501–1512.
Vidal A and Koff A 2000 Cell cycle inhibitors: three families united by a common cause. *Gene* 247: 1–15.

Cell Cycle Effectors at the G1/S Transition in Mammals

Li J et al. 2008 Synergistic function of E2F7 and E2F8 is essential for cell survival and embryonic development. *Dev. Cell* 14: 62–75.
Ogawa H et al. 2002 A complex with chromatin modifiers that occupies E2F- and Myc-responsive genes in G_0 cells. *Science* 296: 1132–1136.
Trimarchi JM and Lees JA 2002 Sibling rivalry in the E2F family. *Nat. Rev. Mol. Cell Biol.* 3: 11–20.

Cell Cycle Effectors at the G2/M Transition in Mammals

Ferrari S 2006 Protein kinases controlling the onset of mitosis. *Cell. Mol. Life Sci.* 63: 781–795.
Nigg EA 2001 Mitotic kinases as regulators of cell division and its checkpoints. *Nat. Rev. Mol. Cell Biol.* 2: 21–32.

The SCF and APC/C

Peters JM 2006 The anaphase promoting complex/cyclosome: a machine designed to destroy. *Nat. Rev. Mol. Cell Biol.* 7: 644–656.
Thornton BR and Toczyski DP 2006 Precise destruction: an emerging picture of the APC. *Genes Dev.* 20: 3069–3078.
Zheng N et al. 2002 Structure of the Cui1-Rbx1-Skp1-F Box^{Skp2} SCF ubiquitin ligase complex. *Nature* 416: 703–709.

Cell Cycle Dynamics

Angeli D, Ferrell JE Jr and Sontag ED 2004 Detection of multistability, bifurcations, and hysteresis in a large class of biological positive-feedback systems. *Proc. Natl. Acad. Sci. USA* 101: 1822–1827.
Ferrell JE Jr 2002 Self-perpetuating states in signal transduction: positive feedback, double-negative feedback and bistability. *Curr. Opin. Chem. Biol.* 6: 140–148.
Goldbeter A 1991 A minimal cascade model for the mitotic oscillator involving cyclin and cdk2 kinase. *Proc. Natl. Acad. Sci. USA* 88: 9107–9111.
Holt LJ, Krutchinsky AN and Morgan DO 2008 Positive feedback sharpens the anaphase switch. *Nature* 454: 353–357.
Novak B and Tyson JJ 1993 Numerical analysis of a comprehensive model of M-phase control in *Xenopus* oocyte extracts and intact embryos. *J. Cell Sci.* 106: 1153–1168.
Pomerening JR, Kim SY and Ferrell JE Jr 2005 Systems-level dissection of the cell cycle oscillator: bypassing positive feedback produces damped oscillations. *Cell* 122: 565–578.
Pomerening JR, Sontag ED and Ferrell JE Jr 2003 Building a cell cycle oscillator: hysteresis and bistability in the activation of Cdc2. *Nat. Cell Biol.* 5: 346–351.
Sha W et al. 2003 Hysteresis drives cell-cycle transitions in *Xenopus laevis* egg extracts. *Proc. Natl. Acad. Sci. USA* 100: 975–980.
Tsai TYC et al. 2008 Robust, tunable biological oscillations from interlocked positive and negative feedback loops. *Science* 321: 126–129.
Tyson JJ, Chen KC and Noval B 2003 Sniffers, buzzers, toggles and blinkers: dynamics of regulatory and signaling pathways in the cell. *Curr. Opin. Cell Biol.* 15: 221–231.

The Timing of Events

Nash P, et al. 2001 Multisite phosphprylation of a Cdk inhibitor sets a threshold for the onset of DNA replication. *Nature* 414: 514–521.
Rape M, Reddy SK and Kirschner MW 2006 The processivity of multiubiquitination by the APC determines the order of substrate degradation. *Cell* 124: 89–103.

Traversing the Cell Cycle with the APC and SCF

Hsu JY et al. 2002 E2F-dependent accumulation of hEmi1 regulates S phase entry by inhibiting APCCdh1. *Nat. Cell Biol.* 4: 358–366.

Rape M and Kirschner MW 2004 Autonomous regulation of the anaphase-promoting complex couples mitosis to S-phase entry. *Nature* 432: 588–595.

Chapter 10
Cell Cycle Checkpoints and DNA Damage Repair

Oncogene-driven, abnormal patterns of cell growth and division generate DNA damage, and this damage triggers a cellular response aimed at preventing inappropriate cell proliferation. Even in the absence of oncogene-driven processes, cells are continually exposed to DNA damaging agents such as ionizing radiation, environmental chemicals, and mutagens, and carcinogens generated endogeneously. As a result, DNA is continually monitored for damage and quickly repaired when damage is detected. This same system responds to the oncogene-driven damage, as well.

Because of its importance there are five distinct DNA damage repair systems. The base excision repair (BER) system removes bases damaged by both exogenous and endogenous agents. The latter causes of damage such as oxygen radicals, alkylating chemicals, and deaminating agents predominate. Injuries brought on by these causes do not, in general, distort the DNA helix and are remedied by specialized repair glycosylases. Nucleotide excision repair (NER) handles bulky DNA lesions brought on by ultraviolet radiation and by environmental agents such as polycyclic aromatic hydrocarbons compounds contained in cigarette smoke. Damage produced by these agents does produce distortions in the DNA helix. Portions of chromosomes bearing genes that are actively undergoing transcription receive greater NER attention than DNA segments containing genes that are only rarely transcribed. The third type of repair, mismatch repair (MMR), monitors and treats damage occurring during DNA replication and meiotic recombination. It corrects A-G and T-C mismatches, and improper sequence insertions and deletions.

Homologous recombination (HR) and nonhomologous end joining (NHEJ) are responsible for repairing DNA double-strand breaks. Homologous recombination utilizes regions of DNA sequence homology from the sister chromatids to repair damaged DNA in an error-free manner. Nonhomologous end joining does not use extensive regions of sequence homology to effect repairs and is not necessarily error-free. Double-strand breaks can be generated endogenously by oxidative agents and environmentally by ionizing radiation. This form of damage, while not as common as the other forms of DNA injury, can be extremely harmful. Double-strand breaks can lead to chromosomal

M. Beckerman, *Cellular Signaling in Health and Disease*,
Biological and Medical Physics, Biomedical Engineering,
DOI 10.1007/978-0-387-98173-4_10,

translocations, deletions, and fragmentation and other, even more dramatic forms of chromosome instability.

Two interlocking cellular signaling pathways lie at the heart of a rapid cellular response to DNA damage. The checkpoint pathway temporarily halts the cell cycle when such pauses are needed to ensure a proper temporal sequence of events. In the all-important cases of double-strand breaks, the regulatory pathways provide time for the orderly repair of damage by the DNA repair machinery and for the shunting aside or destruction of permanently damaged cells. At the same time, the repair pathway is activated. The main role of this pathway is to activate and recruit the factors needed for the repair to the site of damage. These factors include enzymes that affect the repair and enzymes that manipulate chromatin and expose the damage site to the repair enzymes.

10.1 The G1/S Checkpoint Pathway

Checkpoint pathways are activated during the progression through the cell cycle at several critical points just before, during, and after DNA replication. The first place where checkpointing is done is at the G1/S transition. If double-strand breaks are sensed, the cell cycle is halted to allow time for repair. If the damage to the DNA cannot be repaired in a reasonable time period, the cell is not allowed to proceed to S phase and mitosis but instead is directed towards senescence or apoptosis. S phase is a particularly critical time as this is when the DNA is replicated. Damage sensing, signaling, and repair are carried out throughout this phase of the cell cycle in a manner coordinated with replication. These sensing and signaling operations are referred to as the intra S-phase checkpoint. Lastly, the DNA is checked for damage and is repaired subsequent to completion of replication at the G2/M boundary.

The G1/S checkpoint pathway is depicted schematically in Fig. 10.1. This pathway has three components. The first of these is the serine/threonine kinase ataxia telangeictesia mutated (ATM). This kinase is a member of the phosphoinositide-3-kinase-related kinase (PIKK) family. Three members of this family participate in checkpoint and repair processes – ATM, the ATM, and Rad3 related (ATR) protein, and the catalytic subunit of DNA-PK, designated as DNA-PKcs. These proteins launch not only checkpointing but also repair. The second element of the checkpoint pathway is checkpoint kinase 2 (Chk2), another serine/threonine kinase. The third component is p53, the guardian of the genome, so-called for reasons that will become clear shortly.

ATM becomes activated in response to double-strand breaks. The mechanisms by which DSBs are sensed and signaled to ATM will be discussed in the next section. Once activated, ATM phosphorylates Chk2 and activates it. Both ATM and Chk2 then phosphorylate p53 and its negative regulator Mdm2. As a result of these phosphorylations, the binding of Mdm2 to p53 is weakened, and p53 is free to interact with downstream targets. If the damage to the DNA is not

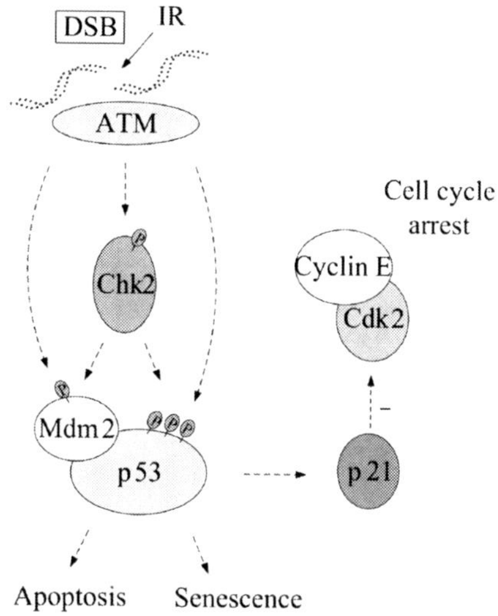

Fig. 10.1 The G1-S phase checkpoint. Signaling from the site of a double-strand break leading to either cell cycle arrest, apoptosis, or senescence

too severe, p53 will upregulate p21 through its role as a transcription factor. Recall from the previous chapter that p21 (p21Cip1) is a cyclin-dependent kinase inhibitor. Once upregulated by p53 this protein will interact with and inhibit the G1/S transition-promoting activities of cyclin E/Cdk2.

10.2 Formation of IRIFs and Activation of ATM

In responding to replicative stresses during S-phase, ATM and ATR localize to specific damage-defined chromatin subcompartments. They are recruited to these locations by, and associate with, protein complexes that recognize the DNA damage and sever as damage sensors. The ATM protein responds to double-strand breaks brought on by agents such as ionizing radiation. These proteins respond to IR-induced double-strand breaks by localizing to chromatin subcompartments called IR-induced foci (IRIFs). The ATR protein responds to stalled replication forks resulting from, for example, ultraviolet light and reactive chemical-induced DNA damage. ATR is recruited to chromatin subcompartments delineated by the presence of single-strand DNA (ssDNA) and these sites are referred to ssDNA-associated chromatin subcompartments (SDCSs). The formation and activation of the checkpoint- and repair-initiating complexes at these locales is depicted in Figs. 10.2 and 10.3.

In launching its response to double-strand breaks, ATM works together with a DNA damage sensor, the Mre11-Rad50-Nbs1 (MRN) complex, and a number of DNA damage mediators. These proteins assist ATM in initiating checkpoint signaling and the repair process. They do so by recruiting proteins involved in these actions to the damage site, and facilitating their access, attachment, and signaling. The recruitment and activation processes occur in an orderly fashion.

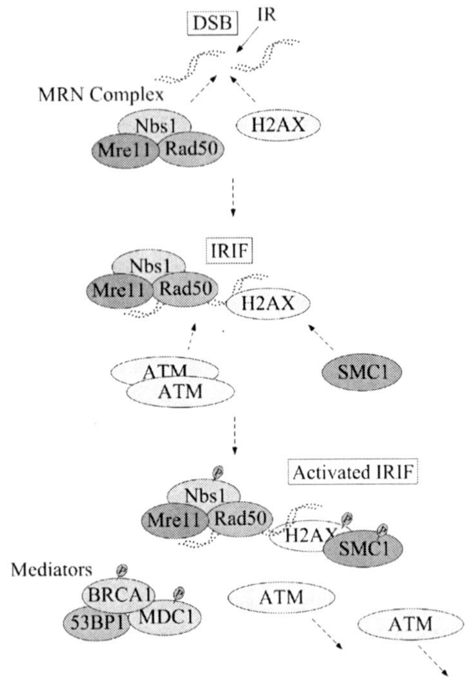

Fig. 10.2 Formation and activation of IR-induced foci (IRIFs) at sites of double-strand breaks

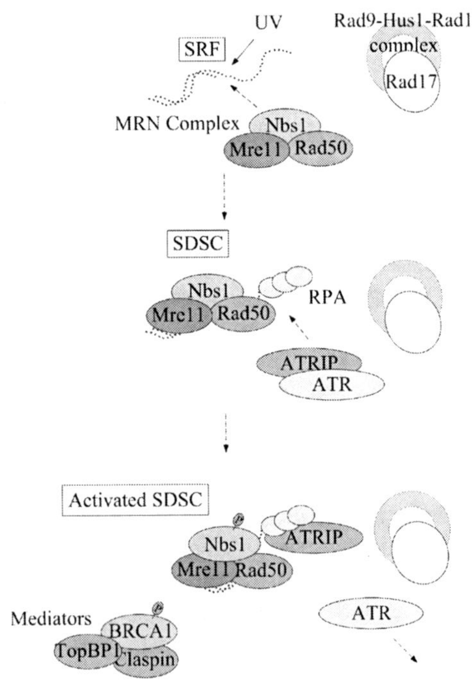

Fig. 10.3 Intra-S and G2/M checkpoint pathways. Different Cdc25s and Cyclin/Cdks are active during different phases of the cell cycle. Signaling from stalled replication forks and double-strand breaks leading to temporary arrest of the cell cycle

The first step is the recruitment of ATM to the IRIFs by the MRN complex. This complex functions as the sensor of DNA double-strand breaks. In the absence of DNA damage, ATM molecules form inactive dimers, but dissociate into active monomers upon stimulation by the MRN complex and DNA.

The MRN complex together with DNA stimulates ATM autophosphorylation on Ser1981 and subsequent dissociation into monomers. These initial but as yet weak responses are greatly amplified when early acting mediators relocate to the IRIF. The first to become involved is the histone H2A variant H2AX. In mammalian cells about 10% of the histone H2A is of the H2AX variant form. The histone H2AX variant differs from the predominant histone H2A by the presence of several additional residues in the C-terminal. These residues are centered on one or two SQ motifs (SQ followed by TQ in humans) that are strongly phosphorylated by ATM. The resulting phosphorylated H2AX, designated as γ-H2AX, spreads out over a 50-kbp region adjacent to the damage foci acting as a recruiting marker for other checkpoint and repair participants including additional ATM molecules.

10.3 Mediators Amplify the ATM Signal

Three mediators are depicted in Fig. 10.2. The mediator of DNA damage checkpoint protein 1 (MDC1) is an early arrival at the damage foci. MDC1, like nibrin (Nbs1), functions as an adapter that, together with γ-H2AX, recruits repair and signaling proteins to DSB damage sites. The ATM, γ-H2AX, MDC1, and Nbs1 proteins operate in a positive feedback loop to rapidly increase the number of activated ATM molecules at γ-H2AX marked damage sites – ATMs make γ-H2AXs; γ-H2AXs recruit ATMs; MDC1s bind phosphorylated H2AXs and recruit Nbs1s, which recruit the other members of the MRNs. In this signal amplification, MDC1 not only mediates the positive feedback between H2AX and ATM, but also interacts with Chk2 via phosphorylated Thr68 and facilitates its activation as well.

Another early acting mediator is the p53 binding protein 1 (53BP1). Like MDC1, it is recruited to damage foci tagged by γ-H2AX. This protein recognizes methylated lysine 79 residues of histone H3. These histone residues are constitutively in a methylated state, and binding sites for 53BP1 become accessible when DSBs occurs. The bound 53BP1s are then stabilized by γ-H2AX. As its name implies 53BP1 interacts with p53 as well as other later acting factors such as BRCA1, and facilitates ATM signaling when the numbers of DSBs are low.

The third mediator, the breast cancer associated gene 1 (BRCA1) protein, is regulated through phosphorylation by ATR, ATM, and Chk2. Once it is phosphorylated and activated by these kinases, the BRCA1 protein binds p53 and acts as a transcriptional coactivator at the p21 promoter and other sites involved in growth arrest thereby shifting p53's actions away from sites connected to apoptosis.

Another one of the proteins recruited to the IRIFs by γ-H2AX is SMC1. Structural maintenance of chromosome (SMC) proteins facilitates sister chromatid cohesion during S-phase and maintains proper segregation of chromosomes during mitosis. They are also required for DSB repair subsequent to completion of DNA replication. There are six SMC protein family members. These proteins form heterodimeric complexes with one another. In particular, SMC1 associates with SMC3. In DSB repair, SMC1/SMC3 binds γ-H2AX; it is phosphorylated by ATM assisted by Nbs1 and BRCA1, and the resulting cohesion complex mediates postreplication DSB repair in which the presence of sister chromatids is required.

10.4 Intra-S Phase and G2/M Checkpoints

Mediators and other signaling intermediaries that operate at the apex of the checkpoint and repair pathways remain in close association with DNA and chromatin. In contrast to these, Chk1 and Cdk2 checkpoint kinases and Cdc25 phosphatases are far more mobile. These enzymes are the key signaling elements in the intra-S phase and G2/M checkpoint pathways. As shown in Fig. 10.4, ATM signals through Chk2 while ATR signals via Chk1. The checkpoint kinases, in turn, target members of the Cdc25 family of protein phosphatases. Recall from the last chapter that different Cyclin/Cyclin-dependent kinase combinations are active at different stages of the cell cycle.

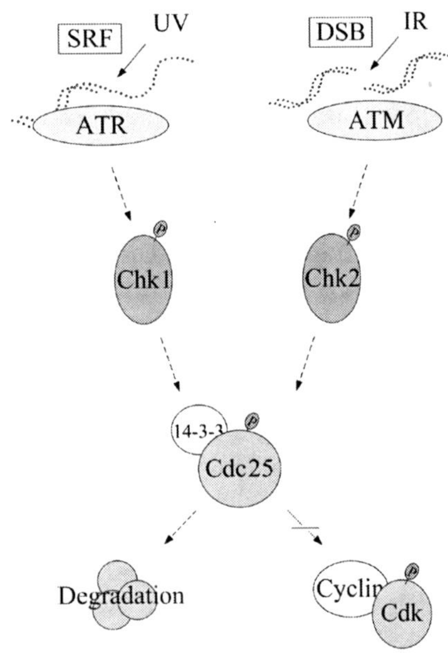

Fig. 10.4 Formation and activation of ssDNA-associated chromatin subcompartments (SDSCs) at sites of stalled replication forks

Cyclin E/Cdk2 predominates during the G1/S transition; Cyclin A then takes over and binds Cdk2 early in S phase and then Cdk1 later in S phase. During G2 and mitosis, Cdk1 forms complexes with Cyclin B. Also recall from Chapter 9 that there are three members of the Cdc25 protein phosphatase family – Cdc25A, Cdc25B, and Cdc25C. Each of these participates in the checkpoint pathway, but Cdc25A has emerged as the general contributor, active at both the G1/S and G2/M transitions, while Cdc25B and Cdc25C are more active at mitosis.

The downstream signaling partners and actions that emerge from these considerations are, as follows. During the intra-S phase, Cdc25A normally dephosphorylates Cdk2, which partners with Cyclins E and A. Dephosphorylation activates the Cdk2s thereby enabling passage through S phase. The onset of DSBs and/or SFAs triggers activation of the checkpoint. In response, Cdk2 and/or Cdk1 phosphorylates Cdc25A. This action leads to its enhanced degradation of Cdc25A; the Cdks remain phosphorylated, and the cell cycle is halted. Later, in G2/M and M phase, Cdc25C is active and in the absence of replicative stresses dephosphorylates Cdk1 that forms complexes with cyclin B. DSBs and replication fork arrest lead to enhanced degradation of Cdc25C resulting in the temporary arrest of the cell cycle.

10.5 Formation of SDSCs and Activation of ATR

ATR forms stable complex with a regulatory subunit known as the ATR-interacting protein, or ATRIP. These complexes respond to stalled replication forks by generating checkpoint signals using Chk1 as the key signal transducer. As noted earlier, the machinery for doing this includes DNA manipulation complexes referred to as sliding clamps and clam loaders, replication protein A, and ATR mediators that assist in shaping and conveying checkpoint signals from ATR/ATRIP to Chk1.

DNA sliding clamps are heterotrimeric complexes that encircle DNA molecules and tether the DNA polymerases. These molecular machines slide along the dsDNA, organizing the various protein complexes involved in replication, cell cycle checkpointing, and DNA damage repair. They are loaded onto DNA by complexes called clamp loaders. In eukaryotic cells, the complex functioning during replication as the sliding clamp is the proliferating cell nuclear antigen (PCNA) complex and the clamp loader is replication factor C (RFC).

A second DNA-encircling complex assists in DNA damage checkpointing and repair. When replicative stresses arise, Rad9, Hus1, and Rad1 proteins form heterotrimeric, PCNA-like complexes at stalled replication forks. These 9-1-1 complexes have a clamp loader that is slightly different from RFC. The RFC clamp loader contains five subunits, RFC1 to RFC5. In the DNA damage clamp loader, referred to as the RFC-hRad17 complex, the hRad17 protein replaces RFC1. The DNA damage clamp loader binds single-stranded DNA generated by stalled replication forks and also binds gapped DNA produced by DSBs.

Replication protein A (RPA) is another early participant in the DNA damage response. This protein exists as a stable complex of three subunits. It, like BRCA2, possesses a number of oligonucleotide binding (OB) folds, and attaches to single-stranded DNA and DNA gaps. A model of how RPA assists in the damage response is, as follows. When replication forks stall as a result of the DSB-generated lesions, the DNA polymerase activity is halted but not helicases that still carry out their DNA unwinding activities. As a result there is a buildup of unwound RPA-coated ssDNA. The RPA-ssDNA facilitates the loading of the ATR-ATRIP and 9-1-1 complexes leading to an increased phosphorylation of Chk1. The longer the segment of unwound DNA, the more ATR-ATRIPs can accumulate and the greater the phosphorylation activity on Ser317 and Ser345 of Chk1s.

As was the case for ATM checkpoint signaling, ATR checkpoint signaling is facilitated by a set of mediators. These include BRCA1, DNA Topoisomerase II-binding protein 1 (TopBP1), and Claspin. Both TopBP1 and Claspin mediate activation of Chk1 by ATR. Claspin is loaded onto origins during DNA replication initiation. The protein binds branched DNA with high affinity, forms a ring-like structure that encircles the DNA, and serves as a substrate for ATR and Chk1 recruitment to SDSCs. TopBP1 has a role in the initiation of replication at origins through its loading of Cdc45, an essential replication factor, and also has a role as a checkpoint mediator by its ability to activate ATRIP/ATR leading to the latter's phosphorylation and activation of Chk1. Thus, both ATM and ATR rely strongly on a set of mediators for assistance in activation of the downstream kinases Chk1 and Chk2.

10.6 Structure and Posttranslational Modifications of Checkpoint Proteins

The ATM and ATR kinases that lie at the apex of the checkpoint and repair pathways phosphorylate a number of substrates. Their kinase domains are situated in their C-terminal regions as shown in Fig. 10.5(a). They occupy only a small portion of these rather large proteins, no more than 5–10%, the remainder consisting primarily of a large number of HEAT (huntingtin, elongation factor 3, A subunit of protein phosphatase 2A, and TOR1) repeats. A HEAT repeat consists of a pair of alpha helices connected by a flexible loop and shaped like a hairpin. HEAT repeats are found in a number of proteins that associate with condensins and cohesins, and other proteins involved in chromosome dynamics. The HEAT repeats in ATM and ATR enable these proteins to form a superhelical scaffolding structure that bend and fold about

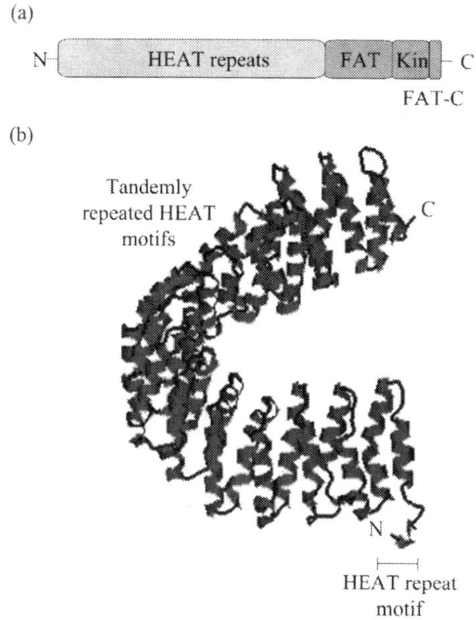

Fig. 10.5 ATR and ATM structure. (**a**) Shown is the ATR protein. The ATM protein is organized in the same way with a slightly larger N-terminal HEAT repeat region. (**b**) Heat repeats. X-ray crystal structure of the A subunit of protein phosphatase 2A (PP2A). The bar in the *lower right hand* corner of the figure delineates the boundaries of a single helix-turn-helix, or helical hairpin, HEAT repeat. Situated between the N- and C-termini are 15 repeats. The figure was prepared using Protein Explorer with atomic coordinates deposited in the PDB under accession number 1b3u

the chromosomes. ATM contains 49 of these motifs while ATR possesses 41 repeats. The X-ray crystal structure of the A subunit of PP2A is presented in Fig. 10.5(b). This figure shows how HEAT motifs can link together to form flexible scaffolds. Returning to Fig. 10.5(a), the kinase domains of ATR and ATM are flanked by a pair of FAT (Frap, ATM, Trrap) domains. These domains mediate protein–protein interactions.

The two checkpoint kinases, Chk1 and Chk2, are the main checkpoint signal transducers. The domain structures of the kinases are depicted in Fig. 10.6(a) and (b). The ATM and ATR kinases prefer to phosphorylate substrates serine (S) and threonine (T) residues that are followed by a glutamine (Q) residue. ATR kinase responds to stalled replication forks by phosphorylating Chk1 on Ser317 and Ser345 situated in that protein's SQ/TQ region. In response to DNA damage, the ATM kinase activates Chk2 by phosphorylating it on Thr68 located in its SQ/TQ rich region. Phosphorylated Chk2 kinases respond by forming dimers, which then undergo autophosphorylation on Thr383 and Thr387 residues located in the kinase domains. The Forkhead-associated

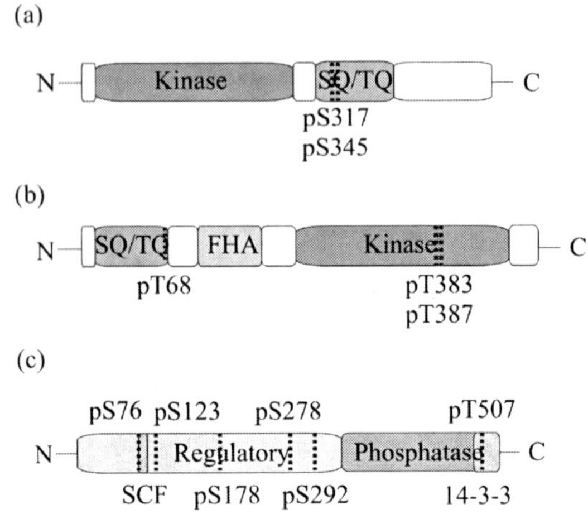

Fig. 10.6 Checkpoint kinases and phosphatases. (**a**) Chk1, (**b**) Chk2, and (**c**) Cdc25A protein phosphatase

(FHA) domains recognize peptide sequences containing phosphorylated threonines, phosphorylated serines, and phosphorylated tyrosines with a pronounced affinity for phosphothreonines. They are found in Chk2 and several other proteins involved in detecting and repairing DNA damage.

The domain structure and key regulatory sites of Cdc25A are depicted in Fig. 10.6(c). 14-3-3 proteins are small adapter proteins, 28–30 kDa in size. They bind with high affinity to peptide sequences containing phosphoserine residues followed by a proline two positions toward the C-terminal. The 14-3-3 proteins bind and localize their substrates within the cytoplasmic compartment and keep them sequestered from the nucleus. These proteins operate in the growth and apoptosis pathways as well as in cell cycle control. In the Cdc25A protein, the 14-3-3 protein binds to C-terminal region in response to phosphorylation on T507. By doing so the 14-3-3 protein blocks binding to Cyclin B/Cdk1 complexes and promotes sequestration of Cdc25A in the cytoplasm away from the nucleus. Another key regulatory site is Ser76. Phosphoprylation of Cdc25A at this site targets the phosphatase for ubiquitin-mediated degradation by the SCF.

10.7 p53 Structure and Function

The p53 protein contains 393 amino acid residues. It possesses an N-terminal transactivation domain, followed by a proline-rich region, a centrally situated DNA-binding domain, and two C-terminal regions – a teteramerization domain and a C-terminal negative regulatory region (Fig. 10.7). The extreme N- and C-terminal regions are natively unfolded. This important property

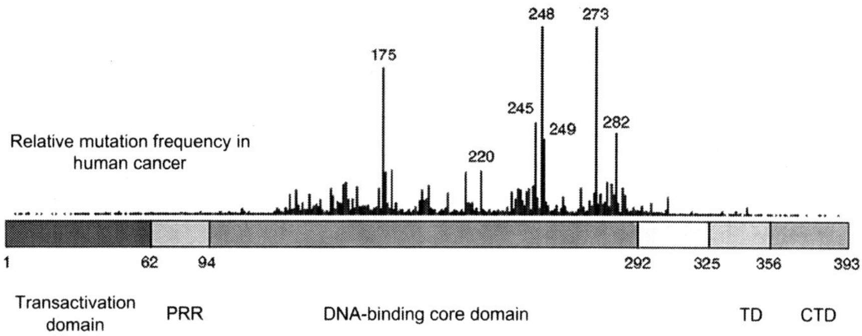

Fig. 10.7 Domain structure of p53. Superimposed on the p53 domain organization is a plot of the relative frequencies of mutations encountered in human cancers. Reprinted by permission from Macmillan Publishers Ltd.: [Oncogene] Joerger AC and Fersht AR 2007 *Oncogene* 26: 2226

facilitates interactions of p53 with multiple binding partners, as will be discussed shortly, while the proline-rich region provides binding sites for proteins bearing SH3 domains. The central core domain binds doubly stranded DNA (dsDNA) cooperatively. The protein self-assembles on dsDNA sites into a transcription-competent tetrameric form. These actions are mediated by the C-terminal tetramerization domain. In contrast to the N- and C-terminal unstructured regions, the DNA-binding core domain and the tetramerization domain have well-defined secondary and tertiary structures.

The three-dimensional structure of the core domain in contact with DNA is depicted in Fig. 10.8. As can be seen, the p53 DNA-binding domain is organized as a beta sandwich. This structure is formed by pair of antiparallel β-sheets that serves as a platform for two loops, designated in Fig. 10.8 as L2 and L3, and a zinc ion; these three elements constitute the DNA binding surface. Superimposed upon the overall domain structure of p53 in Fig. 10.7 are the relative frequencies of common cancer-causing mutations in p53. As can be seen in Figs. 10.7 and 10.8, six out of the ten most frequently encountered mutations are situated in or near the DNA-binding surface.

The DNA-binding domain is only marginally stable at physiological temperatures. Mutations in the DNA-binding domain may directly impair DNA binding or act more indirectly to thermodynamically destabilize the DNA-binding domain (and full-length protein) leading to misfolding and aggregation. Most of the cancer-causing p53 mutations in this region are point mutations affecting a single amino acid residue. (Point mutations leading to misfolding, aggregation, and functional impairment enter prominently in the etiology of neurodegeneration as will be discussed in Part III of CSHD.)

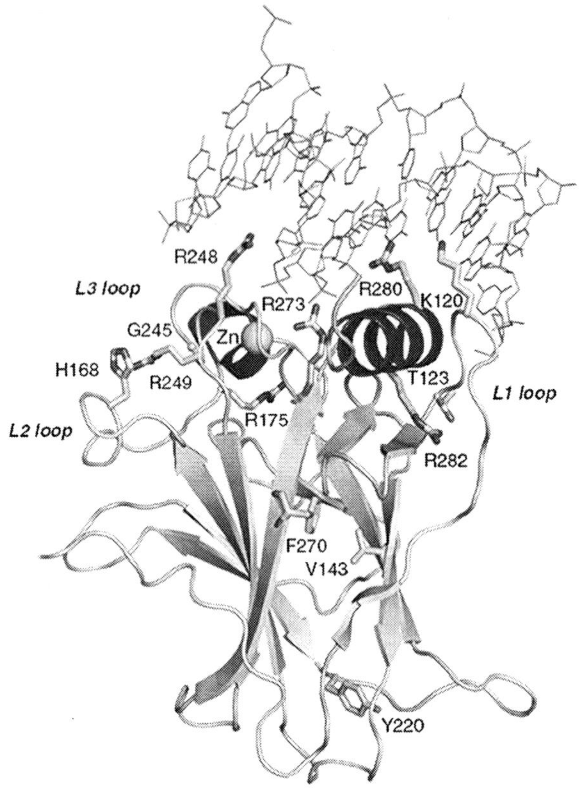

Fig. 10.8 p53 DNA-bound p53 core domain determined by means of X-ray crystallography. Critical residues and structural elements are highlighted in the figure. Reprinted by permission from Macmillan Publishers Ltd.: [Oncogene] Joerger AC and Fersht AR 2007 *Oncogene* 26: 2226

10.8 Restoration of p53 Function by Second-Site Suppressors

The negative effects of some of the point mutations on p53 can be reversed by a second mutation at a different site. The effect of a restorative, or second-site suppressor, mutation can be mimicked by a designed stabilizing drug, a small peptide that binds the folded state of p53 and not the unfolded one, thereby stabilizing the protein. Interventions of this type if successful in restoring function to p53 would reestablish the anticancer barriers and return a crucial portion of the internal networks of the cell to a healthy state. This goal is a challenging one; different mutations give rise to a spectrum of effects on the folding state of the protein, its tetramerization properties, its affinity for DNA target sequences, and its interactions with binding partners. All of these details need to be dealt with in designing restorative drugs. The p53 protein is the key to barrier function but not all cancers are associated with mutated p53. In those cancers that do not have p53 mutations, other proteins that interact with p53 and influence its stability or act downstream of p53 in the apoptosis and senescence pathways are mutated and inactivated. These other mutations have to be found and characterized at the biophysical level and then treated in a way that does not produce excessive toxicity through overactivation of the barrier pathways.

10.9 Special Domains Mediate Protein–Protein Interactions and Chromatin Binding by Proteins that Function at the Apex of the Checkpoint and Repair Pathways

The mediators that assist in the DNA damage checkpoint and repair have several noteworthy structural features. These have been summarized briefly in Table 10.1. The first of these is the presence of BRCA1 C-terminal (BRCT) repeats in most if not all adapters and other proteins with roles in DNA checkpointing and repair. These domains are phosphopeptide-recognition modules. The signaling and repair proteins are recruited and become organized into complexes at sites of DNA damage foci through phosphorylation by kinases such as ATM and ATR, and BRCT-mediated phosphopeptide binding.

Several other domains have prominent roles in DNA damage signaling (Table 10.1). Tudor domains recognize methylated peptides present in histones H3 and H4. The double Tudor domains of 53BP1, for example, bind methylated lysines of histone H3. Forkhead associated (FHA) domains are phosphothreonine-binding modules. They are used, for instance, by MDC1 to bind ATM. Members of the MRN complex possess these domains as well. Both MDC1 and nibrin possess FHA and BRCT domains. MDC1 also uses an FHA domain to interact with Chk2 via phosphorylated Thr68 and facilitates its activation, as well.

The domain structure of the proteins encoded by breast cancer associated genes 1 and 2 (BRCA1 and BRCA2) is presented in Fig. 10.9. These two large proteins are unrelated both structurally and functionally. The BRCA1 protein has roles in checkpointing as well as repair whereas BRCA2 is solely a participant in homologous recombination repair. The BRCA1 protein has a

Table 10.1 Checkpoint mediators and their special structural properties

Mediator	Structure–function
Breast cancer associated protein 1 (BRCA1)	BRCT and RING domains; binds BARD1 leading to activation of Chk1 at the DNA damage-induced intra-S phase and G2/M checkpoints
Claspin	Chk1-binding domain (CKBD) plus two P-Ser sites within this domain mediate binding to CHK1
DNA Topoisomerase II binding protein 1 (TopBP1)	Eight BRCT repeats; mediates Chk1 phosphorylation by ATR
Mediator of DNA damage checkpoint protein 1 (MDC1)	Tandem BRCT domain directly binds tails of phosphorylated histone variant H2AX, regulates its phosphorylation. FHA domain used to bind ATM
P53 binding protein 1 (53BP1)	Two BRCT domains; tandem Tudor domains; Attaches to IRIFs along with MRN and γ-H2AX; binds to p53, BRCA1, and Chk2

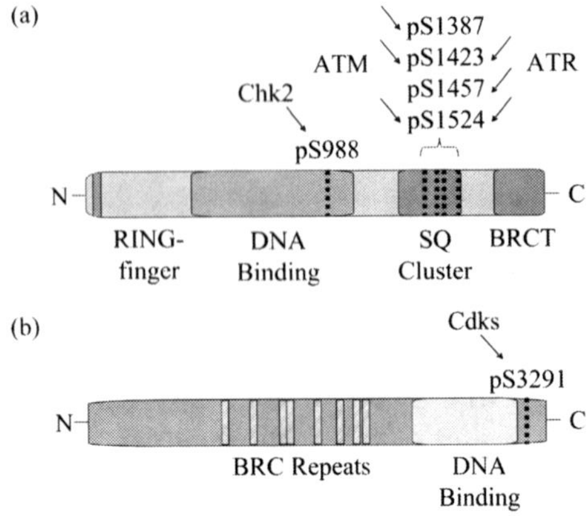

Fig. 10.9 (**a**) Structure of the 1863 amino acid residue BRCA1 protein. The overlapping ATM and ATR phosphorylation sites in the SQ cluster are indicated in the figure along with the Ser988 site phosphorylated by Chk2. (**b**) Structure of the 3418 amino acid residue BRCA2 protein. The 8 BRC repeats situated in the central portion of the protein function somewhat like Velcro and bind and sequester Rad51 until needed. The COOH terminal DNA binding domain consists of five domains – three oligonucleotide/oligosaccharide binding (OB) domains that bind ssDNA, a tower domain that binds DNA and is the locus of a large number of cancer-inducing mutations, and a globular domain at the N-terminal-most end. The DNA binding domain targets the protein and bound Rad51 protein(s) to single-strand and double-strand DNA

RING domain in its N-terminal regions and thus has E3 ubiquitin ligase activity. It binds DNA by means of its DNA-binding domain and has a C-terminal BRCT domain. As indicated in Fig. 10.9, it is regulated through phosphorylation by ATR, ATM, and Chk2. The BRCA2 protein possesses a set of eight BRC repeats along with a C-terminal DNA-binding domain. The extreme C-terminus of BRCA2 binds Rad51, a homologous recombination repair protein; phosphorylation at Ser3291 by the Cdks negatively regulates this binding.

10.10 Base Excision Repair

Specialized enzymes respond to damaged DNA by operating sequentially one after the other to first remove and then replace the damaged bases, nucleotides, and sequences and restore the original structure. As noted in the introduction to this chapter, there are five repair systems. These are briefly characterized in Table 10.2.

Environmental toxins such as UV radiation and cigarette smoke and endogeneous metabolic by-products can damage DNA, causing alterations in the base pairings. The base excision repair system is the main barrier

Table 10.2 The Five DNA Damage Repair Systems

Repair system	Description
Base excision repair (BER)	Removes/replaces damaged bases brought on exogenously (e.g., UV, cigarette smoke) and endogenously (e.g., ROS)
Nucleotide excision repair (NER)	Removes/replaces nucleotide stretches, 25–30 nt in length, damaged by environmental agents
Mismatch repair (MMR)	Corrects mispaired nucleotides and improper nucleotide insertions and deletions during DNA replication and meiotic recombination
Nonhomologous end joining (NHEJ)	Error-prone repair of double-stand breaks brought on by ionizing radiation, oxidative stresses, etc.; main method of DSB repair
Homologous recombination (HR)	Error-free repair of double-stand breaks brought on by ionizing radiation, oxidative stresses, etc.; requires complementary DNA sequences

against cancer-causing ROS-driven mutations. Oxidized guanine bases produced endogenously, generated by inflammatory responses, arising exogenously from IR and UV radiation, from an excessive buildup of heavy metals, and from DNA alkylation are particularly troubling because of the low redox potential of guanine. If the 8-oxoGs are not removed prior to DNA replication, these will lead to G → T mutations.

In base excision repair, the BER glycosylase OGG1 catalyzes the removal of abnormal 8-oxoG bases from the 8-oxoG-C base pairs. Other enzymes then catalyze the replacement of the base to restore the correct G-C base pairing. If this repair is not carried out prior to DNA replication, then another enzyme, the DNA glycosylase MUTYH acts to remove the A from the resulting 8-oxoG-A pairing, thereby setting up a situation where OGG1 can again act. A third enzyme, MTH1, handles the conversion of the oxidized guanine donor 8-oxo-dGTP, incorporated into DNA during DNA synthesis, to the monophosphate form 8-oxo-dGMP thereby preventing this potential source of G → T mutations.

Several different types of damage to DNA bases can occur. Besides the production of oxidized guanines, there is, for example, the hydrolytic deamination of cytosine to generate uracil. If this alteration is not repaired prior to DNA synthesis, the U-G error generates C-G → T-A mutations. This type of alteration is removed by the uracil-DNA glycosylase UNG. Similarly, other forms of damage are handled by the glycosylase specific for that kind of alteration. One of the most common alterations is the loss of a base from a particular nucleotide resulting in the creation of an apurinic/apyrimidinic (AP) site. Repair of this type of damage is carried out either by a set of repair enzymes through the so-called short patch repair pathway in instances where a single base is lost and by means of long-patch repair wherever multiple bases are lost.

10.11 Nucleotide Excision Repair

Nucleotide excision repair (NER) is the main protection against DNA damage caused by ultraviolet (UV) radiation. In nucleotide excision repair, lesions induced by UV and DNA adducts generated by environmental carcinogens such as benzo[a]pyrene, are removed. The importance of this repair system in protecting against cancer is exemplified by defects in NER that lead to xeroderma pigmentosum (XP), a hereditary disease in which skin cancers develop from exposure to UV radiation.

The NER repair process operates differently from BER. In the case of BER, single free bases are excised and replaced. In contrast, in NER oligonucleotide tracts, 25–30 nt in length, are excised and replaced using the opposite strand as a template. There are two NER routes – global genomic repair and transcription-coupled repair. The former removes lesions across the entire genome while the latter, as its name suggests, specifically targets DNA actively undergoing transcription. Initiation of the repair process in the case of global genomic repair is through the actions of the XPC-hR23B dimer that recognizes the DNA damage. The XPC unit recognizes helical distortions while hR23B stabilizes the XPC protein. Initiation in the case of transcription-coupled repair is through the arrest of RNA Poly II at the damage site mediated by two other proteins, called CSA and CSB. These initiating events are followed by the sequential arrival of the repair enzymes and their assembly into functional repair complexes.

10.12 Mismatch Repair

The fidelity of DNA replication is extremely high. Error rates, that is, the frequency of incorrect base-pairings, are on the order of 10^{-10}. These low rates are achieved through a combination of proofreading and mismatch repair. The DNA polymerase holoenzyme has as one of its subunits a 3′–5′ proofreading exonuclease. This subunit checks for erroneous base pairing, and when found the DNA polymerase replaces the incorrect pairings with correct ones. The proofreading operation reduces the core error rate from 10^{-5} to 10^{-7}. In mismatch repair, errors missed by the proofreading exonuclease are corrected. Errors targeted by the MMR system include not only mispaired nucleotides but also inappropriate nucleotide insertions and deletions. These are most likely to occur when multiple tandem repetitive sequences, or microsatellites, are encountered. The mismatch repair system reduces the error rate even further, from 10^{-7} to 10^{-10}.

There are four basic steps in MMR: recognition, recruitment, excision, and resynthesis. Recognition is carried out by human MutS homologs (MSH) of the bacterial MutS proteins. In humans, the MutS proteins form heterodimers. There are two kinds. One heterodimer, termed Mutα, consists of MSH2 plus

MSH6; the other, called Mutβ, contains MSH2 and MSH3. The Mutα complex recognizes base mismatches and short insertions and deletions while Mutβ recognizes larger insertions and deletions that form loops. These complexes bind ATP forming a sliding clamp that translocates along the DNA molecule. In the recruitment step, MutS heterodimers recruit MutL heterodimers consisting of MLH1 plus PMS2 to the damage site. The MutL complexes prepare the site for attachment of exonucleases by displacing the DNA polymerase and PCNA. Once Exo1 is recruited to and binds the sliding clamp, the erroneous nucleotides are excised. DNA polymerases then carry out the resynthesis of the DNA.

Information on how the MutS mismatch repair proteins recognize mispairs is provided by X-ray crystallographic studies of the bacterial MutS homodimer. The three-dimensional structure of the protein is that of an asymmetric homodimer, the counterpart to the eukaryotic MSH2-MSH6 heterodimer. The two chains comprising the dimer form a pair of hands that grip the DNA in between the "thumb" and "fingers" and form a ring about the DNA (Fig. 10.10). The mismatch recognition module sits at the end of one of the thumbs, and the

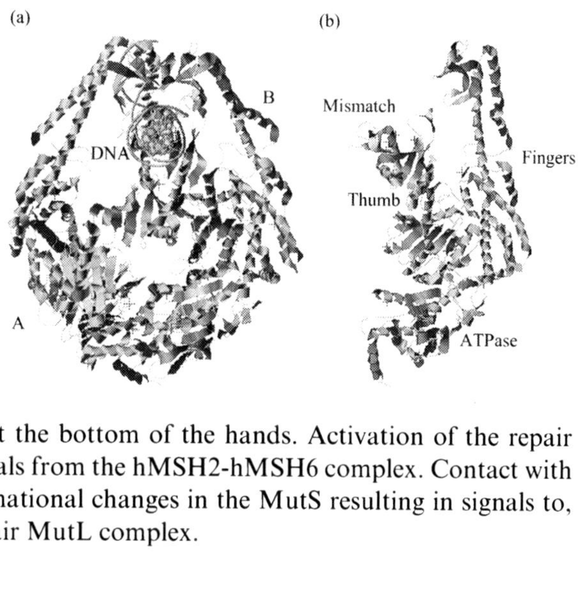

Fig. 10.10 Three-dimensional structure of the (**a**) MutS homodimer from *E. coli* in contact with a DNA G·T mismatch determined through X-ray crystallography. PDB 1e3m and (**b**) MutS monomer (from chain B)

ATPase domains reside at the bottom of the hands. Activation of the repair machinery is through signals from the hMSH2-hMSH6 complex. Contact with a mispair induces conformational changes in the MutS resulting in signals to, and activation of, the repair MutL complex.

10.13 Repair Proteins Diffuse Laterally in One-Dimension Along DNA

One of the key biophysical principles underlying signal transduction is that lateral diffusion in two dimensions is far more efficient in bringing collision partners together than three-dimensional diffusion. For that reason, signal transduction following ligand-receptor binding is dominated by encounters between collision partners that diffuse laterally in two dimensions at and just below the plasma membrane. A nuclear, DNA-centric counterpart to plasma-membrane dominated signaling is lateral diffusion in one dimension along DNA. This allows for far more rapidly occurring encounters between signaling partners than would be the case if the DNA-associated collision partners diffused in two (or worse yet, three) dimensions.

The MutS complex functions as a sliding clamp that moves by means of one-dimensional diffusion in either direction in contact with and along the helical DNA phosphate backbone. This movement is believed to continue until a lesion is encountered. The increase in dwell time would then allow for signaling and recruitment of the MutL repair complex. The movement does not require ATP hydrolysis. The switch from ADP-bound to ATP-bound forms induces conformational changes leading to MSH2-MSH6 release from the DNA and eventual continuation of scanning.

Other critical repair factors exhibit one-dimensional diffusion along DNA, as well. The base-excision glycosylase OGG1 does so as does Rad51, involved in repair of double-strand breaks by homologous recombination (to be discussed in the next section). Finally, not only repair proteins but also p53 operating as a transcription factor and RNA polymerase II have been found to diffuse in one dimension along DNA.

10.14 There Are Two Double-StrandBreak Repair Systems

The proteins responsible for repairing and rejoining broken strands of DNA are organized into several complexes, each consisting of proteins that come into physical contact with one another to carry out their repair functions. Proteins comprising the modules have been grouped together in Table 10.3. As noted several times there are two double-strand break repair systems, homologous recombination (HR) and nonhomologous end joining (NHEJ). The HR repair process requires a complementary strand that serves as a template to guide the repair. These may be supplied by, for example, sister chromatids. More often than not these are not available in the close vicinity of the damaged strand and the repair proceeds by NHEJ. This method does not require a complementary strand, but is not error free.

One of the key findings in studies of the DSB repair machinery is that some of the proteins are centrally involved in a number of rare, inherited genetic disorders whose study reveal important details. These participants in DSB repair are named for the disorders in which they were discovered. Examples of this naming convention include ataxia-telangiectasia mutated (ATM) and Nijmegen breakage syndrome (NBS) proteins. Other rare, cancer-associated diseases involving DSB proteins are Fanconi anemia and Bloom syndrome.

The first entry in the table is ATM/ATR, two proteins that function as sensors and as signaling elements that convey damage signals to the p53 cellular

Table 10.3 Double-strand break sensing and repair

Component	Function	Description
ATM/ATR	Signaling	Phosphorylates NBS1 and BRCA1 in response to DSBs
Rad50	HR, NHEJ	Forms an end-processing complex with Mre11 and NBS1
Mre11	HR, NHEJ	Endonuclease activity
Nbs1	HR, NHEJ	Regulatory
Rad51	HR	Searches for DNA homology; Forms a complex with Rad52, Rad 54, and BRCA1,2
Rad52	HR	Stimulates Rad51 activity
Rad54	HR	Chromatin remodeling
BRCA1,2	HR	Regulatory; Chromatin remodeling
Ku70	NHEJ	Forms a heterodimer with Ku80; Recruits and activates DNA-PKcs
Ku80	NHEJ	KU70 and KU80 bind to the free ends of the DSB
DNA-PKcs	NHEJ	Regulatory
XRCC4	NHEJ	Forms a complex with DNA ligase IV
DNA ligase IV	NHEJ	End reattachment

controller and to the regulatory units of the repair modules. The ATM/ATR proteins are kinases and convey their signals by catalyzing the transfer of phosphoryl groups to their substrates. Three protein complexes follow the ATM/ATR entry. Each complex is made up of proteins with some proteins with regulatory functions and others with repair functions. The modules listed in Table 10.3 work synergistically and sequentially to make the repairs to the DNA.

The first module, referred to as the Mre11 complex, is composed of the Rad50, Mre11, and Nbs1 proteins. This module participates in both homologous recombination and nonhomologous end joining along with several other DNA-strand manipulations involved in mitosis and meiosis. The key function of this complex is to form flexible bridges between ends of broken DNA strands. The Rad50 proteins contain hooks, which join opposing Rad50 proteins protruding from the ends of broken DNA strands. The hooks join pairs of Rad50 proteins in the middle of the break while the ends of the Rad50s remain attached to the DNA strands via the Mre11 proteins.

The next module is the Rad52 group. The Rad52 family of proteins is responsible for homologous recombination. The first step in HR is the coating of the single-strand DNA (ssDNA) with molecules of replication protein A (RPA). In the next step, molecules of Rad51 and BRCA2 are loaded onto the strands so that Rad51 displaces RPA. Both RPA and Rad51 remove secondary structure from the DNA to facilitate the pairing of sister chromatids. As indicated in Table 10.3, Rad52 facilitates the replacement of RPA by Rad51. The Rad51 and BRCA2 proteins form nucleprotein filaments – BRCA2 anchors the filaments to the DNA and Rad51 proteins form the body of the filaments. The pairing of sister chromatids follows growth of the filaments.

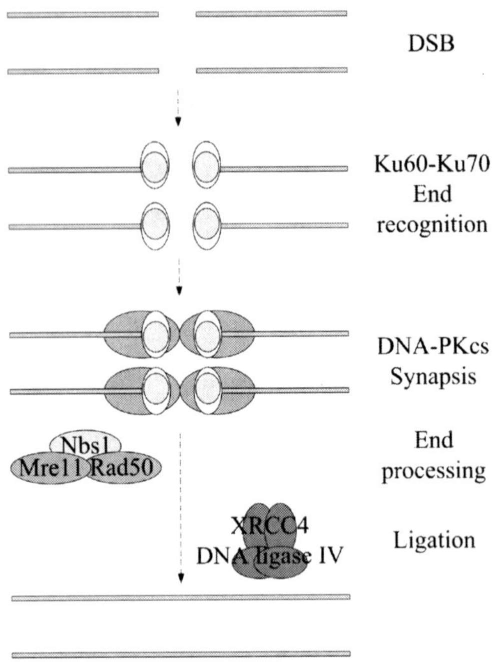

Fig. 10.11 The repair of double-strand breaks by means of nonhomologous end joining

The nonhomologous end joining (NHEJ) proteins complete the list. The last two entries in the table, XRCC4 and DNA Ligase IV, are needed in the final steps of DSB repair. These proteins, operating synergistically with members of the Ku module, rejoin the two ends of the DNA to complete V(D)J recombination and NHEJ. This type of repair is better understood at present than homologous recombination. The steps taken in NHEJ, in a rough order of execution, are depicted in Fig. 10.11.

10.15 The Mre11–Rad50–Nbs1 (MRN) Complex Is Involved in DNA Damage Sensing, Signaling, and Repair

The MRN complex consists of three proteins – Mre11, Rad50, and nibrin (Nbs1), and has a central role in both checkpoint signaling and DNA repair. In its role as a repair protein, Mre11 processes DNA strands and ends. As shown in Fig. 10.12(a), the enzyme possesses a nuclease domain and a pair of DNA-binding domains. The enzyme functions as (i) an exonuclease removing nucleotides from the ends of dsDNA, (ii) as an endonuclease removing phosphodiester bonds from ss and dsDNA molecules, and (iii) as a strand manipulator/winder that separates complementary strands of DNA duplexes and form duplexes from ssDNA.

The second member of the MRN complex, Rad50, has a striking overall structure. As depicted in Fig. 10.12(b), the N- and C-terminals are occupied by

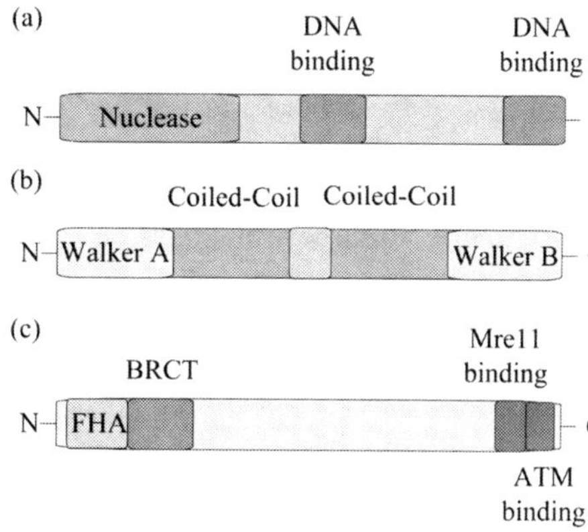

Fig. 10.12 Structure of the components of the Mre11-Rad50-Nbs1 (MRN) complex. (**a**) Mre11; (**b**) Rad50; and (**c**) Nbs1

ATP and DNA binding sequences, Walker A and Walker B motifs. The middle portion of the protein consists of a pair of coiled-coil regions separated from one another by a flexible hinge. This protein assembles into homodimers and in that form is able to bind DNA ends and bring them into close proximity.

The third member of the complex, Nbs1, possesses N-terminal FHA and BRCT domains and C-terminal Mre11 and ATM binding motifs (Fig. 10.12(c)). The Nbs1 protein interacts with ATM through the latter's HEAT repeats and stimulates its activation. Nbs1 recruits other MRN members as well as ATM to sites of damage, which have been marked by γ-H2AX mediator proteins.

10.16 Completing the Repair and Terminating the Checkpoint

The complete DSB damage sensing, signaling, and repair response involves:

- Identifying locations, large and small, where damage has occurred,
- sending a vigorous pause signal to Cdc25 and p53 checkpoint effectors,
- maintaining the pause while the damage is repaired,
- directing repair effectors to the damage sites,
- facilitating access to damage sites,
- repairing the damage, and
- restoring histone and chromatin structure at the end.

The first entry in the above list is sensing step in which damaged foci are sensed and marked. As discussed earlier in this chapter, the MRN complex plays a large role in this step along with the mediators. The next step, generating the pause signal, has been discussed in part already. In the checkpoint pathway, the participants that have been discussed such as Chk1, Chk2, Cdc25s, and p53 diffuse to and from the sites of damage. Other participants are less mobile and tend to collect and remain at the damage foci. The ATM molecules that anchor the pathway at damage sites must generate a checkpoint pause that is necessary and sufficient. The checkpoint participants that localize at the damage site amplify and help generate the appropriate ATM checkpoint signals. Repair of damage to DNA is carried out in the context of chromatin. Because of the tightly wound chromatin structure damage sites may be difficult to access. Inaccessible damage sites must be opened up for repair proteins to converge; a repair process must be selected and carried out, and histone and chromatin structure must be restored at the end.

Chromatin manipulation and reorganization is part of the DNA damage repair process. Posttranslational modifications of histones and chromatin remodeling both take place and are coordinated. These activities facilitate checkpoint signaling and access to damage sites. They provide protection and resistance against further damage, facilitate restoration of the nucleosome structure at the end of the repair, and resumption of the interrupted cell cycle progression.

Histone modification occurs first followed by ATP-dependent chromatin remodeling. As discussed earlier, the histone variant, H2AX, is phosphorylated by the PIKKs. The phosphorylated histone variants are then acetylated by the NuA4 histone acetyltransferase (HAT). The γ-H2AX acetylation is a signal for chromatin remodeling complexes such as the INO80 to converge to the DNA damage sites. The chromatin remodeling complexes then open the nucleosome to allow for easier access to damage sites by repair proteins. The γ-H2AXs also recruit large-scaffold proteins known as cohesions to the damage sites. These proteins bind sister chromatids bringing them into close proximity and thus assisting homologous recombination repair.

Just as phosphorylation and acetylation of H2AX initiates the convergence of checkpointing and repair proteins to the sites of DNA damage, dephosphorylation of γ-H2AX functions as a termination signal that restores cell cycle progression. At the beginning of DNA damage repair, INO80, MRN, and other complexes manipulate and open the nucleosomes. These operations expose the ssDNA segments requiring repair and rejoining. These sites are identified by the presence of γ-H2AX. The pool of γ-H2AX that is created about all the break-points is depleted as the ssDNA binds donor templates, either by actions of resident protein phosphatase 2A (PP2A) dephosphorylation or, alternatively, by means of Tip60-mediated exchange of variant histones. When these events (and deacetylation) occur, signaling completion of repair, the repair foci dissolve and checkpointing terminates.

Further Reading

Ataian Y and Krebs JE 2006 Five repair pathways in one context: chromatin modification during DNA repair. *Biochem. Cell Biol.* 84: 490–504.

Bekker-Jensen S, et al. 2006 Spatial organization of the mammalian genome surveillance machinery in response to DNA strand breaks. *J. Cell Biol.* 173: 195–206.

Lindahl T 1993 Instability and decay of the primary structure of DNA. *Nature* 362: 709–715.

Cell Cycle Checkpointing

Abraham RT 2001 Cell cycle checkpoint signaling through the ATM and ATR kinases. *Genes Dev.* 15: 2177–2196.
Bartek J and Lukas J 2003 Chk1 and Chk2 kinases in checkpoint control and cancer. *Cancer Cell* 3: 421–429.
Chen Ms, Ryan CE and Piwnica-Worms H 2003 Chk1 kinase negatively regulates mitotic function of Cdc25A phosphatase through 14-3-3 binding. *Mol. Cell. Biol.* 23: 7488–7497.
Hingorani MM and O'Donnell M 2000 A tale of toroids in DNA metabolism. *Nat. Rev. Mol. Cell Biol.* 1: 22–29.
Majka J and Burgers PMJ 2004 The PCNA-RFC families of DNA clamps and clamp loaders. *Prog. Nucl. Acid Res. Mol. Biol.* 78: 227–260.
Perry J and Kleckner N 2003 The ATRs, ATMs and TORs are giant HEAT repeat proteins. *Cell* 112: 151–155.
Shiloh Y 2003 ATM and related protein kinases: safeguarding genome integrity. *Nat. Rev. Cancer* 3: 155–168.
Sørensen CS, et al. 2003 Chk1 regulates the S phase checkpoint by coupling the physiological turnover and ionizing radiation-induced accelerated proteolysis of Cdc24A. *Cancer Cell* 3: 247–258.
Sørensen CS, et al. 2004 ATR, Claspin and the Rad9-Rad1-Hus1 complex regulates Chk1 and Cdc25A in the absence of DNA damage. *Cell Cycle* 3: 941–945.

p53

Joerger AC and Fersht AR 2007 Structure–function-rescue: the diverse nature of common p53 cancer mutants. *Oncogene* 26: 2226–2242.
Kitayner M et al. 2006 Structural basis of DNA recognition by p53 tetramers. *Mol. Cell* 22: 741–753.
Tidow H et al. 2007 Quaternary structures of tumor suppressor p53 and a specific p53-DNA complex. *Proc. Natl. Acad. Sci. USA* 104: 12324–12329.

The MRN Complex

D'Amours D and Jackson Sp 2002 The Mre11 complex: at the crossroads of DNA repair and checkpoint signaling. *Nat. Rev. Mol. Cell Biol.* 3: 317–327.
Lee JH and Paull TT 2005 ATM activation by DNA double-strand breaks through the Mre11-rad50-Nbs2 complex. *Science* 308: 551–554.
van Gent DC, Hoeijmakers JHJ and Kanaar R 2001 Chromosomal stability and the DNA double-stranded break connection. *Nat. Rev. Genet.* 2: 196–206.

Mediators

Ehrenhofer-Murray AE 2004 Chromatin dynamics at DNA replication, transcription and repair. *Eur. J. Biochem.* 271: 2335–2349.
Glover JNM, Williams RS and Lee MS 2004 Interactions between BRCT repeats and phosphoproteins. *Trends Biochem. Sci.* 29: 579–585.

Greenberg RA et al. 2006 Multifactorial contributions to an acute DNA damage response by BRCA1/BARD1-containing complexes. *Genes Dev.* 20: 34–46.

Lee J et al. 2006 Roles of replication fork-interacting and Chk1-activating domains from claspin in a DNA replication checkpoint response. *Mol. Biol. Cell* 16: 5269–5282.

Liu SZ et al. 2006 Caspin operates downstream of TopBP1 to direct ATR signaling towards Chk1 activation. *Mol. Cell. Biol.* 26: 6056–6064.

Lou Z et al. 2006 Mdc1 maintains genomic stability by participating in the amplification of ATM-dependent DNA damage repair. *Mol. Cell* 21: 187–200.

Redon C et al. 2002 Histone H2A variants H2AX and H2AZ. *Curr. Opin. Genet. Dev.* 12: 162–169.

Base-Excision Repair

David SS, O'Shea VL and Kundu S 2007 Base-excision repair off oxidative DNA damage. *Nature* 447: 941–950.

Neeley WL and Essigmann JM 2006 Mechanism of formation, genotoxicity, and mutation of guanine oxidation products. *Chem. Res. Toxicol.* 19: 491–505.

Nucleotide-Excision Repair

Friedberg EC 2001 How nucleotide excision repair protects against cancer. *Nat. Rev. Cancer* 1: 22–33.

Mitchell JR, Hoeijmakers JHJ and Niedernhofer LJ 2003 Divide and conquer: nucleotide excision repair battles cancer and ageing. *Curr. Opin. Cell Biol.* 15: 232–240.

Mismatch Repair

Marti TM, Kunz C and Fleck O 2002 DNA mismatch repair and mutation avoidance pathways. *J. Cell. Physiol.* 191: 28–41.

Stojic L, Brun R and Jiricny 2004 Mismatch repair and DNA damage signalling. *DNA Repair* 3: 1091–1101.

Sliding Clamps

Gradia S et al. 1999 hMSH2–hMSH6 forms a hydrolysis-independent sliding clamp on mismatched DNA. *Mol. Cell* 3: 255–261.

Lamers MH et al. 2000 The crystal structure of DNA mismatch repair protein MutS binding to a G·T mismatch. *Nature* 407: 711–717.

Obmolova G et al. 2000 Crystal structures of mismatch repair protein MutS and its complex with a substrate DNA. *Nature* 407: 703–710.

One-Dimensional DNA Scanning Along DNA

Blainey PC et al. 2006 A base-excision DNA-repair protein finds intrahelical lesion bases by fast scanning in contact with DNA. *Proc. Natl. Acad. Sci. USA* 103: 5752–5757.
Gorman J et al. 2007 Dynamic basis for one-dimensional DNA scanning by the mismatch repair complex Msh2–Msh6. *Mol. Cell* 28: 359–370.
Granéli A et al. 2006 Long-distance lateral diffusion of human Rad51 on double-stranded DNA. *Proc. Natl. Acad. Sci. USA* 103: 1221–1226.
Harada Y et al. 1999 Single-molecule imaging of RNA polymerase-DNA interactions in real time. *Biophys. J.* 76: 709–715.
Tafvizi A et al. 2008 Tumor suppressor p53 slides on DNA with low friction and high stability. *Biophys. J.* 95: L01–L03.

DSB Repair

Bekker-Jensen S et al. 2006 Spatial organization of the mammalian genome surveillance machinery in response to DNA strand breaks. *J. Cell. Biol.* 173: 195–206.
Rouse J and Jackson SP 2002 Interfaces between the detection, signaling, and repair of DNA damage. *Science* 297: 547–551.
Valerie K and Povirk LF 2003 Regulation and mechanisms of mammalian double-strand break repair. *Oncogene* 22: 5792–5812.

Histones and Completing the Repair

Chowdhury D et al. 2005 γ-H2AX dephosphorylation by protein phosphatase 2A facilitates DNA double-strand break repair. *Mol. Cell* 20: 801–809.
Fernandez-Capetillo O et al. 2004 H2AX: the histone guardian of the genome. *DNA Repair* 3: 959–967.
Keogh MC et al. 2006 A phosphatase complex that dephosphorylates γH2AX regulates DNA damage checkpoint recovery. *Nature* 439: 497–501.
Murr R et al. 2006 Histone acetylation by Trrap-Tip60 modulates loading of repair proteins and repair of double-strand breaks. *Nat. Cell Biol.* 8: 91–99.
Tsukuda T et al. 2005 Chromatin remodeling at a DNA double-strand break site in *Saccharomyces cerevisiae*. *Nature* 438: 379–383.
van Attikum H and Gasser SM 2005 The histone code at DNA breaks: a guide to repair? *Nat. Rev. Mol. Cell Biol.* 6: 757–765.

Chapter 11
Apoptosis and Senescence

The DNA damage checkpointing and repair responses discussed in the last chapter act as the first barrier to unrestrained cellular growth and proliferation. This barrier is raised early whenever oncogenic stimuli, that is, inappropriate cellular growth conditions, arise. Mechanistically, the unscheduled growth stimuli give rise to DNA damage and stalled replication forks. These stresses activate checkpoint pathways that halt the cell cycle to allow for time to repair the damage and restart DNA replication. If the damage is not repaired in a timely fashion, the same checkpoint-signaling elements, for example, p53 and Chk2, route the cells away from growth toward apoptosis and senescence, the later-acting barriers to cancer.

The term apoptosis was coined by Kerr, Wyllie, and Currie in 1972 to denote the process whereby cells that are damaged or no longer needed are destroyed in a systematic way through cell shrinkage, chromatin condensation, DNA fragmentation, and membrane blebbing. This systematic process of deconstruction prevented damage to surrounding cells from leakage of the cellular contents. Apoptosis is heavily utilized during development to sculpt tissues and organs and remove transient cell populations. It is used throughout early and adult life by the immune system to maintain optimally responsive B and T cell populations, and to kill virally infected and cancerous cells. Anticancer cytotoxic drugs inhibit cellular proliferation by inducing apoptosis. They accomplish this by damaging DNA and membrane lipids, and causing cellular stress and loss of cellular homeostasis. The goal in these therapeutic actions is to activate the cell cycle checkpoint and DNA damage signaling pathways, creating damage that is severe enough to trigger apoptosis thereby removing the cancerous cells.

Apoptosis is not the only form of regulated cell death. Another form is necrosis. Necrosis, derived from the Greek term for corpse, differs from apoptosis in two essential ways. First, it is characterized by membrane rupture, which does not occur during apoptosis. Second, it occurs under conditions of impaired cellular production of ATP. Necrosis has been long thought to be unregulated, but that view has changed recently. The process can occur in a regulated fashion and, in addition, the release of cellular contents can serve a

M. Beckerman, *Cellular Signaling in Health and Disease*,
Biological and Medical Physics, Biomedical Engineering,
DOI 10.1007/978-0-387-98173-4_11, © Springer Science+Business Media, LLC 2009

useful purpose, namely, to alert the immune system of a danger and help launch an immune response.

Cells that enter a state of permanent growth arrest where they no longer divide and proliferate are senescent. This state of cell cycle arrest can arise in several ways. It can be brought about as a result of telomere shortening, resulting in replicative senescence. Alternatively it can arise through cumulative cellular stresses arising from damage to DNA by radiation or reactive oxygen species. Most interesting from the point of view of disease and health is that it can be produced as a protective response to oncogenic influences. This form of cellular senescence is referred to as oncogene-induced senescence (OIS).

In this chapter, apoptosis and senescence will be examined in their guise as anticancer barriers. Apoptosis will be examined first followed by senescence.

11.1 Pathways to Apoptosis – Extrinsic and Intrinsic

It is customary to think of apoptosis in terms of its two major signaling routes – the extrinsic pathway and the intrinsic pathway. The extrinsic pathway handles death signals originating from nearby cells. Death ligands bind death receptors, both belonging to the death-inducing branch of the tumor necrosis factor alpha (TNF-α) family, introduced and discussed in Chapter 8. Prominent members of this subfamily are the Fas ligand and receptor. Ligand binding by members of this subfamily stimulates assembly of a death-inducing signaling complex, or DISC, at and just below the plasma membrane. This assembly step triggers the recruitment and activation of initiator Caspases 8 or 10, which lead to activation of effector Caspase 3, a prime mediator of apoptosis. This series of steps is sketched in Fig. 11.1(a).

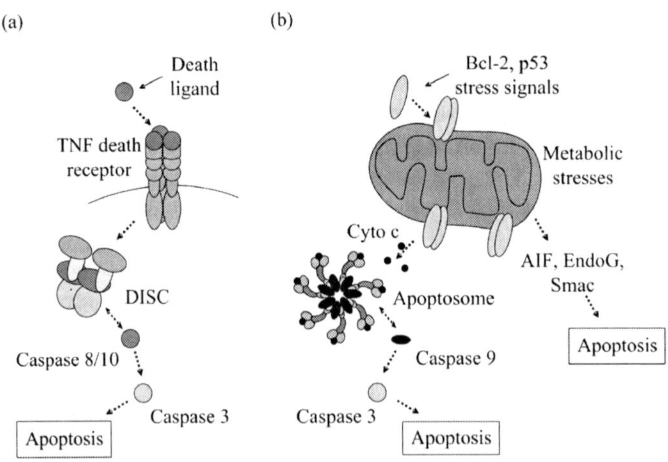

Fig. 11.1 Pathways to apoptosis. (**a**) The extrinsic pathway in which external death signals are transduced by death receptors and the DISC; and (**b**) the intrinsic pathway in which internal stress signals are transduced by mitochondria and the apoptosome

The intrinsic pathway is activated by stresses generated within the cell. Apoptosis-triggering stress signals can be generated by damage to DNA and conveyed via p53, by stresses to the metabolic machinery within the mitochondria, and by endoplasmic reticulum stresses. The internal, or intrinsic, pathway is centered about the mitochondria and utilizes Bcl2 signaling proteins as central agents. At its most basic level the pathway operates, as follows. Cellular stress signaling and actions by p53 and the Bcl2 proteins give rise to an increase in the permeability of the outer mitochondrial membrane allowing for release of apoptotic intermediaries, most notably cytochrome c, stored in the spaces between the inner and outer mitochondrial membranes. This release leads to the assembly of an apoptosome, a signaling complex that serves as the intrinsic pathway counterpart to the DISC. Initiator Caspase 9 is activated along with Caspase 3, and apoptosis ensues as depicted in Fig. 11.1(b).

Caspases are cysteine aspartate-specific proteases; that is, they cleave peptide bonds after Asp residues and possess a highly conserved cysteine residue in their catalytic site. Once activated caspases target repair, splicing, and replication components, cut up membranes and cytoskeleton regulators, and break up cellular DNA. They also stimulate the expression of markers on the cell surface that tag the cell for orderly destruction and engulfment by neighboring cells. Like many other proteolytic enzymes, caspases are synthesized as inactive zymogens. They contains a prodomain in the amino terminal region that regulates the pro-enzyme, followed by a large domain, approximately 20 kDa in size, and then a small domain, about 10 kDa in size. The pro-enzyme is activated by proteolytic cleavage at two Asp-x sites, one situated at the end of the prodomain and the other separating the large and small domains. These operations are accompanied by formation of tetramers and induction of conformational changes that render the assembled and rearranged unit catalytically active.

Two classes of caspases participate in the extrinsic and intrinsic pathways. Caspases 8, 9, and 10 are known as initiator caspases. These caspases respond to pro-apoptotic signals and once activated respond by activating effector Caspase 3. The latter caspase (others in the effector category are Caspases 6 and 7) carry out the cleavage operations on critical cellular components as described above.

The electron carrier cytochrome c is a small protein consisting of 104 amino acid residues plus a covalently attached heme group. Under normal conditions, it functions as an electron carrier shuttling electrons between Complex III and Complex IV of the electron transport chain. Under apoptotic conditions, it carries out a second function that of a signaling intermediary that triggers assembly of the apoptosome as indicated in Fig. 11.1(b).

Once it is released from the mitochondria, cytochrome c interacts with Apaf-1, an adapter protein located in the cytoplasm just outside the mitochondria. These interactions stimulate the assembly of the apoptosome consisting of cytochrome c-bound Apaf-1 molecules. The apoptosome is a seven-fold symmetric wheel-shaped platform that activates Pro-caspase 9 when these zymogens are incorporated into it. When fully saturated seven Pro-caspase

9 molecules are bound to the central Apaf-1 domains and seven cytochrome c molecules are bound to the outermost domains. The formation of an apoptosome containing activated Caspase-9 molecules initiates the apoptosis process by interacting with and activating effector Caspase 3.

11.2 Bcl2 Proteins Mediate the Apoptotic Balance

Members of the Bcl2 family of proteins are key apoptotic signal transducers. The Bcl2 proteins act as stress sensors and as regulators of outer mitochondrial membrane permeability. There are three different categories of Bcl2 proteins. One group is made up of Bcl2 proteins that inhibit apoptosis while the other two groups contain proteins that promote apoptosis. The three groups of Bcl2 proteins are listed in Table 11.1.

Table 11.1 Bcl2 family of pro- and antiapoptotic proteins

Bcl2 subfamily	Bax subfamily	BH3-only subfamily
A1	Bak	Bad
Bcl2	Bax	Bid
Bcl-xL	Bok	Bik/Blk/Nbk
Bcl-w		Bim
Mcl-1		Bmf
		Hrk
		Mule
		Noxa
		Puma

Members of the Bcl2 antiapoptotic subfamily fold in a compact globular domain formed by eight alpha helices. In this fold, the core helix, $\alpha 5$, is surrounded by the seven other helices and connecting loops. These helices form a set of four domains, known as BH1, BH2, BH3, and BH4. Members of this subfamily function as negative regulators of apoptosis. Proteins belonging to the Bax family have a similar arrangement of α helices and multiple BH domains. These proteins are apoptosis effectors. When activated they form pores in the OMM thereby increasing membrane permeability to cytochrome c and other apoptotic agents. In the absence of internal stresses such as irreparable DNA damage, the inhibitory Bcl2 proteins bind and sequester the pro-apoptotic Bax proteins in the OMM. As a result, the pore-forming proteins are maintained in an inactive state, and the OMM remains relatively impermeable.

The third family is characterized by the presence of a single BH3 domain and these proteins are termed BH3-only proteins. BH3-only proteins interact with proteins conveying the upstream stress signals. Once activated by these signals they translocate to the mitochondria where they function as ligands that interact with the mitochondrial Bcl2 apoptotic inhibitors. This interaction forces the antiapoptotic Bcl2s to release the pore forming bax proteins. The aforementioned interaction is made possible by the presence of a hydrophobic groove formed by helices $\alpha 2$–$\alpha 5$ of the full-length Bcl2 proteins that provides binding site for the α helical ligand domain of the BH3-only proteins such as Bim and Bad. Finally, most, but not all, of the multiple BH domain proteins contain a C-terminal hydrophobic tail that mediates membrane attachment. The known interactions between Bcl2s and BH3-only proteins are summarized in Table 11.2.

Table 11.2 Properties of members of the BH3 subfamily of Bcl2 proteins

BH3 protein	Binding partners	Sentinel role	Activation
Bad	Bcl2, Bcl-xL Bcl-w	Growth factor deprivation	Dephosphorylation
Bid (tBid)	all	Death signals	Caspase 8 cleavage
Bim	all	Growth factor deprivation; ER stress	Dephosphorylation; transcription
Noxa	A1, Mcl1	UV-induced DNA damage	p53-mediated transcription
Puma	all	γ-radiation-induced DNA damage	p53-mediated transcription

The BH3-only proteins function as nonredundant cellular stress sensors. Some are upregulated in response to DNA damage and ER stresses while others are positioned in strategic locations in the cytosol. Once they are activated, they translocate to the mitochondrial OMM where they de-repress pro-apoptotic Bcl2 proteins. The specific sensor functions of the BH3-only proteins are listed in Table 11.2. Of these Bim and PUMA are perhaps the most essential. Some BH3-only ligands bind only certain specific subsets of the antiapoptotic Bcl2 proteins while others are less specific in selection of a binding partner.

The balance between pro- and antiapoptotic Bcl2s is controlled by the combined actions of upstream stress signals. The BH3-only proteins that respond to stress signals are situated at strategic locations in the cell. In unstressed cells, Bad responds to growth/survival signals relayed by Akt. When phosphorylated by that protein, it is bound to 14-3-3 proteins and remains inactive and immobilized in the cytosol. Withdrawal of Akt survival signals results in translocation of these molecules to the mitochondria where they promote apoptosis.

The BH3-only protein Bid is sited at the DISC. It functions as a sensor and as part of the circuitry that integrates externally and internally generated death signals. Caspase 8 cleaves the 22 kDa Bid sensor protein to create the 15 kDa tBid truncated form. Once formed, the tBids translocate to the mitochondria. Bim is sequestered at microtubule-associated myosin V motors where it awaits activation by cytokine and other stress signals. Bmf is also immobilized at myosin V motors where it responds to loss of cell attachment (anoikis) signals. Two other BH3-only proteins, Bik and Blk, function in the endoplasmic reticulum as sensors of cellular stress. The remaining BH3-only proteins are regulated at the transcriptional level. Noxa and Puma are transcribed in a p53-dependent manner and may be regarded as DNA damage sensors while Hrk and Bim are upregulated in response to growth-factor deprivation and cytokine withdrawal.

When Bcl2 proteins are not expressed at the proper levels, the option of triggering apoptosis to remove cells that are growing out of control is no longer possible. Aberrant Bcl2 genes are present in 90% of follicular lymphomas or B-cell lymphomas for which the Bcl2 protein is named. In these cells, a translocation has occurred. The gene encoding Bcl2 located on chromosome 18 has been translocated and fused with an immunoglobulin gene located on chromosome 14. The result of this translocation, designated as t(14:18), is that the Bcl2 gene is positioned next to the enhancer for the antibody gene leading to its continual overexpression. The decision between proliferation and apoptosis is shifted toward the former, and the B-cells do not die off as they should when they age. Worse, the B-cells are resistant to radiation therapy and chemotherapy since these forms of treatment work in large measure by stimulating apoptosis.

11.3 Sequestration and Release of Cytochrome c

Unlike most other components of the electron transport chain, cytochrome c is small, water soluble, and mobile. In healthy, nonapoptotic cells, cytochrome c localizes to the outer face of the inner mitochondrial membrane. It is immobilized at these locations through binding to the anionic phospholipid cardiolipin and is sequestered within the cristae. The inner mitochondrial membrane is highly invaginated and although the materials located within the cristae appear contiguous with the matrix, they are, in fact effectively isolated within these compartments, with small bottlenecks and certain proteins serving as diffusion barriers to migration of the contents of the cristae into the matrix. Contributions to apoptosis occur from changes in cristae structure allowing for release of cytochrome c from this compartment. These structural changes are referred to as cristae remodeling.

The BH3-only proteins aid in the release of cytochrome c from the mitochondria in at least two distinct ways. As discussed above, they stimulate the pore-forming activities of Bax, Bak, and other pore-forming Bcl2s. And independently of this activity they promote the freeing of the cytochrome c molecules from cardiolipin and the remodeling of the mitochondrial cristae. This last activity may be thought of as an amplification step. While some cytochrome c can be released rapidly in response to the arrival of pro-apoptotic signals, most of the cytochrome c molecules are sequestered in the mitochondrial cristae. Remodeling of the cristae is needed to free up the bulk of the cytochrome c and produce a more vigorous, longer lasting apoptotic response. These responses are mediated by the integral membrane protease Presenilin-associated rhomboid-like (PARL) acting on the inner membrane protein Optic atropy 1 (OPA1). At least one BH3-only protein, Bid, triggers remodeling by disrupting OPA1 oligomers required to maintain cristae-mediated sequestration.

An important additional mechanism for the freeing of cytochrome c from its binding to cardiolipin is oxidation. The binding and release of cytochrome c is regulated by the cardiolipin oxidation state. Cardiolipin is an unsaturated phospholipid, and it is highly susceptible to lipid peroxidation brought on by, for example, reactive oxygen species generated by the electron transport chain. When oxidized cardiolipin is no longer able to bind cytochrome c and the latter are released into the intramembrane spaces. The cytochrome c molecules are then available for passage out of the mitochondria into the cytosol in response to apoptosis-activating signals and permeabilization of the OMM by Bcl2 proteins.

11.4 Damage-Induced Apoptosis via p53 Transcription and Mitochondrial Actions

In its function as the director of the anticancer barriers, p53 relays DNA damage checkpoint signals and when the damage is severe and irreparable directs the cell toward apoptosis or senescence. The decision between these different responses is based on p53 expression levels, and target promoter properties and cofactors. In particular, high-affinity sites are present in promoters of genes involved in DNA damage/growth responses whereas low-affinity sites are found in genes associated with apoptosis. This mechanism couples sustained DNA damage leading to increased levels of stabilized p53 to apoptotic outcomes. Coactivators present at p53 responsive promoter sites have a role in the decision, some favoring cell cycle arrest and others promoting apoptosis.

p53 acts in several ways to shift the balance between life and death toward the latter. One way is through its actions as a transcription factor. In its role as a transcription factor, p53 activates transcription of a number of pro-apoptotic Bcl2 family members including pore-forming Bax, and BH3-only proteins NOXA and PUMA. The latter BH3-only proteins bind to antiapoptotic Bcl2 proteins such as Bcl-xL thereby promoting the dissociation of the pore-forming Bcl2s from their inhibitors and enabling them to oligomerize and form pores. The pores enable cytochrome c to pass through the OMM into the cytosol.

The p53 protein not only acts in the nucleus but also translocates to mitochondria. It interacts with Bcl-xL and Bax disrupting the inhibition of Bax by Bcl-XL and enabling the former to oligomerize and form pores in the OMM. This p53 mitochondrial role is not limited to Bcl-xL and Bax. It also extends to Mcl1 and Bak.

11.5 Cells That Are Healthy Do Not Have an Unlimited Capacity to Divide

After a certain number of cell divisions, called the Hayflick limit, cells no longer grow and divide but rather permanently arrest in G_0 and become senescent. The concept of senescence goes back to 1881 when the German biologist August Weismann posited that cells are not immortal, but instead tissues wear out when their cells lose the ability to replicate. This limitation was rediscovered by Leonard Hayflick in 1961–1965; cells that avoid the Hayflick limit are said to be immortal.

The Hayflick limit is intimately connected to telomere shortening. Telomeres, the capping structures that protect the ends of chromosomes, are shortened by about 100 base pairs every cell division. When the capping structure is degraded sufficiently the cell's DNA repair machinery is able to sense the ends of the DNA molecules and interprets the ends as double-strand breaks. If the cell keeps dividing several negative outcomes become possible because of imperfect replication and repair. These include degradation, chromosome recombination leading to loss of genetic information, aberrant rearrangements of chromosomes, and genomic instability. To avoid such dangerous situations, cells cease to divide after several kilo base pairs of telomere are lost. The cells have reached Hayflick limit, and instead of replicating further enter a nondividing stage called replicative senescence.

Cells become senescent not only through telomere shortening but also as a result of cumulative DNA damage brought on by cellular stresses such as a buildup in reactive oxygen species. In addition, cells can go into a senescent state when subjected to inappropriate growth signals. Progress has been made in understanding just how cells distinguish between appropriate and inappropriate growth and proliferation. An emerging and unifying view is that DNA damage responses underlie oncogene-induced senescence just as they do for telomere shortening and DNA damage-induced senescence.

11.6 Telomere Structure and Capping Proteins

The ends of chromosomes are capped by telomeres. These structures were first discovered by Barbara McClintock in 1941. In her studies of corn (*Zea mays*) genetics, she found that these capping structures protected the ends of the DNA molecules. As is now known they protect the ends of the DNA molecule from enzymatic attack and fusions, and from recognition as breaks by the resident DNA repair machinery. Telomeres consist of a series of TTAGGG repeats and associated proteins. They are terminated at their very end by a lasso-like structure (Fig. 11.2) possessing a protrusion or overhang formed by its 3′ strand G-rich in content. Two of the proteins that associated with the telomeres, namely, telomeric repeat binding factors 1 and 2 (TRF1 and TRF2), bind

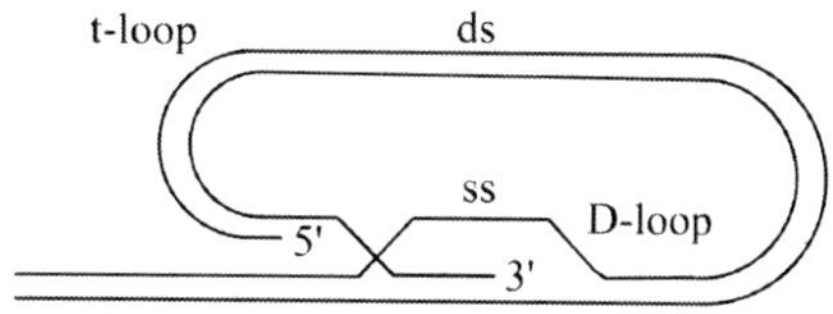

Fig. 11.2 Telomere termination (t) and displacement (D) loops

double-stranded telomeric DNA. Another, the protection of telomeres 1 (POT1) protein binds single-stranded telomeric DNA. There are three other telomeric-binding proteins besides TRF1, TRF2, and POT1. The six capping proteins have been listed in Table 11.3 where it can be noted that the others serve as bridging factors between the TRFs and POT1s, and as protein–protein binding adaptors.

Table 11.3 Telomeric capping proteins and their functions

Telomere capping proteins	Function
TRF1	Binds dsDNA, bends, loops, pairs DNA
TRF2	Binds dsDNA, mediates t-loop formation
TIN2	Tethers TPP1/POT1 to TRF1 and TRF2
hRAP1	Interacts with TRF2
TPP1	Interacts with POT1
POT1	Binds ssDNA

DNA repair proteins are present in the vicinity of the telomeres, and under normal circumstances are shielded from the telomeres by the capping proteins. This DNA repair machinery includes members of the Mre11/Rad50/NBS1 double-strand break repair (MRN) complex, and the Ku70/Ku86 (Ku80) heterodimer and DNA-PK$_{cs}$ involved in NHEJ. Members of these complexes convey pro-senescence signals to the p53 circuitry when telomeres become excessively shortened and exposed, and consequently are sensed by this machinery.

11.7 Cancer Cells Increase Their Production of Telomerase, an Enzyme That Immortalizes the Cells

Cancer cells avoid senescence by increasing the production of the telomerase chromosome-capping enzyme. This activity greatly expands the number of cell divisions possible, effectively immortalizing the cells. The cells avoid passing into senescence, and instead continue to divide and progress toward more lethal states. Mutations in the p53 and pRb that disable these proteins can contribute to the immortalization by disrupting the conveyance of pro-senescence signals.

Telomeres prevent chromosomal rearrangements and fusions. They shorten at each cell division due to incomplete replication at the 3′ end by the DNA polymerases. Under normal situations, one or a few mutations will be sufficient to activate the cell's senescence barrier. Mortality stage-1 (M1), or senescence, is triggered by stresses associated with the mutations and accompanying excessive rate of cell division. These stresses activate the p53–pRb checkpoint signals that lead to arrest of the cell cycle in G1. If the p53–pRb elements are disabled through mutations then the checkpoint and associated arrest is bypassed and the cells can continue growing and dividing and proceed to mortality stage-2 (M2), or crisis, where the telomeres have reached a critical shortening. During further cell divisions, chromosomal rearrangements and fusion may occur leading to genomic instability. Crisis can be bypassed in one of two ways, either through telomerase activation or by alternative lengthening of telomeres (ALT) activation, and resulting in immortalized cells.

Immortalization is a hallmark of cancer. It is established through actions by c-Myc, a growth signal-activated transcription factor that is overexpressed in a high percentage of cancers. Telomeres are maintained by the (human) telomerase reverse transcriptase (hTERT), which catalyzes the addition of multiple TTAGGG repeats at the ends of chromosomes. The hTERT catalytic subunit is assisted by other subunits such as of the human telomerase RNA, or hTR, which contains the template for adding telomeric repeats. However, hTERT is the rate-limiting enzyme in the process. The c-Myc protein upregulated expression of hTERT thereby maintaining the telomeres during repeated cycles of growth and division.

11.8 Regulation of Replicative Senescence by p53 and pRb

The core circuit-mediating senescence in the cell is depicted in Fig. 11.3. Two key tumor suppressors discussed in previous chapters – p53 and pRb – lie at the heart of the tumor barrier network. Signals sent through these proteins ensure

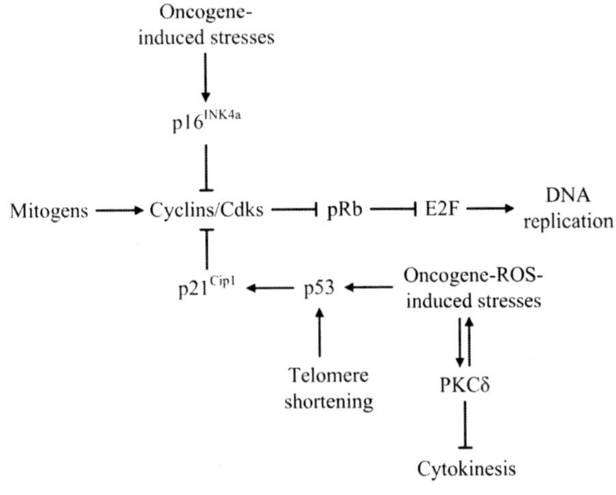

Fig. 11.3 Cancer barrier network responsible for senescence program in cells

that the cell ceases to divide. In the absence of cellular stresses or telomere shortening, mitogenic signals stimulate the activity of G1 phase and G1/S transition cyclin/Cdk complexes. As discussed earlier, once activated the cyclin-dependent kinases hyperphosphorylate pRb. These posttranslational modifications reduce pRb's ability to bind and inhibit the E2F transcription factors. The E2Fs are then able to stimulate the transcription of genes required for DNA replication during S phase; the cell can then pass through the remainder of the cell cycle and divide.

Two other key elements present in the cancer barrier circuit, p16Ink4a and p21Cip1, negatively regulate the cyclin/Cdks. Recall from Chapter 9 that the former is a member of the inhibitors of Cdk4, or Ink4, family, and the latter is a member of the cyclin dependent kinase inhibitor, or Cip/Kip, family. The p21 protein is upregulated by p53 in response to signals sent to p53 in response to excessive and potentially dangerous telomere shortening. As discussed in Chapter 10, ATM is activated in response to double-strand breaks and appears to be the key signaling link between shortened telomeres and p53.

11.9 DNA Damage and Oncogene-Induced Senescence

Excessive and uncontrolled proliferative signals are hallmarks of cancer. Oncogenic Ras is a prototypic example of mutations that produce signals of this type. A seminal finding in study of these signals is that oncogenic Ras by itself does not lead to cancer but instead induces just the opposite. Healthy cells into which oncogenic Ras is introduced become senescent rather than cancerous. This occurrence is not only observed in the laboratory but it is also seen in premalignant and benign tumors. This remarkable observation first made in 1997 led to the recognition that cells possess intrinsic barriers to cancer. The requirement that emerges from these studies is that mutations in oncogenes such as Ras need to be accompanied by additional mutations, most notably in tumor suppressors such as p53 or pRb. These other mutations enable cells to overcome the three protective anticancer barriers – DNA damage responses, apoptosis, and senescence. The lowering of these protective barriers is then a key property of the disease state. The process whereby cells halt this progression was originally referred to as premature senescence in recognition that it occurs before cells reach the Hayflick limit. It is more commonly referred to today as oncogene-induced senescence (OIS).

The Ink4a/ARF locus situated in a small region of human chromosome 9p21 is deleted in a wide variety of tumors. It encodes two prominent tumor suppressors – p16Ink4a and the alternative reading frame (p14ARF) protein. The former potentiates actions by pRb and the latter has a similar stimulatory effect upon p53. The core circuitry connecting these proteins to each other is depicted in Fig. 11.4. The p16Ink4a protein is a cell cycle inhibitor and prevents activation of Cyclin E/Cdk2 by targeting the CyclinD/Cdk4 complexes and freeing

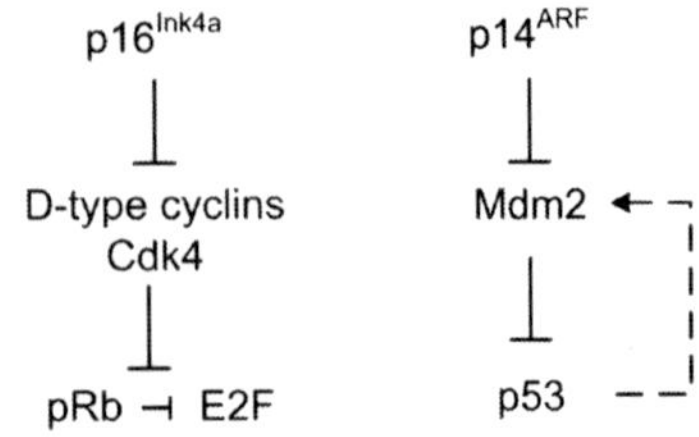

Fig. 11.4 p53 and pRb core circuitry

p21Cip1 and p27Kip1. As a result the cell cycle progression is arrested; pRb is not hyperphosphorylated and is able to bind and prevent the growth-promoting activities of the E2F transcription factors.

The ARF pathway operates through suppression of Mdm2, an E3 ubiquitin ligase that maintains p53 expression at a basal level in nonstressed healthy cells. The p53 protein upregulates Mdm2 gene expression leading to Mdm2-mediated p53 ubiqitination and degradation. This feedback dependence ensures that p53 levels remain fairly constant under nonstressed conditions. At its low basal levels, the protein is able to halt the cell cycle to allow for repair of DNA damage and also fine tunes cellular metabolism to prevent excessive ROS buildup and DNA damage.

ARF and p16Ink4a are upregulated in response to excessive growth activity derived from, for example, overexpressed Ras. Signals relayed from Ras to Raf and through the ERK mitogen activated protein kinase (MAPK) cascade (a signaling route that will be discussed in detail in Chapter 13) activates both of these tumor suppressors. First, these signals lead to upregulation of hDMP1 (human cyclin D binding myb-like protein 1), a transcription factor which, in turn, binds to the ARF promoter and stimulates its transcription. Second, the p16Ink4a protein is believed to be the primary agent in humans mediating oncogene-induced senescence. Overexpressed Ras and signaling by the MAPK cascade lead to activation of the Ets1,2 transcription factors and increased expression of p16Ink4a. When ARF levels are increased, it represses the ability of Mdm2 to limit the p53 lifetime, and the p53 levels are able to rise. As a result the Cdks do not phosphorylate and inhibit pRb; it remains in a hypophosphorylated state, and the growth-promoting activities of the E2Fs are repressed.

11.10 A Model or Two of Oncogene-Induced Stress

The key question in studying oncogen-induced senescence is: how does a cell tell appropriate from inappropriate or excessive growth and proliferation signals? The inappropriate/excessive signals are referred to as producing oncogenic-induced stresses so the questions become just what are these "stresses" that trigger the raising of the apoptotic/necrotic and senescence barriers. Evidence has been found supporting a model in which hyperproliferation generates stresses in DNA replication and these stresses result in DNA damage leading to a DNA damage response as is the case for telomere shortening. Examples of stresses in DNA replication are impaired fork progression, increased DNA replication origin firing, and re-replication.

One appealing answer to the question of the nature of the "oncogenic stress" signals is encapsulated by the following model. In this model, there are two distinct signaling events, and both are required to trigger apoptosis or senescence. The first is sustained mitogenic signaling. DNA repair barriers are erected fairly rapidly in response to DNA damage. The senescence barrier, in contrast, takes much longer to erect – time courses are typically several days and once senescence is established it is usually not reversible. This part of the model is fairly well established whereas the character of the second signal is far more uncertain. The second signal is a stressor that signals in a clear and unambiguous way that there is a serious problem. Several reasonably well-supported candidate stressor-signals have been advanced. These include signals emanating from the sites of DNA damage and, similarly, from shortened telomeres. They also include signals conveyed by reactive oxygen species (ROS).

The generation of ROS with a direct role in senescence is particularly interesting. Under normal cellular conditions, mitogenic signals and Cyclin/cdks are active; the retinoblastoma protein is repressed, and E2Fs stimulate transcription of genes required for growth and proliferation. Among the ensemble of genes upregulated by the E2Fs are genes encoding redox enzymes and this activity further supports normal growth by inhibiting the buildup of DNA damaging ROS. Under abnormal conditions such as excessive Ras-stimulated growth, p16Ink4a is upregulated. This action abrogates the ROS-limiting actions of the E2Fs. Rather, pRb is activated, the E2Fs are not, and the excessive growth activity is now accompanied by a rapid increase in ROS production. This transient effect is converted into a sustained one by protein kinase Cδ (PKCδ), which through a positive feedback loop sustains both the ROS signaling and activity of PKCδ. The latter enforces an irreversible block on the cell cycle progression at G2-M that maintains the senescent state through inhibition of cytokinesis (Fig. 11.3).

It was noted earlier that the high-affinity binding is associated with sites that encode proteins that mediate cell cycle arrest, while low-affinity binding occurs at sites encoding proteins involved in apoptosis. As a consequence the expression levels of p53 can, in part, determine which program will be executed. One facet of this type of encoding is that p53 can direct a program that protects against ROS under normal conditions, yet switch to actions that generate ROS when the cells have sustained irreparable damage, as reflected by high p53 expression levels, and need to undergo apoptosis. The expression levels associated with senescence may lie somewhere intermediate between these low- and high-ROS regimes corresponding to a situation where the cell may still perform useful functions but should not undergo replication.

Another potential "second signal" is one generated through an inflammatory response to DNA damage-induced stresses. Genes upregulated by DNA damage are not restricted to p53 and those at the Ink4a locus. They also include genes encoding secreted factors that act in the microenvironment and modify components of the ECM and signal to cells in a paracrine and autocrine manner. Prominent among these are the chemokine receptor, CXCR2, and inflammatory cytokines IL-1, IL-6, IL-8, and Groα. These collectively trigger amplification processes that reinforce the initial senescence signal conveyed by p53. As shown in Fig. 11.5, the cytokines and their receptors activate pathways that operate through the NF-κB and C/EBPβ transcription factors to further upregulate the aforementioned inflammatory mediators, thereby generating a robust inflammatory response that helps in the clearance of potential tumor cells and debris, and maintains the senescent state.

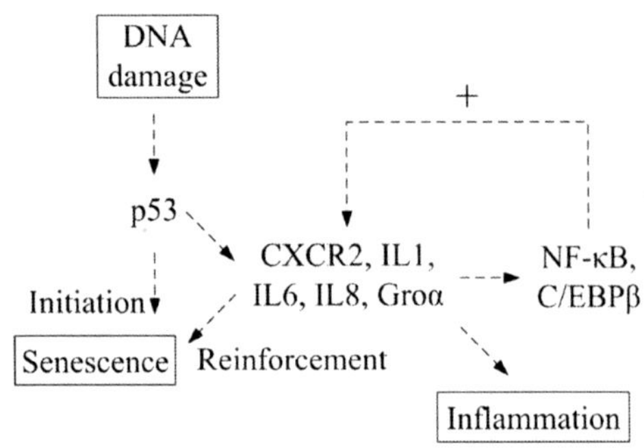

Fig. 11.5 A model of initiation and reinforcement of senescence containing a positive feedback cytokine loop

11.11 p53 Undergoes Posttranslational Modifications Including Phosphorylation, Acetylation, and Ubiquitination at Multiple Sites

The p53 protein receives DNA damage and genotoxic stress signals from a number of upstream signaling elements. Signals conveyed by upstream kinases impinge on serine and threonine residues situated in the N-terminal transactivation domain and also in the C-terminal domain. The N-terminal region is phosphorylated at multiple sites. As is indicated in Fig. 11.6, the PIKK-signaling proteins ATM, ATR, and DNA-PKcs target S6, S9, and S15, Chk2 phosphorylates T18 and S20, and other kinases phosphorylate still other sites in this region.

The N-terminal region serves as the interface between p53 and Mdm2. Phosphorylation of residues in this region, most notably, serine 15, threonine 18, and serine 20, by ATM, ATR, and Chk2 weakens p53's binding and inhibition by Mdm2 leading to stabilization and increased transcriptional activity by p53. Limitations placed on p53 half-life by Mdm2 are further reduced by phosphorylation of Mdm2 by these upstream kinases as shown in Fig. 11.6.

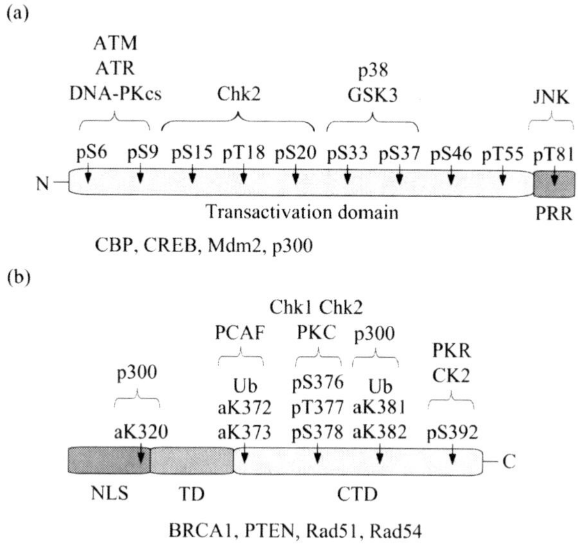

Fig. 11.6 p53 posttranslational modifications, proteins responsible for phosphorylation and acetylation, and binding partners. (**a**) N-terminal multisite phosphorylation region, which supplies sites for Mdm2 and transcriptional cofactor binding (a few of which are shown) and (**b**) C-terminal negative regulatory region which regulates p53 stability, affinity of p53 for DNA, and provides additional sites for binding partners (some of which are shown)

The p53 protein is an example of a protein that undergoes multisite phosphorylation. This class of posttranslational modification was encountered in Chapter 9 with respect to proteins central to the cell cycle progression such as Cdc25 and Cyclin E. One of the main purposes served by this signaling strategy is a dynamic one – it is a mechanism for the precise timing of a series of events so that each one occurs in the correct place at the right time. Other transcription factors besides p53 utilize this process for signal integration, for creating binding sites for cofactors, and for modulating DNA-binding affinity, and strength and duration of transcription. The C-terminal region is thought to play a central role in the latter. This region possesses a number of acetylation and ubiquitination sites thought to influence DNA binding and stability.

11.12 Heterochromatin Formation Provides a Route to Oncogene-Induced Senescence

Entry into the senescent state is facilitated by at least two distinct mechanisms. The first is the direct inhibition of E2Fs by pRb mediated by p53 and p21 as discussed in the preceding sections. There is a second route used by cells to impede the development of cancer through entry into a senescent state. This additional mechanism is gene silencing brought on through alterations in chromatin structure. Recall that heterochromatin is a compacted form of chromatin in which the genes are relatively inaccessible to transcription factors and cofactors. Oncogene-induced senescence has one of its prime characteristics altered chromatin structure typified by the presence of numerous DNA-dense structures called senescence-associated heterochromatin foci (SAHF). Four of the molecular components of the SAHFs are listed in Table 11.4.

Table 11.4 Major components of senescence-associated heterochromatin foci (SAHF)

SAHF component	Function
ASF1: Antisilencing function 1	Histone chaperone
HP1: Heterochromatin protein 1	Nonhistone chromatin-associated transcription silencing proteins
HMGA proteins: High-mobility group A proteins	Nonhistone chromatin-associated hub proteins
HIRA: Histone repressor A	Histone chaperone

Formation of SAHFs proceeds through formation of nuclear structures known as promyelocytic leukemia bodies (PMLs) and involves a variant form of histone H2A known as macro H2A. Promyelocytic leukemia (PML) nuclear bodies are doughnut-shaped macromolecular structures. Their numbers vary throughout the cell cycle, but cells typically contain 10–30 of these structures. PMLs have essential roles in DNA damage responses and in the associated apoptotic and senescent barriers to oncogenic transformation. Central to these roles is the involvement of PMLs in p53 and pRb signaling. In the case of p53, the PNl serves as a signaling node that mediates the colocalization of CBP/p300 and p53. This results in the acetylation of p53, a step needed for p53 transcriptional activity at Bax and p21 promoters, the former promoting apoptosis and the latter supporting senescence. Formation of PMLs is promoted by pRb leading to the assembly of corepressor complexes containing, for example, Ski/Sno and N-CoR/SMRT as depicted in Fig. 11.7; these complexes greatly enhance the transcription-repression activities of pRb and the Myc antagonist Mad.

High-mobility group A (HMGA) proteins are nonhistone chromatin-associated proteins. These proteins bind the minor groove of AT-rich DNA and by altering chromatin structure influence transcription positively or negatively. This family of proteins, of which there are four members (HMGA1a-1c, HMGA2), are highly expressed during embryonic development. They are largely absent in healthy adults, but are present in both benign and malignant neoplasias. HMGA proteins are crucial components of SAHFs. In these situations, the HMGAs together with $p16^{Ink4a}$ proteins promote SAHF formation. They act as structural components of the SAHFs, repress oncogenic-promoting genes, and help stabilize the senescent state thus act in an anticancer mode.

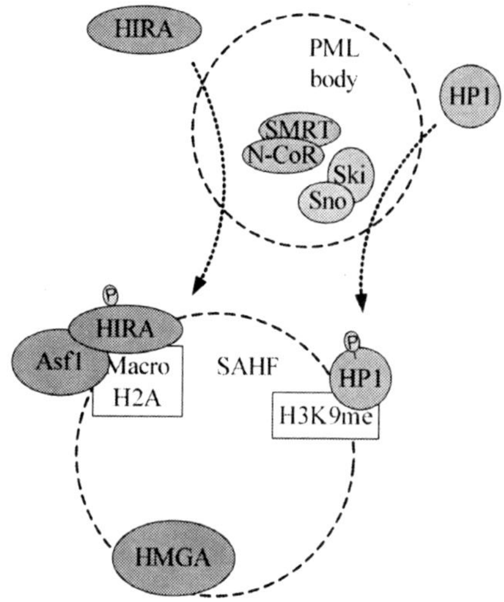

Fig. 11.7 Representative activities at promyelocytic leukemia (PLM) bodies and senescence-associated heterochromatin foci (SAHF) control nodes

11.13 The Retinoblastoma Protein Helps Establish the Senescent State by Mediating Heterochromatin Formation

The role of pRb as a tumor repressor is depicted in Fig. 11.4 as one in which the protein binds and directly represses E2F transcription. This is not the only way that pRb acts to prevent tumor formation. It also recruits histone deacetylates (HDACs) to promoter sites thereby facilitating gene repression through histone deacetylation. In addition, the pRb protein facilitates the association of components of SWI/SNF chromatin-remodeling complexes and Polycomb group (PcG) proteins to promoter sites. Polycomb group proteins, such as Bmi-1 and Ezh2, form multiprotein complexes that repress Hox gene expression during embryonic development. They also repress genes involved in the cell cycle progression and proliferation, and are overexpressed in a variety of tumors.

One of the key factors recruited to SAHFs and central to transcription repression is heterochromatin protein 1 (HP1). These proteins function as adaptors while other crucial factors listed in Table 11.4, ASF1 and HIRA, are chromatin-directed molecular chaperones. The chromatin-compacting HP1 proteins bind to methylated histone H3 at lysine 9. One of the proteins that can catalyze the transfer of methyl groups to histone H3 at lysine 9 is SUV39H1. This histone methyltransferase (HMT) is conveyed to SAHFs along with HP1 by pRb. The ASF1 and HIRA chaperone proteins convey the heterochromatin-forming histone variants macroH2A to chromatin sites and facilitate their incorporation leading to compacted, heterochromatin formation with one SAHF at each locus of compacted chromatin.

Results of the aforementioned recruitment and facilitation activities of pRb are depicted in Fig. 11.8. First, the pRb protein and Polycomb group protein Ezh2 are present at E2F promoter sites. At these sites E2Fs forms heterodimers

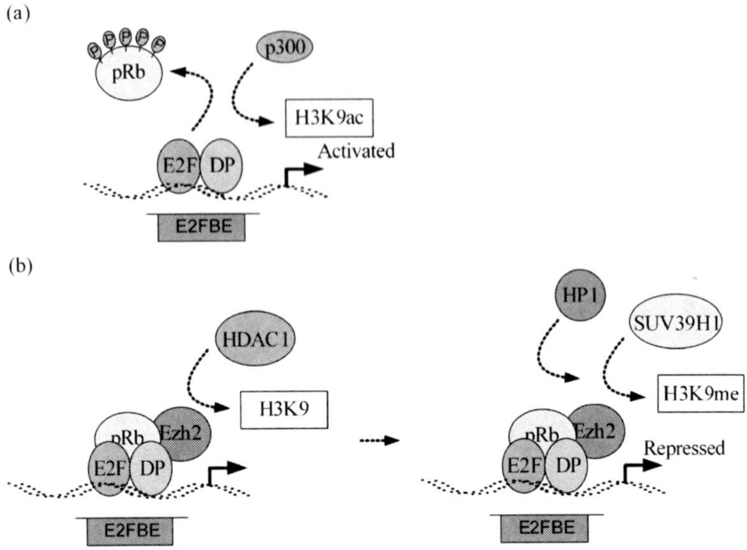

Fig. 11.8 E2F and pRb at promoter sites. (**a**) Activation of transcription through pRB phosphorylation and recruitment of HDAC1. (**b**) Repression of transcription mediated by pRb leading to replacement of the H3K9ac mark with an H3K9me mark and heterochromatin formation

with partner DP proteins. Histone deacetylases such as HDAC1 remove acetyl groups from histone H3, lysine 9. The addition of such groups is associated with transcriptionally active euchromatin. This type of posttranslational modification is catalyzed by histone acetylases such as p300. Once this posttranslational modification is removed, the methyl transferase SUV39H1 catalyzes the addition of a methyl group to that site. This posttranslational modification creates a docking site for HP1 and is characteristic of transcriptionally inactive heterochromatin.

In raising the barrier to cancer in healthy cells, the aforementioned proteins aggregate in SAHFs and upon migrating to promoter sites form transcriptional inactive heterochromatin thereby blocking transcription of genes needed for proliferation. The processes of heterochromatin formation and associated transcription repression and gene silencing are the subject of the next chapter. The various families of proteins introduced in this section and discussed only briefly will be examined in more detail therein.

Further Reading

Kerr JF, Wyllie AH and Currie AR 1972 Apoptosis: a basic biological phenomenon with wide-ranging implications in tissue kinetics. *Br. J. Cancer* 26: 239–257.

Bcl2 Proteins

Chen L et al. 2005 Differential targeting of prosurvival Bcl2 proteins by their BH3-only ligands allows complementary apoptotic function. *Mol. Cell* 17: 393–403.

Kuwana T et al. 2002 Bid, Bax, and lipids cooperate to form supramolecular openings in the outer mitochondrial membrane. *Cell* 111: 331–342.

Willis SN et al. 2007 Apoptosis initiated when BH3 ligands engage multiple Bcl2 homologs, not Bax or Bak. *Science* 315: 856–859.

Youle RJ and Strasser A 2008 The Bcl2 protein family: opposing activities that mediate cell death. *Nat. Rev. Mol. Cell Biol.* 9: 47–59.

Release of Cytochrome c

Cipolat S et al. 2006 Mitochondrial rhomboid PARL regulates cytochrome c release during apoptosis via OPA1-dependent cristae remodeling. *Cell* 126: 163–175.

Frezza C et al. 2006 OPA1 controls apoptotic cristae remodeling independently from mitochondrial fusion. *Cell* 126: 177–189.

Garrido C et al. 2006 Mechanisms of cytochrome c release from mitochondria. *Cell Death Diff.* 13: 1423–1433.

Ott M et al. 2002 Cytochrome c release from mitochondria proceeds by a two-step mechanism. *Proc. Natl. Acad. Sci. USA* 99: 1259–1263.

Scorrano L et al. 2002 A distinct pathway remodels mitochondrial cristae and mobilizes cytochrome c during apoptosis. *Dev. Cell* 2: 55–67.

p53 and Apoptosis

Chipuk JE et al. 2005 PUMA couples the nuclear and cytoplasmic proapoptotic functions of p53. *Science* 309: 1732–1735.

Leu JIJ et al. 2004 Mitochondrial p53 activates Bak and causes disruption of a Bak-Mcl1 complex. *Nat. Cell Biol.* 6: 443–450.

Barriers and Senescence

Bodner AG et al. 1998 Extension of life-span by introduction of telomerase into normal human cells. *Science* 279: 349–352.

Goldstein S 1990 Replicative senescence: the human fibroblast comes of age. *Science* 249: 1129–1133.

Hayflick L 1965 The limited in vitro lifetime of human diploid cell strains. *Exp. Cell Res.* 37: 614–636.

Mallette FA et al. 2007 The DNA damage signaling pathway is a critical mediator of oncogene-induced senescence. *Genes Dev.* 21: 43–48.

Telomeres

D'Adda di Fagagna F, Teo SW and Jackson SP 2004 Functional kinks between telomeres and proteins of the DNA damage response. *Genes Dev.* 18: 1781–1799.

de Lange T 2004 T-loops and the origins of telomeres. *Nat. Rev. Mol. Cell Biol.* 5: 323–329.
de Lange T 2005 Shelterin: the protein complex that shapes and safeguards human telomeres. *Genes Dev.* 19: 2100–2110.
Longhese MP 2008 DNA damage response at functional and dysfunctional telomeres. *Genes Dev.* 22: 125–140.
McClintock B 1941 The stability of broken ends of chromosomes in *Zea mays*. *Genetics* 26: 234–282.

Replicative Senescence

Campisi J 2005 Senescent cells, tumor suppression and organismal aging: good citizens, bad neighbors. *Cell* 120: 513–522.
Herbig U et al. 2004 Telomere shortening triggers senescence of human cells through a pathway involving ATM, p53, and p21, but not $p16^{Ink4a}$. *Mol. Cell* 14: 501–513.

Oncogene-induced Senescence

Bartkova J et al. 2006 Oncogene-induced senescence is part of the tumorigenesis barrier imposed by DNA damage checkpoints. *Nature* 444: 633–637.
Braig M et al. 2005 Oncogene-induced senescence as an initial barrier in lymphoma development. *Nature* 436: 660–665.
Chen Z et al. 2005 Crucial role of p53-dependent cellular senescence in suppression of Pten-deficient tumorigenesis. *Nature* 436: 725–730.
Collado M et al. 2005 Senescence in premalignant tumours. *Nature* 436: 642.
Di Micco R et al. 2006 Oncogene-induced senescence is a DNA damage response triggered by DNA hyper-replication. *Nature* 444: 638–642.
Michaloglou C et al. 2005 BRAFE600-associated senescence-like cell cycle arrest of human naevi. *Nature* 436: 720–724.
Serrano M et al. 1997 Oncogenic *ras* provokes premature cell senescence associated with accumulation of p53 and $p16^{Ink4a}$. *Cell* 88: 593–602.

Cancer and Telomeres

Hahn WC and Weinberg RA 2002 Modelling the molecular circuitry of cancer. *Nat. Rev. Cancer* 2: 331–341.
Henson JD et al. 2002 Alternative lengthening of telomeres in mammalian cells. *Oncogene* 21: 598–610.
Hornsby PJ 2007 Senescence as an anticancer mechanism. *J. Clin. Oncol.* 25: 1852–1857.
Shay JW and Wright WE 2005 Senescence and immortalization: role of telomeres and telomerase. *Carcinogenesis* 26: 867–874.

Second Signals

Acosta JC et al. 2008 Chemokine signaling via the CXCR2 receptor reinforces senescence. *Cell* 133: 1006–1018.

Dimri GP 2005 What has senescence got to do with cancer? *Cancer Cell* 7: 505–512.
Kuilman T et al. 2008 Oncogene-induced senescence relayed by an interleukin-dependent inflammatory network. *Cell* 133: 1019–1031.
Sablina AA et al. 2005 The antioxidant function of the p53 tumor suppressor. *Nat. Med.* 11: 1306–1313.
Takahashi A et al. 2006 Mitogenic signalling and the p16Ink4a-Rb pathway cooperate to enforce irreversible cellular senescence. *Nat. Cell Biol.* 8: 1291–1297.
Xue W et al. 2007 Senescence and tumour clearance is triggered by p53 restoration in murine liver carcinomas. *Nature* 445: 656–660.

p53 Posttranslational Modifications

Appella E and Anderson CW 2001 Post-translational modifications and activation of p53 by genotoxic stresses. *Eur. J. Biochem.* 268: 2764–2772.
Holmberg CI et al. 2002 Multisite phosphorylation provides sophisticated regulation of transcription factors. *Trends Biochem. Sci.* 27: 619–627.
Lavin MF and Gueven N 2006 The complexity of p53 stabilization and activation. *Cell Death Diff.* 13: 941–950.

Senescence-Associated Heterochromatin Foci (SAHF)

Zhang RG et al. 2005 Formation of macroH2A-containing senescence-associated heterochromatin foci and senescence driven by ASF1a and HIRA. *Dev. Cell* 8: 19–30.

PML Nuclear Bodies

Dellaire G and Bazett-Jones DP 2004 PML nuclear bodies: dynamic sensors of DNA damage and cellular stress. *BioEssays* 26: 963–977.
Solomoni P and Pandolfi PP 2002 The role of PML in tumor suppression. *Cell* 108: 165–170.

High-Mobility Group A (HMGA) Proteins

Fusco A and Fedele M 2007 Roles of HMGA proteins in cancer. *Nat. Rev. Cancer* 7: 899–910.
Narita M et al. 2006 A novel role for high-mobility group A proteins in cellular senescence and heterochromatin formation. *Cell* 126: 503–514.

Polycomb Group Proteins

Bracken AP et al. 2003 Ezh2 is downstream of the pRb-E2F pathway, essential for proliferation and amplified in cancer. *EMBO J.* 22: 5323–5335.
Bracken AP et al. 2007 The polycomb group proteins bind throughout the INK4a-Arf locus and are dissociated in senescent cells. *Genes Dev.* 21: 525–530.

Itahana K et al. 2003 Control of the replicative life span of human fibroblasts by p16 and polycomb protein Bmi-1. *Mol. Cell. Biol.* 23: 389–401.
Jacobs JJL et al. 1999 The oncogene and Polycomb-group gene *bmi-1* regulates cell proliferation and senescence through the Ink4a locus. *Nature* 397: 164–168.

Rb-Mediated Heterochromatin Formation

Luo RX, Postigo AA and Dean DC 1998 Rb interacts with histone deacetylase to repress transcription. *Cell* 92: 463–473.
Narita M et al. 2003 Rb-mediated heterochromatin formation and silencing of E2F target genes during cellular senescence. *Cell* 114: 703–716.
Nielsen SJ et al. 2001 Rb targets histone H3 methylation and HP1 to promoters. *Nature* 412: 561–565.
Ogawa H et al. 2002 A complex with chromatin modifiers that occupies E2F- and Myc-responsive genes in G_0 cells. *Science* 296: 1132–1136.

Chapter 12
Epigenetics

The term "epigenetics" has meant different things to different people and has evolved over time. The term was introduced by Conrad Waddington in the early 1940s. He defined this term as the study of the "causal interactions between genes and their products, which bring the phenotype into being." His focus was on development and on how specific groups of genes got turned on and off at different developmental stages. This definition has been taken to mean the study of those phenomena that produce alterations in the outcome of a locus of gene expression without changing the underlying DNA sequence. That is, epigenetic processes produce alterations in chromatin structure and cellular phenotype without changing the primary DNA coding sequence.

Nowadays, several kinds of processes are considered as being epigenetic. Some of these produce changers that are heritable, through mitosis or meiosis, from one generation to the next while others are far more dynamic and can be reversed over time within a particular cell in a regulated fashion. There are two prime types of epigenetic processes leading to long-lasting heritable changes – those generated by the Polycomb-Trithorax systems and those involving DNA methylation. In general, epigenetic changes in chromatin structure are shorter-lasting than heritable changes yet may persist for considerable time and provide a level of control over which genes are expressed and which ones are repressed. Numerous examples of transient modifications in chromatin structure by coactivators and corepressors have been encountered in earlier chapters. Many of these modifications produce heterochromatin silencing and are driven by enzymes that add and remove acetyl, phosphoryl, methyl, ubiquitin, or SUMO groups to histones. Commonly encountered examples are the HATs and HDACs that add or remove acetyl groups.

Abnormal epigenetic patterns of silent and active regions of gene transcription are a hallmark of cancer. These changes, like the alterations in coding sequences, result in disease-causing changes in the normal operation of the cellular signaling networks that maintain homeostasis and prevent cancers from developing. During the past few years, epigenetic changes that contribute to several specific kinds of cancers have been uncovered along with an entire new, and previously unsuspected, layer of control over gene expression

M. Beckerman, *Cellular Signaling in Health and Disease*,
Biological and Medical Physics, Biomedical Engineering,
DOI 10.1007/978-0-387-98173-4_12,

involving small regulatory molecules called microRNAs. These molecules silence gene expression posttranscriptionally by degrading mRNAs and by blocking mRNA translation. MicroRNAs along with abnormal histone modifications and DNA methylation patterns help drive the transformation of healthy cellular states into disease-causing ones. In this chapter, these three kinds of epigenetic processes and their roles in cancer will be explored. In doing so, some of the most dramatic findings, those centered on cell cycle inhibitors, p53, Bcl2, Ras, and Myc, will be examined.

12.1 Nucleosomes and Chromatin Structure

The basic unit of chromatin, the nucleosome, is a repeating unit of 146 base pairs of DNA wrapped about a histone octamer. The duplex DNA is wrapped about a histone core composed of H2A H2B, H3, and H4 histone pairs. The core nucleosome unit is connected to the next nucleosome core unit by a linker segment (called H1) of chromatin. Each nucleosome has the DNA wrapped about 1.75 times about the histone core forming a bead-like structure roughly 11 nm in diameter as depicted in Fig. 12.1.

Chromatin structure plays an important role in determining which genes are to be transcribed at any given moment in time. There are two forms of chromatin. Transcriptionally inactive chromatin, called heterochromatin, is tightly compacted. In heterochromatin, sites where transcription initiation and regulation take place are largely inaccessible to the proteins responsible for these actions. In contrast, transcriptionally active chromatin, or euchromatin, has a far more open shape. In euchromatin, sites where transcription factors and the transcription machinery bind are accessible to the responsible proteins.

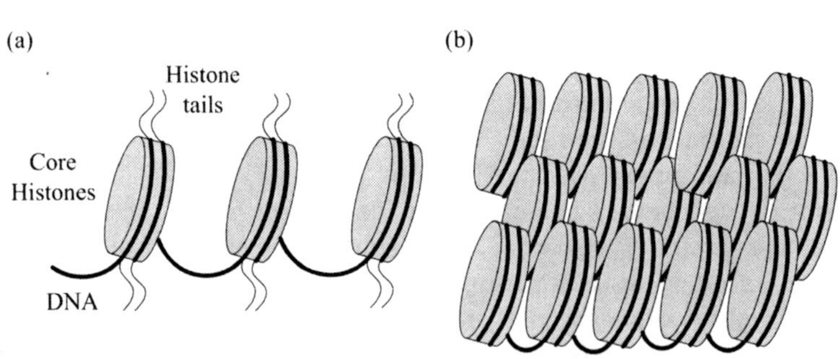

Fig. 12.1 Beads-on-a-string model of chromatin structure. (**a**) Open, transcriptionally active euchromatin. Core histones are depicted as grey discs with DNA as black strings. Histone tails appear as wavy lines. (**b**) Condensed, transcriptionally inactive heterochromatin

12.2 Epigenetic Marks

The amino terminals of the core histones extend out from the nucleosomes to form histone tails. These tails provide a means for regulatory proteins to influence transcription. The histone tails provide sites for attachment of methyl groups, acetyl groups, and SUMO groups. The predominant targets of the posttranslational modifications are the amino groups lying at the end of lysine side chains. The lysine residues, along with arginine residues, can be methylated, acetylated, sumoylated, or ubiquitinated. As shown in Fig. 12.2, the tails of the histones are enriched in these residues, and thus supply multiple sites for attachment of these groups. Attachment of acetyl and other groups neutralizes the net positive charge on the tail regions, and this reduction weakens the attraction between the tails and the DNA. As a consequence the histone-DNA interactions are lessened allowing for a greater access of the DNA to transcription regulators. Similarly, addition of net negative charge through phosphorylation on serine residues also serves to decondense chromatin.

Transcriptionally inactive heterochromatin is tightly compacted. Sites where transcription initiation and regulation take place are inaccessible to

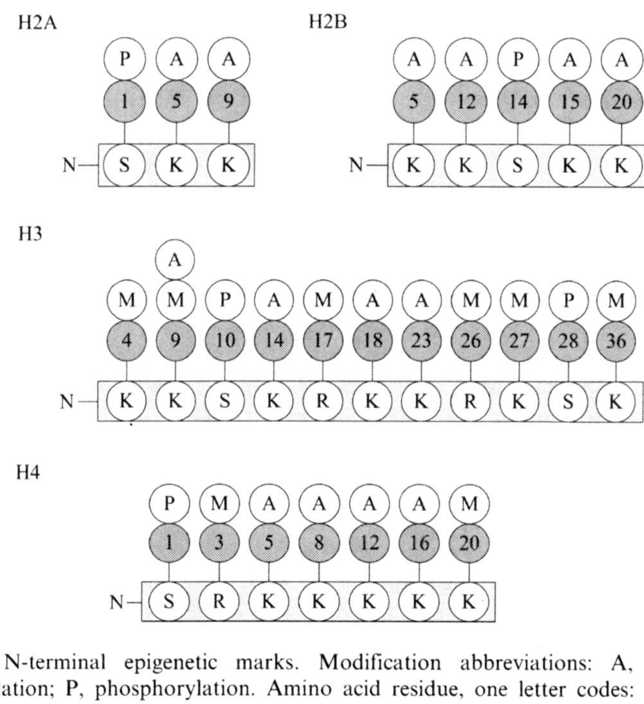

Fig. 12.2 N-terminal epigenetic marks. Modification abbreviations: A, acetylation; M, methylation; P, phosphorylation. Amino acid residue, one letter codes: R, arginine; K, lysine; S, serine

the proteins responsible for these actions. In contrast, transcriptionally active euchromatin has a far more open shape, and sites where transcription factors and the basal transcription machinery bind are accessible to the responsible proteins. The covalent addition of acetyl groups to the side chains reduces the net positive charge on the histone tails thereby weakening their attraction to the DNA-strands. The addition of these groups counteracts the natural tendency for the chromatin fibers to fold into compact nucleosomal units making transcription difficult to impossible. Upon acetylation, the promoter sites become far more accessible to the transcription machinery. The various kinds of additions to the histone tails are referred to as histone marks, and the ensemble of marks serves as an epigenetic profile or pattern of gene activity for the cell.

The enzymes that mediate the modifications on the H3 and H4 tails are quite specific. This aspect, together with the presence of so many potential modification sites, has led to the suggestion that there is a histone code that determined whether a particular gene or cluster of genes is silenced or not. An example of how this might work is provided by data regarding the H3 lysine 9 (K9) site. As indicated in Fig. 12.2, this is a site that can be either acetylated or methylated. The selection process is a competitive one. Acetylation at K9 blocks methylation at that site, and vice versa. The silencing process for that site works in the following way. A HDAC acts on K9 to remove an acetyl group. Next, a histone methyltransferase specific for that site catalyzes the covalent addition of a methyl group. Then, heterochromatin (adaptor) protein 1 (HP1) binds to the methylated K9 site and blocks transcription.

Chromatin tails stick out where they are ideally positioned to influence DNA and protein interactions. The tails provide platforms for chromatin remodeling and DNA-modifying enzymes and they mediate nucleosome-nucleosome interactions. The tails can control the openness of the chromatin to the transcription machinery. Figure 12.1 depicts nucleosomes in their open structure and in a far more compact configuration that prohibits transcription. Chromatin can organize into several states of chromatin compactness; these range from 10 nm fibers to 30 nm fibers to even more tightly compacted 100–400 nm fibers. An archtypic example of how modification to histone tails influences chromatin structure is acetylation of lysine 16 on histone H4 (H4K16ac). This posttranslational chromatin modification is catalyzed by histone acetyltransferases (HATs) such as Tip60 and p300/CBP and hMOF, the human homolog of the *D. melanogaster* H4K16ac-specific enzyme MOF. This modification prevents nucleosome-nucleosome interactions required to form compact chromatin structures. It also alters that character ISWI-chromatin interactions. Overall, it opposes gene silencing, and loss of acetylation at this site and at Lys20 on histone H4 is observed in human cancers.

12.3 DNA Methylation

DNA methylation status has a bearing on human health and disease. In DNA methylation, methyl groups are covalently attached to 5′ position of cytosines just before guanosines in CpG nucleotides. CpG nucleotides are uncommon in mammalian DNA, occurring far more rarely (about 1%) than the nominally expected ¼ × ¼ frequency. The CpG dinucleotides that do occur are concentrated into fairly short stretches, 0.5–5 kb in length, and these are referred to as CpG islands. Inside these stretches the concentration of CpG is high – 60–70%. Under normal, healthy conditions the cytosines are unmethylated; the chromatin is hyperacetylated and in an open chromatin conformation.

CpG islands are found at the 5′ ends of many if not most human genes. In promoters, methylation of the cytosines silences gene expression. Methylation increases as part of the aging process; aberrant CpG methylation in promoters is encountered in a number of neurodevelopmental disorders and cancers. When CpG islands are situated in exons, the 5-methylcytosines (m^5Cs) become subject to C → T transitions arising from deamination. On the opposite strand G → A transitions may occur, as well. The methylated CpG islands in coding regions serve as mutational hot spots leading to mental retardation and cancer progression.

The addition of methyl groups to promoters in CpG islands is catalyzed by a family consisting of three methyltransferases, namely, DNMT1, DNMT3A, and DNMT3B. These enzymes each possess an N-terminal regulatory domain followed by a C-terminal catalytic domain. Once the enzymes have acted, members of a family of methyl-CpG binding proteins are recruited. In mammals, this family too has three members – MeCP2, MBD1, and MBD2. These proteins function as adapters that link the methylated promoters to histone deacetylases (HDACs). The latter along with corepressors such as Sin3 silence transcription activity at those sites. DNA Methylation and histone modification work hand in hand to activate and silence gene expression.

Several prominent examples of aberrant gene hypermethylation in cancers have been uncovered. A short list of these is presented in Table 12.1. As can be seen these genes encode signaling proteins, cell cycle regulation, and DNA damage repair proteins. When silenced these genes are not expressed leading to an increase in cancer susceptibility.

Table 12.1 Cancer-associated hypermethylated gene products

Gene product	Protective role	Associated cancer
APC	Wnt signaling[a]	Colorectal
BRCA1	DNA damage repair	Breast, ovarian
hMLH1	DNA damage repair	Colorectal, gastric
pRb	Cell cycle regulation	Retinoblastoma
p16Ink4a	Cell cycle regulation	(Various)

[a] Wnt signaling will be discussed in Chapter 15

12.4 Polycomb and Trithorax Group Proteins

Polycomp (PcG) and trithorax (trxG) group proteins were first identified in *Drosophila* where they regulate the expression of *Hox* genes during embryonic development by means of chromatin modification. Members of the Polycomb group function as transcription repressors while members of the trithorax group function in the opposite way as transcription activators. These proteins are essential factors during embryonic development and in adult stem cell renewal.

Polycomb proteins associate into two large complexes (Table 12.2). Polycomb repressive complex 1 (PRC1) contains at least 10 subunits including the oncoprotein Bmi1. The other complex, PRC2, contains four subunits including EZH2 which functions as a catalytically active histone methyltransferase specific for Lys27 on H3 (H3K27me3). Several histone-modifying complexes contain trithorax group proteins. The two best characterized and most highly conserved of these are the SWI/SNF and NURF complexes. Their core components have been listed in Table 12.2 together with the core members of the PRC1 and PRC2 Polycomb group complexes.

Table 12.2 Polycomb (PrG) and trithorax (trx) group proteins and the complexes into which they assemble

PcG complex	Members	Trx complex	Members
PRC2	**EZH2**	SWI/SNF	BRM
	EED		BAF250
	SUZ12		BAF170
	RpAp48, RpAp46		BAF47
PRC1	RING1A	NURF	SNF2L
	HPC1-3		BPTF
	HPH1-3		RpAp48, RpAp46
	Bmi1	MLL	MLL1-3
	SCMH1-2		WDR5
			ASH2L
			RbBP5
			CFP1

12.5 Histone Acetylation and Deacetylation

A large number of histone acetyltransferases acting either singly or as catalytic subunits as part of large chromatin-modifying complexes append acetyl groups to histone H2, H3, and H4 lysines. In general, these enzymes are less specific than their methyl counterparts for particular lysines, and these modifications activate transcription. Some of the most frequently encountered of these are listed in Table 12.3 along with the primary histones targeted.

Table 12.3 Histone acetyltransferases (HATs)

HAT	Family members	Histone specificity
GNAT Family	Gcn5	H3
	PCAF	H3
	Hat1	H4
CBP/p300		H3/H4/H2A/H2B
MYST Family	Tip60 (NuA4)	H4
	HBO1	H4
TAFII250		H3/H4

Acetyl groups covalently attached to histone lysines are removed by histone deacetylases. There are 18 known members of this group in humans, and these are divided into four classes (Table 12.4). Classes I, II, and IV family members are zinc-dependent while Class III enzymes are NAD^+-dependent.

Sirt2 is a member of the Sirtuin family central to caloric restriction discussed in Part I of the text. This enzyme has a preference for H4K14Ac, particularly during the G2/M phase of the cell cycle, and thereby sharply reducing H4K16Ac levels during mitosis from the high values present during S-phase.

Table 12.4 Histone deacetylases (HDACs)

HDAC	Family members
Class I (Rpd-like)	HDAC1, -2, -3, -8
Class II (Hda1 like)	HDAC4, -5, -6, -7, -9, -10
Class III (Sir2-like)	Sirt1, -2, -3, -4, -5, -6, -7
Class IV	HDAC11

12.6 Histone Methylation and Demethylation

Lysine and arginine residues on histone tails provide sites for attachment and removal of methyl groups. The importance of these activities is reflected by the large number of lysine methyltransferases – at least 50 – that have been found so far. Like other operations dealing with epigenetic marks, the view of methyl marks has undergone considerable revision from one that regarded these marks as static to one where they are added and removed dynamically just as phosphoryl groups are added and removed from serine, theronine, and tyrosine residues. The turning point in the case of methylation was the discovery in 2004 and 2005 of the lysine demethylase LSD1 followed shortly thereafter with the discovery that enzymes such as JMJD2A function as lysine demethylases through catalytic actions of their jumanji (Japanese for cruciform) C, or JmjC, domains. More recently, in 2007, the JmjC-containing protein JMJD6 was discovered to function as an arginine demethylase specific for methylated histone H3R2 and H4R3 residues (Table 12.5).

Table 12.5 Lysine methyltransferases and demethylases and their histone substrates

Histone specificity	Lysine methyltransferase	Lysine demethylase
H3K4	MLL1-4, Set1, ALR1-2, Set7/9, ASH1	LSD1
H3K9	Suv39H1/2, G9a, GLP, ESET/SETDB1, ASH1	JHDM2A, JHDM2B, JHDM2C, JHDM2D, LSD1
H3K27	EZH1/2, NSD3	
H3K36	HYPB, NSD1, SMYD2	FBXL11, FBXL10, JHDM2A, JHDM2C
H3K79	DOT1	
H4K20	PR-Set7, SETB, ASH1	

The various methyltransferases are not all equivalent in their actions nor are the demethylases. Lysine residues can be mono-, di- and tri-methylated, while arginines can be mono- and di-methylated, either symmetrically or asymmetrically. Different, overlapping subsets of methyltransferases and demethylases are specific for each of the many different combinations of histone lysines and their methylation states. The well-identified arginine methyltransferases are listed in Table 12.6 along with some of the serine-threonine kinases found to target histones. The kinases act downstream of growth factor and stress-related stimulations that are transduced through the actions of MAP kinase modules (These important cellular signaling modules will be examined in Chapter 13).

Table 12.6 Arginine methyltransferases and serine/threonine kinases and their histone substrates

Arginine methyltransferase	Histone specificity	Serine-threonine kinases	Histone specificity
CARMI1	H3R17, H3R26	MSK1	H3S28
PRMT1	H4R3	MSK2	H3S28
PRMT5	H3R8, H4R3	Mst1	H2BS14

12.7 Reading Out Histone Marks by Recognition Modules

The establishment of a histone mark is the first step in a series of binding and modification events leading to changes in the transcriptional status of the region of DNA. In establishing a histone mark, binding sites are created for histone recognition modules. For example, there are three commonly encountered modules that recognize methylated lysines on histones H3 and H4. These are the chromodomains, plant homeodomains (PHDs), and tandem tudor domains. Proteins containing these domains are able to bind the methylated lysine marks and function as their "readouts". Table 12.7 lists a number of the best characterized of these readout effectors.

Table 12.7 Readout of methylated lysines on histones H3 and H4

Effector protein	Recognition module	Epigenetic mark
BPTF	PHD finger	H3K4me3
ING2	PHD finger	H3K4me2,3
WDR5	WD-40	H3K4me2
CHD1	Chromodomain	H3K4me3
JMJD2A	Tandem tudor domain	H3K4me3
HP1	Chromodomain	H3K9me3
Polycomb	Chromodomain	H3K27me3
53BP1	Tandem tudor domain	H4K20me2

Acetylated lysines are recognized by just one kind of readout module – the bromodomain. This module is encountered in a variety of proteins involved in manipulating chromatin, most notably in histone acetyltransferases and other transcription activators. Many of these proteins have been encountered in earlier chapters in which their role of coactivators was emphasized. The prototypic example is the p300/CBP-associated factor, PCAF. This is the homolog of the yeast HAT Gcn5 (Fig. 12.3(a)), which has been found to bind by means of its bromodomain to histone H4 which has been acetylated on lysine 16 (H4K16ac). Another example is human TAF1, a large subunit of the transcription-initiating complex TFIID. This protein possesses a pair of bromodomains that binds H4K5acK12ac.

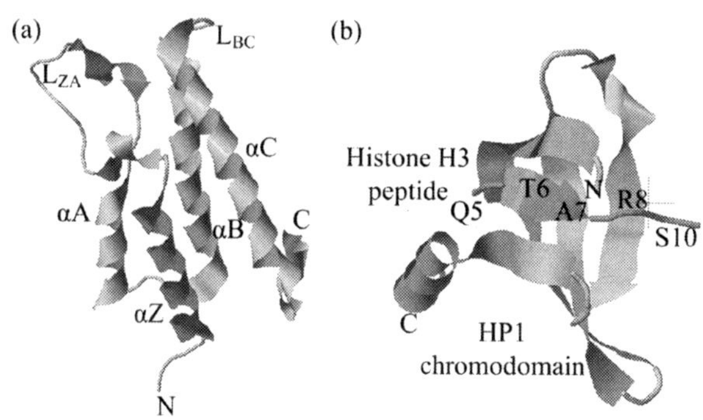

Fig. 12.3 (**a**) Solution structure of the Gcn5 bromodomain determined by means of NMR. The resulting structure is that of an antiparallel four-helix bundle. A binding pocket formed by the LZA and LBC loops and core helices, and lined by hydrophobic residues, reads H4K16ac. (**b**) Structure determined by means of X-ray crystallography of the HP1 chromodomain in complex with histone H3 peptide fragment dimethylated on lysine 9 (Me_2K9). The chromodomain backbone is depicted in this figure along with residues Gln5, Thr6, Ala7, and Arg8 and Ser10 of the histone H3 peptide. The figure was prepared using Jmol with atomic coordinates deposited in the PDB under accession numbers 1 kna (**a**) and 1e6i (**b**)

Another readout module that is broadly distributed is the SANT domain, named for the first letter in four proteins first found to contain this module – switching-defective protein 3 (Swi3), adaptor 2 (Ada2), nuclear receptor corepressor (N-CoR), and transcription factor IIB (TFIIB). These domains are present in numerous chromatin-interacting proteins as is clear from the set of four proteins that lend this module its name. They occur in complexes involved in both activation and repression of transcription. In the Ada2 complex, they promote activation. Recall that cofactors such as N-CoR and SMRT repress transcription from nonliganded nuclear receptors by recruiting HDAC3. Both of these repressors contain a pair of SANT domains.

Readout of phosphorylation marks is carried out by 14-3-3 proteins. The best studied of these activities is readout of H3S10ph, which is associated with activation of transcription and chromatin condensation. Readout effectors of histone marks do not operate in isolation but rather form complexes and synergize with other readout effectors. They work together to modify chromatin structure and either activate or repress activation of specific sets of genes. In its transcription regulatory activities, there is a synergy between readout of H3S10ph by 14-3-3- proteins and effectors of acetylation marks at H3K14 (such as Gcn5) that counteract repression of transcription by methylation at H3K9 thereby activating transcription.

Heterochromatin-associated protein 1 family members, or HP1 proteins, are major components of chromatin with roles in heterochromatin organization, maintenance, and silencing. These proteins possess chromo domains that recognize methylated lysines on histone tails. These proteins are key constituents of heterochromatin. They stabilize the heterochromatin structure thereby silencing gene expression from loci locked off from transcription factors and the basal transcription machinery. A key epigenetic mark associated with gene silencing is methylated lysine 9 on histone H3 (H3K9me2). The mechanism underlying the stable long-term maintenance of this state is revealed by X-ray crystallography (Fig. 12.3(b)). It seems that the histone tail inserts into the chromo domain and functions as a β-strand that completes the formation of a beta sandwich structure.

HP1 proteins contain two domains – an N-terminal chromo domain and a C-terminal chromo shadow domain. These domains are connected to one another through a linker segment. The chromo domain is responsible for binding di- ands tri-methylated lysine 9. The chromo domain folds into a globular structure with a hydrophobic groove that is the site of H3 tail attachment. The peptide tail functions as the third strand of a three-stranded β-sheet. The chromo shadow domain mediates HP1 dimerization and functions as a docking site for nuclear proteins. This domain folds into a globular structure similar to that of the chromo domain. Residues unique to the chromo shadow domain facilitate its function as a dimerization interface.

12.8 Cooperative Actions by Histone Modification Enzymes and DNA Methyltransferases Can Silence Genes and Lead to Cancer

A particularly clear example of how epigenetic dysregulation can lead to cancer is acute promyelocytic leukemia (APL). This form of cancer arises from a genomic translocation resulting in the fusion of the retinoic acid receptor α (RAR α) with one of five other proteins, the most common of which is the promyelocytic leukemia (PML) protein. The RARα protein belongs to the nuclear receptor superfamily of transcription factors. In the absence of ligand binding, this protein heterodimerizes with the RXR coreceptor and represses transcription though association with an N-CorR corepressor that recruits the HDAC protein to the promoter site. Binding to its retinoic acid (RA) ligand provokes the release of HDAC followed by recruitment of coactivators leading to transcription activation and resulting in cellular differentiation and eventual apoptosis.

The fusion product generated by the translocation t(15,17) results in altered response properties in which differentiation is arrested. The fusion protein is still able to bind DNA and recruit N-CoR, which then recruits the HDAC. However, the affinity of the receptor for HDAC is increased and ligand binding is not able to trigger the switch from repression to activation. Instead, as shown in Fig. 12.4, histone deacetylation is followed by recruitment of the DNA methyltransferases DNMT1 and/or DNMT3A (abbreviated in the figure as DNMT) and the histone methyltransferase SUV39H1. Methylation is then followed by the binding of HP1 and methyl-CpG domain (MBD) proteins

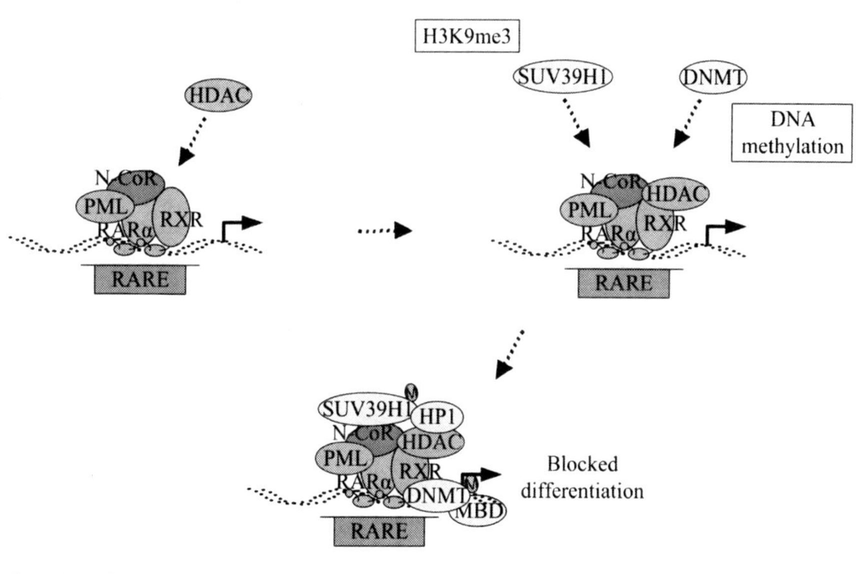

Fig. 12.4 Cooperative actions by histone and DNA-modification enzymes leading to transcription silencing and blocked differentiation. RARE, Retinoic acid response element

that effectively shut down transcription and block differentiation. This sequence of events brings together histone deacetylation, DNA methylation, and histone methylation along with readout units/adapters such as HP1 to repress transcription. This teaming arrangement is not unique to the translocation but rather is encountered in other silenced genes as well.

This form of leukemia has been successfully treated by administering pharmacological levels (i.e., higher doses) of RA resulting in release of HDAC from RARE promoters. In situations where another of the fusion partners, PLZF, is involved, this strategy alone is insufficient to stimulate the switch from repression to activation. However, administering an HDAC inhibitor (HDACi) along with RA has been successful in arresting the leukemia.

12.9 Recently Discovered Small Noncoding RNAs (ncRNAs) Regulate Gene Expression

The term noncoding RNA (ncRNA) encompasses not only translation-associated RNAs such as mRNAs, tRNA, and rRNAs, but also RNAs with a role in RNA processing such as small nuclear RNAs, or snRNAs (splicing) and small nucleolar RNAs, or snoRNAs (RNA modification). During the past few years this world has expanded to include an entire spectrum of regulatory RNAs that promote gene silencing. Among the growing list of newly identified ncRNAs are small interfering RNAs (siRNAs), micro RNAs (miRNAs), piwi RNAs (piRNAs) and trans-acting siRNAs (tasiRNAs). These regulatory RNAs silence genes by preventing mRNA translation.

Short interfering RNAs (siRNAs) are small RNA molecules, 21–23 nucleotides (nt) in length that are exactly homologous to portions of the longer target mRNAs. The silencing of a gene by long strands of dsRNA occurs in two stages (Fig. 12.5). In the dsRNA processing stage, long dsRNA strands, either

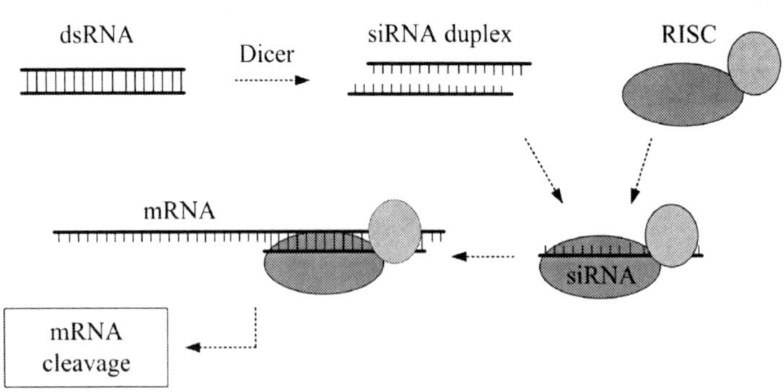

Fig. 12.5 Processing of short-interfering RNAs leading to degradation of the target mRNA. RISC, RNA-induced silencing complex

introduced into the cell or produced endogenously, are cleaved by a protein called Dicer several times resulting in short, duplex siRNA strands with a slight 2 nt overhang structure (Fig. 12.5). In the effector stage, the siRNA strands are unwound. One strand is incorporated into a protein complex called the RNA-induced silencing complex, or RISC. The RISC then binds to the mRNA by means of complementary base pairing, and the mRNA is cleaved by an enzyme belonging to the Argonaute family functioning as the Slicer. This slicing event effectively prevents translation of the mRNA and leads to its destruction.

Heterochromatin silencing is another task carried out by siRNAs. In fission yeast (*Saccharomyces pombe*), this process is associated with centromeres. In place of the RISC complex, the RNA-induced transcriptional silencing (RITS) complex and RNA-directed RMA polymerase complex (RDRC) function as the effectors. In this operation, the siRNAs match the sequences of repetitive DNAs in the outer centromeric repeats. The effector complexes that localize to the centromeres contain the machinery to generate dsRNA, Dicer, and Slicer (Argonaute/Ago1) proteins, a histone lysine methyltransferase that generates a H3K9 methylation mark, and chromodomain-bearing silencing proteins that associate with the H3K9 methyl mark. Evidence has been uncovered supporting a similar mechanism in mammalian cells mediated by a complex containing the DNMT3A H3 lysine methyltransferase among other participants.

MicroRNAs (miRNAs), the second major class of small ncRNAs, are structurally similar to siRNAs. However, there are several differences. These ncRNAs are encoded in the cell's genome. In humans, there are several hundred microRNA-encoding genes. These are transcribed into long RNA precursors termed primary miRNA (pri-miRNA) characterized by the presence of one or more hairpin structures (Fig. 12.6). The pre-miRNAs are processed sequentially by a pair of enzymes. The first enzyme, called Drosha, crops the pri-miRNA in the nucleus into ~65 nt pre-miRNA hairpin-containing intermediates. The intermediates are exported to the cytoplasm. As was the case for siRNAs, Dicer chops the pre-miRNA intermediate into the short ~22 nt miRNA sequences with an overhang structure. Single strands are then incorporated into the RISCs.

The mature miRNAs silence gene expression primarily by inhibiting their translation whereas the siRNAs usually operate by degrading the mRNA (Fig. 12.6). In contrast to the siRNAs, the miRNAs do not have to be exactly homologous to the mRNA being silenced. Instead one kind of miRNA may silence several different mRNAs and different miRNAs may silence the same mRNA. The method of silencing used depends on the degree of homology between the miRNA and the targeted mRNA. If the overlap is high the mRNA will be degraded, and if the overlap is not high mRNA translation will be blocked.

In addition to siRNAs and miRNAs there are one or two other classes of mammalian small negative regulatory RNAs. This grouping includes the Piwi-associated RNAs (piRNAs), consists of 29–30 nt long RNAs that are expressed in the germlines, and carry out germline-specific actions and transposon control

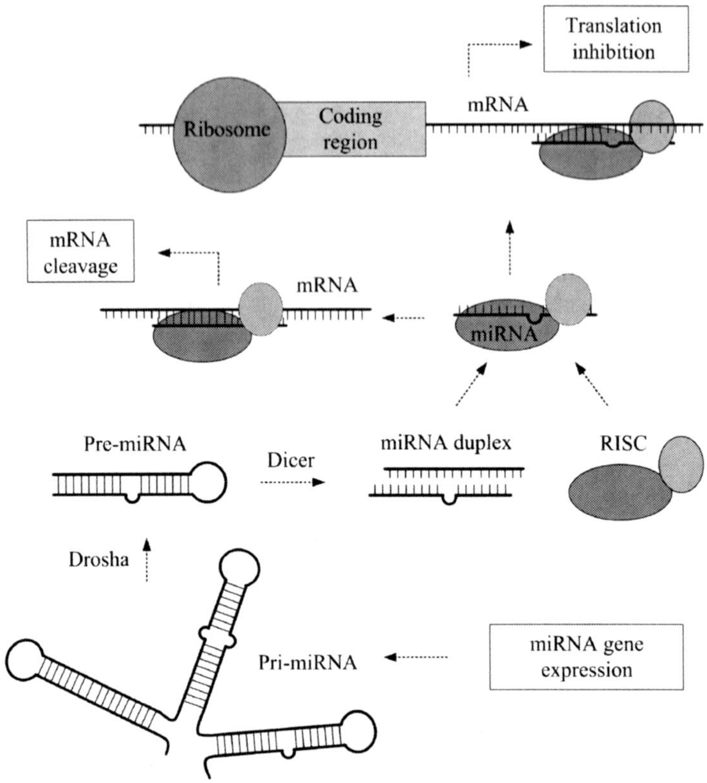

Fig. 12.6 Processing of micro RNAs leading to either degradation or blockage of the target mRNA

and related tasiRNAs. Members of this group of small regulatory RNAs are thought to be derived from ssRNA. They are Dicer independent, and P-element-induced wimpy testis (Piwi), not Ago2, is the slicer. Functions associated with piRNAs in mammalian cells may include gene silencing. Evidence for this regulatory activity has begun to emerge.

12.10 Atomic-Level Studies of Dicer and Slicer Provide Crucial Insights into ncRNA Function

Recent studies of Dicer and Slicer (Argonaute) proteins using biophysical methods have revealed crucial details on how these key enzymes function. The domain structure of the human Dicer protein is depicted schematically in Fig. 12.7(c). The N-terminal helicase domain depicted in the figure encompasses two helicase activities. This domain is followed by a domain

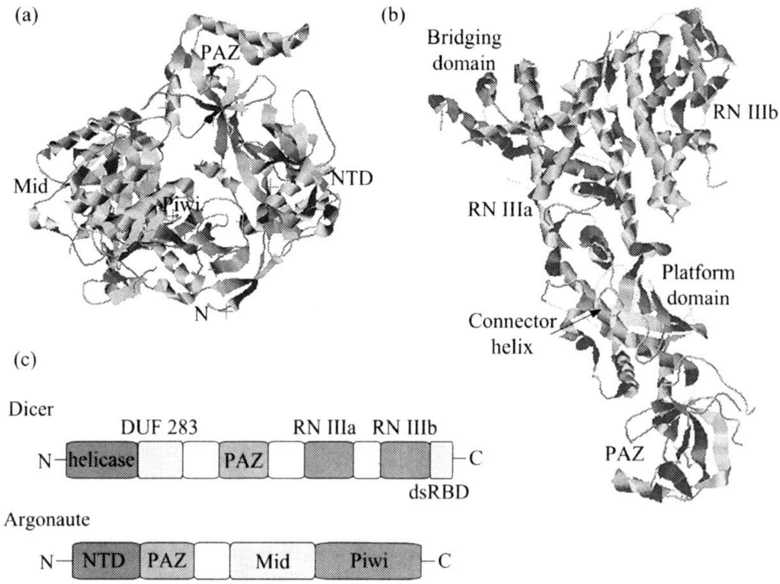

Fig. 12.7 Slicer and Dicer. (**a**) Three-dimensional structure of (**a**) Argonaute from *Pyrococcus furiosus*. PDB accession number 1u04. From Song, 2004 and (**b**) Dicer from *Giardia intestinalis* that lacks the N-terminal helicase domain, dsRBD, and extended interdomain regions found in higher eukaryotes. PDB accession number 2ffl. (**c**) Domain structure of Dicer and human Argonaute 2 (HsAgo2)

of unknown function (DUF 283). PAZ domains bind the dsRNAs, as will be discussed further below. These domains are unique to Dicer and Argonaute proteins. This domain is followed by a pair of ribonuclease III domains that catalyze the dicing of the RNAs into 21–23 nt segments with 2 nt overhang and the ends. The C-terminal-most domain shown in the figure is the double-stranded RNA binding domain (dsRBD) that assists in RNA binding and positioning.

In the three-dimensional crystal structure of the molecule (Fig. 12.7(b)), the PAZ domain is located at one extremity while the catalytic (cleaver) domains sit at the other. The 65 angstrom separation between PAZ and RNase domains serves as a molecules ruler producing measured 21–23 base pair RNA dicer products. The region in between the two end regions is flat and has a positive charge enabling it to bind negatively charged dsRNA molecules.

Argonaute proteins are characterized by the presence of PAZ and Piwi domains. As shown in Fig. 12.7(a), hArgo2 has a PAZ (Piwi-Argonaute-Zwiller) domain in its N-terminal half and a Piwi domain in its C-terminal region. The Piwi domain contains the catalytic core of the protein, which functions as an RNase H enzyme, a class of divalent metal-ion (Mg^{2+})-dependent enzymes similar to RNase III enzymes like Dicer, and in the Argonaute proteins functions as Slicer.

The Argonaute protein is crescent shaped. Its base is built from the NTD, Mid, and Piwi domains. As was the case for the flat-surfaced middle region of the Dicer proteins, the grooves in the Argonaute protein are lined with positive charges that can bind nucleic acids. The PAZ domain supported by a stalk is situated directly above the base. The PAZ domains operate as RNA binding and positioning modules. As can be seen in Figs. 12.5 and 12.6, both Drosha and Dicer generate dsRNAs with 2 nt overhangs at the two 3′ ends. The PAZ domains found in members of the Dicer and Argonaute protein families utilize this special feature (as well as others) in the RNA duplexes. The Mid domain loads small RNAs onto the RNP complex. The Piwi catalytic core is centered about an Asp-Asp-His motif that coordinates one metal ion (Mg^{2+}) binding.

12.11 MicroRNAs and Cancer

MicroRNAs provide yet another layer of control over cellular processes. As is the case for protein-coding regions, coding alterations and inappropriate expression levels of microRNA genes can stimulate cancer development. An example of this kind is a region on human chromosome 13 designed as 13orf25. Expression levels from this region, which encodes a cluster of seven microRNAs, are increased in B-cell lymphomas. The microRNAs in this cluster are activated by c-Myc, which at the same time activates E2F1. When this happens a small circuit is formed – c-Myc turns on the E2F1s but two of the microRNAs activated by c-Myc negatively regulate the E2F1s. The net outcome, under normal circumstances, is that the E2F1s are activated and then turned off after a short while. In the case of the altered microRNA expression levels from 13orf25, coding changes have altered the microRNA actions, converting them from agents that suppress excessive growth to agents that promote it.

MicroRNAs are underexpressed in several kinds of cancer. Since miRNAs are negative regulators of messenger RNA translation, their underexpression results in protein overexpression. An example of this reciprocal relationship is provided by lung cancers in which members of the let-7 family of miRNAs are underexpressed. These miRNAs regulate Ras, and, in accordance with the reciprocal relationship, Ras is overexpressed in the lung tumors.

The microRNAs miR-15a and miR-16-1 regulate the antiapoptotic protein Bcl2. In chronic lymphocytic leukemia (CLL), the Bcl2 protein is overexpressed and in most cases of this disease, the most common form of leukemia, miR-15a and miR-16-1 are either deleted or downregulated. In a mouse model of CLL, repressing Bcl2 action by these microRNAs induces apoptosis thereby pointing to a potential role of these small noncoding RNAs in cancer therapy.

The microRNA miR-34, situated at the 1p36 chromosome locus, is a target of p53 gene expression triggered by cellular stresses. Three miRNAs are encoded at this site – miR-34a, b, and c. Upregulation of these miRNAs enhances p53's protective effects strengthening its ability to direct the cell

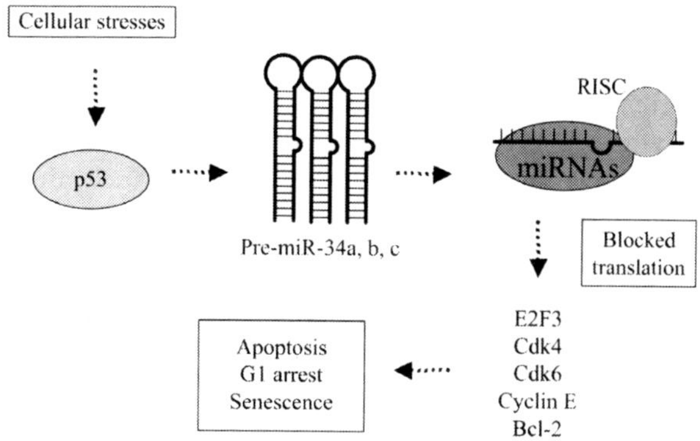

Fig. 12.8 A portion of the miR-34 p53 network. Cellular stresses upregulate miR-34 expression leading to a blocking of growth and survival mRNA translation

toward apoptosis or cell cycle arrest or senescence by blocking expression of pro-growth factors such as Cdk4 and Cdk4 and antiapoptotic Bcl2 (Fig. 12.8). Most recently it has been found that p53 and miR-34 may operate in a positive feedback loop to strengthen this component of the p53 response to cellular stresses and directing p53 toward p21 and PUMA activation, and apoptosis.

Some microRNAs function as tumor suppressors while others operate as oncogenes, or as they are sometimes called, oncomirs. Which of these two roles is followed is determined by which classes of proteins are being repressed by them – oncoproteins or tumor suppressors. Referring back to the just-discussed examples, let-7 suppresses Ras and thus functions as a tumor suppressor. miR-15 and miR-16 also operate as tumor suppressors since they inhibit actions by the anti-apoptotic Bcl2 protein, and thus strengthen the apoptotic barrier to cancer. Two examples of microRNAs that function as oncomirs are miR-21 and miR-155. A general observation is that tumor suppressors such as let-7 are underexpressed and, conversely, oncomirs like miR-155 are overexpressed in human cancers.

The first of the two afore-mentioned oncomirs, miR-21, opposes the protective, apoptotic actions of p53 in glioblastoma cells. It accomplishes this by targeting several factors that stabilize p53 and enhance its ability to orchestrate DNA repair and, if needed, induce apoptosis. Another regulatory route targeted by miR-21 is the TGF-β pathway. This pathway can act either as part of the anti-cancer defense system of the cell or, as will be discussed in the next chapter, as a tumor promoter. The p53-stabilzing factor Daxx, which is one of the target proteins repressed by miR-21, acts as a TGF-β pro-apoptotic and growth arrest effector.

The second above-mentioned oncomir, miR-155, is transcribed from the non-protein-encoding *BIC* (B-cell integration cluster) gene. It is overexpressed in lymphomas and leukemias and is believed to play a causal role in the development of these cancers. The miR-155 microRNA functions as part of the immune system's response to pathogens. It is upregulated in macrophages in response to viral pathogens and in B and T lymphocytes in response to inflammatory cues. In B and T lymphocytes, it helps direct which lineage fates are assumed by the T cells and helps control antibody production.

12.12 Induced Pluripotent Stem Cells

Cells can reprogram themselves. Metabolic reprogramming was encountered in Chapter 4 and, as will be explored in the next few chapters, reprogramming, metabolic and more, is fundamental to cancer development and metastasis. A long-standing belief has been that cells follow a unidirectional path from less-differentiated "primitive" or "simpler" form to more differentiated and eventually terminally differentiated, end-use state. This path is, of course, the one followed during embryonic development. However, it is not, as has been thought, the only one available. Cells can be reprogrammed from one state to another, either less or more highly differentiated, and they can reprogram themselves.

This plastic ability manifests itself in a particularly dramatic fashion in the so-called induced pluripotent stem (iPS) cells. These are pluripotent embryonic-like stem cells derived in the laboratory from more highly differentiated somatic cell lines. These cells can then be made to differentiate into cell types that differ from the original adult cells. This back and forth route is not the only one that can be taken to generate different cell types. Cells can also be induced to transdifferentiate. That is, they can be made to directly transform from one kind of differentiated 'end-use' cell to another without going through the intermediate and simpler pluripotent stem cell stage.

A recently presented example of the iPS-intermediate route involving neurons is as follows. Cells taken from skin of elderly patients suffering from ALS were reprogrammed to become iPS cells and then stimulated to differentiate into motor neurons. In this example, the patients had a mutation in the SOD1 gene (discussed in Part III) but the same approach might be applicable to people suffering from sporadic forms of the disease. By using nuclear reprogramming, motor neurons can be studied to isolate factors and processes responsible for ALS and other diseases. A second recently published example, this time involving nuclear reprogramming of cells from the pancreas, illustrates the use of the alternative, transdifferentiation route. In this situation, pancreatic exocrine cells were directly converted into insulin-secreting beta cells. The potential medical utility of cells generated through these transformations is apparent. Donor cells taken elsewhere in the body of an individual may be used not only to study the disease progression and isolate factors responsible for the disease, but also to replace diseased and lost cells, and if needed to grow and regenerate entire organs and tissue.

Sitting at the center of the cellular regulatory network responsible for establishment and maintenance of the iPS cells is a small set of transcription factors, signaling pathway components, and epigenetic factors. This exclusive club includes most centrally transcription factors, OCT4, SOX2, NANOG, and Tcf3, the latter a downstream member of the Wnt signaling pathway. These four TFs in turn activate a larger group of transcription factors, chromatin remodeling enzymes, histone modifiers, and microRNAs. In particular, the four TFs occupy promoter sites for a number of microRNAs. Some of these are upregulated immediately while other sites remain silent. Promoter sites for the latter group are co-occupied by Polycomb group proteins that maintain the silent state. These microRNAs are thus poised for upregulation later in a cell-specific manner.

Further Reading

Jones PA and Baylin SB 2003 The fundamental role of epigenetic events in cancer. *Nature Rev. Genet.* 3: 415–428.
Yoo CB and Jones PA 2006 Epigenetic therapy of cancer: past, present and future. *Nature Rev. Drug Disc.* 5: 37–50.

Nucleosomes

Khorasanizadeh S 2004 The nucleosome: from genomic organization to genomic regulation. *Cell* 116: 259–272.

Epigenetic Marks

Fraga MF et al. 2005 Loss of acetylation at Lys16 and trimethylation at Lys20 of histone H4 is a common hallmark of human cancer. *Nat. Genet.* 37: 391–400.
Shogren-Knaak M et al. 2006 Histone H4-K16 acetylation controls chromatin structure and protein interactions. *Science* 311: 844–847.

DNA Methylation

Bird A 2002 DNA methylation patterns and epigenetic memory. *Genes Dev.* 16: 6–21.
Robertson KD 2005 DNA methylation and human disease. *Nat. Rev. Genet.* 6: 597–610.

Polycomb and Trithorax Group Proteins

Bracken AP et al. 2006 Genome-wide mapping of Polycomb target genes unravels their roles in cell fate transitions. *Genes Dev.* 20: 1123–1136.
Lee Mg et al. 2007 Demethylation of H3K27 regulates Polycomb recruitment and H2A ubiquitination. *Science* 318: 447–450.
Pasini D et al. 2008 Coordinated regulation of transcriptional repression by the RBP2 H3K4 demethylase and Polycomb-repressive complex 2. *Genes Dev.* 22: 1345–1355.
Schuettengruber B et al. 2007 Genome regulation by Polycomb and Trithorax proteins. *Cell* 128: 735–745.
Schwartz YB and Pirrotta V 2007 Polycomb silencing mechanisms and the management of genomic programmes. *Nat. Rev. Genet.* 8: 9–22.

Histone Acetylation and Deacetylation

Lee KK and Workman JL 2007 Histone acetyltransferase complexes: one size doesn't fit all. *Nat. Rev. Mol. Cell Biol.* 8: 284–295.
Minucci S and Pelicci PG 2006 Histone deacetylase inhibitors and the promise of epigenetic (and more) treatments for cancer. *Nat. Rev. Cancer* 6: 38–51.

Histone Methylation and Demethylation

Barford MT and Richard S 2005 Arginine methylation: an emerging regulator of protein function. *Mol. Cell* 18: 263–272.
Chang S et al. 2007 JMJD6 is a histone arginine demethylase. *Science* 318: 444–447.
Klose RJ and Zhang Y 2007 Regulation of histone methylation by demethylimination and demethylation. *Nat. Rev. Mol. Cell Biol.* 8: 307–318.
Shi Y and Whetstone JR 2007 Dynamic regulation of histone lysine methylation by demethylases. *Mol. Cell* 25: 1–14.

Readout Modules

Bannister AJ et al. 2001 Selective recognition of methylated lysine 9 on histone H3 by the HP1 chromo domain. *Nature* 410: 120–124.
Jacobs SA and Khorasanizadeh S 2002 Structure of HP1 chromodomain bound to a lysine 9-methylated histone H3 tail. *Science* 295: 2080–2083.
Lachner M et al. 2001 Methylation of histone H3 lysine 9 creates a binding site for HP1 proteins. *Nature* 410: 116–120.
Nielsen AL et al. 2001 heterochromatin formation in mammalian cells: interaction between histones and HP1 proteins. *Mol. Cell* 7: 729–939.
Yang XJ 2004 Lysine acetylation and the bromodomain: a new partnership for signaling. *BioEssays* 26: 1076–1087.

Cooperativity in Gene Silencing

Fuks F 2005 DNA methylation and histone modifications: teaming up to silence genes. *Curr. Opin. Genes Dev.* 15: 1–6.
Narlikar GJ, Fan HY and Kingston RE 2002 Cooperation between complexes that regulate chromatin structure and transcription. *Cell* 108: 475–487.
Smallwood A et al. 2007 Functional cooperation between HP1 and DNMT1 mediates gene silencing. *Genes Dev.* 21: 1169–1178.
Zhang Y and Reinberg D 2001 Transcription regulation by histone methylation: interplay between different covalent modifications of the core histone tails. *Genes Dev.* 15: 2343–2360.

MicroRNAs

Bartel DP 2005 MicroRNAs: genomics, biogenesis, mechanism, and function. *Cell* 116: 281–297.
Colmenares SU et al. 2007 Coupling of double-stranded RNA synthesis and siRNA generation in fission yeast RNAi. *Mol. Cell* 27: 449–461.
Han J et al. 2006 Molecular basis for the recognition of primary microRNAs by the Drosha-DGGR8 complex. *Cell* 125: 887–901.
Houwing S et al. 2007 A role for Piwi and piRNAs in germ cell maintenance and transposon silencing in zebrafish. *Cell* 129: 69–82.
Kim VN 2005 MicroRNA biogenesis: coordinated cropping and dicing. *Nat. Rev. Mol. Cell Biol.* 6: 376–385.
Meister G and Tuschl T 2004 Mechanism of gene silencing by double-stranded RNA. *Nature* 431: 343–349.
Morris KV et al. 2004 Small interfering RNA-induced transcriptional gene silencing in human cells. *Science* 305: 1289–1292.
Sontheimer EJ 2005 Assembly and function of RNA silencing complexes. *Nat. Rev. Mol. Cell Biol.* 6: 127–137.

MicroRNAs and Disease

Calin GA et al. 2004 Human microRNA genes are frequently located at fragile sites and genomic regions involved in cancers. *Proc. Natl. Acad. Sci. USA* 101: 2999–3004.
Esquela-Kerscher A and Slack FJ 2006 Oncomirs – microRNAs with a role in cancer. *Nat. Rev. Cancer* 6: 259–269.
Jin P, Alisch RS and Warren ST 2004 RNA and microRNAs in fragile X syndrome. *Nat. Cell Biol.* 6: 1048–1053.

Slicer and Dicer Structure

Hutvagner G and Simard MJ 2008 Argonaute proteins: key players in RNA silencing. *Nat. Rev. Mol. Cell Biol.* 9: 22–32.
MacRae IJ et al. 2006 Structural basis for double-stranded RNA processing by Dicer. *Science* 311: 195–198.
Song JJ et al. 2004 Crystal structure of Argonaute and its implications for RISC Slicer activity. *Science* 305: 1434–1437.

MicroRNAs and Cancer

Chang TC et al. 2008 Widespread microRNA repression by Myc contributes to tumorigenesis. *Nat. Genet.* 40: 43–50.
Chang TC et al. 2008 Transactivation of miR-34a by p53 broadly influences gene expression and promotes apoptosis. *Mol. Cell* 26: 745–752.
Cimmino A et al. 2005 miR-15 and miR-16 induce apoptosis by targeting BCL2. *Proc. Natl. Acad. Sci. USA* 102: 13944–13949.
He L et al. 2007 MicroRNAs join the p53 network – another piece in the tumour-suppression puzzle. *Nat. Rev. Cancer* 7: 819–822.

Johnson SM et al. 2005 *Ras* is regulated by the *let-7* microRNA family. *Cell* 120: 635–647.
O'Donnell KA et al. 2005 c-Myc-regulated microRNAs modulate E2F1 expression. *Nature* 435: 839–843.
Raver-Shapira N et al. 2008 Transcriptional activation of miR-34a contributes to p53-mediated apoptosis. *Mol. Cell* 26: 731–743.
Yamakuchi M, Ferlito M and Lowenstein CJ 2008 MiR-34a repression of SIRT1 regulates apoptosis. *Proc. Natl. Acad. Sci. USA* 105: 13421–13426.

Oncomirs

Papagiannakopoulos T, Shapiro A and Kosik KS 2008 MicroRNA-21 targets a network of key tumor-suppressive pathways in glioblastoma cells. *Cancer Res.* 68: 8164–8172.
Baltimore D et al 2008 MicroRNAs: new regulators of immune cell development and function. *Nature Immunol.* 9: 839–845.
O'Connell RM et al 2007 MicroRNA-155 is induced during the macrophage inflammatory response. *Proc. Natl. Acad. Sci. USA* 104: 1604–1609.

Embryonic SCs and microRNAs

Cole MF et al 2008 Tcf3 is an integral component of the core regulatory circuitry of embryonic stem cells. *Genes Dev.* 22: 746–755.
Marson A et al 2008 Connecting microRNA genes to the core transcriptional regulatory circuitry of embryonic stem cells. *Cell* 134: 521–533.
Niwa H 2007 How is pluripotency determined and maintained? *Development* 134: 635–646.

iPS cells and Disease Remediation

Dimos JT et al 2008 Induced pluripotent stem cells generated from patients with ALS can be differentiated into motor neurons. *Science* 321: 1218–1221.
Hanna J et al 2007 Treatment of sickle cell anemia mouse model with iPS cells generated from autologous skin. *Science* 318: 1920–1923.
Quintana E et al 2008 Efficient tumour formation by single human melanoma cells. *Nature* 456: 593–632.
Zhou Q et al 2008 *In vivo* reprogramming of adult pancreatic exocrine cells to β-cells. *Nature* 455: 627–632.

Chapter 13
Tumor Growth

The focus in the discussions of the preceding two chapters was on the three anticancer barriers – DNA damage repair, apoptosis, and senescence. In those discussions, tumor suppressors such as p53 and pRb had a prominent role along with some of the cell cycle checkpointing and DNA damage repair proteins. The emphasis now shifts in this and the following two chapters to growth and metastasis, and the associated oncogenes. This second phase of the examination of cancer begins with tyrosine kinases required for cell-to-cell signaling and coordination. It then shifts to the downstream components of the central growth and survival pathways. Cancerous growth can only be sustained through modifications in cellular metabolism. The discussion of tumor growth continues into the next chapter with an exploration of how tumor cells adjust their metabolism to support and sustain rapid growth and proliferation.

The microenvironment plays a central role in development of a cancer. That involvement is not obvious from the titles of the two chapters just mentioned. However, tyrosine kinase signaling occurs between cells of the tumor microenvironment and is a key to some of the hallmarks of cancer listed in Fig. 9.1 such as sustained angiogenesis. In addition, and most interestingly, the shifts in cellular metabolism by cancerous cells are intimately connected to the cellular microenvironment. The preparation for metastasis, the stage associated with most cancer deaths, begins with these early interactions. Metastasis, and its connections to the microenvironment, and especially to inflammation, is the subject of the concluding chapter of Part II.

13.1 Growth and Survival Signaling Pathways

Cancers arise when cells undergo uncontrolled growth and proliferation. The dysregulated growth pathways that trigger the diseases convey local autocrine and paracrine signals resulting in alterations in expression levels of genes. The cell-to-cell signals, in the form of polypeptide growth factors, bind plasma membrane receptors belonging to the protein tyrosine kinase family, found

M. Beckerman, *Cellular Signaling in Health and Disease*,
Biological and Medical Physics, Biomedical Engineering,
DOI 10.1007/978-0-387-98173-4_13,

only (mostly) in metazoans. Cytoplasmic signals adapters and relays then convey the upstream signals to mitogen-activated protein kinase (MAPK) modules and from there activate transcription factors and alter gene expression patterns (Fig. 13.1). Prominent among the malfunctioning signaling elements associated with cancers are the receptor tyrosine kinases, and the RasGTPase and Myc transcription factors mentioned in earlier chapters. Mutations in these elements are second only to p53 and pRb in their cancer-associated frequencies.

Growth signals are relayed from the receptor tyrosine kinase to the ERK MAP kinase modules through two distinct routes. One of these is Ras independent and the other is Ras dependent. The Ras-independent route utilizes phospholipase C and protein kinase C as key signaling intermediates. Activated RTKs stimulate PLC to produce DAG and IP_3 from PIP_2 leading to activation of protein kinase C, which can then phosphorylate and activate Raf. The Ras-dependent route begins with activation of the adaptor Grb2 by the RTKs. This step leads to activation of the Ras GEF Son-of-sevenless (Sos), which catalyzes the exchange of GDP for GTP. Activated (GTP-bound) Ras is able to recruit Raf to the plasma membrane and activate it. The MAP kinase module has a number of downstream targets among which are immediate early genes and c-Myc.

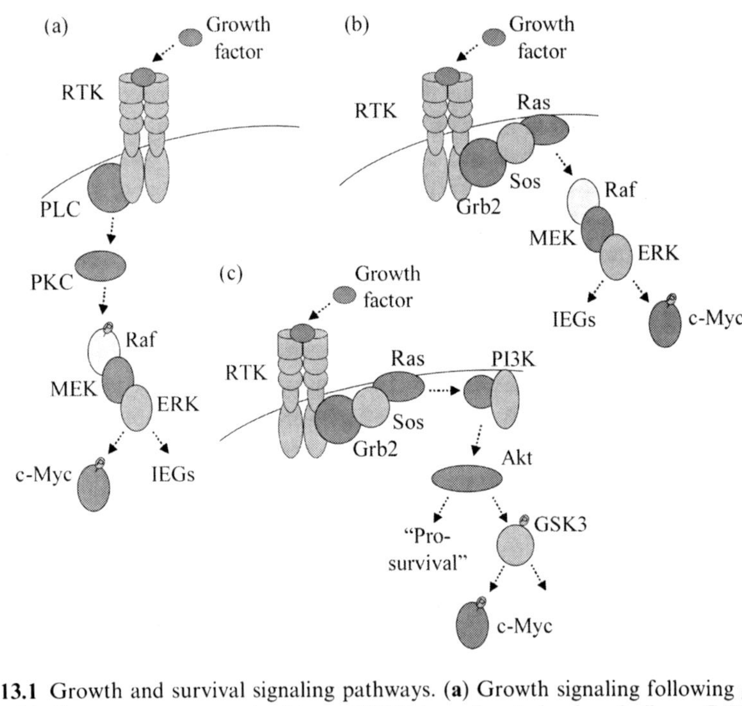

Fig. 13.1 Growth and survival signaling pathways. (**a**) Growth signaling following growth factor binding to receptor tyrosine kinase (RTK) transduced via phospholipase C (PLC) to the central mitogen-activated protein (MAP) kinase module and from there to c-Myc and immediate early gene (IEG) expression. (**b**) Signaling following growth factor binding to RTK transduced via Grb2, Sos, and Ras. (**c**) Survival signaling pathway leading from growth-factor/RTK binding via Grb2/SoS/Ras to protein kinase B/Akt, GSK3 and c-Myc

Growth factor signals activate the PI3K/Akt survival pathway in addition to the MAP kinase module-dependent growth pathways. The key participants in this second major signaling route are depicted in Fig. 13.1 along with the two alternative growth signal routes. Activated Akt promotes survival in part by negatively regulating GSK3, a negative regulator of c-Myc. Protein kinase B/ Akt also promotes survival by inhibiting apoptosis and influencing tumor cell metabolism. In this chapter, and the one following, these pathways and components will be examined in detail.

13.2 Receptor Activation Leads to Recruitment of Molecular Adaptors to Docking Sites

Growth factors and hormones such epidermal growth factor (EGF) and insulin bind to receptor belonging to the receptor tyrosine kinase (RTK) superfamily. In humans, there are twenty different families of receptor tyrosine kinases encoded by 59 genes. As was the case for the insulin receptor, these receptors possess cytoplasmic kinase domains similar in structure to other members of the serine, threonine, and tyrosine kinase superfamily. Five families of RTKs have especially prominent roles in carcinogenesis, and have become primary targets of anticancer drugs. These cancer-associated RTKs are listed in Table 13.1. As can be seen in their names, these ligands and receptors mediate signaling between the cellular constituents of the microenvironment.

Table 13.1 Receptor tyrosine kinase families with prominent roles in carcinogenesis and as small molecule kinase inhibitor targets

Receptor tyrosine kinase family	Receptors	Ligands
Epidermal growth factor	EGFR, ErbB2, ErbB3, ErbB4	EGFR: EGF, TGFα, amphiregulin; EGFR, ErbB4: β-cellulin, epiregulin, HB-EGF; ErbB3, ErbB4: neuregulins
Fibroblast growth factor	FGFR1, FGFR2, FGFR3, FGFR4	22 FGF family members
Hepatocyte growth factor/MET	MET	HGF/SF
Platelet-derived growth factor	PDGFR-AA, PDGFR-AB, PDGFR-BB	α-PDGF, β-PDGF
Vascular endothelial growth factor	VEGFR1, VEGFR2, VEGFR3	VEGFA, VEGFB, VEGFC, VEGFD, PLGF

Signaling begins with ligand binding and the resulting placement of the cytoplasmic domains of two RTK chains in close proximity to one another. The resulting cross or autophosphorylation of tyrosine residues situated in their activation loops leads to full activation of the kinase. The activated tyrosine kinase then catalyzes additional phosphorylations on tyrosine resides outside the activation loop opening up docking sites for cytoplasmic signaling molecules possessing Src homology 2 (SH2) and phosphotyrosine binding (PTB) domains. These initial events trigger the recruitment of additional signaling elements and seed the assembly of signaling complexes in the vicinity of the activated receptors as depicted in Fig. 13.1. This assembly operation is facilitated by lipid binding and protein–protein interaction modules. The particular mix of docking sites for each receptor type leads to the recruitment of particular sets of adapters and other signaling intermediaries leading to activation of downstream serine/threonine kinases and their associated cassettes and modules.

The first family listed in Table 13.1 is the epidermal growth factor (EGF) family, a major target of anticancer drug development. There are four receptors in this family – the EGF receptor (EGFR), ErbB2 (HER-2), ErbB3 (HER-3), and ERBB4 (HER-4). The second member of this family, ErbB2, has no ligand while ErbB3 is almost completely devoid of kinase activity. These receptors heterodimerize with other family members when transducing signals into the cell. Phosphorylation of their cytoplasmic segments exposes docking sites for adapter proteins such as Shc and Grb2, for phospholipase Cγ, and for the transcription factor STAT1.

Fibroblast growth factors activate signaling pathways similar to those stimulated by EGF. These receptors are activated by fibroblast growth factor and heparan sulfate proteoglycan (HSPG) heterodimers. Like the insulin and insulin-like growth factor receptors, signaling is mediated by docking site intermediates. Whereas insulin and IGF signaling requires IRS proteins, fibroblast growth factor signaling requires membrane-associated adaptor proteins such as FRS2α,β, which upon tyrosine phosphorylated by the receptors provides docking sites for SH2 and PTB domain-bearing signaling elements. In addition to differences in docking sites, there are differences in the feedback loops with a somewhat more complex circuitry present for FGF signaling compared to EGF signaling.

Hepatocyte growth factor, or scatter factor, binds to its receptor called Met. Activation and phosphorylation provide docking sites for a number of adapters, prominent among which is Gab1, which possesses a unique Met-binding site distinct from the typical SH2 and PTB domains. The Gab1 proteins function as a scaffold for binding of other signaling elements. Chief among these are the protein tyrosine phosphatase 2 (Shp2), phospholipase C, PI3K, and the adapter protein Crk. Once activated through interactions with Gab1, Shp2 activates ERK MAPK signaling. In epithelial cells, PI3K and ERK signaling induces cell dispersion, or scattering. Signals relayed through Ras, Rac1, and PAK regulate cytoskeleton remodeling, while Crk signaling induces the dissolution of adherens junctions.

Vascular endothelial growth factors signal through VEGFR1, VEGFR2, and VEGFR3. VEGFR1 is another example of a receptor with greatly reduced tyrosine kinase activity, and the most important of these receptors with regard to angiogenesis is VEGFR2. A central feature of VEGFR2 signaling is there is lack of growth stimulation through standard Shc – Grb2 – Sos to the ERK MAPK module (Fig. 13.1(b)). In its place, there is a signaling route involving the sequential activation of phospholipase C, protein kinase C, Ras, and c-Raf leading to ERK1/2-mediated gene expression and cellular proliferation (Fig. 13.1(a)). Other signals relayed by means of FAX/Paxillin and p38 MAPKs stimulate migration. This situation is not the case for PDGF signaling, which utilizes signaling routes resembling those for EGF.

Platelet-derived growth factors are growth factors for fibroblasts, smooth muscle cells, and glia. They are composed of homo- and hetero-dimers of two chains, an A-chain, and a B-chain. These bind PDGFRs, also homo- and hetero-dimeric constructions from PDGFR-A and PDGFR-B. These members of the PRK family have roles in driving mesenchymal cell migration and proliferation. Activation and signaling are tightly controlled through joint positive and negative stimulation, feedback loops, and endocytosis. In fibroblasts, signal intensity is important. Low-intensity signals induce migration while high signal intensity triggers a shift to proliferation. The reason for this dependence can be explained by a simple chemotactic model. When the fibroblasts are far from the site of secretion concentration is low, and the cells migrate up the concentration gradient. When they arrive near the site of secretion the concentration is high and this conveys the message that they have arrived at their destination and should begin to proliferate.

13.3 Ras and Other Small GTPases Link Adaptors to Downstream Signaling Elements

One of the main functions of the adaptors, once they are tyrosine phosphorylated and activated, is to recruit and activate small GTPases. Ras is one of the most frequently mutated proteins encountered in human cancers. It lies upstream of the ERK MAPK signaling module that regulates cellular growth, and also proliferation and differentiation. As shown in Fig. 13.1, the Grb2 adaptor recruits Son-of-sevenless (Sos), the GEF for Ras, to the plasma membrane. Once activated by Grb2, Sos catalyzes the exchange of GDP for GTP thereby activating Ras, which then activates Raf leading to ERK1/2 signaling.

Another key family of small GTPase involved in relaying growth factor signals is the Rho family, which includes RhoA, Rac1, and Cdc42. Members of this family of small GTPases are activated by a large number of GEFs collectively referred to as Dbls, so-named for their earliest identification as a transforming gene in diffuse B-cell lymphomas. Rho GTPases regulate the actin cytoskeleton and by that means regulate cell adhesion and motility.

Signaling by Ras and Rac is coordinated by Sos-1. This protein acts as a GEF for Ras and Rho. These alternative activities are associated with formation and dissolution of two plasma membrane-associated complexes. In one, Sos-1 and

Grb2 form a complex that activates Ras. In the other, Sos-1 complexes with a RTK substrate called Eps8 and with a scaffolding protein termed E3b1. This second complex along with PI3K and ERK mediation activates Rac. The two modes of signaling differ in their duration; Ras signaling is short lived, while Rac signaling, which ensues upon dissolution of Sos-1/Grb2 complexes, is prolonged.

13.4 Many of the Growth Signaling Proteins Function as Oncogenes

Two other pathways activated by growth factor and hormone binding have as their triggering events the recruitment of c-Src and STAT proteins to the docking sites on the RTKs. c-Src is a member of the Src family of nonreceptor tyrosine kinases. It has a prominent role as an organizer of signaling complexes involving the cytoskeleton that are important for motility and metastasis. It will be discussed in detail in Chapter 15. Signal transducer and activator of transcription (STAT) proteins function as upstream cytoplasmic signaling elements and as downstream transcription factors. They are activated by cytokine and growth factors receptors. In the former instance, ligand binding to cytokine receptors results in the activation of Janus kinases situated at the cytoplasmic face of the cytokine receptors. These kinases, in turn, recruit and activate the STATs, which then trranslocate to the nucleus where they induce gene expression. In the latter case, STATs are recruited to the plasma membrane and activated directly through the tyrosine kinase activity of the growth factor receptors. Alternatively, these transcription factors can undergo phosphorylation by NRTKs in c-Src-organized complexes.

Ras and Raf, the uppermost members of the MAP kinase signaling module, are prominent members of the growth signaling pathways. There are three vertebrate Raf proteins: A-Raf, B-Raf, and C-Raf (Raf-1). The domain organization of B-Raf, depicted in Fig. 13.2(b), reveals the presence of three conserved regions designated as CR1, CR2, and CR3. The N-terminal CR1 region has within it a Ras-binding domain and also cysteine-rich regions needed for membrane recruitment while the CR3 regions contain the catalytic site (activation segment). As indicated in the table, ASBC mutations in B-Raf are prominently associated with several cancers. The most frequently encountered mutation is the V599E (Val 599 → Glu) mutation in the kinase domain. This mutation disturbs the protein's inactive conformation and renders the protein 460 times more catalytically active.

The domain organization of K-Ras is depicted in Fig. 13.2(a). Shown in the figure are the locations of the predominant mutations – to codons 12, 13, and 61. These are found in the P-loop (12 and 13) and Switch II region (61). Mutations in codons 12, 13, 61 impair the intrinsic GTPase activity of the protein and render the protein insensitive to GAP-catalyzed hydrolysis. The protein thus remains bound to GTP and stuck perpetually in the "ON" position.

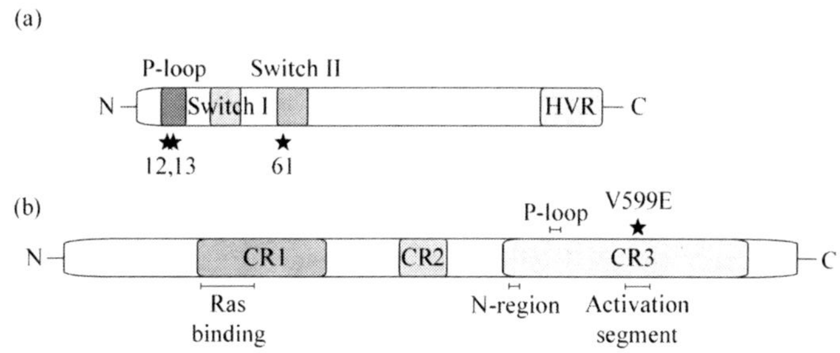

Fig. 13.2 Domain organization of (**a**) K-Ras and (**b**) B-Raf

There are three Ras family proteins: H-Ras, K-Ras, and N-Ras. Several of the most prominent K-Ras mutations and the cancers affected are listed in Table 13.2. In addition to these, N-Ras mutations are encountered with high frequencies in melanomas and H-Ras mutations in bladder cancer.

Table 13.2 Upstream components of the growth pathway overexpressed (EGFR and ErbB2) or mutated (K-Ras and B-Raf) in cancers. The altered signaling proteins (oncoproteins) and the frequency at which these alterations are encountered in the cancers are listed

Oncoprotein	Cancer (frequency)	Oncoprotein	Cancer (frequency)
EGFR	Breast (30%)	K-Ras	Bilary tract (30%)
	Glioblastoma (40%)		Pancreas (60%)
	SCCHN (80–90%)		Colon (30–40%)
	NSCLC (40–80%)		Liver (30%)
ErbB2	Breast (30–60%)	B-Raf	Melanoma (30–60%)
	Lung (40%)		Thyroid (30–50%)
	Pancreas (90%)		Colon (5–20%)
	Thyroid (60%)		Ovarian (30%)

NSCLC, Non-small cell lung cancer; SCCHN, Squamous-cell carcinoma of the head and neck.

The GEF for plasma membrane associated Ras is Sos (Son of sevenless). It works in the following way to speed up the dissociation of GDP from Ras. The Ras molecule consisting of 188 amino acids residues contains two flexible regions on its surface. These surface regions can alternate between several conformational substates, and they are referred to as Switch I and Switch II. In its inactive state, the Ras molecule binds GDP tightly. Sos binding produces a tertiary Sos-Ras-GDP complex that stabilizes Ras in a conformation in which Switch I has swung out to open the binding pocket. A portion of Sos stabilizes Switch II in an alternative conformation, and this change plus an altered electrostatic environment further weakens the binding of the phosphate group of GDP and its associated magnesium ion to Ras. The changes in the switch regions increase the dissociation rate by several orders of magnitude and allow GDP to speedily exit the binding pocket.

The RasGAP works in the following manner. Arginine and lysine residues have long side chains and are positively charged under physiological conditions. When bound to the Ras-GTP complex, the RasGAPs extend an "arginine finger" into the active site that neutralizes negative charges in its vicinity. A network of hydrogen bonds forms. These stabilize the transition state and promote cleavage of the phosphodiester bond linking the gamma phosphate group to the remaining GDP molecule. The result is an increase in the rate of hydrolysis by a factor of a thousand. These hydrolysis-promoting actions fail to occur in the presence of mutations to the three aforementioned codons.

The ErbB2 receptor differs from the other ErbBs. As noted earlier, it does not require a ligand, is the favored binding partners of the other ErbBs, and prefers the others as binding partners over forming homodimeric complexes with itself. Examination of the three-dimensional crystal structure of the chain reveals the reasons for this unusual behavior. In its native form, the protein assumes an activated conformation resembling the other receptors bound to their ligands. In more detail, ErbB receptor dimerization is mediated by a dimerization arm that is raised and primed to mediate dimerization upon ligand binding. The ErbB2 receptor also has a dimerization arm, but in this case the arm is oriented in an active conformation in the absence of ligand binding. This active conformation is maintained by structural features that inhibit ligand binding (and homodimerization).

13.5 MAP Kinase Signaling Modules

Two major cytoplasmic signaling nodes have been introduced and described in the previous chapters. The first of these was the NF-κB signaling node. This module has as its central elements the NF-κB proteins serving as output units, IκB inhibitors, and activating IKK proteins that function as input units. This node is strongly associated with the regulation of inflammation. The second node introduced was the mTOR cassette consisting of mTOR and S6K proteins operating as output units, Rheb GTPase linker, and the tuberous sclerosis complex TSC as its input unit. This module was discussed earlier with an emphasis on insulin signaling. When all inputs and outputs are taken into account this node becomes a central regulator of metabolism and growth. The mitogen activated protein (MAP) kinase cassette constitutes a third major intracellular signaling node. This cassette, presented in Fig. 13.3, is intimately associated with cellular growth and proliferation, and with stress responses and inflammation, and for that reason will now be introduced and discussed.

The existence of the MAP kinase signaling module and its cellular functions were uncovered starting in the mid 1980s. Originally referred to as microtubule-associated kinases, these signaling units were first identified by Sturgill and Ray in 1986–1988 as insulin-stimulated signal transducers. Similar units were found in both yeasts (the budding yeast *Saccharomyces cerevisiae* and the fission yeast *Schizosaccharomyces pombe*) and mammals during the 1989–1991 time period

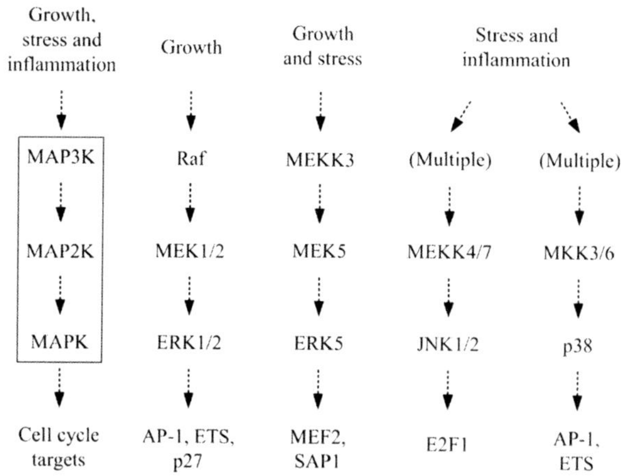

Fig. 13.3 Four major mammalian MAP kinase pathways. The stereotypic organization of the MAP kinase pathways is depicted in the leftmost column. The three elements comprising the core MAP kinase module are shown in the *boxed area*

with the yeast MAP kinases providing key insights into the biophysical properties of the signaling module.

Yeasts respond to mating pheromones and stresses by altering their patterns of gene expression. Environmental and pheromone signals are sent from the plasma membrane to the nucleus through a set of parallel pathways named for the last in a series of protein kinases that serve as the central signal transducers. The kinases that lend their name to the pathways are the mitogen-activated protein (MAP) kinases. Six of these parallel pathways have been found in yeasts, and these findings helped spur the discovery of a similar set of parallel pathways operating in mammalian cells. The mammalian pathways convey cytokine, stress, and growth signals that are sent to them from the plasma membrane to the nucleus where they influence transcription of target genes. The four mammalian MAP kinase pathways are depicted in Fig. 13.3. One of these, the extracellular signal-regulated kinase (ERK) pathway, carries growth signals. The second pathway is known as the ERK5 pathway and also as the big mitogen-activated protein kinase 1 (BMK1) pathway. It is not as well characterized as the others; it conveys both growth and stress signals. The other two pathways, the c-Jun NH_2-terminal kinase (JNK) pathway and the p38 MAP kinase pathway relay environmental stress signals (e.g., from ionizing radiation and osmotic shock) and inflammatory signals transmitted by cytokines.

Three kinases, activated sequentially, comprise the MAP kinase module. Their stereotypic arrangement is illustrated in the leftmost portion of Fig. 13.3. The first or uppermost kinase in the cascade is referred to as a MAP kinase kinase kinase, or MAP3K. It is a serine/threonine kinase that phosphorylates a pair of serine and threonine residues on the middle element of the MAP kinase module. The middle kinase, the MAP kinase kinase, or MAP2K, is a dual specificity kinase. It phosphorylates a tyrosine residue and a threonine residue on the third component of the cascade, the MAP kinase. The targeted residues are arranged as Thr-X-Tyr, where X is either Glu, Pro, or

Lys for the ERK, JNK, and p38 pathways, respectively. Once they are phorphorylated, the third and last kinases in the module typically stimulate the transcriptional activity of members of several families of transcription factors. Some of these transcriptional targets are shown in Fig. 13.3.

In most instances, additional residues located well away from target phosphorylation sites help confer substrate specificity. These sites are referred to as docking motifs or docking domains. The presence of docking sites on kinases and their substrates increases the signaling specificity. These sites are typically located far from the catalytic site of the kinase and far from the phosphoacceptor site of the substrate. An example of a kinase family that uses this form of matching is the MAP kinase family. In this family, the docking sites are located in a C-terminal domain of the kinase. The docking site is characterized by the presence of several acidic amino acids and is matched to a cluster of basic amino acids in the matching docking sites used by upstream activators, deactivators, and downstream substrates.

13.6 The MAP Kinase Modules and Their Substrates Function as Dynamical Circuits

Recall from Chapter 9 that the cell cycle progression depends upon several key dynamical properties of the modules and component units. In particular, feedback – positive and negative – was essential for the orderly progression from one phase to the next. Negative feedback produces ultrasensitive responses, positive (and double negative) feedback generates bistability and hysteresis, and multisite phosphorylation serves as a timing mechanism that orders events and set sharp thresholds between them.

Paralleling the cell cycle examinations discussed in Chapter 9, the MAP kinase module has been subjected to studies of its dynamical properties. Examples of some of the same underlying principles have been found in the case of the MAP kinase cascades that drive the maturation of frog (*Xenopus*) oocytes. Two MAP kinase modules were found to exhibit bistable responses. The first of these is the p42-MAP kinase cascade in which ultrasensitivity and positive feedback generate bistability. The second was the JNK cascade, which seems to produce bistability, but by an entirely different mechanism from that utilized by the p42-MAP cascade.

Mammalian MAP kinase modules exhibit not only bistable properties but also considerable flexibility in their responses. MAP kinase modules such as the ERK module can change its behavior from monostability to ultrasensitivity to bistability in accordance with changing protein kinase C and protein phosphatase feedback signal strengths. Bistable behavior can also be generated by multisite phosphorylation arising from dual phosphorylation-dephosphorylation cycles.

The oocyte maturation circuit serves as an example of how MAP kinase circuits operate. Progesterone is a steroid hormone. In the circuit depicted schematically in Fig. 13.4, the progesterone hormone signals through a cell surface receptor. In Chapter 4, steroid signaling through nuclear receptors was discussed. That route

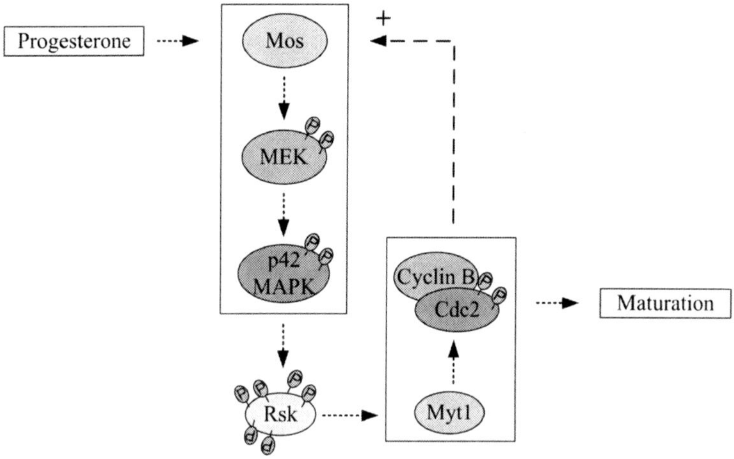

Fig. 13.4 A MAP kinase – cell cycle regulatory circuit that drives the maturation of *Xenopus* oocytes

of signaling involves ligand-activated gene transcription mediated by the nuclear receptors. It is complemented by rapid signaling through cell surface (extra-nuclear) steroid receptors and serine/threonine kinases as depicted above.

The circuit presented in Fig. 13.4 incorporates several of the design principles. The transition of the oocyte from an immature egg into a fertilizable one is known as maturation. This transition is triggered by progesterone, and in it a number of positive feedback loops ensure that the *Xenopus* maturation transition occurs in a sharp, all-or-nothing one manner. The circuitry can be decomposed into four parts. The first is the upstream cell-surface signaling taking place at and near the PR leading to activation of the Map kinase cascade, the second component of the circuit. The MAP kinase module outputs signals to downstream serine/threonine kinases belonging to the Rsk family which, in turn, influences and changes the operation of the cell cycle regulatory machinery centered about cyclin B/Cdc2.

The progesterone receptor is a putative 7TM receptor and signals through the G-protein subunits Gα and Gβγ leading to inhibition of adenylyl cyclase leading to reduced cAMP production and protein kinase A signaling. Downstream events from PKA signaling reduction are quite interesting and still being uncovered. Progesterone and also growth factors such as insulin convey signals that result in increased translation of Mos and cyclin B mRNAs thereby activating both modules.

13.7 Active and Inactive Conformations of Protein Kinases

Studies of how catalysis is achieved by proteins kinases have revealed that the active conformations for all serine, threonine, and tyrosine kinases are remarkably similar to one another. This is not the case for the inactive conformations. In this situation, binding to a specific set of substrates becomes the paramount technical challenge. As a result the inactive conformations of the catalytic domains of the kinases exhibit considerable plasticity and are broadly different from one another.

The catalytic core of MEK1, the middle element in the ERK pathway, is shown in Fig. 13.5. The structure revealed by X-ray crystallography consists of a small, N-terminal lobe, a linker, and a large C-terminal lobe. The small lobe is composed mostly of β-sheets while the large lobe is α helical. A cleft formed by elements of the small and large lobes functions as the catalytic site. The large lobe binds the substrate peptide or protein, while the small lobe supplies the main site for attachment of the ATP molecule. The ATP molecule sits at the base of the cleft and provides structural support by helping to fix and maintain the orientations of the large and small lobes with respect to one another. The interface between the two lobes also provides the binding surface for substrates. In the crystal structure presented in Fig. 13.5, there is a unique inhibitor-binding pocket adjacent to the MgATP binding site. Binding of the (noncompetitive) inhibitor induces conformational changes leading to a closed but catalytically inactive conformation of the enzyme.

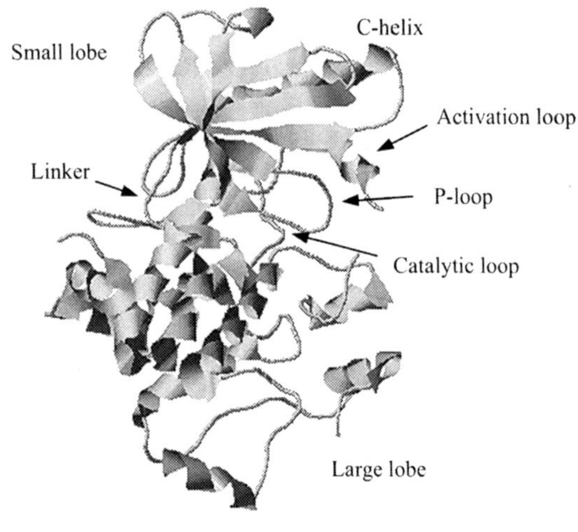

Fig. 13.5 Three-dimensional structure of the catalytic domain of dual-specificity MEK1 enzyme in an inactive conformation determined by means of X-ray crystallography. The figure was prepared using Jmol with atomic coordinates deposited in the PDB under accession code 1s9j

13.8 Oncogene Addiction

Oncogene addiction, a term coined by Bernard Wienstein in 2000–2002, encapsulates the observation that in escaping the barriers to cancer the cellular circuitry is altered from a normal one to one in which there is a stronger than normal dependence on one or a few components. Under normal healthy conditions, multiple redundant circuits and attendant positive and negative feedback loops regulate growth, proliferation, and apoptosis. The circuitry ensures cellular and tissue homeostasis and robustness against normal perturbations in concentrations of the components. In a cancer cell, the circuitry established and maintained by oncogenes and tumor suppressors is different from that of a healthy cell, just as the wiring of blood vessels in the tumor stroma is different from the normal one. In particular, the cancer cells can develop a strong dependence on one or a few gene products for their survival, and these proteins may potentially serve as an "Achilles heal" to cancer.

The targeting of the Bcr-Abl gene fusion and resulting constitutive activation is the first of a number of success stories in treating cancers through molecular targeting. Another more challenging example is provided by the epidermal growth factor receptor (EGFR). Overexpression of this receptor is encountered in ovarian, breast, colon, lung, and brain tumors. Roughly 10% of patients with non small cell lung cancers (NSCLCs) respond dramatically to treatment with the small molecule inhibitor gefitinib. The tumors successfully attacked by this agent are those with activating mutations in the EGFR centered about the ATP binding pocket. Gefitinib binding disrupts the catalytic activity of the kinase domain thereby inhibiting signaling. The mutated receptors selectively activate Akt and activator of transcription (STAT) pathways promoting survival and rendering the cancerous cells resistance to conventional chemotherapy.

A common feature of the tumors treated with tyrosine kinase inhibitors (TKIs) is the development over time of resistance to the inhibitors leading to regrowth of the tumors. The acquired resistance is arrived at in several ways. One of the ways of evading the effects of the TKIs is to develop an additional mutation in the activation loop that prevents binding by the TKI developed to bind to sites containing the initial mutation. The way that this may be overcome necessitates the engineering of a new TKI that selectively binds the modified kinase. Acquisition of new mutations in the catalytic domain of the kinase target is not the only route to TKI evasion. Another is for the treated cancerous cells to develop an additional mutation in a signaling element that promotes the rerouting of growth-sustaining signals so that protein targeted for treatment is no longer essential for the survival and growth.

13.9 Target-Based Anticancer Therapies

Anticancer drugs have traditionally been developed though empirical, random screening approaches resulting in identification and testing of agents that are cytotoxic and kill cancer cells. During the past few years, a new approach has been gaining favor and is believed to hold promise for more successful eradication of cancer and less damaging to the body. The target-based approaches are mechanistic in nature and rather than being cytotoxic are cytostatic. They are based on development of inhibitors of specific cell-signaling pathways and components thereof.

An example where this approach has been used with great success is in treating chromic myeloid leukemia (CML). In this disease, the Abl gene has translocated from chromosome 9 to chromosome 22, where it fuses with the Bcr gene. The altered form of chromosome 22 is known as the Philadelphia chromosome. It results in the constitutive activation of the Abl tyrosine kinase and leukemia. The objective in the target-based approach to treatment of CML is to devise a drug that will bind and turn off the inappropriate Abl signaling and only that signaling.

A short Asp-Phe-Gly motif that is critical for catalysis lies at the base of the activation loop of the Abl catalytic domain. It is flipped by about 180% in an inactive conformation compared to the active conformation in which the activation loop is open and extended. The drug imatinib recognizes the flipped motif and stabilizes the kinase in the inactive conformation. Most importantly, it does not bind the closely related tyrosine kinase Src or other similar kinases. These others lack the flipped Asp-Phe-Gly motif, and thus the designed inhibitor is highly specific to Abl.

13.10 Myc Protein Structure and Function

Myc proteins function as throttles for growth and proliferation. Members of this family include c-Myc (Myc), N-Myc, and L-Myc. Signals relayed into the nucleus by these transcription factors influence growth, proliferation, apoptosis, and angiogenesis. They often act in concert with other oncoproteins such as Ras to drive tumorigenesis. The Myc family of transcription factors is part of a larger family of transcription factors known as basic, helix-loop-helix, leucine zipper (bHLH-LZ) proteins. c-Myc possesses an N-terminal globular transactivation domain, a central unstructured region, and a C-terminal bHLH-LZ domain (Fig. 13.6). This structure supports the formation of heterodimers between Myc and other transcription factors, and in that form Myc binds DNA and influences transcription, positively or negatively. A network of

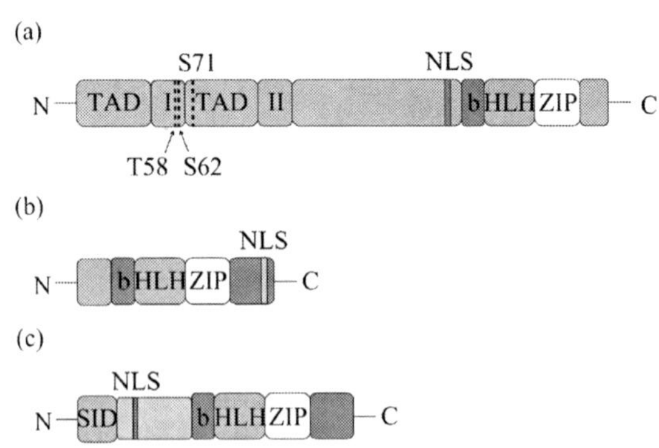

Fig. 13.6 Domain structure of the Myc network proteins. (**a**) Myc (c-Myc), (**b**) Max, and (**c**) Mad

bHLH-LZ binding partners has been identified. These include the closely related Myc-associated factor x, or Max, and Mad proteins.

The Myc, Max, and Mad transcription factors presented in Fig. 13.6 influence transcription in large measure by recruiting proteins that interact with DNA and histones thereby making transcription either easier or harder. The transcriptional activation region (TAD) of Myc and the Sin3 interaction domain (SID) of Mad provide recruitment/attachment sites for these factors. Generally speaking the factors recruited to the TAD positively influence transcription while those recruited to the SID negatively affect transcription. Correspondingly, complexes formed about Myc-Max activate transcription while complexes build from Mad-Max heterodimers repress it. Myc box-1 (MB1) and Myc box-2 (MB2) are two highly conserved regions within the Myc transactivation domain.

The basic regions of these proteins interact with the DNA, and the leucine zippers mediate dimer formation. The heterodimers establish multiple contacts with segments of DNA known as E-boxes. These DNA sequences are of the form CACGTG. E-boxes are found in promoters of a large number of proteins. While a cell is quiescent and not growing, the cellular levels of Myc are low. Myc is a short-lived protein. When the cell enters a growth phase triggered, for example, by the receipt of mitogenic signals, Myc expression is increased and its lifetime is extended. The Myc proteins, in turn, stimulate the transcription of numerous proteins.

13.11 Phosphorylation and Polyubiquitination Sculpt Myc-Mediated Gene Transcription

Myc proteins are regulated by posttranslational modifications, most notably, by phosphorylation and ubiquitination. They are phosphoproteins and like many others discussed in this text possess multiple phosphorylation sites distributed throughout the protein; these are listed in Table 13.3. Several of these phosphorylation sites (e.g., S71, S82, and S293) contain proline residues at the + 1 position and are candidate sites for proline-directed kinases.

Table 13.3 Potential Myc phosphorylation sites. The two-bolded residues are discussed in detail in the text

Location	Phosphorylated residues
N-terminal region	**T58**, **S62**, *S71*, *S82*, and S164
Acidic region (240–262)	5 sites
C-terminal region	*S293*, T343, S344, S347, and S348

The best understood of these phosphorylation sites are T58 and S62. These sites are phosphorylated by kinases associated with growth and cell cycle control. The Ser 62 site is phosphorylated by proline-directed kinases such as the mitogen-activated protein (MAP) kinases, while GSK3 catalyzes phosphorylation at Thr 58. Phosphorylation of Myc at Ser 62 by the mitogens stabilizes the protein and extends its lifetime, while phosphorylation of Myc at Thr 58 by GSK3 targets the protein for ubiquitination and proteolytic destruction. The PI3K-Akt survival pathway has as one its downstream targets the GSK3 protein. Phosphorylation of GSK3 by Akt redirects the former away from Myc thereby extending its lifetime. Thus Myc serves as a point of convergence for growth and survival signals.

In response to phosphorylation by Erk and GSK3, nearby lysine residues Lys 48 and Lys 63 are polyubiquitinated by E3 ubiquitin ligases. Three families of E3s participate in these regulatory activities – SCF-Fwb7, SCF-Skp2, and HECTH9. Phosphorylation at Ser 62 occurs before c-Myc can be phosphorylated at Thr 58, so that a sequence of actions takes place that first turns on and then turns off c-Myc. In response to MAPK phosphorylation at Ser 62 and the subsequent GSK3 phosphorylation at Thr 58, Pin1 binds followed by PP2A binding resulting in the deephosphorylation at Ser 62, recruitment of SCF-Fbw7, and Lys 48 polyubiquitination and degradation. Alternatively, recruitment of the Skp2 to Myc-Max at promoters results in Lys 48 polyubiquitination that enhances gene expression. That is, Skp2 acts as a coactivator that first stimulates transcription and then terminates it through polyubiquitination and proteasomal degradation. The third E3 ligase complex, HECTH9, also stimulates transcription by polyubiquitinating Myc at Lys 63. This activity results in recruitment of coactivators such as CBP and p300.

13.12 Regulation of Cellular Growth by Ras, Erk, and Myc

During its growth phase a cell doubles its protein content. In order to support this level of cellular growth, ribosomal RNA genes are transcribed with high efficiency from multiple copies. RNA polymerase I and III transcriptional complexes are assembled and then shipped to nucleoli where transcription takes place. Mature rRNAs, tRNAs, and ribosomal proteins are assembled into ribosomes, which are then exported to the cytoplasm. Nucleoli are dynamic structures. They disappear when transcription is halted, and become more pronounced when ribosomal transcription is at high levels. One of the indicators of cancerous growth is the swelling of nucleoli. It is therefore not surprising that some of the primary control points for growth signaling are located at the sites for transcription of ribosome components by RNAP I and RNAP III.

The machinery responsible for RNAP I gene transcriptions has four core components: (1) RNA polymerase I; (2) Selectivity factor 1 (SF1) complex, which consists of the TATA binding protein (TBP) and a set of three RNAP I-specific TBP associated factors – TAFs 41, 63, and 110; (3) Upstream binding factor (UBF) protein; and (4) Transcription intermediary factor (TIF) IA, a RNAP I-specific transcription factor that recruits the polymerase. The

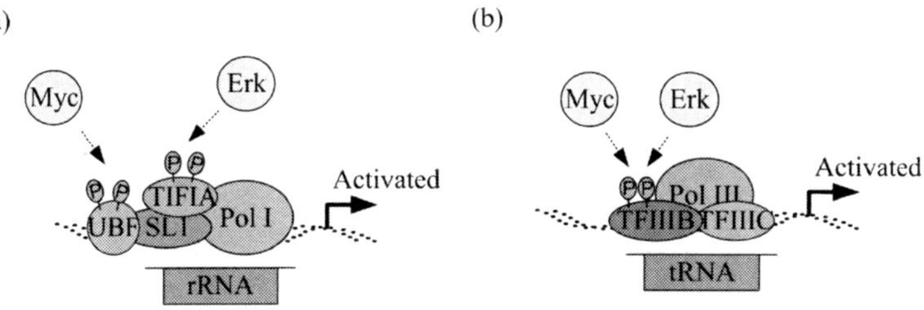

Fig. 13.7 RNAP I and RNAP III gene transcription

arrangement of these factors at rRNA promoters is depicted in Fig. 13.7(a). Two of these components, TIFIA and UBF, function as nodes for receipt of upstream growth signals.

The RNAP I machinery transcribes rRNA genes while the RNAP III machinery transcribes the 5S RNA and tRNA genes. The core factors responsible for transcribing tRNA genes includes TFIIB and TFIIC, as depicted in Fig. 13.7(b). In RNAP III gene transcription, TFIIIB functions as a regulatory signal node. Key growth signals impinging on TIFIA, UBF, and TFIIIB are supplied by Erk and Myc.

Myc regulates ribosome production in a number of ways. It binds E-boxes situated in promoters of genes encoding ribosomal and nucleolar proteins thereby stimulating their transcription by RNAP II. It binds directly to TFIIIB thus stimulating RNAP III activity, and it binds to E-boxes in promoter sites for rRNA genes thereby stimulating RNAP I activity. E-box sequences are found in an appreciable number of protein-coding genes, perhaps as many as 10% in flies and humans. Among the proteins upregulated by Myc are not only ribosomal and nucleolar proteins required for ribosome biogenesis, but also regulators of translation initiation such as eIF4A and eIF4E. Thus, Myc coordinates and regulates growth at multiple levels.

13.13 Regulation of Cellular Proliferation by Myc

Myc promotes cellular proliferation through its positive and negative influences on the transcription of cell cycle regulators. In particular, Myc stimulates transcription of genes that control the G1-S cell cycle transition such as Cyclin D and Cdk4. Their kinase activity targets the pRB protein. In response, the E2Fs that had been inhibited by pRb are able to escape this control and activate transcription of Cyclin E and Cdk2 genes. This step leads sequentially to further phosphorylation of pRb, full activation of the E2Fs, and entry into S phase.

Myc and Myc-associated proteins repress transcription in two ways (Fig. 13.8(b) and (c)). As discussed earlier, Mad-Max dimers repress transcription at promoters containing E-boxes. Alternatively, Myc proteins can directly

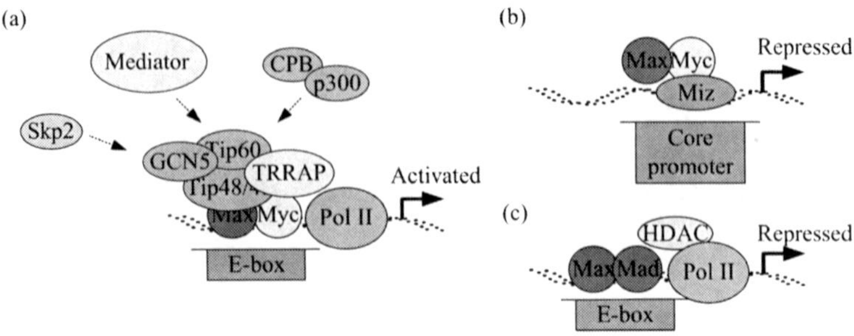

Fig. 13.8 Regulation of gene transcription at RNAP II promoters by Myc proteins. (**a**) Recruitment of transcription activators to E-Box sites, (**b**) repression of transcription through binding to Miz, and (**c**) recruitment of transcription repressors at E-boxes

function as transcriptional repressors through binding to the Myc-interacting zinc-finger 1 (Miz-1) protein. Myc-Max heterodimers can bind to Miz-1, which binds to the core promoter. In the absence of Myc, Miz-1 functions as a coactivator recruiting HATs such as p300 to the promoter. This ability is blocked by Myc-Max. This inhibitory mechanism is used by Myc to inhibit transcription of the cell cycle negative regulators p15Ink4b and p21Cip1. Thus, Myc promotes cellular proliferation by stimulating the transcription of G1-S cyclins/Cdks and by inhibiting their inhibitors.

Max-Myc stimulates transcription by recruiting and mediating the assembly of several complexes at promoters containing E-boxes. Prominent among these cofactors are several proteins and protein complexes that modify chromatin structure in the vicinity of the E-box. Among these are histone acetyltransferases such as TRRAP, hGCN5, and Tip60 and the ATPases/helicases Tip48/49. These factors modify the chromatin structure necessary for recruitment of the core transcription machinery and initiation of transcription. This mode of action is illustrated in Fig. 13.8(a).

Further Reading

Growth Factors and Their Receptors

Birchmeier C et al. 2003 Met, metastasis, motility and more. *Nat. Rev. Mol. Cell Biol.* 4: 915–925.

Burgess AW et al. 2003 An open and shut case? Recent insights into the activation of EGF/ErbB receptors. *Mol. Cell* 12: 541–552.

Eswarakumar VP, Lax I and Schlessinger J 2005 Cellular signaling by fibroblast growth factor receptors. *Cytokine Growth Factor Rev.* 16: 139–149.

Hynes NE and Lane HA 2005 ERBB receptors and cancer: the complexity of targeted inhibitors. *Nat. Rev. Cancer* 5: 341–354.

Schubbert S, Shannon K and Bollag G 2007 Hyperactive Ras in developmental disorders and cancer. *Nat. Rev. Cancer* 7: 295–308.

Wan PTC et al. 2004 Mechanism of activation of the Raf-ERK signaling pathway by oncogenic mutations in B-Raf. *Cell* 116: 855–867.
Wellbrock C, Karasarides M and Marais R 2004 The Raf proteins take centre stage. *Nat. Rev. Mol. Cell Biol.* 5: 875–885.
Yu H and Jove R 2004 The STATs of cancer – new molecular targets come of age. *Nat. Rev. Cancer* 4: 97–105.

MAP Kinases

Biondi RM and Nrbreda AR 2003 Signalling specificity of Ser/Thr protein kinases through docking-site-mediated interactions. *Biochem. J.* 372: 1–13.
Boulton TG et al. 1991 ERKS – a family of protein-serine threonine kinases that are activated and tyrosine phosphorylated in response to insulin and NGF. *Cell*65: 663–675.
Goldsmith EJ et al. 2007 Substrate and docking interactions in serine/threonine protein kinases. *Chem. Rev.*107: 5065–5081.
Ray LB and Sturgill TW 1987 Rapid stimulation by insulin of a serine/threonine kinase in 3T3-L1 adipocytes that phosphorylates microtubule-associated protein 2 in vitro. *Proc. Natl. Acad. Sci. USA* 84: 1502–1506.

MAP Kinase Circuits

Bagowski CP and Ferrell JE Jr 2001 Bistability in the JNK cascade. *Curr. Biol.* 11: 1176–1182.
Bhalla US, Ram PT and Iyengar R 2002 MAP kinase phosphotase as a locus of flexibility in a mitogen-activated protein kinase signaling network. *Science* 297: 1018–1023.
Ferrell JE Jr and Machleder EM 1998 The biochemical basis of an all-or-nothing cell fate switch in Xenopus oocytes. *Science* 280: 895–898.
Hammes SR and Levin ER 2007 Extra-nuclear steroid receptors: nature and actions. *Endocr. Rev.* 28: 726–741.
Markevich NI, Hoek JB and Kholodenko BN 2004 Signaling switches and bistability arising from multisite phosphorylation in protein kinase cascades. *J. Cell Biol.* 164: 353–359.
Wilkinson MG and Millar JBA 2000 Control of the eukaryotic cell cycle by MAP kinase signaling pathways. *FASEB J.* 14: 2147–2157.

Kinase Catalysis

Canagarajan BJ et al. 1997 Activation mechanism of the MAP kinase ERK2 by dual phosphorylation. *Cell* 90: 859–869.
Huse M and Kuriyan J 2002 The conformational plasticity of protein kinases. *Cell* 109: 275–282.
Kim C, Xuong NH and Taylor SS 2005 Crystal structure of a complex between the catalytic and regulatory (RIα) subunits of PKA. *Science* 307Z: 690–696.
Ohren JF et al. 2004 Structure of human MAP kinase kinase 1 (MEK1) and MEK2 describe novel noncompetitive kinase inhibition. *Nat. Struct. Mol. Biol.* 11: 1192–1197.
Taylor SS et al. 2004 PKA: a portrait of protein kinase dynamics. *Biochem. Biophys. Acta* 1697: 259–269.
Zhang X et al. 2006 An allosteric mechanism for activation of the kinase domain of epidermal growth factor receptor. *Cell* 125: 1137–1149.

Oncogene Addiction

Baselga J 2006 Targeting tyrosine kinases in cancer: the second wave. *Science* 312: 1175–1178.
Engelman JA et al. 2007 MET amplification leads to gefitinib resistance in lung cancer by activating ERBB3 signaling. *Science* 316: 1039–1043.
Gorre ME et al. 2001 Clinical resistance to STI-571 cancer therapy caused by BCR-ABL gene mutation or amplification. *Science* 293: 876–880.
Noble MEM, Endicott JA and Johnson LN 2004 Protein kinase inhibitors: insights into drug design from structure. *Science* 303: 1800–1805.
Sordella R et al. 2004 Gefitinib-sensitizing *EGFR* mutations in lung cancer activate anti-apoptotic pathways. *Science* 305: 1163–1166.
Weinstein IB and Joe AK 2006 Mechanisms of disease: oncogene addiction – a rationale for molecular targeting in cancer therapy. *Nat. Clin. Pract. Oncol.* 3: 448–457.

Target-Based Anticancer Therapies

Sebolt-Leopold JS and Herrera R 2004 Targeting the mitogen-activated protein kinase cascade to treat cancer. *Nat. Rev. Cancer* 4: 937–947.

c-Myc Is a Transcription Factor

Fernandez PC et al. 2003 Genomic targets of the human c-Myc protein. *Genes Dev.* 17: 1115–1129.
Grummt I 2003 Life on a planet of its own: regulation of RNA polymerase I transcription in the nucleolus. *Genes Dev.*17: 1691–1702.
Levens DL 2003 Reconstructing Myc. *Genes Dev.* 17: 1071–1077.
Li Z et al. 2003 A global transcriptional regulatory role for c-Myc in Burkitt's lymphoma cells. *Proc. Natl. Acad. Sci. USA* 100: 8164–8169.
Orian A et al. 2003 Genomic binding by the Drosophila Myc, Max, Mad/Mnt transcription factor network. *Genes Dev.* 17: 1011–1014.
White RJ 2005 RNA polymerases I and III, growth control and cancer. *Nat. Rev. Mol. Cell Biol.* 6: 69–78.

Chapter 14
Tumor Metabolism

Changes in metabolic strategy must accompany the transition from modest growth and proliferation that characterizes normal, healthy cells to the rapid expansion that typifies cancerous growth and proliferation. These changes are made possible by alterations in the signaling pathways that govern and synchronize metabolism and growth, protect against inappropriate growth, and integrate the interlocking growth and metabolic demands with attendant intracellular stress responses and conditions prevalent in the tumor microenvironment.

Signals from at least four pathways coordinately regulate growth and metabolism. Growth factor receptor activation stimulates signaling through the MAP kinase growth and Akt survival pathways as was discussed in the last chapter. Energy balance is maintained by signaling through the AMPK pathway as was discussed in Chapter 2, and yet another pathway conveys information on amino acid status. These signals all converge onto the mTOR signaling cassette. That signaling node, the central component of the growth signaling network, was introduced in Chapter 3 in the discussion of insulin signaling and Type 2 diabetes. Key components of the networks regulating growth were discussed further in the last chapter from the perspective of cancer. Those discussions are continued in this chapter with an emphasis on cancer metabolism.

A growing tumor is supported by the continual generation of new vasculature. In contrast to vasculature generated during wound healing, the vasculature that supports a growing tumor is irregular in its morphology, and cells situated at a distance from the nearest blood vessels must contend with a reduced oxygen supply. Overall, tumor microenvironments are hypoxic and acidic. The low oxygen conditions present in tumors is one of the hallmarks of cancer. It is intimately related to another of the cancer hallmarks – an increased reliance on aerobic glycolysis.

The shift toward aerobic glycolysis by cancerous cells was first noted and remarked upon by the noted chemist Otto Warburg in the mid to late 1920s. He underscored the importance of this shift in metabolic strategy by asserting that it was the key causal event in the onset of a cancer. That central status has since then been assigned to mutations, genetic and somatic. However, a metabolic shift does occur, is connected to hypoxia, and is advantageous to tumors in ways that are still actively being explored.

M. Beckerman, *Cellular Signaling in Health and Disease*,
Biological and Medical Physics, Biomedical Engineering,
DOI 10.1007/978-0-387-98173-4_14,

In examining the signaling pathways that regulate growth and metabolism, one fact stands out above all others. That fact is: the interlocking network that regulates growth and metabolism encompasses an extraordinary large number of the known oncoproteins and tumor suppressors. The goal of this chapter is to explore the mTOR signaling node, examine the interrelationships between hypoxia and metabolism, and look at how the p53 keeper of the barrier contributes to the alterations in metabolism.

14.1 The Central Growth Network of the Cell Is Organized About the mTOR Cassette

The mTOR cassette consists of three component modules. The upstream unit is the TSC1/TSC2 complex; the downstream component is the mTOR modules consisting of TORC1 and TORC2, and these are connected to one another by the Rheb GTPase. Recall from Chapter 3 that TORC1 promotes cellular growth by increasing the rate of protein synthesis and ribosome biogenesis. TORC2 signals to Akt through a feedback loop and functioning as an activating kinase. The balance between TORC1 and TORC2 signaling is regulated by the GTPase. When bound to ATP, Rheb shifts the balance toward TORC1 signaling. Otherwise, TORC2 remains active and not TORC1. Ordinarily, Rheb remains bound to GDP and is maintained in that state by TSC1/TSC2. These actions are summarized in Fig. 14.1 by the negative signaling path from TSC1/TSC2 to Rbeb and a positive signaling path from Rheb to TORC1.

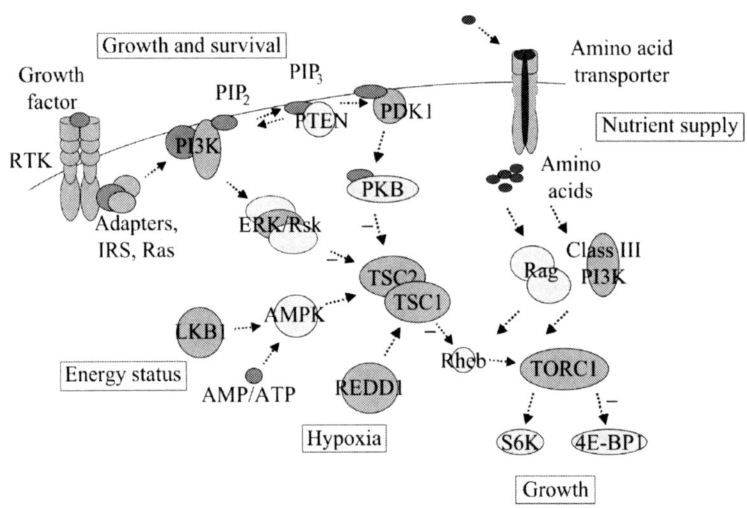

Fig. 14.1 mTOR integration and regulation of growth and metabolism. Shown is the convergence of growth, survival, energy status, hypoxia, and nutrient supply pathways onto the mTOR signaling cassette, which integrates these signals to regulate growth

The mTOR cassette coordinates and regulates growth and metabolism in the cell. It receives growth signals initiated by the binding of growth factors and cytokines to cell surface receptors and these signals converge upon mTOR through the growth pathways described in the last chapter. Additional growth-promoting signals are sent to TSC1/TSC2 by Akt/PKB. These signals are not passed through the mTOR cassette unless adequate metabolic resources are present to support the growth. Three sets of signals, one representing energy supply, the second amino acid availability, and the third oxygen supply, gate the growth signals.

14.2 AMPK Supplies a Gating Signal Indicative of Energy Balance

Recall from Chapter 2 that APM-activated protein kinase (AMPK) is the master regulator of energy balance. This kinase is activated when energy supplies are depleted, and helps to restore energy balance by stimulating energy-producing catabolic processes and throttling back energy-using anabolic ones. Under normal conditions, this is accomplished, in part, by shutting off growth and survival stimuli that would otherwise ensure continued cell growth. AMPK is activated by upstream kinases such as LKB1 and by depleted energy supplies signaled through elevated AMP:ATP ratios. These events occur whenever the supply of glucose is severely depleted.

Glucose starvation puts the cell under considerable stress – low ATP levels, low ribosome biogenesis rate, low oxygen levels. These stresses, in turn, activate p53, which responds by upregulating the transcription of genes for:

- β1 and β2 subunits of AMPK,
- PTEN, negative regulator of Akt signaling, and
- TSC2.

The AMPK-p53 route thus provides a means for several forms of cellular stress to coordinately shut off growth when necessary. This closure is maintained by a positive feedback loop in which AMPK phosphorylates p53 on serine 15, an early step in its stabilization and activation. Recovery from glucose starvation leads to a reopening of the gate through mTOR-mediated activation of protein phosphatase 2A, which dephosphorylates p53 on serine 15.

14.3 Cells Halt Growth in Response to Hypoxia and Other Cellular Stresses

The amount of oxygenation varies widely from cell-to-cell both within a given solid tumor and across the spectrum of tumor types and individuals. Some regions may be close to normal (normoxic) in the extent of its oxygen supply while other regions may be entirely lacking in available molecular oxygen (anoxic). A general statement, then, is that portions of most solid tumors are hypoxic, and cells within those regions must deal with reduced oxygenation. Normal, healthy cells when exposed to low oxygen conditions respond by ceasing protein translation since that process consumes an enormous amount of energy. Low energy, inadequate nutrient supply, and low oxygen conditions all lead to a cessation of protein translation and growth, and the cells further respond by shifting toward certain catabolic processes at the expense of anabolic ones. Cancerous cells exploit some of these responses and bypass others, and take advantage of the hypoxic microenvironment. Most strikingly, and oddly enough on the surface, they increasingly rely on aerobic glycolysis for bioenergetics and biosynthesis.

How cells respond to hypoxia depends on the degree and duration of the hypoxic conditions. There are two predominant responses, one occurring almost immediately and the other more gradually over a longer time span. Both are directed at shutting down protein synthesis. The time course of these responses under conditions of severe hypoxia is depicted in Fig. 14.2.

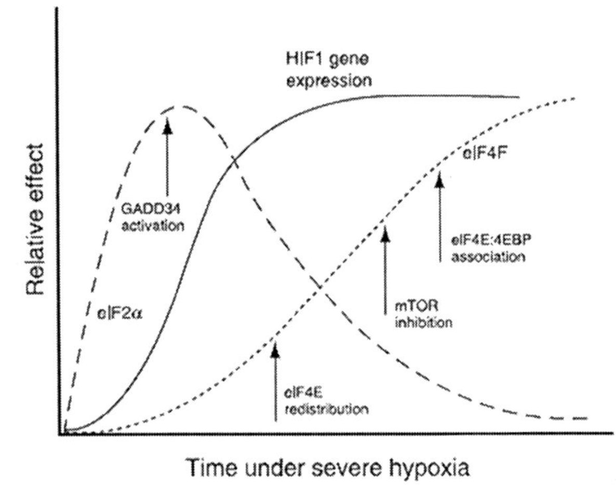

Fig. 14.2 Time course of hypoxic responses. Reprinted by permission from the American Associated for Cancer Research from Koumenis C and Wouters BG 2006 *Mol. Cancer Res.* 4: 423–436

Low oxygen tension, nutrient deprivation, and low energy supplies are cellular stresses that are sensed almost immediately through a cell's unfolded protein response (UPR). This response is triggered by an increase in the number of proteins in the endoplasmic reticulum that experience either misfolding or unfolding as a result of the aforementioned stresses. In response, protein translation and growth are transiently halted to allow time for recovery and restoration of protein homeostasis. Protein initiation factor eIF2α is the signaling target. Endoplasmic reticulum resident kinases such as the general control non-derepressible-2 (GCN2) protein and RNA-dependent protein kinase (PKR)-like ER kinase, or PERK, are activated by the stresses and phosphorylate eIF2α on Serine 51. This posttranslational modification leads to eIF2α binding and inhibition of eIF2B and a halt to the assembly of the translation preinitiation complex. This response is terminated through a negative feedback loop mediated by the protein phosphatase GADD34, which dephosphorylates eIF2α and terminates the inhibition.

Normally, in response to growth factor and survival stimuli, mTOR and S6K1 are activated and target mRNA required for sustained growth and proliferation. mTOR targets 4E-BP1 while S6K1 interacts with eIF4B. In the former instance, mRNA containing highly structured 5′UTR involved in growth and proliferation such as cyclin D1 and c-Myc are strongly affected. In the latter case, mRNAs are targeted containing a characteristic 5′TOP (terminal oligopyrimidine tract) sequence in their mRNA regulatory region, which are translated into ribosomal proteins and eEFs. The two S6K proteins, S6K1 and S6K2, also regulate cellular growth at the transcriptional level. The targets in these instances are ribosomal genes, which are encoded by rDNA. The timely production of ribosomes is a crucial rate-limiting step in rapidly growing and proliferating cells. Regulation by mTOR and S6K appears to occur though S6K-mediated phosphorylation of rDNA transcription factors.

Longer-term responses are initiated under conditions of prolonged hypoxia and other stresses. In these instances, mTOR growth stimulation is prevented. The translation initiation factor targeted in this case is eIF4E, the rate-limiting factor in protein translation. This factor is repressed by unphosphorylated 4E-BP1, and this inhibition is alleviated by activated mTOR. Prevention of mTOR activation results in the sequestering of eIF4E by 4E-BP1 and cessation of translation of the target mRNAs. The increasing shutdown of translation is represented in Fig. 14.2 by the curve labeled eIF4F.

Persistent low oxygen conditions lead to upregulation of the transcription factor HIF-1, and a curve representing this increase has been included in Fig. 14.2. One of the ways that HIF-1 influences metabolism is by inhibiting growth. This is accomplished by HIF-1 though its upregulation of REDD1, a factor that controls TSC2/TSC1 signaling by binding 14-3-3 proteins that would otherwise bind and inhibit TSC2/TSC1. By binding the 14-3-3 protein, TSC2 is free to inhibit TORC1 signaling and growth promotion.

14.4 Regulation of Cell Growth by Amino Acid Starvation Signaling to mTOR

The decision to growth or not to grow takes into account the availability of nutrients, in particular, whether or not there is an adequate supply of amino acids. The most abundant of these is glutamine. This amino acid like glucose is consumed at a high rate by rapidly growing cells with an attendant secretion of lactate. These processes complement one another with glutamine metabolism supplementing and supporting glucose metabolism.

The pathway through which information on amino acid status is integrated into the growth decision begins with the import of amino acids into the cell through amino acid transporters. Amino acid signals are generated by two signaling routes. One of these routes is initiated when amino acids bind Class III PI3Ks, and the second is triggered when the amino acids bind members of the Rag family of GTPases (discussed shortly).

Recall that PI3Ks, or more precisely Class I PI3Ks, consisting of p85 and p110 subunits, are localized in the vicinity of the plasma membrane where they transduce signals conveyed into the cell by receptor tyrosine kinases (Class IA) and G protein-coupled receptors (Class IB). Class III PI3Ks, in contrast, are cytosolic proteins that function as regulators of membrane trafficking and vesicle formation. In this case, a Class III PI3K designated as hVps34 acts as a cytoplasmic amino acid receptor. Upon ligand binding hVps34 facilitates downstream ATP-bound Rheb promotion of TORC1 signaling as depicted in Fig. 14.1. hVsp34 acts in a second way, too. In response to amino acid deprivation, a role that will be discussed shortly, it associates with other regulators and promotes autophagy.

The second route of amino acid signaling involves a family of GTPases known as Rags. These GTPases are closely related to members of the Ras family. Members of this group of which there are four – RagA, RagB, RagC, and RagD – bind Raptor and facilitate the relocation of TORC1 to sites where Rheb is present. In doing so, they mediate interaction between the two and activation of TORC1. Once TORC1 is activated, it can activate S6K and inhibit 4E-BP1 leading to increased protein synthesis and cellular growth. The control of protein synthesis by TORC1 and S6K is now examined in greater detail.

14.5 Regulation of the Translation Initiation Complex by mTOR

Recall that messenger RNA (mRNA) molecules contain special structures at their 5′ and 3′ termini. First, a cap is added to its 5′ end. This structure consists of a 7-methylguanylate (a CH_3 group appended to the 7 position of the G base) plus a triphosphate group that links the 7-methylguanylate to the sugars at the 5′ end of the RNA molecule. Second, a string of up to 200 adenylates is appended to the 3′ terminus of the mRNA. This addition is referred to as a polyadenylate tail or poly (A) tail. The cap and poly (A) tail are separated from the mRNA-coding region by flanking noncoding regions and start and stop codons.

A set of proteins termed eukaryotic initiation factors (eIFs) is responsible for assembling the ribosomal 40S and 60S subunits at the mRNA molecule in the correct location and orientation near and at the start codon. These proteins are listed in Table 14.1. Translation is regulated through phosphorylation by several kinases at key assembly control points. There are two of these, one involving the GTP-binding protein eIF2 and its GEF eIF2B and the other

Table 14.1 Eukaryotic initiation factors

Initiation factor	Function
eIF1A	Promotes Met-tRNA, 40S, mRNA binding
eIF2	GTPase, guides Met-tRNA onto 40S
eIF2B	GEF for eIF2
eIF3	Scaffold; binds to 40S, prevents 60S binding
eIF4A	RNA-dependent ATPase and an RNA helicase
eIF4B	Facilitates ribosome-mRNA binding
eIF4E	Cap binding protein
eIF4G	Augments binding of the preinitiation complex
eIF5	GAP for eIF2

centered about the cap-binding protein eIF4E and an associated regulatory protein called 4E-BP1.

Translation initiation begins at the 5′ end of the mRNA. An initiator tRNA is a tRNA molecule loaded with a methionine (Met), the amino acid encoded by the AUG start codon. For translation to begin, the Met-tRNA initiator, and the small and large ribosomal subunits, must be recruited and positioned at the AUG start site of the mRNA. In the first main step, the loaded tRNA molecule and small ribosomal subunit associate. The Met-tRNA molecule is guided to its correct position by the initiation factor eIF2 that forms a complex with the Met-tRNA and a GTP molecule. This tertiary complex associates with the small (40S) ribosomal subunit to form a 43S preinitiation complex. The GTP molecule is hydrolyzed once Met-tRNA engages the ribosome, and eIF2 plus the GDP molecule is recycled back to its active eIF2-GTP form by eIF2B.

The initiation factor eIF2 and its GEF eIF2B are both regulated by kinases. The first of these, eIF2, is deactivated by phosphorylation and activated by dephosphorylation. When eIF2 is phosphorylated it binds eIF2B tightly, thereby sequestering both and preventing their translational activities. A family of eIF2 kinases responds to cellular stresses such amino acid deprivation and the unfolded proteins response (UPR) by phosphorylating eIF2 and halting protein synthesis to allow for the cell to recover and restore cellular homeostasis. Examples of these are the general control nonderepressible-2 (GCN2) protein (amino acid deprivation) and RNA-dependent protein kinase (PKR)-like ER kinase, or PERK (ER stress). The second component, eIF2B, can be phosphorylated by GSK3, and other kinases, to prevent its activities. In response to insulin and other growth and survival signals, Akt is activated, and when this kinase phosphorylates GSK3, eIF2B is freed of its repression by GSK3.

The preinitiation complex attaches to the mRNA near its 5′ end. An initiation factor complex, eIF4F (consisting of eIF4A, eIF4E, and eIF4G), binds to the m^7GpppN cap and prepares the way for the attachment of the preinitiation complex to the mRNA. Upon attachment the preinitiation complex scans in the 5′–3′ direction for the start AUG codon and halts upon arrival. Another initiation factor, eIF5, stimulates the hydrolysis by eIF2 of GTP to GDP

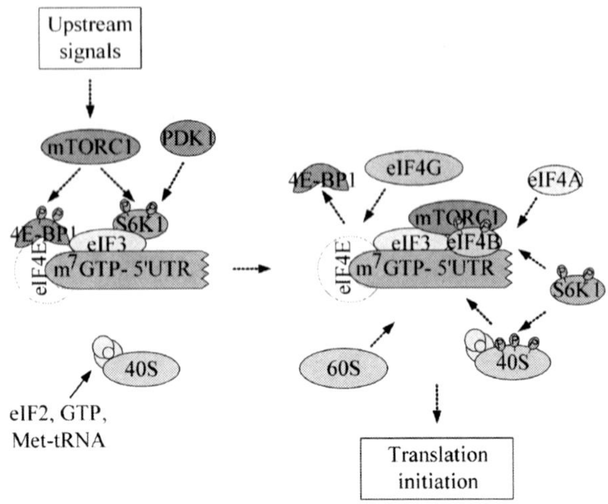

Fig. 14.3 Regulated assembly of the preinitiation complex (PIC)

resulting in the dissociation of the initiating factors from the assemblage leading to the attachment of the large (60S) ribosomal subunit and translation elongation begins.

Regulation of this stage of the assembly process is depicted in Fig. 14.3. The cap-binding protein eIF4E is negatively regulated by eIF4E-binding protein 1 (4E-BP1). This small protein competes with eIF4G for binding sites. The binding affinity of the 4E-BP1 for eIF4E is dependent upon its phosphorylation status – binding affinity decreases with increasing phosphorylation. As shown in the figure, mTOR regulates assembly through its ability to stimulate the phosphorylation and dissociation of 4E-BP1 from eIF3. Once the 4E-BP1 proteins dissociate, eIF4G can bind and the assembly process can continue past this potential stop point.

The initiation factor eIF3 functions as a chaperone/scaffold. Binding by S6K1 blocks further assembly onto this platform. This blockage is relieved when S6K1 is phosphorylated in two crucial locations – in T229 located within the catalytic loop and T389 inside the hydrophobic motif. Phosphorylation at T389 is mediated by mTOR/raptor and occurs first; it leads to recruitment of PDK1 and phosphorylation by it at T229. These actions lead to dissociation and activation of S6K1 enabling it to phosphorylate eIF4B and the 40S ribosomal subunit. Assembly then runs to completion.

14.6 Starvation and Autophagy

Autophagy (Greek for self-eating) is a survival metabolic strategy used by cells under starvation conditions to maintain cellular homeostasis. In this process, long-lived proteins and organelles that are either damaged or can be dispensed with are packaged into compartments called autophagosomes and sent for lysosomal degradation and recycling. By this means the demands on cellular resources are reduced, and the material so recovered can be utilized for energy production and macromolecule biosynthesis.

Beclin1 is a member of the Bcl2 family of proteins discussed earlier with respect to apoptosis. Recall that full length Bcl2 pro- and antiapoptotic proteins fold in a compact globular domain formed by eight α helices. In this fold, the core helix, α5, is surrounded by the seven other helices and connecting loops. A hydrophobic groove formed by helices α2–α5 provides the binding site for the BH3-only proteins such as Bim and Bad. The three-dimensional structure of Beclin1 in contact with Bcl-X_L (Fig. 14.4) is characteristic of the way full-length Bcl2 proteins interact with BH3-only family members. In these interactions, residues from Beclin1 form a network of van der Waals interactions with the residues lining the hydrophobic groove and these interactions are further stabilized by flanking hydrogen bonds.

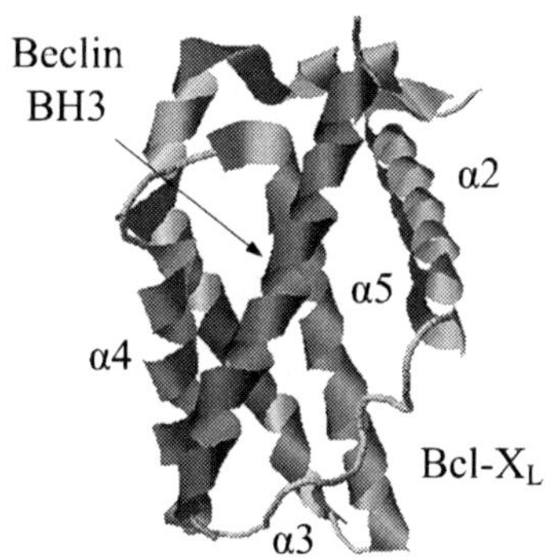

Fig. 14.4 The three-dimensional structure of the Beclin1BH3 peptide, amino acid residues 112–123, bound to the negative-regulatory, full length Bcl2 protein, Bcl-X_L determined by means of X-ray crystallography. *Highlighted* in the figure are Bcl-X_L helices α2–α5; Bcl-X_L helix α1 has been omitted for clarity. The figure was prepared using Jmol with atomic coordinates deposited in the PDB under accession code 2p1l

Autophagic responses to starvation conditions are mediated by a complex containing Class III PI3Ks and Bcl2 family members such as Bcl-X_L. The core components of the decision complex consist of vacuolar protein sorting 34 (Vps34), the PI3K3 discussed earlier, the full-length negative regulator (antiapoptotic) proteins Bcl2 or Bcl-X_L, UV radiation resistance-associated gene (UVRAG) protein, the endophilin Bif-1, and Beclin1. A tentative and highly schematic picture of how this complex might operate is presented in Fig. 14.5. As indicated in the figure, the output from the module is PthIns3P, which nucleates the formation of autophagosomes by recruiting autophagy (Atg) proteins required for their biogenesis (represented in this figure by hVps15). The Bif-1 endophilin has a role in controlling membrane curvature. It possesses a BAR domain central to its lipid manipulations, an oligomerization domain, and an

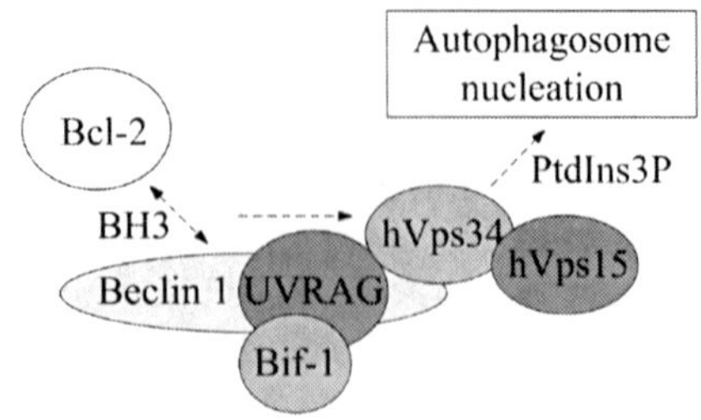

Fig. 14.5 The autophagy decision complex that nucleates formation of the autophagosome

SH3 domain that mediates its interactions with UVRAG. The UVRAG protein interacts with Beclin through their mutual CCD domains. Just as was the case for apoptosis at mitochondrial membranes, Bcl2 and Bcl-X_L negatively regulate autophagy when strongly bound to, in this situation, the pro-autophagy protein Beclin. Weakening of this binding leads to activation of the Class III PI3Ks.

14.7 p53 Modulation of Metabolism Is One of Its Barrier Functions

The p53 protein not only orchestrates DNA damage repair, cell cycle arrest, apoptosis, and senescence but also influences cellular metabolism. In normal cells exposed to typical rather than pathological levels of DNA damage, p53 upregulates the gene for TP53-induced glycolysis and apoptosis regulator (TIGAR). The TIGAR protein inhibits glycolysis by redirecting the early steps in the glycolytic pathway away from producing Fructose-6-P2 and toward the Pentose phosphates shunt resulting in an increased production of NADPH and critical nucleotides (Fig. 14.6). The increased production of NADPH leads to an increase in the amount of available glutathione, which scavenges ROS. This activity assists in the repair process by reducing the amount of ROS available to cause more damage to the cell's DNA.

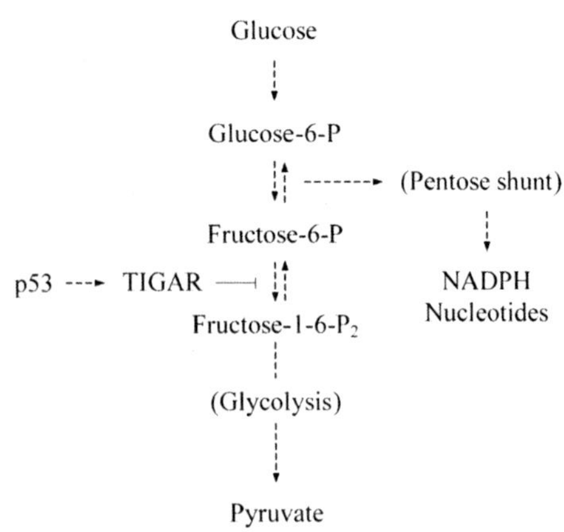

Fig. 14.6 Regulation of glucose metabolism by TIGAR results in a shift away from glycolysis and toward the pentose phosphate shunt (pentose shunt) leading to increased production of NADPH and synthesis of critical nucleotide-containing biomolecules

The gene encoding the synthesis of cytochrome c oxidase 2 (SCO2) protein is another p53 target. In response to normal levels of DNA damage, this gene is upregulated by p53. The SCO2 protein is required for assembly of the cytochrome c oxidase (COX) complex, Complex IV of the electron transport chain and the primary user of molecular oxygen in the cell. Stepping up production of this protein, a copper transporter, alters the balance between oxidative phosphorylation and glycolysis. It leads to an increase in respiration that may compensate for the reductions brought on by TIGAR gene expression. Conversely, in diseased cells in which p53 is mutated in ways that prevent SCO2 synthesis, respiration declines and energy production shifts in the direction of glycolysis.

A third target of p53 in its role as a metabolic modulator is p53-mediated upregulation of antioxidant genes including the glutathione peroxidase and sestrins. Recall from Chapter 8 that sestrins are an important component of hydrogen peroxide signaling. They mediate the reversibility of the protein modifications and a buildup of ROS that would otherwise damage DNA. At low levels of stress in healthy cells p53 is expressed at low levels and not only halts the cell cycle to allow for repair of DNA damage but also stimulates the increased expression of antioxidants to limit and alleviate damage to DNA.

Sustained and irreparable DNA damage is reflected in the amount of stabilized p53 present in the cell. In normal cells, the low amounts of p53 produced help orchestrate, and participate in, the DNA damage response and cell cycle arrest. This response changes in badly stressed cells. In these situations, p53 rather than assisting in repair by transactivating antioxidants and repair proteins stimulates apoptosis by upregulating apoptotic Bcl2s and oxidants. The latter triggers positive feedback, whereby further damage to the DNA by the oxidants leads to a further production of p53 and apoptotic factors.

One of the main downstream targets of p53-mediated transcription that mediates p53's role as a stress-induced elevator of oxidant production and apoptosis is p66Shc. Shc proteins function as molecular adapters. There are three alternatively spliced Shc isoforms – of 46, 52, and 66 kDa mass. These proteins have an SH2 domain, a collagen-homology (CH1) motif, and a phosphotyrosine binding (PTB) domain. In addition, the 66 kDa form, p66Shc, possesses a 14-kDa N-terminal CH2 domain. The two smaller isoforms undergo tyrosine phosphorylation upon activation of growth factor receptors, and form complexes with Grb2, the Ras/SOS adapter. Unlike the other Shc proteins, p66Shc does not respond to growth factor/cell cycle signals but instead responds to oxidative stresses and promotes apoptosis.

The pro-apoptotic actions of p66Shc depend on phosphorylation of Ser36 in the CH2 domain. This action is triggered by H_2O_2 and UV radiation. These stresses activate protein kinase Cβ, which then phosphorylates p66Shc. This action is then followed by Pin1-induced prolyl isomerization, which induces the translocation of p66Shc to the mitochondria where it accumulates and stimulates calcium release and further oxidative stresses that stimulate cytochrome c release from its sequestration by cardiolipin leading to apoptosis.

In carrying out this role, p66Shc operates as a redox enzyme that generates mitochondrial ROS using electrons from the electron transport chain. The overall system operates as follows. Under normal cellular conditions, p66Shc is sequestered in the cytosol where it is maintained in a reduced and inactive state by resident Trx1 and GSH. Cellular stresses related via PKCβ result in phosphorylation and translocation to the mitochondria. Apoptosis is normally inhibited by the mitochondrial Trx and GSH systems but large oxidative stresses overwhelm this system allowing for formation of an oxidized active state characterized by formation of a disulfide bond between thiols located in the N-terminal CH2 domain. Once activated another feature of the CH2 domain, a copper-dependent redox center, stimulates the ROS production and apoptosis.

14.8 The PTEN Tumor Suppressor Acts at the Plasma Membrane and in the Nucleus

A large fraction of all the major cancer-associated proteins participate in the pathway presented in Fig. 14.1 The list of oncoproteins includes receptor tyrosine kinases (RTKs), PI3K, Akt, Ras, and Raf. The corresponding list of tumor suppressors contains LKB1, TSC1 and 2, and PTEN. If p53 is included, and also pRb and c-Myc, the latter two as honorary members of the list, then just about all the major players are present in one way or another. More often, not just one of these but rather several are mutated in a given cancer, and this represents a roadblock to effective treatment using either cytotoxic approaches or cytostatic, target-based ones. In addition to this list of widely encountered oncogenes and tumor suppressors, each type of cancer seems to have associated with it any number of specific additional mutated proteins.

The phosphatase and tensin homolog deleted on chromosome 10, or PTEN, protein is a tumor suppressor that is mutated in a large number of cancers. This phosphoinositide phosphatase is the primary negative regulator of the PI3K – Akt pathway. It carries out this role by catalyzing the conversion of PIP_3 to PIP_2 thereby opposing the actions of PI3K and shutting down signaling to Akt. This protein is composed of two domains, an N-terminal phosphatase domain and a C-terminal C2 domain followed by an extreme C-terminal PDZ-binding motif. The C2 domain targets the enzyme to the inner leaflet of the plasma membrane, which is rich in acidic lipids. The three-dimensional structure of this protein determined through X-ray crystallography is presented in Fig. 14.7.

Recall from Chapter 8 that PTEN is a target of hydrogen peroxide-induced oxidation. In healthy cells, this activity serves a useful signaling function. PTEN is constitutively active and in order to allow for tyrosine phosphorylation in response to growth factors, PTEN is temporarily shut down. PTEN contains target cysteines in its active site that function as nucleophiles. When oxidized these can no longer function as a nucleophile and the phosphatase loses its

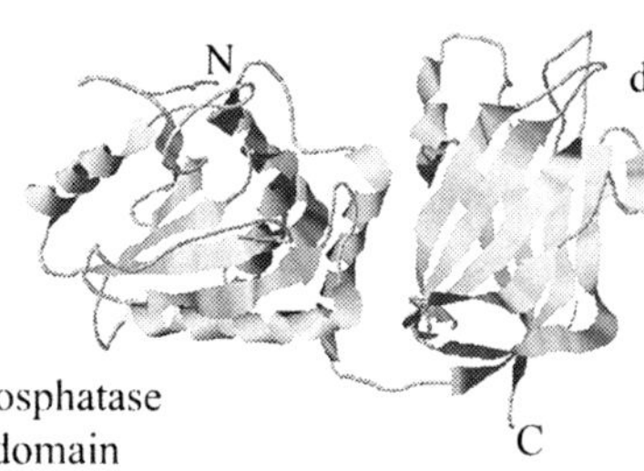

Fig. 14.7 The three-dimensional structure of PTEN determined by means of X-ray crystallography. The figure was prepared using Jmol with atomic coordinates deposited in the PDB under accession code 1d5r

catalytic ability. Ligand binding to a growth factor receptor stimulates NADPH oxidase to produce hydrogen peroxide and allow for phosphorylation unimpeded over the short term by PTEN. In PTEN, a second cysteine (Cys^{71}) participates in the inactivation process. Oxidation of the active site cysteine (Cys^{124}) triggers the formation of a disulfide bond between the two cysteines that prevents further oxidations and ensures reversibility of the oxidative modification.

PTEN contains a pair of PEST motifs, a typical sequence that mediates ubiquitin-mediated degradation. The E3 ubiqutin ligase responsible for activation of this route is the neural precursor cell expressed, developmentally down-regulated 4-1 (NEDD4-1) protein. Two distinct sets of outcomes are possible. NEDD4-1 may polyubiquitinate PTEN thereby targeting it for degradation by the proteasome, or NEDD4-1 may monoubiquitinate PTEN resulting in its translocation to the nucleus. Once in the nucleus, PTEN may further inhibit Akt or it may interact with a number of factors to help maintain chromosome stability. In the latter instance, PTEN interacts with Cenp-C, a component of contromeres. Mutations that disrupt the function of the C-terminal disable this capability resulting in an increased number of chromosomal instabilities. In addition, PTEN interacts with chromatin and acts as a cofactor for the E2Fs at Rad51 promoters and through this mechanism promotes DSB repair. This protein thus acts in the nucleus as a guardian of the genome.

14.9 Mutations and Disturbed Redox Balance Deactivate PTEN

The PTEN tumor suppressor is one of the most prominent of the cancer-associated mutated proteins; it is encountered with high frequency in many different cancers. In these mutated forms, the protein is no longer able to catalyze the conversion of PIP_3 to PIP_2 resulting in overactivation of the PI3K/Akt pathway. As a result, in cells expressing these mutated forms, the balance is shifted away from apoptosis and toward growth and survival.

The shift toward excessive growth and survival can be triggered and exacerbated by dysregulated oxidative balance. Hydrogen peroxide and metabolites such as NADH/NADPH released from mitochondria under glycolytic and hypoxic conditions and/or in the presence of mutations in mtDNA can deactivate PTEN leading again to a persistent elevation in Akt activity. Additional positive feedback loops can further stabilize the tumorigenic signaling route and associated cancer disease state. For example, if ROS (hydrogen peroxide) is increasingly being generated by the mitochondria, it would lead through HIF1α stabilization to enhanced glycolysis and metabolic shifts that maintain PTEN inhibition through maintenance of elevated NADH/NADPH ratios.

PTEN and p53 reciprocally regulate each other. The CTD of p53 contains a binding site for PTEN along with numerous Lys acetylation sites (Fig. 11.6). PTEN enhances p53 transcription by acting as a coactivator that recruits p300 and promotes high p53 acetylation leading to p53 tetrameter formation and gene transcription. The mutual protective actions of both PTEN and p53 are disabled by the cancer-causing mutation to either of these tumor suppressors.

14.10 HIF Transcription Factors Sense and Respond to Low Oxygen Conditions

Hypoxia-inducible factors (HIFs) are transcription factors. When oxygen levels are low, HIF proteins stimulate the transcription of a host of genes that enable the body to cope with low oxygen supply. HIFs upregulate proteins that promote red blood cell production and angiogenesis, and also stimulate increased production of glycolytic enzymes that can produce ATP under low oxygen conditions. Genes upregulated by HIFs include those encoding erythropoietin [Epo] responsible for stimulating survival and proliferation of red blood cell progenitors, growth factors such as VEGF, and glycolytic enzymes such as HKI and HKII.

The HIF-1 protein is a heterodimer, consisting of a HIF-1α subunit and a HIF-1β subunit. The latter is constitutively expressed while the former is regulated by cellular O_2 levels. Each subunit has an N-terminal bHLH domain, and a PAS domain. (PAS domains are named after the first three proteins found to contain them, namely, the *Drosophila* clock protein PER, the mammalian basic helix-loop-helix containing transcription factor ARNT, and the insect SIM protein.) PAS domains function as sensor units detecting physical properties such as oxygen tension, redox potential, or light intensity. They also mediate protein–protein interactions. The bHLH and N-terminal half of the PAS domains of HIF-1α and HIF-1β mediate DNA binding while the C-terminal portion of the PAS domains enable formation of DNA-binding competent HIF heterodimers. The C-terminal segments of the HIF proteins contain transactivation domains (TADs) responsible for recruiting coactivators.

HIF-1α contains several crucial amino acid residues that undergo posttranslational modifications according to cellular oxygen status (Fig. 14.8). Under normal oxygen conditions, or normoxia, proline residues situated in the

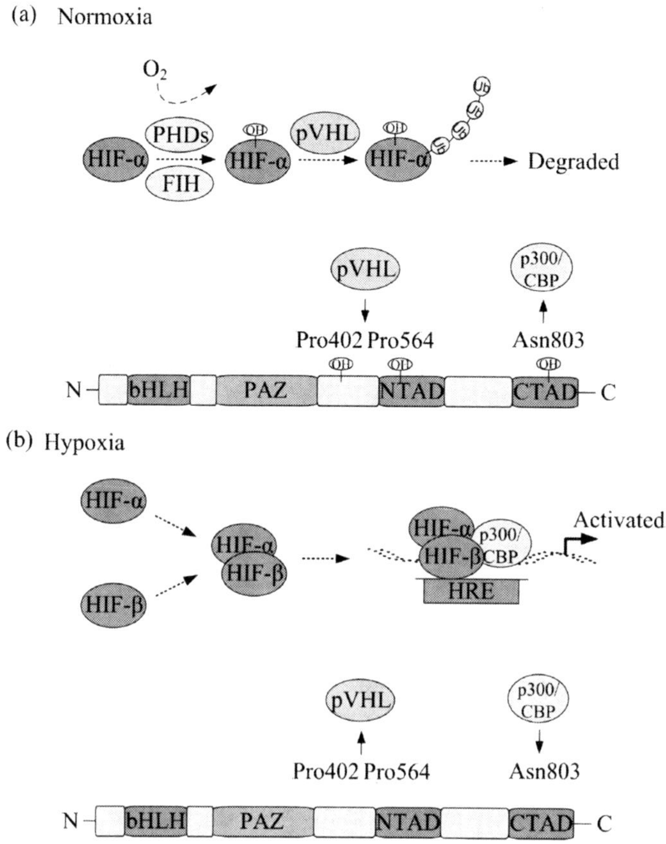

Fig. 14.8 HIF actions and HIF domain organization under (**a**) normoxia and (**b**) hypoxia/ROS conditions

N-terminal TAD of HIF-1α are modified by oxygen-sensitive enzymes leading to rapid degradation of the subunit. Under low oxygen conditions, hypoxia, or in the presence of ROS, the oxygen-sensitive enzymes are unable to hydroxylate HIF-1α. Its lifetime is extended and it is able to translocate to the nucleus and interact with its HIF-1β subunit and CPB/p300 and other coactivators to stimulate gene transcription from promoters bearing hypoxia-responsive element (HRE).

Three enzymes have been identified that modify proline residues in HIF-1α according to oxygen status. These enzymes are referred to as prolyl-hydroxylase domain-containing proteins, or PHDs. The enzymes, PHD1, PHD2, and PHD3, hydroxylate HIF-1α proline residues Pro402 and Pro564 in the presence of molecular oxygen. The degradation triggered by hydroxylation begins with binding of the von Hppel-Lindau tumor suppressor protein, pVHL, to the hydroxylated oxygen sensitive region. The pVHL protein is a component of an

E3 ubiquitin-ligase complex. In the steps that follow, the remaining components of the complex are recruited; HIF-1α is ubiquitinated and is then degraded by the 26S proteosome.

Another enzyme, factor inhibiting HIF-1 (FIH-1), functions as an asparaginyl hydroxylase and in the presence of molecular oxygen it catalyzes the transfer of a hydroxyl group to Asn803 situated in the C-terminal TAD. This operation blocks the binding site for transcriptional coactivators p300/CBP, thereby preventing activation of transcription. The contrasting situations of normoxia versus hypoxia are depicted in Fig. 14.7.

Not only hypoxia but also elevated ROS levels activate the HIFs. The increased ROS interact with the ferrous ion that resides in the PHD active site thereby preventing the PHD from hyroxylating the HIFs. The HIFs are stabilized and as was the case for hypoxic conditions they initiate transcription of genes directed at restoring homeostatis.

14.11 HIFs Regulate Cellular Metabolism and Drive the Glycolytic Shift

Cells can become deprived of oxygen under both normal and abnormal conditions. Examples of normal conditions under which hypoxia may arise include exercise and embryonic development. Abnormal conditions include not only tumor growth but also inflammation and myocardial infarction. In response to oxygen deprivation, HIF-1 gene expression is triggered leading to an increase in glycolytic enzymes and a shift toward glycolytic ATP generation and away from oxidative phosphorylation, which requires a ready supply of molecular oxygen.

Substrates needed for oxidative phosphorylation include molecules oxygen and pyruvate acting as carbon sources. Pryuvate is an end product of glycolysis. It is enzymatically converted to acetyl-CoA and in that form enters the TCA cycle. HIF-1 upregulates pyruvate dehydrogenase kinase 1 (PDK1), an enzyme that inhibits the enzymes that convert pyruvate to acetyl-CoA. By upregulating PDK1, HIF-1 shunts pyruvate away from the TCA cycle and oxidative phosphorylation, which occurs downstream from the TCA cycle. In its place, free pyruvate is converted to lactate by lactate dehydrogenase (LDH), and the lactate is released from the cell into the extracellular spaces. In the conversion of pyruvate to lactate, NAD^+ is recovered from NADH. These actions enhance cell survival; they conserve O_2 and reduce ROS production by the electron transport chain.

Under hypoxic conditions, oxidative phosphorylation is reduced in favor of glycolysis and cell proliferation is halted. Myc and HIF interactions contribute to the cellular responsiveness to low oxygen by stimulating the expression of cyclin-dependent kinase inhibitors p27 and p21, which inhibit the cell cycle. Under healthy, normal or hypoxic conditions, HIF-1 and HIF-2 regulate Myc both positively and negatively. In particular, HIF-1 inhibits Myc while HIF-2 with a more limited tissue distribution cooperates with Myc. Under tumorigenic conditions in which Myc is overexpressed, the negative influences are no longer present, and both HIFs cooperate with Myc and jointly promote the Warburg effect and tumorigenesis.

Increases in glycolysis by tumor cells is not driven by malfunctions in the oxidative phosphorylation machinery. The latter is intact. Rather, rapidly growing cancer cells require large amounts of ATP, and while the numbers of ATP molecules generated per mole of glucose are far greater in OXPHOS – 32 versus 2, the rate of production via glycolysis is far higher. OXPHOS is limited by the relatively slow NADH shuttling rate from the cytosol to the mitochondrial electron transport chain. This gives the tumor cells a selective advantage in hypoxic environments and enables them to out-compete normal cells for resources.

In mammalian cells, the expression of cytochrome oxidase (COX) 4-1 and 4-2 isoforms is regulated by means of alterations in gene expression by the HIFs. In particular, HIF-1 adjusts the subunit composition of the COX complex in order to optimize the efficiency of OXPHOS at different oxygen concentrations. It does so by stimulating transcription of genes that promote the switching of COX4-1 (normoxia) and COX4-2 (hypoxia) isoforms. HIF-1 activates transcription of COX4-2 and the mitochondrial protease LON. The latter stimulates increased proteolyitc degradation of COX4-1s.

14.12 Hexokinase II and Akt Drive the Glycolytic Shift and Prevent Apoptosis in Tumors

One of the main components supporting glycolysis in cancerous cells is Hexokinase II. This enzyme along with Hexokinase I is frequently overexpressed in cancer. These proteins are the leadoff enzymes in glycolysis. They catalyze the transfer of a phosphate group to glucose, forming glucose-6-phosphate. They use ATP as the donor and by attaching to the cytosolic facing surface of the outer mitochondrial membrane next to the voltage-dependent anion channel (VDAC) are well positioned to utilize newly synthesized ATP molecules, and under normal conditions to coordinate glycolysis with oxidative phosphorylation (Fig. 14.9).

Recall from Chapter 7 that ADP is supplied to complex V, ATP synthase, of the electron transport chain through the adenine nucleotide translocator (ANT) situated nearby in the IMM. The ANT translocates ADP from the IMS to the matrix and translocates newly formed ATP from the matrix to the IMS. The glycolytic machinery is positioned just outside the OMM, and the enzymatic machinery comprising the tricarboxylic acid, or TCA, cycle (also known as the Krebs or citric acid cycle) resides in the mitochondrial matrix (Fig. 14.9). The VDAC resides in the outer mitochondrial membrane in close proximity to the ANT. It is a porin type of ion channel that mediates transport of adenine nucleotides, ions (both anions and cations), and metabolites across the OMM. The channel opening is fairly small, less than 3 nm, and cytochrome c is too large to pass out through the VDAC.

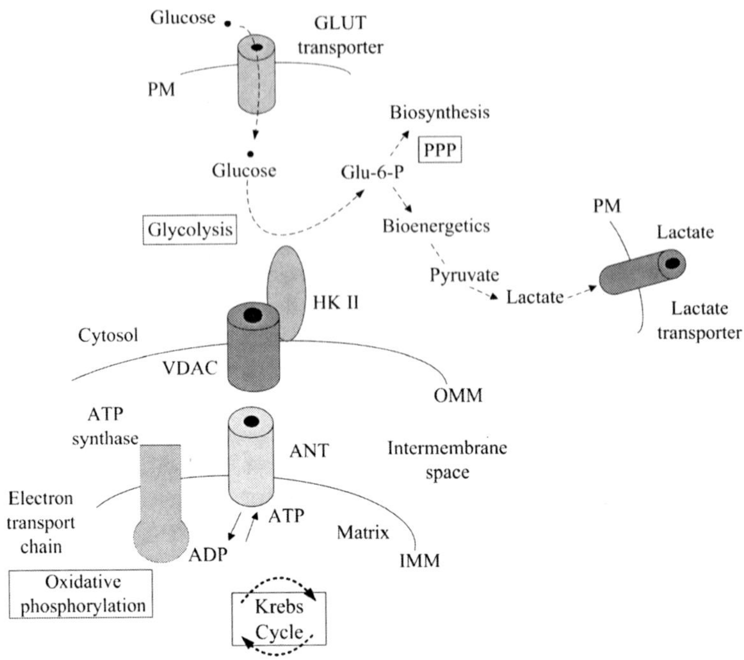

Fig. 14.9 Hexokinase II and related regulatory elements and steps. VDAC, voltage-dependent anion channel; ANT, adenine nucleotide translocator; IMM, inner mitochondrial membrane; OMM, outer mitochondrial membrane; PM, plasma membrane

The conversion of glucose to glucose-6-phosphate by the hexokinases is an important step; by appending negatively charged phosphoryl groups to glucose, the glucose molecules are unable to pass through the plasma membrane and exit the cell. There are four hexokinase isozymes. Hexokinases I and II are broadly distributed throughout many different cell types whereas hexokinase IV, also known as glucokinase, is found mostly in pancreatic β-cells and liver. Recall from Chapter 3 that in healthy pancreatic β-cells, glucose enters through GLUT2 transporters. Glucokinase converts glucose to gluco-6 phosphate which then undergoes further steps of glycolysis and oxidative metabolism. The ratio ATP to ADP ratio goes up, and this increase activates the Kir2 potassium channel embedded in the plasma membrane. The resulting depolarization of the plasma membrane triggers an opening of voltage-dependent calcium channels (VDCCs) through which calcium enters the cell leading to insulin release from the cells.

Hexokinase II-mitochondria interactions are believed to have a major role in decisions whether or not to undergo apoptosis. In particular, VDAC-bound hexokinase II is though to inhibit apoptosis through mechanisms that are not entirely clear as yet. There is evidence that it prevents apoptosis by disrupting the binding of channel-forming proteins Bax and Bak to the OMM as well as preventing the relocation of BH3-only proteins such as Bad. Overall, it prevents increases in mitochondrial membrane permeability to cytochrome c efflux, which would otherwise trigger apoptosis. Its influence on apoptosis is mediated by Akt, which stimulates HKII-VDAC binding and thus promotes survival. This prevention of apoptosis is further enhanced by overexpression of Akt, a common event in cancers. It is also enhanced by the HKII promoter, which is activated by hypoxia (HIF-1) and by growth factor signaling. All of these antiapoptotic processes are of special importance since the main mode of action of cytotoxic drugs is induction of apoptosis.

Further Reading

The mTOR Cassette and Signaling Pathways

Fingar DC and Blenis J 2004 Target of rapamycin (TOR): an integrator of nutrient and growth factor signals and coordinator of cell growth and cell cycle progression. *Oncogene* 23: 3151–3171.

Wullschleger S, Loewith R and Hall MN 2006 TOR signaling in growth and metabolism. *Cell* 124: 471–484.

AMPK and Energy Balance

Feng Z et al. 2005 The coordinate regulation of the p53 and mTOR pathways in cells. *Proc. Natl. Acad. Sci. USA* 102: 8204–8209.

Inoki K et al. 2003 TSC2 mediates cellular energy response to control cell growth and survival. *Cell* 115: 577–590.

Jones RG et al. 2005 AMP-activated protein kinase induces a p53-dependent metabolic checkpoint. *Mol. Cell* 18: 283–293.

Hypoxia and Growth Regulation

deYoung MP et al. 2008 Hypoxia regulates TSC1/2-mTOR signaling and tumor suppression through REDD1-mediated 14-3-3 shuttling. *Genes Dev.* 22: 239–251.

Koritzinsky M et al. 2006 Gene expression during acute and prolonged hypoxia is regulated by distinct mechanisms of translation control. *EMBO J.* 25: 1114–1125.

Liu L et al. 2006 Hypoxia-induced energy stress regulates mRNA translation and cell growth. *Mol. Cell* 21: 521–531.

Starvation/Amino Acid Supply and Growth

DeBerardinis RJ et al. 2007 Beyond aerobic glycolysis: transformed cells can engage in glutamate metabolism that exceeds the requirement for protein and nucleotide synthesis. *Proc. Natl. Acad. Sci. USA* 104: 19345–19350.

Kim E et al. 2008 Regulation of TORC1 by RagGTPases in nutrient response. *Nat. Cell Biol.* 10: 935–945.
Liang C et al. 2006 Autophagic and tumor suppressor activity of a novel Beclin 1-binding protein UVRAG. Nat. Cell Biol. 8: 688–699.
Nobukuni T et al. 2005 Amino acids mediate mTOR/raptor signaling through activation of class 3 phosphatidylinositol 3OH kinase. *Proc. Natl. Acad. Sci. USA* 102: 14238–14243.
Sancak Y et al. 2008 The Rag GTPases bind Raptor and mediate amino acid signaling to mTORC1. *Science* 320: 1496–1501.

Control of Protein Synthesis

Holz MK et al. 2005 mTOR and S6K1 mediate assembly of the translation preinitiation complex through dynamic protein interchange and ordered phosphorylation events. *Cell* 123: 569–580.
Inoki K, Zhu T and Guan KL 2003 TSC2 mediates cellular energy response to control cell growth and survival. *Cell* 115: 577–590.
Jacinto E and Hall MN 2003 Tor signaling in bugs, brain and brawn. *Nat. Rev. Mol. Cell Biol.* 4: 117–126.

Starvation and Autophagy

Lum JJ et al. 2005 Growth factor regulation of autophagy and cell survival in the absence of apoptosis. *Cell* 120: 237–248.
Lum JJ, DeBerardinis RJ and Thompson CB 2005 Autophagy in metazoans: cell survival in the land of plenty. *Nat. Rev. Mol. Cell Biol.* 6: 439–447.
Maiuri MC et al. 2007 Self-eating and self-killing: crosstalk between autophagy and apoptosis. *Nat. Rev. Mol. Cell Biol.* 8: 741–752.
Mizushima N 2007 Autophagy: process and function. *Genes Dev.* 21: 2861–2873.
Oberstein A, Jeffrey PD and Shi Y 2007 Crystal structure of the Bcl-XL-Beclin 1 peptide complex. *J. Biol. Chem.* 282: 13123–13132.
Pattingre S et al. 2005 Bcl2 antiapoptotis proteins inhibit Beclin 1-dependent autophagy. *Cell* 122: 927–939.

Regulation of Metabolism by p53

Bensaad K et al. 2006 TIGAR, a p53 inducible regulator of glycolysis and apoptosis. *Cell* 126: 107–120.
Gertz M et al. 2008 Activation of the lifespan regulator p66Shc through reversible disulfide bond formation. *Proc. Natl. Acad. Sci. USA* 105: 5705–5709.
Giorgio M et al. 2005 Electron transfer between cytochrome c and p66Shc generates reactive oxygen species that trigger mitochondrial apoptosis. *Cell* 122: 221–233.
Matoba S et al. 2006 p53 regulates mitochondrial respiration. *Science* 312: 1650–1653.
Ott M et al. 2007 Mitochondria, oxidative stress and cell death. *Apoptosis* 12: 913–922.
Pinton P et al. 2007 Protein kinase Cβ and prolyl isomerase 1 regulate mitochondrial effects of the life-span determinant p66Shc. *Science* 315: 659–663.
Sablina AA et al. 2005 The antioxidant function of the p53 tumor suppressor. *Nat. Med.* 11: 1306–1313.

The PTEN Tumor Suppressor

Li AG et al. 2006 Mechanistic insights into maintenance of high p53 acetylation by PTEN. *Mol. Cell* 23: 575–587.
Shen WH et al. 2007 Essential role for nuclear PTEN in maintaining chromosomal integrity. *Cell* 128: 157–170.
Trotman LC et al. 2007 Ubiquitination regulates PTEN nuclear import and tumor suppression. *Cell* 128: 141–156.
Wang X et al. 2007 NEDD4-1 is a proto-oncogenic ubiquitin ligase for PTEN. *Cell* 128: 129–139.

PTEN Structure and Function

Lee JO et al. 1999 Crystal structure of the PTEN tumor suppressor: implications for its phosphoinositide phosphatase activity and membrane association. *Cell* 99: 323–334.
Leslie NR and Downes CP 2004 PTEN function: how normal cell control it and tumour cells lose it. *Biochem. J.* 382: 1–11.

HIFs and the Glycolysis

Fantin VR, St-Pierre J and Leder P 2006 Attenuation of LDH-A expression uncovers a link between glycolysis, mitochondrial physiology, and tumor maintenance. *Cancer Cell* 9: 425–434.
Fukuda R et al. 2007 HIF-1 regulates cytochrome oxidase subunits to optimize efficiency of respiration in hypoxic cells. *Cell* 129: 111–122.
Gordan JD et al. 2007 HIF-2α promotes hypoxic cell proliferation by enhancing c-Myc transcriptional activity. *Cancer Cell* 11: 335–347.
Kim JW et al. 2006 HIF-1-mediated expression of pyruvate dehydrogenase kinase: a metabolic switch required for cellular adaptation to hypoxia. *Cell Metab.* 3: 177–185.
Lum JJ et al. 2007 The transcription factor HIF-1α plays a critical role in the growth factor-dependent regulation of both aerobic and anaerobic glycolysis. *Genes Dev.* 21: 1037–1049.
Papandreous I et al. 2006 HIF-1 mediates adaptation to hypoxia by actively downregulating mitochondrial oxygen consumption. *Cell Metab.* 3: 187–197.
Zhang H et al. 2007 HIF-1 inhibits mitochondrial biogenesis and cellular respiration in VHL-deficient renal cell carcinoma by repression of c-Myc activity. *Cancer Cell* 11: 407–420.

Hexokinases and Akt

Mathupala SP, Ko YH and Pedersen PL 2006 Hexokinase II: cancer's double-edged sword acting as both facilitator and gatekeeper of malignancy when bound to mitochondria. *Oncogene* 25: 4777–4786.
Majewski N et al. 2004 Akt inhibits apoptosis downstream of BID cleavage via a glucose-dependent mechanism involving mitochondrial hexokinases. *Mol. Cell. Biol.* 24: 730–740.
Wilson JE 2003 Isozymes of mammalian hexokinase: structure, subcellular localization and metabolic function. *J. Exp. Biol.* 206: 2049–2057.

Elstrom R et al. 2004 Akt stimulates aerobic glycolysis in cancer cells. *Cancer Res.* 64: 3892–3899.

Gottlob K et al. 2001 Inhibition of early apoptotic events by Akt/PKB is dependent on the first committed step of glycolysis and mitochondrial hexokinase. *Genes Dev.* 15: 1406–1418.

Plas DR and Thompson CB 2005 Akt-dependent transformation: there is more to growth than just surviving. *Oncogene* 24: 7435–7442.

Majewski N et al. 2004 Hexokinase-mitochondria interaction mediated by Akt is required to inhibit apoptosis in the presence or absence of Max and Bak. *Mol. Cell* 16: 819–830.

Rathmell JC et al. 2003 Akt-directed glucose metabolism can prevent Bax conformation change and promote growth factor-independent survival. *Mol. Cell. Biol.* 23: 7315–7328.

Chapter 15
Metastasis

Metastasis begins with the invasion of tumor cells into its surrounding tissue, the stroma. This initial event is followed by migration into the circulation, or intravasation, homing to distant sites, followed by extravasation, seeding and colonization. Deciphering the mechanisms of how, step by step, a cancer cell can leave the site of a primary tumor to seed a colony at a secondary site represents one of the great challenges of science today. It is estimated that about 90% of all cancer deaths arise from cancers that have metastasized. At present there are few successes in treatment interspersed among the far larger number of failures to substantially impede the spread and ensuing deaths.

Given its importance it is not surprising that considerable efforts have been made to better understand metastasis. Two major themes have emerged from these studies. The first is the ascendency of the microenvironment to a place where it must be considered co-equally with the primary tumor cell in devising new therapeutic approaches. This rise has as its underpinnings the twin bulwarks of cancer – sustained angiogenesis and chronic inflammation – that link together in a mechanistic fashion tumor growth and metastasis.

The second leap forward is the demise of the notion of a linear, one-way progression of cellular phenotypes from less differentiated to more highly differentiated and eventually terminally differentiated forms. In its place are the concepts of phenotype plasticity and nuclear reprogramming; that is, the ability of cells not only to differentiate, but also to dedifferentiate and even transdifferentiate in response to right cellular signals. Macrophage and fibroblast phenotype switching, and metabolic reprogramming in response to oxygen supply and nutrient availability, are two of an ever-growing list of reversible changes that cells can initiate when needed. This chapter will highlight another set of even more dramatic plastic changes – those associated with the reawakening of a developmental program called the epithelial-to-mesenchymal transition, or EMT.

In an epithelial-to-mesenchymal transition, epithelial cells, the main type of cells likely to form solid tumors, lose their polarized shape and adhesiveness. In place of this morphology, they adopt a mesenchymal form suitable for migration out of the tissue into circulation, transit in the circulation, homing to a secondary site, migration out of circulation, and colonization, the latter

M. Beckerman, *Cellular Signaling in Health and Disease*,
Biological and Medical Physics, Biomedical Engineering,
DOI 10.1007/978-0-387-98173-4_15,

supported by the reverse process, a mesenchymal-to-epithelial transition (MET). Several hallmarks of cancer as listed in Fig. 9.1 are associated with an EMT including downregulation of adhesive molecules, activation of extracellular proteases and, of course, inflammation. The EMT and, more generally, tumor growth and metastasis are community affairs. The two aforementioned themes are deeply intertwined; how they and their associated hallmarks influence metastasis are the focus in this chapter.

15.1 Tumor Growth and Metastasis Are Community Affairs

In order for a tumor to grow beyond 1–2 mm in size, it requires a ready supply of nutrients and oxygen. Tumor cells respond to this need by secreting factors such as the polypeptide VEGF that stimulate the continual growth of new vasculature. The resulting vasculature differs from that produced under normal conditions of wound healing and tissue repair. It is highly irregular in its organization and leaky. As discussed in the last chapter, the tumor microenvironment surrounding a growing tumor is hypoxic and acidic. Angiogenesis is one of a number of modifications that are continually made to the tumor microenvironment all leading to further tumor growth and metastasis.

A key event in tumor growth and metastasis is the angiogenic switch. This term refers to a change in tumor cell signaling strategy. In place of antiangiogenic factors, tumor cells begin to secrete VEGF and other agents that stimulate growth of new vasculature. They accomplish this in part by secreting cytokines such as macrophage-colony stimulating factor (M-CSF), TNF-α, and IL-6 that attract macrophages. Once these factors succeed in attracting macrophages to the developing tumor and vasculature, the macrophages (and CAFs) assist by secreting a mix of angiogenic growth factors that further stimulate proliferation of endothelial cells thereby amplifying the signaling efforts.

A broad spectrum of growth factors belonging to the receptor tyrosine kinase (RTK) and cytokine superfamilies are secreted by the tumor cells, CAFs, and TAMs. The most prominent of the RTKs, from the point of view of cancer, are listed in Table 15.1 along with some of their cancer-promoting activities. Their role in activating growth pathways was examined in the last chapter. As will be discussed shortly, many of these agents (e.g., EGF) also operate coordinately with other receptor types, most notably the integrins, to promote migration.

Several classes of cells reside in the tumor microenvironment and work together to promote tumor growth and metastasis. In addition to tumor cells, there are fibroblasts, adipocytes, blood and lymph vessels, and macrophages. Under the influence of tumor cells and the factors they secrete, activated stromal fibroblasts, commonly referred to as carcinoma-associated fibroblasts (CAFs), secrete factors that stimulate tumor growth and the production of new blood vessels, and recruit endothelial progenitor cells (EPCs) to the tumor site.

Table 15.1 Receptor tyrosine kinases in health and disease

Receptor tyrosine kinase family	Actions in health and disease
Epidermal growth factor	Growth and survival
Fibroblast growth factor	Tissue repair, stem cell proliferation
Hepatocyte growth factor/MET	Migration and invasiveness
Platelet derived growth factor	Angiogenesis, fibrogenesis
Vascular endothelial growth factor	Angiogenesis, vasodilation

Some of these factors and their contributions to tumor growth and metastasis are presented in Fig. 15.1. Tumor-associated macrophages (TAMs) are essential members of the tumor stroma. They interact with tumor cells and CAFs and have profound influences upon tumor growth and metastasis. Some of their main activities are summarized in Fig. 15.1

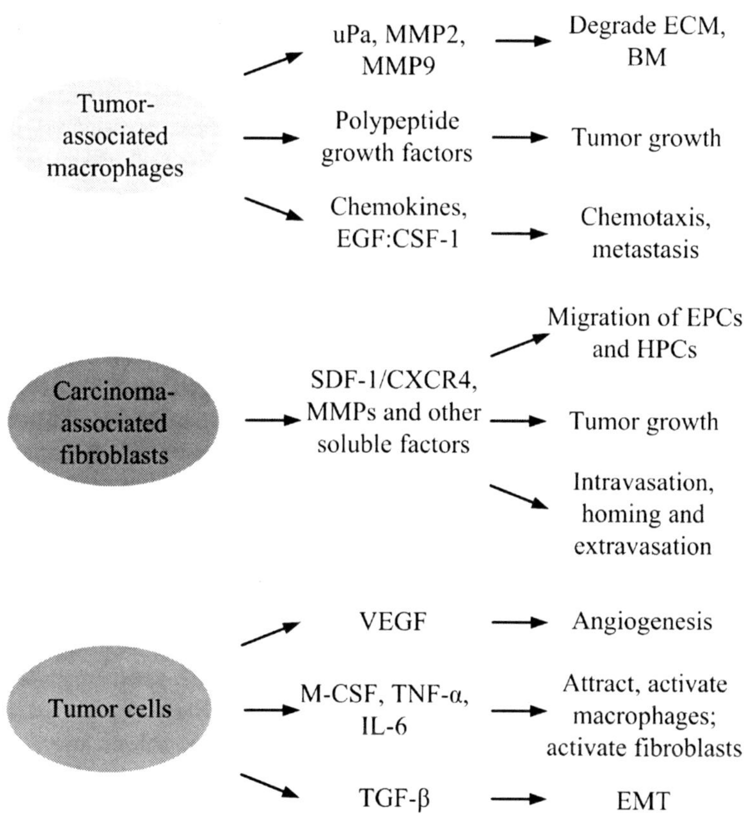

Fig. 15.1 A community affair in which tumor cells, fibroblasts, and macrophages send and receive signals to one another that remodel the microenvironment in the vicinity of the tumor and promote tumor growth and metastasis

15.2 Macrophages and Fibroblasts Direct Invasion and Intravasation

Macrophages facilitate invasion and intravasation in several ways. Macrophages secrete chemotactic agents; they release ECM-remodeling proteases such as MMP2, MMP9, and uPA, and they secrete growth factors as discussed above. Tumor cells and macrophages migrate together toward blood vessels. Intravasation is accomplished by means of a paracrine-signaling loop that generates a self-reinforcing migratory signal for the invading tumor cells to follow. The macrophages express CSF-1 receptors and secrete EGF ligands. The tumor cells express EGF receptors and secrete CSF-1 ligands. Ligand binding by either initiates intracellular signaling leading to secretion of its ligands. Those ligands bind receptors on the surface of the initial signaling partner, again resulting in intracellular signaling and secretion of ligands.

Directionality is supplied by the network of collagen fibers that emanate from the tumor-associated blood vessels. Both tumor cells and macrophages migrate along these fibers to the blood vessel. The EGF/CSF-1 signals direct both cell types to reorganize their actin cytoskeletons in a way that supports the migration. Several key signaling pathways direct these changes in morphology. In particular, they direct the development of membrane protrusions on the leading edge of the cell such as filopodia and lamellipodia that are required for movement.

Fibroblasts promote tumor cell invasion and intravasation by secreting chemokines, most notably, SDF-1 that stimulate and guide tumor cell migration. They also secrete proteolytic factors such as MT1-MMP/MMP-14 that degrade collagen and generate microtracks for tumor cells to follow. Instances have been uncovered in which tumor cells undergo collective migration with the fibroblasts leading the way. Tumor cells, macrophages, and fibroblasts all secrete VEGF that increases the permeability of the microvasculature and helps attract fibroblasts, macrophages, and stem/precursor cells to the stroma. The expression of SDF-1 is not only induced by VEGF but also by HIF-1 upregulated in endothelial cells under hypoxic conditions. This chemokine is laid down by endothelial cells and fibroblasts in gradients that guide CXCR4-expressing progenitors and tumor cells to vascular niches.

Collective migration and preparation of niches for tumor cells to colonize are two of the several operations carried out by tumor-associated macrophages and carcinoma-associated fibroblasts. These cells along with resident endothelial cells, pericytes and smooth muscle cells, and infiltrating precursor/stem cells form a pre-metastatic niche, a potential site for metastatic colonization. As was the case for the primary site of tumor growth, inflammatory and angiogenic conditions generated through joint interactions between these cell populations support colony formation at these sites.

Integrins along with the chemokines, growth factors, and proteases mediate the interactions between cells. Cooperativity between ($\alpha_v\beta_3$) integrins and

VEGF receptors (i.e., VEGFR-2), like that depicted in Figure 15.4 (a), is one of the key elements that trigger the sprouting of new blood vessels at metastatic niches. The integrins are expressed on the surfaces of a variety of cell types in these niches including endothelial cells, pericytes, and the precursor/stem cells. Another integrin with important roles in neovascularization is $\alpha_4\beta_1$ (VLA-4). Binding and signaling mediated by expression of this integrin on proliferating endothelial cells bind VCAM expressed on pericytes, and this interaction is important for angiogenesis. This integrin also binds fibronectin secreted by fibroblasts, and thus supports migration of $\alpha_4\beta_1$-bearing cells into the niche.

15.3 The SDF-1/CXCR4 Axis Is a Central Participant in Metastasis

Under normal tissue conditions, CXCR4 receptors are expressed at a low level while CXCL12 ligands are constitutively expressed. The hypoxic conditions characteristics of the altered tumor microenvironment trigger the upregulation of CXCR4 receptors aided by the presence of a hypoxia-responsive element in its promoter. This upregulation leads to activation of several signaling pathways. These are represented schematically in Fig. 15.2.

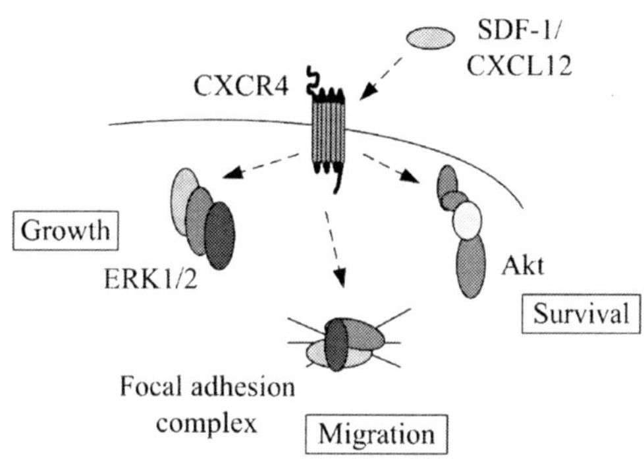

Fig. 15.2 SDF-1/CXCL12/CXCR4 signaling in response to hypoxic conditions in the tumor microenvironment

Like other chemokines SDF-1 binds receptors belonging to the 7-pass, G-protein-coupled receptor superfamily. In response, the CXCR4 receptors activate several intracellular signaling pathways. One of these is the ERK MAP kinase module that mediates the growth and proliferation effects of SDF-1. These actions are further supported by signaling to Akt, which upon activation enhances survival and reduces the chances for apoptosis.

Another signaling route activated leads to the focal adhesion complexes. The associated signaling events promote assembly for components of the focal adhesions such as Fak, paxillin, and p130 CAS. These complexes have a major role in cytoskeleton adjustments, cellular movements, and migration.

Several metastasis-related activities are facilitated by the SDF-1/CXCR4 axis among which are intravasation, homing, and extravasation much as a similar set of activities would under different conditions (1) assist leukocytes in arriving at sites where they are needed in response to infection or injury and (2) contribute to stem cell trafficking.

15.4 Focal Adhesions and Metastasic Migration

Recall from Chapter 6 that integrins connect the extracellular matrix to the cellular cytoskeleton and anchor cells in fixed locations while cadherins organize cell-to-cell adhesion junctions. Both integrins and cadherins convey signals outside-in and inside-out. Cadherins associate with catenins and form complexes tied to the actin cytoskeleton. Integrins form focal adhesions, points of contact and adhesion between the cell and its supporting membranes. Focal adhesions are control points where growth and adhesion signals are integrated together and coordinated across multiple points of ECM-to-cell surface contact, to govern the overall growth and movement of the cell. These control points not only regulate the assembly and disassembly of the focal adhesions but also convey signals that control cellular growth, proliferation, differentiation, and survival.

Extracellular matrix proteins, transmembrane proteins, and the actin cytoskeleton proteins participate in the adhesive contacts. Integrins and growth factor receptor colocalize at focal adhesions. The integrins bind to ECM proteins such as laminin and the growth factor receptors bind to growth factor ligands. In response to ligand binding by these receptors, a number of nonreceptor tyrosine kinases and adaptor/scaffold proteins are recruited to the plasma membrane. Among the nonreceptor tyrosine kinases recruited are Src, its negative regulator Csk, and focal adhesion kinase (FAK). These kinases along with a key adapter called paxillin link the integrin and growth factor receptor signaling to the actin cytoskeleton.

c-Src (Src) is a nonreceptor tyrosine kinase and is the cellular counterpart to the viral protein v-Src encoded by the Roux sarcoma virus. Its discovery in the late 1970s by Michael Bishop and Harold Varmus led to a profound change in thinking about cancer from that of a disease caused by external agents to one caused by mutations. This protein occupies a central place in focal adhesion complexes from which it regulates the actin cytoskeleton. Upstream Src activators include receptor tyrosine kinases, integrins, and cytokine receptors. It thus sits in a central signaling node, and in dysregulated systems coordinates cancer-related cellular adhesion, motility, and invasiveness.

The domain structure of Src is presented in Fig. 15.3. In support of its functions, the protein contains N-terminal SH2 and SH3 domains and a C-terminal kinase domain. Embedded in the C-terminal region are two critical tyrosine phosphorylation sites. Phosphorylation at Y416 in chickens and

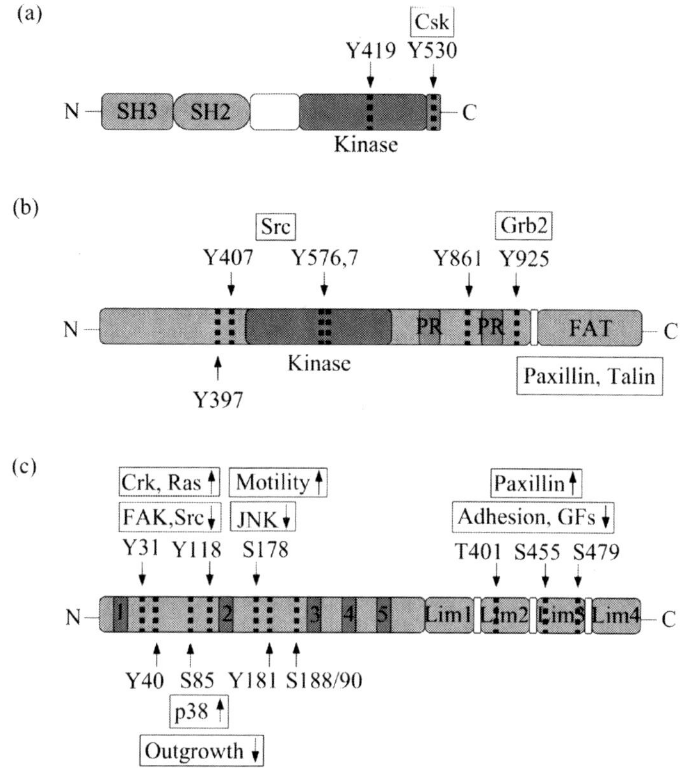

Fig. 15.3 Components of the FAK/c-Src signaling complex. (**a**) c-Src, (**b**) Focal adhesion kinase (FAK), and (**c**) Paxillin. *Arrows* pointing toward the structure (inward) denote kinases and factors responsible for phosphorylation; *arrows* pointing outward indicate the cellular actions and mediators assisted by the phosphorylated residues

Y419 in humans by an upstream kinase activates the protein. The second site, at Y527 in chickens and Y530 in humans in an extreme C-terminal regulatory region, is targeted by the negative regulator Csk. Phosphorylation at this site induces a conformation rearrangement of the domains leading to Src deactivation.

Recall from Chapter 8 that the ECM provides sites for attachment of signaling molecules, sequesters them, supplies ligands for cell-surface adhesion molecules, and regulates cell survival through interactions with the cell surface and cytoskeleton. The extracellular matrix is remodeled during wound healing, and ECM remodeling is crucial for the transition from the normal to the disease states characteristic of atherosclerosis and cancer.

Paxillin is a fairly small protein, only 68 kDa in mass, but it contains a large number of binding sites. Its structure is shown in Fig. 15.3(c). It possesses two tyrosine phosphorylation sites that are targeted by the nonreceptor tyrosine kinases such as Src, Csk, and FAK, and bound by SH2 domain-bearing proteins subsequent to phosphorylation. A proline-rich region serving as an attachment site for SH3 domains is located in the same vicinity. These N-terminal sites provide a linkage to upstream integrins and growth factor receptors, and also downstream to proteins associated with the actin cytoskeleton through the five LD repeats. The C-terminal LIM domains anchor the paxillin protein at the plasma membrane.

Proliferation is a highly regulated process. For proliferation to occur, the appropriate adhesive and growth factor signals must be sent and received. Integrin receptors and receptor tyrosine kinases are localized in the plasma membrane close to one another and together work in signaling growth. The adhesive signals confirm that the cell remains in adhesive contact with the extracellular matrix, and thus growth is permitted. Paxillin associates with focal adhesions; it contains multiple binding sites and serves as a platform for gathering adhesive signals relayed through integrin receptors and growth factor signals sent by receptor tyrosine kinases. Several NRTKs participate in the relay of messages from the transmembrane receptors to paxillin; prominent among these is the focal adhesion kinase (FAK).

The domain composition of FAK is presented in Fig. 15.3(b). It does not have any SH2 or SH3 domains but instead provides phosphorylation and anchoring sites for proteins with these domains. In place of the SH2 and SH3 domains, FAK has two large domains of about 400 amino acids each, one on either side of the catalytic domain. FAK possesses six tyrosine phosphorylation sites. Two of these, Tyr397 and 407, lie just N-terminal to the kinase domain; two others, Tyr576 and 577, lie inside the kinase domain, and the last two, Tyr 861 and 925, lie in the COOH terminal region N-terminal to the FAT. Autophosphorylation at Tyr397 exposes an SH2 docking site for Src. Phosphorylation by Src at Tyr407, 576, and 577 maximally activates the kinase domain, and phosphorylation at the sixth site, Tyr925, provides a Grb2 docking site. The domain structure of Grb2 was presented in Fig. 15.3(b). It is an adaptor protein that links FAK to the MAP kinase pathway. There are also two proline-rich regions in the COOH terminal domain that provide docking sites for adaptor proteins bearing SH3 domains. Thus, like paxillin, FAK serves as integrator of adhesive and growth signals. The COOH-terminal region of FAK contains a focal adhesion targeting (FAT) sequence of about 160 amino acids that provides binding sites for paxillin and talin.

15.5 Receptor Cooperativity and Src Signaling

Integrin signaling is central to tumor growth and metastasis. These receptors together with growth factor receptors regulate cellular growth and survival. In oncogenic circuits, integrin signaling to focal adhesion complexes is disrupted as is cadherin maintenance of adherens junctions. In the signaling pathways activated by RTKs and integrins (Fig. 15.4(a)), the FAK/c-Src complex signals via the JNK pathway to trigger the transcription of matrix metalloproteinases

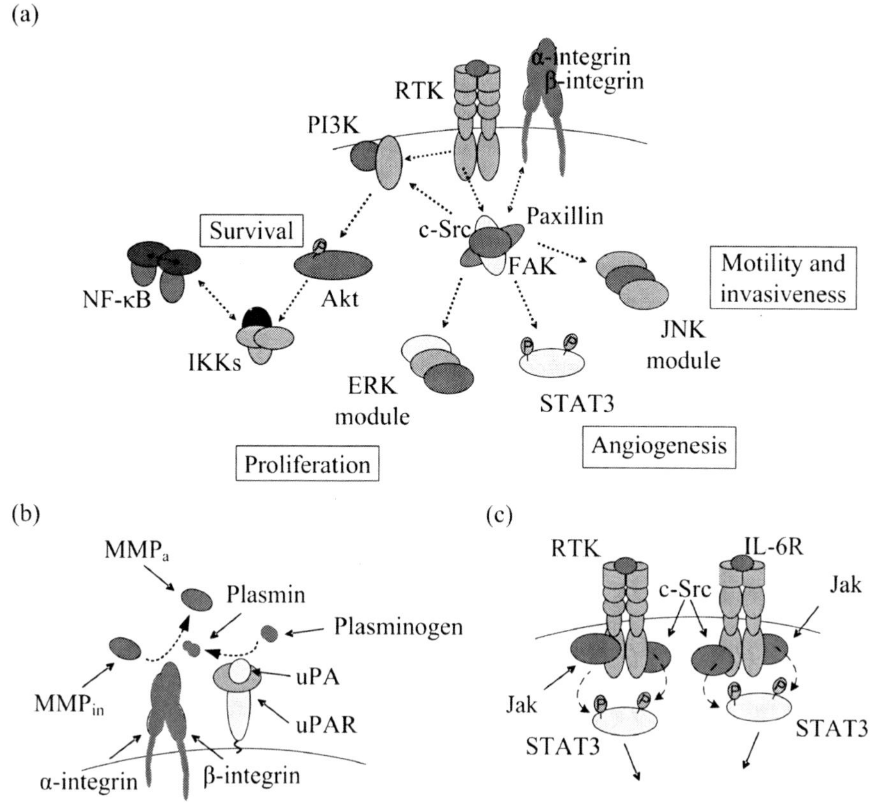

Fig. 15.4 Receptor cooperativity and Src signaling. (**a**) Joint RTK and integrin signaling to focal adhesions resulting in activation of several metastasis-promoting pathways. (**b**) Urokinase-type plasminogen activator (uPA) – integrin regulation of MMP processing and activation. (**c**) RTK and cytokine receptor IL-6R cooperativity resulting in Jak-STAT3 transcription factor signaling to the nucleus

MMP2 and MMP9 that degrade ECM components. It also activates STAT3, which stimulates the expression of VEGF leading to increased angiogenesis. Additonal FAK/c-Src signals activate the ERK MAP kinase pathway and promote the epithelial-to-mesenchymal transition (to be discussed in the sections that follow this one). Overall, signals sent out via this complex act (1) near the cells surface to disable the integrin-focal adhesions and cadherins-adherens junctions that immobilize cells and maintain tissue homeostasis and (2) downstream in the nucleus to upregulate genes needed for motility and invasiveness.

Cytokine receptors such as IL-6R, and growth factor receptors, recruit Janus tyrosine kinases and NRTKs such as c-Src to their cytoplasmic regions. Janus nonreceptor tyrosine kinases, or Jaks, and signal transducer and activator of transcription, or STATs, form a signaling pathway to the nucleus from cytokine and growth factor receptors in the plasma membrane. In response to ligand binding, cytokine receptors dimerize and activate Jaks, which in turn phosphorylate and activate their associated STATs. The latter are transcription factors that once activated translocate to the nucleus where they stimulate the transcription of genes. The Src kinases, in turn, phosphorylate and activate STATs such as STAT3. This protein is notably overexpressed and dysregulated in human cancers. The STAT3s not only stimulate the expression of vascular endothelial growth factor, but also upregulate antiapoptotic BCL-xL proteins and growth-promoting c-Myc. The STAT3 oncogenic pathway is depicted in Fig. 15.4(c).

Urokinase-type plasminogen activator (uPA) and its receptor, UPAR, operate jointly with integrin receptors such as $\alpha_v\beta_3$ to activate proteolytic enzymes that degrade the ECM to promote tumor growth and metastasis. Two actions occur in response to activation of the UPAR by uPA ligand binding (Fig. 15.4(b)). First, the activated UPAR catalyzes the conversion of the proteolytic enzyme plasminogen from its inactive proenzyme (zymogen) to its active form, plasmin, which acts as a serine protease to degrade ECM components. Second, it stimulates the conversion of MMPs recruited to the cell surface by the integrins from their inactive form to their proteolytically active one. In particular, $\alpha_v\beta_3$ integrins and plasmin together activate MMP2 and MMP9, which degrade Type IV collagens, the primary component of the ECM.

15.6 The Transforming Growth Factor-β Pathway

The EMT generates cells with the ability to migrate and as such is an early critical event in cancer metastasis. It logically precedes and initiates the series of events described in the preceding sections. For some time metastasis has been regarded as a late process, one that occurs long after tumor growth has begun. That view is challenged by numerous observations that support a far earlier onset. Two of the developmental signaling pathways closely associated with the EMT are the transforming growth factor-β (TGF-β) pathway and the Wnt pathway. Significantly, the transforming growth factor β (TGF-β) is capable of acting both as an anticancer agent or as a pro-cancer one. As indicated in Fig. 15.1, it is secreted by tumor cells and also it turns out by all of the others present in the tumor stroma.

The transforming growth factor-β (TGF-β) pathway is involved in developmental patterning, wound healing, bone formation, ECM fabrication, and homeostasis. There are three core components in a TGF-β pathway – ligands, transmembrane receptors, and cytoplasmic signal transducers and transcription factors called Smads. The elements of this pathway organize into two complexes. One complex is formed at the cell membrane. Ligand binding results in receptor activation, clustering, phosphorylation, and activation of the Smads. These signal transducers dissociate from the first, cell surface complex, oligomerize, and translocate to the nucleus where they join with a variety of cofactors to form a second set of complexes that mediate transcription activation and repression.

The TGF-β family is a large one consistent with its role in regulating multiple cellular processes. In mammals there are 12 different receptors, 29 distinct ligands, and 8 different Smads. The TGF-β pathway is named for the TGF-β subfamily of diffusible polypeptide ligands. Other prominent ligands involved in signaling through the TGF-β pathway are the bone morphogenetic proteins (BMPs), growth and differentiation factors (GDFs), members of the Activin subfamily, and Nodal. These signaling systems are summarized in Table 15.2.

Table 15.2 The TGF-β signaling pathway

Pathway	Activin	TGF-β	TGF-β	BMP
Ligands	Activins	TGFβs	TGFβs	BMPs
Type 2 Rs	ActII/IIB	TβRII	TβRII	BMPRII, ActRII/IIB, MISRII
Type 1 Rs	ALK4	ALK5	ALK1	ALK3, ALK6, ALK2
R-Smad	Smad2, Smad3	Smad2, Smad3	Smad1, Smad5, Smad8	Smad1, Smad5, Smad8
Co-Smad	Smad4	Smad4	Smad4	Smad4
I-Smad	Smad7	Smad7	Smad6, Smad7	Smad6, Smad7

As indicated in the table, the TGF-β pathway branches at the Type 1 receptor (Type 1 Rs) level into an activin-like branch and a BMP-like branch. ALK, activin receptor-like kinase; BMP, bone morphogenetic protein; TGF, transforming growth factor; R, receptor-activated; I, inhibitory.

TGF-β receptors are receptor serine/threonine kinases possessing cytoplasmic domains that catalyze the phosphorylation of serine and threonine residues of targets proteins. There are two types of receptors; these are designated Type 1 and Type 2 receptors, and are abbreviated as TβRI and TβRII. These receptors possess an extracellular ligand binding domain and a cytoplasmic kinase domain. Type 1 receptors have an additional cytoplasmic domain referred to as a Gly-Ser (GS) motif immediately N-terminal to the kinase domain that functions as a key regulatory region. TGF-β ligands prefer to bind to Type 2 receptors. These binding events recruit Type 1 receptor molecules to the assemblage, resulting in the formation of a complex consisting ligand dimer bound to a receptor foursome. As a consequence, the cytoplasmic domains of the Type 2 and Type 1 proteins are brought into close proximity to one another and are stabilized.

The next step in transducing the signal into the cell is transphosphorylation of serine and threonine residues in the GS domains of the Type 1 receptors, assisted by the Type 2 receptors. This step creates docking sites for a number of different adaptor proteins. One of these, the Smad anchor for receptor activation (SARA), couples Smad2/3 to the activated receptor complex. The SARA protein possesses a lipid-binding FYVE domain that enables it to tether to the plasma membrane. Two other domains permit it to bind simultaneously to the receptor and to the Smad protein. Once the Smad proteins have been recruited to the receptor complex they are phosphorylated, triggering their release into the cytosol as depicted in Fig. 15.5.

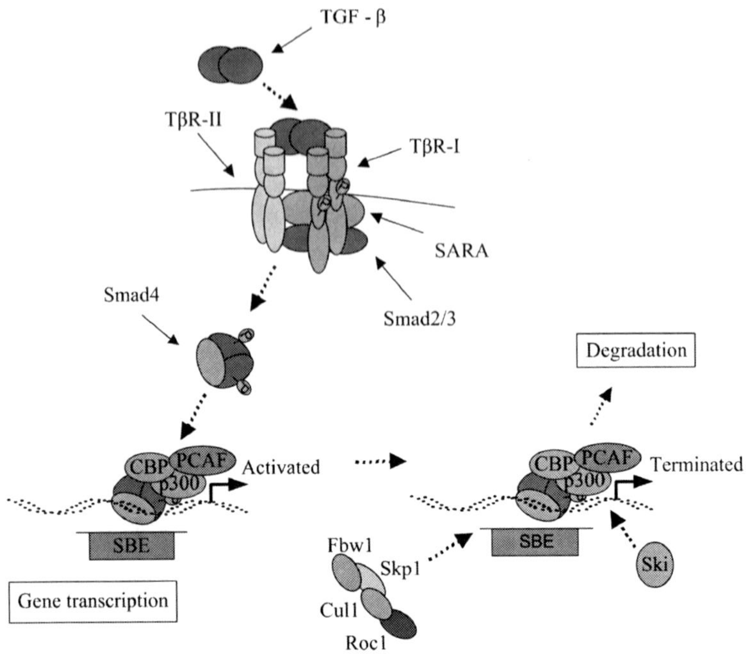

Fig. 15.5 TGFβ signaling pathway from ligand binding to transcription activation and then termination. The corepressor Ski is recruited to the promoter site where it helps dislodge the p300 coactivator. A Roc1-Scf^{Fbw1} E3 ubiquitin ligase complex binds and catalyzes the poly-ubiquitination of Smad3 leading to degradation of Smad3 by the 26S proteasome and termination of transcription. Abbreviations: c-Ski, Sloan-Kettering Institute proto-oncogene

Smad proteins convey messages from the cell surface to the nucleus. The name Smad is a contraction of the names of the first two Smad-type proteins to be discovered – the *C. elegans* Sma protein and the *Drosophila* mothers against dpp, or Mad, protein. There are at eight known Smads, designated Smad1 through Smad8, and they fall into three categories (Table 15.2). Five of the Smads are receptor activated and they are called R-Smads. Smad4 is required for signaling through all pathways and is called a common Smad, or co-Smad. The other two Smads, Smad6 and Smad7, are inhibitory. They turn off signaling, and form an anti-Smad grouping. Smad 6 inhibits signaling through the BMP branch while Smad7 inhibits signaling through all of the branches of the TGF-β signaling pathway. R-Smad proteins contain a pair of globular signaling domains connected by a linker region. Once activated by TβRI the R-Smads translocate to the nucleus. On the way to the nucleus the R-Smads associate with the co-Smad (Smad4) as illustrated in Fig. 15.5 for the case of Smad3. The specific cell fate decision arrived at through the TGF-β pathway depends on context. The steps outlined above are general and are neither cell specific or developmental-stage dependent. Partner and accessory signaling molecules that combine with the Smads to form the active transcriptional complexes supply the contextural information. Coactivators, corepressors, and other partners provide cell type and developmental stage inputs, and they fix parameters such as the duration of transcription.

15.7 TGF-β Promotes Cytostasis

TGF-βs inhibit cancer development by promoting cytostasis, the arrest of cellular growth and proliferation. Central to TGF-βs' cytostatic program is the inhibition of c-Myc and stimulation of cell cycle inhibitors such as p15 and p21. These actions are depicted in Fig. 15.6. The first part (a) illustrated the formation of a c-Myc inhibitory complex composed of E2F family members E2F4/5, the pRb family member p107, Smad3, and Smad4. This complex translocates from the cytoplasm to the cell nucleus where it binds the c-Myc promoter and functions as a transcription repressor that inhibits expression of c-Myc. As a result, c-Myc is not available to bind the p15 and p21 promoters. This is one of the key ways that c-Myc promotes cellular growth and proliferations. When activated it joins with Miz-1 and/or Sp1/Sp3 to form a repressor complex that prevents expression of p15 and p21 genes.

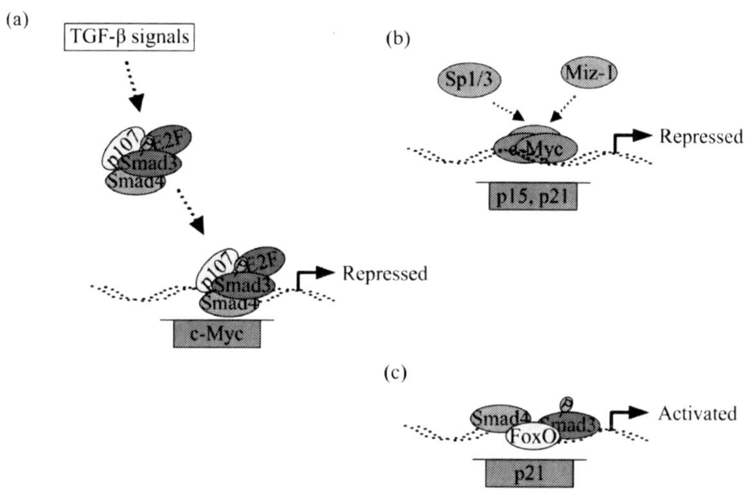

Fig. 15.6 Cytostatic actions of TGFβ. (**a**) Repression of the c-Myc oncogene resulting in the inability of c-Myc to repress p15 and p21 in the manner shown in (**b**). Further actions by the Smads activate p21 gene expression as shown in (**c**)

The cytostatic actions of the Smads include activation of cell cycle inhibitors. This facet is illustrated in Fig. 15.6(c) depicting the Smad-mediated activation of p21. The presence of FoxO proteins in the p21 transcription-activating complex is noteworthy. Recall that FoxO proteins are negatively regulated by

Akt. This kinase phosphorylates the FoxOs, which has the consequence of sequestering these transcription factors in the cytoplasm unable to promote gene expression. Growth factor signals relayed through PI3K to Akt thereby turn off p21-mediated repression of Cdk2 and Cdk4 by disabling p21 gene expression, and thus Ckd2 and Cdk4 are available to drive the cell cycle progression. Thus, TGF-β inhibitory and growth factor stimulatory signals converge and are coordinated under normal cellular conditions.

15.8 The Wnt Pathway

There are three major Wnt signaling pathways – canonical, planar cell polarity (PCP), and Ca^{2+}. These pathways are activated when Wnt ligands bind to Frizzled receptors, which along with an expanding list of coreceptors transduce signals across the plasma membrane into the cell. Wnt ligands are extracellular lipid-modified glycoproteins. There are 19 Wnts in mammalian species, 8 in *Drosophila*, and 7 in *C. elegans*. The Wnt receptors, members of the Frizzled family, are somewhat unusual G-protein-coupled receptors. Their unusual character manifests itself in the difficulty in identifying the specific heterotrimeric G-proteins to which they are coupled. There is good evidence that these proteins are present, but they have largely eluded identification to date. Each of the three pathways has associated with it one or more coreceptors. Coreceptors for the canonical pathway include the single-pass protein low-density lipoprotein (LDL)-receptor-related proteins 5 and 6 (LRP5/6). Another mammalian coreceptor, the atypical receptor tyrosine kinase Ryk, is present in neurites where it provides a link between Wnt and the cytoplasmic scaffolding protein Disheveled, a common element in all three pathways.

In response to ligand binding, Disheveled (Dsh) is recruited to the receptor complex. Interactions between the receptor, coreceptor(s), and Dsh determine whether ligand binding activates the canonical pathway or one of the alternative signaling routes. The exact set of steps leading to this decision is not yet known with certainty. At this point in time it is clear that Dsh is heavily phosphorylated in response to ligand binding.

The canonical Wnt signaling cascade has major roles in embryonic development, stem cell proliferation, and self-renewal, and when dysregulated in cancer progression. It is the best understood of the three pathways and will be the focus of attention in this chapter. The steps leading to activation of signaling in the canonical Wnt pathway are illustrated in Fig. 15.7.

The key event taking place when Wnt ligands are not present is the assembly and activation of the Wnt destruction complex. This complex contains several proteins with prominent roles in stem cell fate determinations and carcinogenesis. Two of the proteins belonging to this complex, the adenomatous polyposis coli (APC) protein and the cytoskeleton protein axin, function as scaffolds that bind the central member of the complex, β-catenin. Another protein present in destruction complexes is the β-TrCP E3 ubiquitin ligase. In the absence of

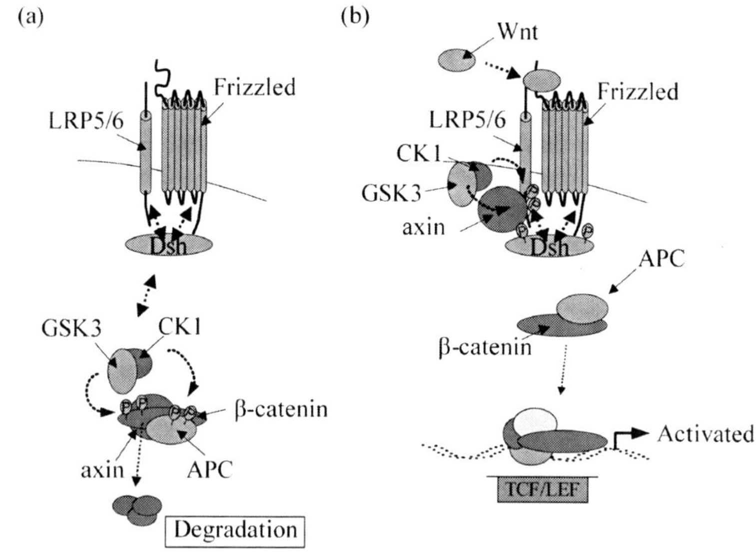

Fig. 15.7 The Wnt canonical pathway. (**a**) Assembly and activation of the destruction complex in the absence of ligand binding. (**b**) Steps leading to TCF/LEF gene expression in the presence of ligand binding

ligand binding, CK1 phosphorylates β-catenin (and also APC). This action primes β-catenin for phosphorylation again, this time by GSK3. These phosphorylations tag β-catenin for proteolytic destruction.

This situation is altered when Wnt binds Frizzled and LRP5/6. When Wnt binding occurs the assembly of the destruction complex is inhibited and β-catenin does not undergo ubiquitin-mediated proteolysis. As shown in Fig. 15.7, the cytoplasmic tail of the LRP5/6 coreceptor is hyperphosphorylated. This site and perhaps those on Dsh as well become binding sites for axin. This protein is present in limited amounts and sequestration at these sites is a possible mechanism for prevention of destruction complex assembly. When this happens β-catenin is dephosphorylated and not degraded; instead it can translocate to the nucleus where it stimulates transcription by acting as activating cofactors for T-cell factor/Lymphoid enhancer factors (TCF/LEF) transcription factors. Several other components of destruction complexes such as APC possess nuclear localization signals and these proteins may shuttle back and forth to the nucleus where they dislodge transcription coinhibitors and help recruit transcription coactivators.

15.9 The Epithelial to Mesenchymal Transition

Recall from the discussion of atherosclerosis that cadherins mediate homophilic attachment of adjacent cells to one another through formation of adherens junctions. Like the vascular endothelial (VE) cadherins discussed with regard to atherosclerosis, epithelial (E) cadherins attach to the actin cytoskeleton by means of catenins. The β-catenins bind to the catenin domain of the E-cadherins and to α-catenin. The α-catenin, in turn, connects to the actin cytoskeleton.

Loss of E-cadherin function can be brought on through genetic mutations, epigenetic (DNA hypermethylation) actions, and most often at the transcriptional levels through downregulation by its transcription repressors. Expression of E-cadherin is regulated by a number of transcription repressors, among which are the E-box-binding proteins SNAI1 (Snail), SNAI2 (Slug), zinc-finger E-box-binding homeobox 1 (ZEB1), and Smad interacting protein 1 (SIP1/ ZEB2), and by the bHLH protein Twist. These negative regulators of E-cadherin gene expression are upregulated and activated by upstream signals conveyed by TGF-βs, Wnts, RTKs, and integrins (Fig. 15.8).

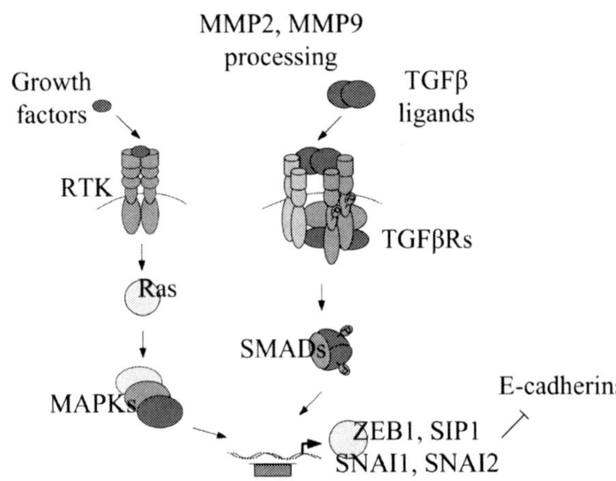

Fig. 15.8 Joint action through the RTK/Ras/ MAPK pathway and TGFβ pathways activates gene expression of one or more of an ensemble of E-cadherin repressors

The loss of cadherin-mediated attachment between adjacent cells is a critical step in conversion of an immobilized, polarized epithelial cell into a mobile, nonpolarized mesenchymal cell capable of relocating to secondary, distant sites and seeding new tumor growth. This action leads to dissolution of adherens junction complexes and releases β-catenin into the cytosol where it can accumulate in complexes with axin and APC involved in Wnt signaling. Once phosphorylated by CK1 and GSK3, β-catenin is targeted for proteolytic destruction by the E3 ubiquitin ligase, βTrCP. As discussed above, activation of the Wnt signaling pathway (or alternatively inactivation of APC by mutations) abrogates that activity. When this happens β-catenin can translocate to the nucleus and stimulate transcription of TCF/LEF genes involved in proliferation and tumor progression (e.g., c-Myc and D1-type cyclins).

In several prominent cancer cell types, E-cadherin is downregulated through joint actions of TGF-β and Ras. The onset of tumor growth leads to overexpression and activity of ECM metalloproteases such as MMP2 and MMP9. These proteases activate resident TGF-β ligands that would otherwise remain inactive.

The Smads translocate to the nucleus and together with cofactors stimulate transcription of proteins such as Snail and SIP1 that bind E-boxes in E-cadherin promoters and suppress transcription. This process requires the concomitant hyperactivation of the Ras-Raf-MAPK kinase module. Under these conditions, E-cadherin genes are silenced and the EMT is induced. In these situations, TGF-β no longer functions as a growth inhibitor but instead as a metastasis promoter. Mutations or other inactivating events in Smad cofactors required for TGF-β's cytostatic actions may be responsible for turning off of the antigrowth program.

15.10 MicroRNAs and Transcription Repressors Jointly Regulate E-Cadherin Expression

Downregulation of E-cadherin expression initiates the EMT, and conversely, restoration of E-cadherin expression initiates the reverse transition, aptly named the mesenchynal-to-epithelial transition, or MET. The need for this alteration would naturally arise at the end of a migratory stage where an epithelial phenotype is once again required. A regulatory circuit composed of TGFβ, the E-cadherin negative regulators ZEB1 and SIP1/ZEB2, and members of the miR-200 family of microRNAs tightly regulate these transitions.

The miR-200 family of microRNAs includes miR-141, miR-200b, and miR-200c. These function as negative regulators of mRNAs for the E-cadherin repressors ZEB1 and SIP1. In more detail, miR-200c targets ZEB1 while miR-200b inhibits SIP1. The third family members in this regulatory circuit, miR-141 inhibits TGFβ. These miRNAS and their targets form a circuit which has two stable states, one of which corresponds to an epithelial phenotype and the other to a mesenchymal form. A simplified view of how this circuit operates is presented in Fig. 15.9.

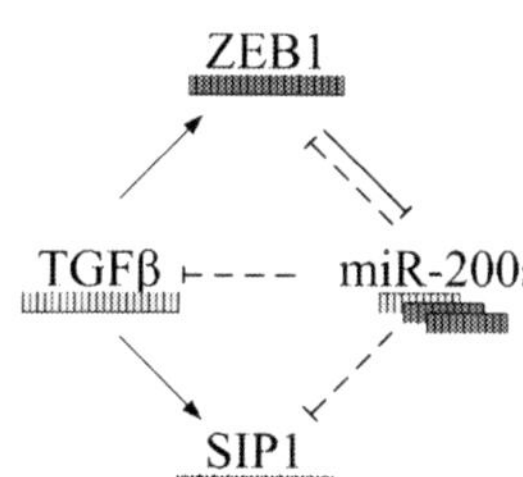

Fig. 15.9 The two-state system that regulates E-cadherin expression. *Dashed lines* represent inhibition by the three miR-200s on the E-cadherin repressors and on TGFβ. In this state, the cell has an epithelial phenotype. When the signal strength from EMT activators is sufficient to overcome the repression, the second state denoted by the solid lines ensues in which TGFβ signaling is the dominant influence and the cell reverts to the simpler mesenchymal form

In the absence of external input to the TGFβ pathway, miR-141, miR-200c, and miR-200b block translation of the mRNAs for TGFβ, ZEB1, and SIP1, respectively. The E-cadherin repressors are inhibited, E-cadherin expression occurs, and the cells exhibit an epithelial phenotype. Population of the alternative state arises through stimulation of the TGFβ signaling pathway that overrides the repression from the three miR-200s; ZEB1 and SIP1 are activated, E-cadherins are downregulated, and the cell undergoes an EMT.

This two-state system has the same kind of dynamical behavior (bistability and hysteresis) as that described in Chapter 9 dealing with the cell cycle transitions. A certain amount of TGFβ signaling, enhanced by signals relayed through the RTK, and supported further in many instances by Wnt and integrin-mediated pathways, is needed to initiate the EMT. This level of signaling is needed to override the inhibition to dissolution of E-cadherins, focal adhesions, and junctional complexes that serve as the underpinnings of the epithelial state. Once the system undergoes the EMT, it must stabilize and maintain the mesenchymal state during migration in the circulation and extravasation at the secondary sites.

Similarly, the transition from mesenchymal to epithelial forms, the MET, requires sustained signaling to override the blocks on restoring the epithelial form. The MET reactivates E-cadherins, reestablishes the junctional complexes, and remodels the actin cytoskeleton. The cells thus exhibit bistability and through hysteresis prevent inappropriate transitions. In the current context of metastasis, the behavior of the system would follow that of Fig. 15.10 in which "Signal strength" and "E-cadherin activity" are shorthand for the totality of expression and structural changes that must occur.

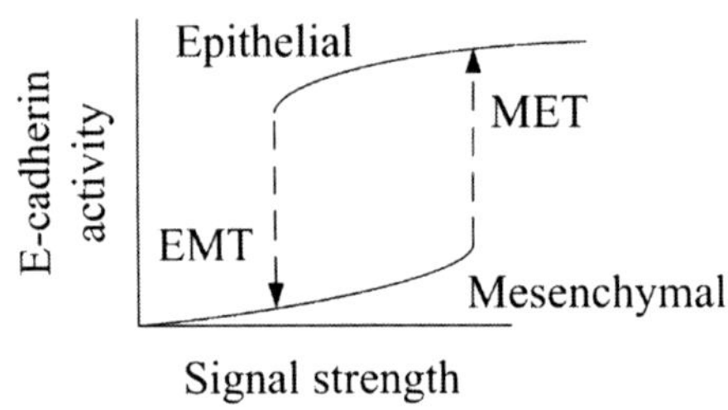

Fig. 15.10 The metastatic cycle. Signal strength denotes the sum of E-cadherin and related pro-adhesion signals and supporting positive feedback loops that induce a MET. E-cadherin activity represents the corresponding buildup in the cell of the machinery that supports a nonmotile, polarized phenotype

Minimally, "signal strength" refers to the ensemble of E-cadherin transcription activators that sustain E-cadherin expression and prevent inappropriate activation of E-cadherin repressors such as the SNAIs and ZEBs. When TGFβ, RTK, and other signals become strong enough to reduce this signal strength, E-cadherin activity rapidly drops off and the system returns to the mesenchymal state that is active during embryonic development and tissue repair. More globally, "signal strength" and "E-cadherin activity" encompass the totality of changes to adherens and tight junctions, focal adhesions, and cytoskeleton with E-cadherins serving a prominent role as initiators.

15.11 MicroRNAs Act as Metastasis Repressors and Activators

Twist functions as an inducer of EMT. It does so, in part, by repressing E-cadherin gene expression. Another of its targets is the microRNA miR10b. In this instance, Twist functions as a transcription activator. Once activated by Twist, miR10b binds and blocks translation of mRNAs for the homeobox D10 protein. This step, in turn, leads to increased expression of RhoC, which functions as a mediator of invasion and metastasis. Twist expression can be upregulated by hypoxic conditions present in tumors through actions of Hif-1, and these signaling connections link hypoxic environmental conditions to metastatic responses.

An emerging theme in the studies of MiR-200s and their repression of the ZEBs and of the alignment of Twist and miR-10b in a signaling pathway is that microRNAs contribute in various ways to the epithelial and mesenchymal cellular states. Some microRNAs act as metastasis suppressors while others function as metastasis promoters. Additional examples of microRNAs functioning as metastasis repressors are miR335, miR-126, and miR-206. These three microRNAs suppress breast cancer metastasis whereas miR10b acting through Twist promotes it.

15.12 Stem Cells and Cancer Stem Cells

There are several models that describe how tumor cells grow and mestatasize. One model is centered on the observation that, in order to metastasize, tumor cells must detach from the tissues in which they are embedded and migrate to remote secondary sites. The EMT model which results is based on the epithelial-to-mesenchymal transition (EMT); it emphasizes the changes in morphology from a sedentary to a mobile form that play prominent roles during embryonic development. That model has been a main subject of this chapter.

A second model, called the stem cell (SC) model, arises from the observation that cells that migrate to remote sites will only produce macroscopic-sized secondary growth if these cells have the ability to self-renew as do stem cells (SCs). This observation leads to the SC model that also emphasizes processes occurring during embryogenesis, in this case those governing self-renewal and differentiation. The SC model has as a second supporting observation that a tumor consists of a heterogeneous population of cells. A small number of these are adult stem cells and some of these cells become cancer stem cells (CSCs). It is posited that these cells and not the main populations of tumor cells are responsible for the ability of cancers to prosper and grow in secondary sites and to restart growth in primary sites after chemotherapy. In the latter situations, the stem cells are not killed by anticancer treatments that target fast-growing cell populations.

In its original formulation, the cancer stem cell model asserts that only small populations of cells are resistant to anti-cancer therapies and capable of reforming tumors and forming colonies. These cancer stem cells, identified by the presence of a characteristic set of surface markers, differ from both normal, non-cancer-causing stem cells and from the majority of non-stem cells residing in the tissue or organ. The cancer stem cells are derived both from the pool of stem cells and through a process called dedifferentiation from the non-stem-cell population.

The first steps in a convergence of EMT and stem cell models have occurred recently. In a landmark study by members of the Weinberg group, similarities between mesenchymal cells derived from tumor cells through an EMT have been found to possess properties of stem cells. Thus, a single population of cells may exhibit properties of both stemness and EMT. To generate this population of metastatic cells TGFβ or SNAIL/Twist was overexpressed in breast cells for a period of about 12 days. The targeted cells not only underwent an EMT but also expressed surface markers indicative of stemness.

In other recent studies, questions of the rareness and homogeneity of cancer stem cells were raised. It seems that, depending on the type of cancer, the cancer stem cells may be rare, perhaps one cell in 10,000 satisfying the criteria, or may be quite common, ranging up to 10 or even 25% of the overall tumor population. Furthermore, it may not follow that all cells in a given cancer stem cell group are identical with respect to their molecular profile and therapeutic targets, and heterogeneity may be the norm for this class of cells.

15.13 Changing Views About Metastatic Spread

Views as to the onset of metastatic spread have undergone a profound shift during the last few years as new findings emerge. Previously, metastasis was thought of as exclusively a late event, triggered long after growth of a tumor in the primary site began. In a recent study from the Varmus laboratory, normal mammary epithelial cells were shown to survive in secondary sites for long periods of time awaiting activation of oncogenes such as Ras and c-Myc that lead to proliferation and colonization. The possibility raised by these data is of an early spread to many potential secondary sites, followed by eventual colonization at a subset of these sites leading to tumors that may significantly differ in their therapeutic potential from the cells at the primary site.

Metastatic spread may occur in more than one way. Just as there are several routes to generating new cells from adult cells metastatic spread may occur late or early, from cancer stem cells or not. The identification of which cells in a heterogeneous population of tumor cells are the seeds for metastasis is an exceptionally challenging problem and the answer may depend on the particular cancer being explored. . In yet another striking and provocative study, surmounting the senescence barrier to tumor development, an early event, and passage through an EMT were found to occur at the same time through actions by Twist. Specifically, Twist proteins were able to cooperate with oncogenic Ras to overcome the p53- and pRb-mediated anti-cancer barriers.

Lastly, changes in the microenvironment that support tumor growth promote metastasis as well so that growth and metastasis are not mechanistically distinct. Alterations in the tumor microenvironment that promote both tumor growth and metastasis were examined in the earlier sections of this chapter. The act of metastasizing involves (i) generating or finding supportive microenvironments and (ii) reactivation of developmental programs, especially those controlled by the TGFbeta and Wnt pathways and by the Notch and Hedgehog pathways. This chapter will end with a brief look at the two not-yet-discussed developmental pathways – Notch and Hedgehog.

The Notch pathway directs cell fate; it is active during embryonic development and in cancer metastasis. Its readily seen effects are its stimulation of the epithelial-to-mesenchymal transition during cardiac development and in the decision whether or not to undergo differentiation during development of the central nervous system. Developmental pathways such as Wnt and Notch remain active in the adult as regulators of self-renewing tissues and especially of their progenitors/stem cells. In addition, Notch signaling along with transcription factors such as Oct4 mediates the influence of hypoxia on precursor and stem cell differentiation and, jointly with TGFβ, direct epithelial-to-mesenchymal transitions through the induction of SNAIL and repression of E-cadherin expression.

Hedgehog is important for adult stem cell renewal. This signaling pathway along with the Polycomb group protein Bmi-1 regulates self-renewal of mammary stem cells and this type of finding lends support to the stem cell picture of tumorigenesis. Hedgehog is also strongly implicated in pancreatic cancer metastasis where excessive pathway activation generates an aggressive metastatic phenotype in pancreatic epithelial cells. The chief feature is the downregulation of E-cadherin and an attendant epithelial-to-mesenchymal transition.

15.14 The Notch Pathway

The Notch protein is synthesized as a single-chain, 300-kDa precursor molecule. Signaling is initiated by ligand binding in which a ligand, a counter-receptor on an adjacent cell, binds the Notch extracellular ligand-binding segment. Two cleavage operations occur in response. The first cut, by a metalloprotease, takes place in the region just outside the membrane resulting in separation of the N-terminal, extracellular portion of the molecule from the membrane-bound remainder. This process is referred to as ectodomain shedding. The second cut, by proteolytic enzymes termed γ-secretases, occurs within the membrane and frees the C-terminal part of the protein from the membrane (Fig. 15.11(a)). This latter sequence of events parallels those taking place in SREBP activation as discussed in Chapter 6, and is another example of (ligand) regulated intramembrane proteolysis (RIP). The catalytic core of the γ-secretase is the protein Presenilin, and this same γ-secretase complex is responsible for cleavage of the amyloid precursor protein (APP), the central element in Alzheimer's disease.

Cleavage following ligand binding produces two fragments, an N-terminal 180-kDa protein and a smaller C-terminal 120-kDa molecule. The N-terminal chain contains from 29 to 36 tandem epidermal growth factor (EGF) repeats and three cysteine-rich Lin/Notch repeats (LNRs) as indicated in Fig. 15.11(b). These EGF and LNR repeats mediate ligand binding and Notch activation.

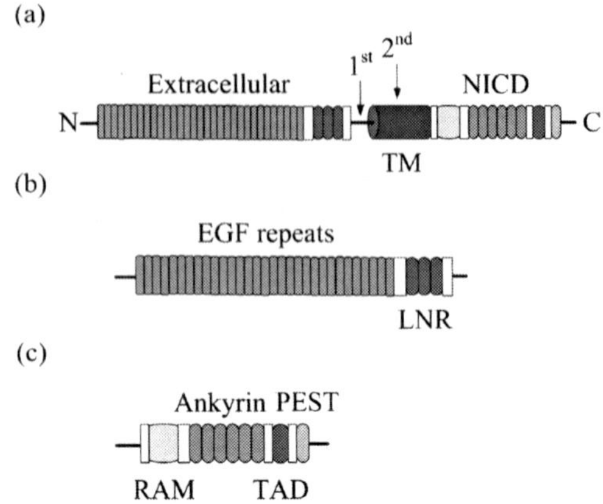

Fig. 15.11 Notch processing. (**a**) Full length Notch showing the locations of the two proteolytic cleavage sites, (**b**) The Notch extracellular domain, and (**c**) the Notch intracellular domain (NICD)

The Notch intracellular domain (NICD) functions as a transcriptional cofactor for members of the CBF1/RBP-Jκ [vertebrates], Su (H) [*Drosophila*], Lag-1 [*C. elegans*], or CSL family of transcription factors. In the absence of NICD translocation and binding, the CSL proteins function as transcriptional repressors through regulatory actions of corepressor proteins. The NICD displaces the corepressor thereby converting the CSL protein complex from a repressor to an activator of gene transcription. The NICD contains a RBP-Jκ-associated molecule (RAM) domain, 6 Cdc10/Ankyrin repeats, a transactivation domain (TAD), and a C-terminal PEST – proline (P), glutamate (E), serine (S), threonine (T) rich motif (Fig. 15.11(c)).

Cooperative interactions between multiple transcription factors and cofactors characterize gene transcription. One of the most striking examples of cooperativity in gene transcription is that of the bacteriophage lambda decision circuit that controls the whether the phage adopts a lytic or lysogenic life cycle. The promoter site architecture and cooperativity between transcription factors produces an environmentally responsive, switch-like behavior from one to the other. Notch provides another equally striking yet different example of bioengineering in which the goal is not creation of a switch but rather ensuring graded responses in which target selectivity, transcription intensity, and transcription duration depend upon the strength of the signaling and the architecture of the particular promoter site.

Canonical Notch signaling occurs in the following manner. In the absence of Notch ligand binding, the transcription factor CBF1/RBP-Jκ forms a repressor complex at Notch transcription sites. These complexes contain corepressors such as SMRT, HDAC-1, SMART/HDCA-1-associated repressor protein (SMART), and CtIP/CtBP. The arriving NICDs dislodge the repressors and initiate the formation of an activator complex in its place. In more detail, NICD/RBP-Jκ recruits the coactivator Mastermind to the promoter site. This complex then recruits several types of chromatin-modifying enzymes that comprise the transcription activating complex. The RAM and Ankyrin domains together with the C-terminal domain of RBP-Jκ mediate assembly of the activator complex. This begins when allosteric changes in RBP-Jκ triggered by the RAM domain are followed by docking of the Ankyrin domain to create a high-affinity binding site for the Mastermind coactivator, displacing SMRT and other corepressors.

An architectural feature of particular importance with respect to target selectivity is that of dual sequence-paired binding sites (SPSs). These consist of two CSL binding sites separated from one another by a spacer region of 16 or 17 nt. Both directionality and distance between the two binding sites influence selectivity. This promoter architecture may support cooperativity and provide a mechanism for the observed dose and context sensitivity to Notch signals.

15.15 The Hedgehog Pathway in *Drosophila*

Studies of *Drosophila* have provided a large body of information about hedgehog signaling; the signaling pathway is somewhat simpler in the fly, exhibits many of the same features as found in humans, and will serve as a starting point for the discussion of the more elaborate mammalian system. The principal upstream components of the Hedgehog signaling pathways are (i) hedgehog (Hh) ligands, (ii) ligand-binding Patched receptors, and (iii) signal-transducing Smoothened (Smo) receptors. Hedgehog ligands are morphogens with important roles in embryonic development and during adult life in stem cell proliferation and migration. Morphogens are proteins secreted and laid down to form concentration gradients either on cell surfaces or immobilized in the extracellular spaces in the close vicinity of the cells secreting them. The signaling pathway activated by Patched and Smoothened in response to Hh ligands reads and transduces this information into appropriate cellular responses.

The key signaling events in the *Drosophila* Hh pathway are depicted in Fig. 15.12. Patched is a 12-pass transmembrane protein while Smoothened is a seven-pass GPCR-like receptor similar to the Wnt receptor, Frizzled. In the absence of ligand binding, Patched inhibits the translocation of Smoothened to the plasma membrane from sites in the cytoplasm. Ligand binding relieves this inhibition and Smoothened receptors are able to accumulate at the plasma membrane. The cytoplasmic tail of Smoothened is crucial for its signaling activities. In the absence of Hh, the C-terminal tail is stabilized in a closed, no-signaling conformation. When Hh is present, Smo is phosphorylated on multiple serine/threonine residues and accumulates at the cell surface in a open signaling competent conformation. The conformational switch is induced by the electrostatic antagonism of multiple Arg clusters by the addition of the phosphoryl groups. It is believed (but not yet proven) that the number of activated Smo proteins at the surface, and degree of to which they are phosphorylated, serves as the intracellular reflection of the gradient concentration.

Several serine/threonine kinases have crucial roles in Hh signaling among which are protein kinase A, casein kinase 1 (CK1), and glycogen synthase kinase 3 (GSK3). These kinases phosphorylate sites in the cytoplasmic tail of Smo and also the downstream signaling effector, the transcription factor Cubitus interruptus (Ci). This protein is part of a complex consisting of the kinesin-like

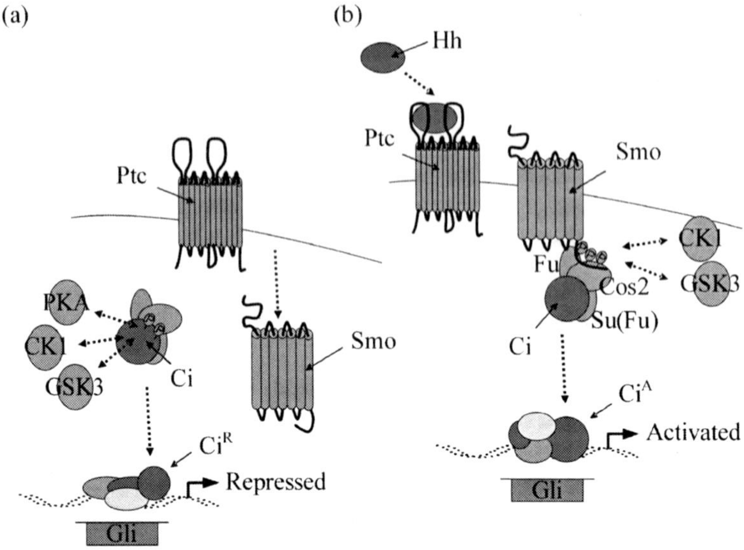

Fig. 15.12 Hedgehog signaling in *Drosophila*. (**a**) Transcription repression by truncated cubitus interruptus. (**b**) Transcription activation by full-length cubitus interruptus

protein, Costal-2, which binds to microtubules, and a second protein, Fused, a serine/threonine kinase. Once recruited to Smoothened cytoplasmic tail in its open conformation, these two proteins function as scaffolds for recruitment of several other proteins to form the signaling node. These others are the Suppressor of Fused (Su(Fu)), Cubitus interruptus, and the aforementioned protein kinases.

In the absence of Hh signaling, protein kinase A [PKA] phosphorylates Ci. This phosphorylation event tags Ci for proteolytic cleavage. The result of this process is the formation of a 75-kDa Ci^{R} protein, which translocates to the nucleus where it functions as a repressor of the transcription of several genes. When Hh is present, phosphorylation of Ci by PKA does not occur, the complex dissociates, and the full-length 155-kDa protein is left intact and free to move into the nucleus. This intact Ci^{A} protein functions as an activator of gene transcription.

15.16 The Hedgehog Pathway in Mammals

Vertebrate hedgehog signaling takes place in a special sensory organelle, the primary cilium. The organelle protrudes out from the surface of most vertebrate cells and serves as a sensory organ that probes the microenvironment for the presence of signals in the form of hedgehog ligands. Primary cilia are immotile relatives of the motile cilia; they consist of a microtubule-based transport system referred to as an axoneme, which through intraflagellar transport (IFT) moves signaling molecules in and out of the structure. The organelle is enclosed in an extension of the plasma membrane and anchored to the cell by a basal body.

Many of the features uncovered in studies of *Drosophila* have counterparts, but in expanded form, in vertebrates. In place of a single Ci transcription factor that is proteolytically processed into activator and repressor forms, there are three proteolytically processed forms, namely, Gli1, Gli2, and Gli3. In mammals, at least, Cos2 and Fu do not seem to be present, and there are several other signaling and mechanistic differences. Mammalian activator and repressor situations are depicted in Fig. 15.13. In the absence of ligand binding, mammalian Patched localizes at the base and inside the primary cilium and impedes entry of Smo to the organelle. The Gli transcription factors move into the organelle, interact with its residents, and are processed into repressor forms. Ligand binding by Sonic Hedgehog (Shh) results in processing of the Glis into

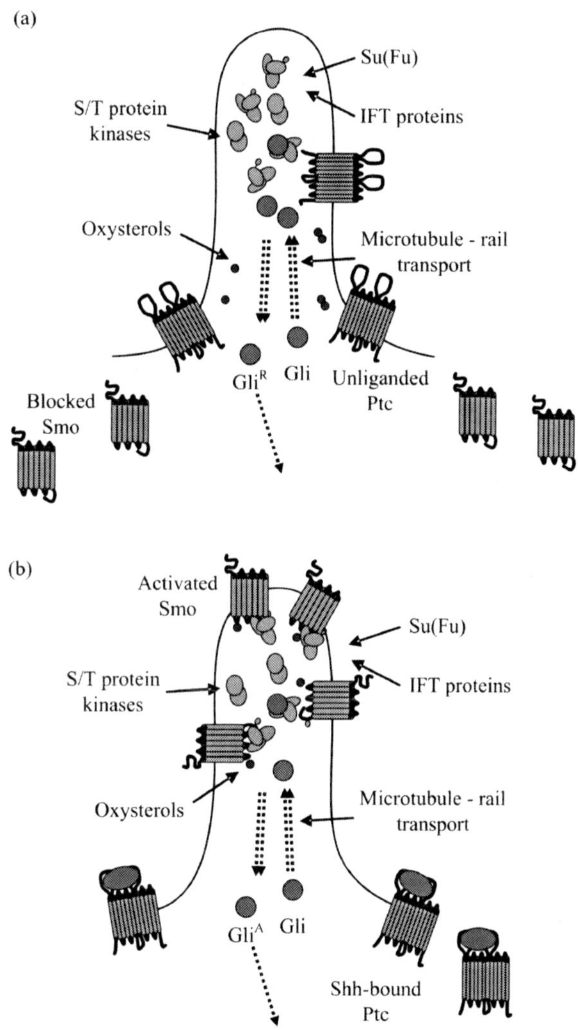

Fig. 15.13 Sonic hedgehog signaling in mammals occurs at the primary cilium. (**a**) In the absence of Shh ligand binding, Ptc1 (Ptc) prevents entry of Smo into the primary cilium resulting in the processing of Gli into Gli^R. (**b**) In response to Shh binding, Ptc no longer blocks entry of Smo leading to the processing of Gli into Gli^A

activator forms. The Patched receptors move away from the primary cilium and no longer prevent entry of Smo to the structure. The Smo interacts with the residents and Glis leading to production of the activator forms. In mammals, IFT proteins assist in the transport in and out of the Glis. How Patched and Smo interact with one another is somewhat mysterious, since they do not directly interact with one another. It is believed that oxysterols may mediate their interactions and contribute to Smo activation.

15.17 Bone Metastasis Is a Seed-and-Soil Exemplar

In 1889, Stephen Paget, in his examinations of cancer metastasis, reached the conclusion that the spread of cancer to distant, secondary sites was not a random process, but instead specific organs were being targeted because they had properties that favored further cancer growth. This led him to introduce the notion of seed and soil. The 'seed' referred to the cells of the primary tumor that were attempting to migrate and the 'soil' denoted sites found by them that supported colonization. There is good evidence today more than a century later in support of this depiction, which has as its basis the notion that tumor cells and cells in the microenvironment of secondary sites converse with one another just as they do at the primary site of tumor growth. A dramatic example of this type of conversation is provided by bone metastasis.

Bone is a preferred site of metastasis of cancer cells from breast, prostate, and to a lesser extent lung and kidney. The bone microenvironment contains mineralized bone matrix and several cell types, prominent among which are osteoclasts that degrade bone matrix and osteoblasts that do the opposite, generating new bone matrix. These opposing activities are coordinated – osteoclast differentiation and activation of bone remodeling are controlled by signals from the osteoblasts. Imbalances in these two activities accompany metastatic spread. In breast metastases, there is too little bone while in prostate there is too much.

Three cell populations participate in the breast-bone metastasis conversation – breast tumor cells, the osteoblasts, and osteoclast/osteoclast precursors. Interactions between these three cell types generate a self-reinforcing positive feedback loop. First, the tumor cells secrete osteolytic factors (Table 15.3) such as parathyroid-hormone-related peptide (PTHrP) that bind to receptors on the

Table 15.3 Positive feedback in bone metastasis

Sending cells	Receiving cells	Signals
Breast tumor cells	Osteoblasts	PTHrP, IL-11, CTGF*
Osteoblasts	Osteoclasts	RANKL
Osteoclasts	Breast tumor cells	TGFβ, IGF, PDGF, FGF

*CTGF: Connective tissue growth factor.

osteoblasts. The PTHrP receptor belongs to the G protein-coupled receptor family and in response to ligand binding the osteoblasts upregulate and secrete receptor activator of nuclear factor κB ligand (RANKL), a TNFα family member. These ligands bind to RANK receptors expressed on osteoclasts and on osteoclast precursors. The resulting RANKL→TRAF6→NF-κB signal transduction leads to osteoclast differentiation and activation. The activated osteoclasts, in turn, release factors such as TGFβ (Table 15.X) from the matrix. The TGFβ ligands bind to receptors on the tumor cells and the resulting activation of the TGFβ→SMAD signal transduction pathway further stimulates PTHrP expression and downregulates the decoy receptor osteoprotegrin (OPG) that functions as a negative regulator of PTHrP signaling. The positive feedback loop is thereby closed leading to sustained colonization and continued bone loss. This exemplar thus highlights the stabilizing influence on the disease state of positive feedback loops generated by cell-to-cell signaling.

Further Reading

Metastasis and Invasion Is a Community Affair

Balkwill F, Charles KA and Mantovani A 2005 Smoldering and polarized inflammation and promotion of malignant disease. *Cancer Cell* 7: 211–217.
Bhomick NA et al. 2004 Stromal fibroblasts in cancer initiation and progression. *Nature* 432: 332–337.
Deryugina EI and Quigley JP 2006 Matrix metalloproteinases and tumor metastasis. *Cancer Metastasis Res.* 25: 9–34.
Grunewald M et al. 2006 VEGF-induced adult neovascularization: recruitment, retention, and role of accessory cells. *Cell* 124: 175–189.
Gupta GP and Massagué J 2006 Cancer metastasis: building a framework. *Cell* 127: 679–695.
Kalluri R and Zeisberg M 2006 Fibroblasts in cancer. *Nat. Rev. Cancer* 6: 392–401.
Kaplan RN et al. 2005 VEGFR1-positive hematopoietic bone marrow progenitors initiate the pre-metastatic niche. *Nature* 438: 820–827.
Kim S et al. 2009 Carcinoma-produced factors activate myeloid cells through TLR2 to stimulate metastasis. *Nature* 457: 102–106.
Sica A and Bronte V 2007 Altered macrophage differentiation and immune dysfunction in tumor development. *J. Clin. Invest.* 117: 1155–1166.
Wycoff J et al. 2004 A paracrine loop between tumor cells and macrophages is required for tumor cell migration in mammary tumors. *Cancer Res.* 64: 7022–7029.
Yamaguchi H and Condeelis J 2007 Regulation of the actin cytoskeleton in cancer cell migration and invasion. *Biochim. Biophys. Acta* 1773: 642–652.

Angiogenic Switch in Cancer

Bergers G and Benjanin LE 2002 Tumorigenesis and the angiogenic switch. *Nat. Rev. Cancer* 3: 401–410.
Carmeliet P 2005 Angiogenesis in life, disease and medicine. *Nature* 438: 932–936.
Lin EY et al. 2006 Macrophages regulate the angiogenic switch in a mouse model of breast cancer. Cancer Res. 66: 11238–11246.

The SDF-1/CXCR4 Axis

Ceradini DJ et al. 2004 Progenitor cell trafficking is regulated by hypoxic gradients through HIF-1 induction of SDF-1. *Nat. Med.* 10: 858–864.
Kucia M et al. 2005 Trafficking of normal stem cells and metastasis of cancer stem cells involve similar mechanisms: pivotal role of the SDF-1/CXCR4 axis. *Stem Cells* 23: 879–894.
Orimo A et al. 2005 Stromal fibroblasts present in invasive human breast carcinomas promote tumor growth and angiogenesis through elevated SDF-1/CXCL12 secretion. *Cell* 121: 335–348.
Schioppa T et al. 2003 Regulation of the chemokine receptor CXCR4 by hypoxia. *J. Exp. Med.* 198: 1391–1402.

C-Src and Focal Adhesions

Frame MC et al. 2002 v-Src's hold over actin and cell adhesions. *Nat. Rev. Mol. Cell Biol.* 3: 233–245.

Wnt Signaling

Clevers H 2006 Wnt/β-catenin signaling in development and disease. *Cell* 127: 469–480.
Gottardi CJ and Gumbiner BM 2004 Distinct molecular forms of β-catenin are targeted to adhesive or transcriptional complexes. *J. Cell Biol.* 167: 339–349.
Nelson WJ and Nusse R 2004 Convergence of Wnt, β-catenin, and cadherin pathways. *Science* 303: 1483–1487.

TGF-β Signaling

Elliott RL and Blobe GC 2005 Role of transforming growth factor beta in human cancer. *J. Clin. Oncol.* 23: 2078–2093.
Schmierer B and Hill CS 2007 TGFβ-SMAD signal transduction: molecular specificity and functional flexibility. *Nat. Rev. Mol. Cell Biol.* 8: 970–982.
Massagué J, Seoane J and Wotton D 2005 Smad transcription factors. *Genes Dev.* 19: 2783–2810.

The Epithelial–Mesenchymal Transition

Hay ED 2005 The mesenchymal cell, its role in the embryo, and remarkable signaling mechanisms that create it. *Dev. Dyn.* 233: 706–720.
Janda E et al. 2002 Ras and TGFβ cooperatively regulate epithelial cell plasticity and metastasis: dissection of Ras signaling pathways. *J. Cell Biol.* 156: 299–313.
Savagner P 2001 Leaving the neighborhood: molecular mechanisms involved during epithelial-mesenchymal transition. *BioEssays* 23: 912–923.
Thiery JP and Sleeman JP 2006 Complex networks orchestrate epithelial-mesenchymal transitions. *Nat. Rev. Mol. Cell. Biol.* 7: 131–142.
Zavadil J and Böttinger EP 2005 TGF-β and epithelial-to-mesenchymal transitions. Oncogene 24: 5764–5774.

MicroRNAs and E-Cadherin Suppressors

Burk U et al. 2008 A reciprocal repression between ZEB1 and members of the miR-200 family promotes EMT and invasion in cancer cells. *EMBO Rep.* 9: 582–589.

Gregory PA et al. 2008 The miR-200 family and miR-205 regulate epithelial to mesenchymal transition by targeting ZEB1 and SIP1. *Nat. Cell Biol.* 10: 593–601.

Park SM et al. 2008 The miR-200 family determines the epithelial phenotype of cancer cells by targeting the E-cadherin repressors ZEB1 and ZEB2. *Genes Dev.* 22: 894–907.

Tsang J, Zhu J and van Oudenaarden A 2007 MicroRNA-mediated feedback and feedforward loops are recurrent network motifs in mammals. *Mol. Cell* 26: 753–767.

Activities of MicroRNAs

Ma L, Teruya-Feldstein J and Weinberg RA 2007 Tumour invasion and metastasis initiated by microRNA-10b in breast cancer. *Nature* 449: 682–688.

Tavazoie SF et al. 2007 Endogenous human microRNAs that suppress breast cancer metastasis. *Nature* 451: 147–152.

Yang J et al. 2004 Twist, a master regulator of morphogenesis, plays an essential role in tumor metastasis. *Cell* 117: 927–939.

Yang MH et al. 2008 Direct regulation of *TWIST* by HIF-1α promotes metastasis. *Nat. Cell Biol.* 10: 295–305.

Cancer Stem Cells

Brabletz T et al. 2005 Migrating cancer stem cells – an integrated concept of malignant tumour progression. *Nat. Rev. Cancer* 5: 744–749.

Liu S et al. 2006 Hedgehog signaling and Bmi-1 regulate self-renewal of normal and malignant human mammary stem cells. Cancer Res. 66: 6063–6071.

Mani SA et al. 2008 The epithelial-mesenchymal transition generates cells with properties of stem cells. *Cell* 133: 704–715.

Polyak K and Weinberg RA 2009 Transitions between epithelial and mesenchymal states: acquisition of malignant and stem cells traits. *Nat. Rev. Cancer* 9: 265–273.

Reya T and Clevers H 2005 Wnt signaling in stem cells and cancer. *Nature* 434: 843–850.

Wicha MS, Liu S and Dontu G 2006 Cancer stem cells: an old idea – a paradign shift. *Cancer Res.* 66: 1883–1890.

Metastatic Spread

Ansieau S et al. 2008 Induction of EMT by Twist proteins as a collateral effect of tumor-promoting inactivation of premature senescence. *Cancer Cell* 14: 79–89.

Podsypania K et al. 2008 Seeding and propagation of untransformed mouse mammary cells in the lung. *Science* 321: 1841–1844.

Notch Signaling

Duncan AW et al. 2005 Integration of Notch and Wnt signaling in hematopoietic stem cell maintenance. *Nat. Immunol.* 6: 314–322.

Radtke F and Raj K 2003 The role of Notch in tumorigenesis: oncogene or tumour suppressor? *Nat. Rev. Cancer* 3: 756–767.

Hedgehog Signaling in Drosophila

Hooper JE and Scott MP 2005 Communicating with hedgehogs. *Nat. Rev. Mol. Cell Biol.* 6: 306–317.

Hedgehog Signaling in Mammals

Christensen ST et al. 2007 Sensory cilia and integration of signal transduction in human health and disease. *Traffic* 8: 97–109.

Liu A, Wang B and Niswander LA 2005 Mouse intraflagellar transport proteins regulate both the activator and repressor functions of Gli transcription factors. *Development* 132: 3103–3111.

Rohatgi R, Milenkovic L and Scott MP 2007 Patched1 regulates hedgehog signaling at the primary cilium. *Science* 317: 372–376.

Rosenbaum JL and Witman GB 2002 Intraflagellar transport. *Nat. Rev. Mol. Cell Biol.* 3: 813–825.

Singla V and Reiter JF 2006 The primary cilium as the cell's antenna: signaling at a sensory organelle. *Science* 313: 629–633.

Seed and Soil

Padua D and Massague J 2009 Roles of TGFβ in metastasis. *Cell Res.* 19: 89–102.

Paget S 1889 The distribution of secondary growths in cancer of the breast. *Lancet* 133:571–573.

Part III
Neurodegeneration

Chapter 16
Protein Folding, Misfolding, and Aggregation

Proteins begin their cellular lives as linear chains of amino acid residues exiting the translation apparatus. Based on the particular sequence of amino acids, each protein goes through a set of conformational changes and arrives at a three-dimensional shape, its native state, which enables it to carry out its cellular roles. The series of steps from a newly synthesized polypeptide chain (called the denatured state) to the native state is called protein folding. Each type of amino acid has a unique set of biophysical and biochemical properties arising from differences in composition of their side chains. The particular sequences of amino acids are not random but instead are selected by evolution to allow the protein to fold into their operational three-dimensional forms in a physiologically meaningful time frame.

Occasionally, proteins misfold resulting in the onset of a disease. That is, they fold into conformations other then the ones that enable them to carry out their normal cellular functions. And, as a consequence, the protein does something it should not do or fails to do something that it ought to be doing leading to the development of a disease state in place of a healthy one. One of the main ways this happens is through mutations as in the case of cancer. Another way this happens is through posttranslational modifications that alter the folding environment either during the initial folding stage or later leading to unfolding and refolding.

One of the most important distinctions among the twenty types of amino acids is their hydrophobicity. Amino acids with nonpolar side chains are hydrophobic and those with polar side chains are hydrophilic. The distinction between the two arises from the binding preference of water for the amino acid. In the case of a hydrophilic (water loving) amino acid, a water molecule would rather bind to the amino acid than to another water molecule. In the case of a hydrophobic amino acid, a water molecule prefers another water molecule to the amino acid.

As will be seen in the following chapters, the distinction between hydrophobic and hydrophilic plays a key role in neurodegenerative disorders arising from protein misfolding. Proteins that misfold quite often do so in a way the exposes hydrophobic patches on their outer surfaces, and this fundamental aspect drives

M. Beckerman, *Cellular Signaling in Health and Disease*,
Biological and Medical Physics, Biomedical Engineering,
DOI 10.1007/978-0-387-98173-4_16,

conformational changes that lead to formation of amyloids and other classes of aggregates. This type of misfolding can arise from mutations and from post-translational modifications. The diseases that arise from deposits of misfolded proteins are referred to as protein-folding diseases.

Neurodegenerative diseases are the third major group of age-related, non-infectious diseases affecting millions of people worldwide. As was discussed in the Introduction, Chapter 1, the likelihood that an individual will develop a neurodegenerative disorder increases sharply with age. Table 16.1 lists some of the most prominent and widespread of these age-dependent disorders. The list is headed by Alzheimer's disease and includes Parkinson's disease, Huntington's disease, and amyotrophic lateral sclerosis, or Lou Gehrig's disease. In all of these instances, the causal agent of the disease is a misfolded protein, that is, a protein that for one of a number of reasons has altered its three-dimensional folded shape from one that supports its normal function in healthy tissue to one that leads to disease.

Table 16.1 Neurodegenerative protein-folding diseases

Neuodegenerative disease	Protein
Alzheimer's disease	Amyloid beta (Aβ), Tau
Taupathies	Tau
Down's syndrome	
FTDP-17	
Pick's disease	
α-Synucleinopathies	α-Synuclein
Parkinson's disease	
Dementia with Lewy bodies	
Multiple system atrophy	
Unstable repeat disorders	PolyQ
Huntington's disease	Htt
Kennedy's disease (SBMA)	Androgen receptor
DRPLA	Atrophin-1
Amyotrophic lateral sclerosis	SOD1
Spongiform encephalopathies	Prions
Mad cow disease (BSE)	
Scrapie	
Creutzfeldt-Jacob disease	

FTDP-17, Fronto-temporal dementia with parkinsonism linked to chromosome 17; SBMA, Spinal and bulbar muscular atrophy; DRPLA, DentatoRubral and PallidoLuysian atrophy; BSE, Bovine spongiform encephalopathy; SOD1, Superoxide dismutase 1.

The diseases presented in Table 16.1 have been grouped according to the most prominent misfolded protein found therein. In the case of Alzheimer's disease, the causal agent is a truncated form of the amyloid precursor protein called amyloid beta. Tau is a microtubule-associated protein that dissociates from the microtubules, becomes hyperphosphorylated, is often truncated, misfolds, and forms aggregates called tangles. These aggregates are a prominent feature of Alzheimer's disease as well as the diseases presented in the short list of taupathies. Intracellular aggregates of α-Synuclein are referred to as Lewy bodies. In the case of Parkinson's disease, an ever-growing number of disease-causing misfolded proteins have been found, and a major challenge in understanding the disease is to understand what this particular collection of agents says about the disease. There are nine known diseases brought on by misfolding-causing mutations in polyglutamine-encoding (CAG) tracts. The most prominent of these are presented in Table 16.1. The last set of entries in this table is the prion diseases. These revealing, paradigm-altering diseases affect mammalian species ranging from sheep and cattle, to elk and deer to humans.

This ensemble of diseases has a number of striking properties in common. First, the proteins that misfold do so in a particular way – most undergo a transition from a shape containing a mixture of secondary structure elements, α-helices, and β-sheets, to a highly ordered arrangement of crossed β-sheets. Second, the altered β-sheet structures support aggregation. That is, the altered proteins tend to form aggregates of various sizes from small oligomers to large amorphous clumps. The aggregates may form outside the cell, in the extracellular spaces, or they may form inside the cell. Prominent among these are the highly regular quaternary structures in which each monomer contributes two or more β-sheets. These are the amyloid fibrils first described by Virchow 150 years ago. Third, larger, less regular heterogeneous aggregates of misfolded and other proteins may build up within the cells. These deposits are referred to as inclusion bodies (IBs) or aggresomes or some neuron-specific name such as the Lewy bodies mentioned earlier. In all of these situations, a specific population of neurons dies leading to progressive deterioration of the associated neural function and eventual death of the individual.

That certain proteins when misfolded can cause disease is not restricted to neurons. Table 16.2 lists a number of protein-folding diseases that affect non-neural tissues. This grouping like that presented at the beginning in Table 1.1

Table 16.2 Nonneurodegenerative protein-folding diseases

Nonneuodegenerative disease	Protein
Amyloidoses	
AA amyloidosis	Serum amyloid A protein
Light-chain amyloidosis	Ig light chain
Dialysis-related amyloidosis	β2-microglobulin
ApoAI, AII, AIV amyloidosis	Apolipoprotein AI, AII, AIV
TTR amyloidosis	Transthyretin
Type 2 diabetes	Amylin (hIAPP)
Cystic fibrosis	CFTR
Serpinopathies	
Cirrhosis, emphysema	α1-antitrypsin
Thrombosis	Antithrombin

TTR, Transthyretin; hIAPP, human islet amyloid polypeptide; CFTR, Cystic fibrosis transmembrane conductance regulator.

includes a broad spectrum of amyloidoses. At the present time, twenty-three distinct amyloid diseases have been identified. In the amyloidoses, the misfolded proteins form amyloids of different sorts depending on the specific protein. Some of these diseases are tissue specific while others are more widespread in the selection of locale. The list of nonneural-specific protein-folding diseases includes cystic fibrosis and also a set of diseases referred to as serpinopathies. Serpins are serine proteinase inhibitors. These enzymes inhibit serine and cysteine proteases thereby preventing excessive proteolytic destruction within and outside the cell. These enzymes regulate extracellular proteolytic cascades and intracellular proteolytic events. When misfolded these enzymes tend to polymerize and cause diseases such as cirrhosis of the liver and emphysema in the lungs.

Some of these diseases are strongly age-dependent. For example, transthyretin is a carrier protein secreted by the liver into the bloodstream and cerebrospinal fluid. It functions as a transporter of the thyroid hormone thyroxine (T4) and retinol (vitamin A). Buildups of amyloid fibrils containing misfolded TTR and reassembled TTR monomers and dimers are responsible for senile systemic amyloidosis (SSA), familial amyloid polyneuropathy (FAP), and familial amyloid cardiopathy (FAC). These disorders affect 25% of the population over 80 years of age.

16.1 Proteins Spontaneously Fold into Their Native State Based Solely on Their Primary Amino Acid Sequence

The central dogma of protein folding, that a protein folds into its three-dimensional, native state based solely on its primary structure, was first articulated and demonstrated by Christian Anfinsen. It was based on an elegant series of studies which he carried out in the 1950s and early 1960s. In his experiments, he observed the spontaneous unfolding and refolding of bovine pancreatic ribonuclease, a small enzyme that facilitates DNA-RNA interactions. His "thermodynamic hypothesis" of protein folding which followed from his observations states that the native conformation is determined exclusively by the amino acid sequence of that protein. Anfinsen reasoned that the native state is favored over alternative states because it has the lowest energy and is therefore thermodynamically stable. This work was inspired in part by the work of Frederick Sanger on insulin and was accompanied by research by Stanford Moore and William Stein who determined the complete primary structure of ribonuclease. Anfinsen, along with Moore and Stein, established that the gene encodes the primary structure of the protein; the latter determines its native conformation which, in turn, defines the protein's functionality.

A question that immediately follows from this seminal observation is: just how is this feat accomplished? Even a small protein has a prohibitively large number of possible conformations. It would take far too long to sample all of these conformations in order to find the lowest energy ones. So the search by a protein for the native state must be directed in some manner. In addition, the folding process must be done in a way that avoids side paths that lead to aggregation. This is the second part of the question – how is it done in a way that avoids misfolding and aggregation. The need to answer this question is driven by the prominent role of aggregation in human diseases ranging from diabetes to prion diseases to Alzheimer's disease and other fatal disorders of the brain.

What about large multidomain proteins? Bovine pancreatic ribonuclease is a 124 amino acid residue, single-domain protein that folds into a compact globular shape. CFTR, the proteins central to cystic fibrosis (Table 16.2), in contrast, contains 1480 amino acid residues. It possesses two membrane-spanning domains along with two cytoplasmic domains and a regulatory domain. These domains interact with one another in the process of forming a Cl^- ion channel. So how does this protein fold? This question is pertinent to many of the diseases as they are caused by proteins involved in signaling and related activities. These proteins are large multidomain proteins, far more complex than small single-domain proteins.

The short answer to the above-mentioned question(s) is that Anfinsen's thermodynamic hypothesis is alive and well. It is supported by an impressive amount of cellular machinery that ensures the validity of this biophysical principle and in doing so preserves protein homeostasis. It includes interlocking networks of molecular chaperones including some that form Anfinsen cages for protein folding, a sophisticated endoplasmic reticulum (ER) quality-control system, and several pathways that collect and remove aberrant misfolded proteins from the cell.

How proteins fold is the subject of this chapter. This is one of the most fundamental questions in all of science. Numerous experimental and theoretical studies into this question over the past 50 or so years have not only revealed a number of key protein-folding principles, but also provided an overall conceptual framework for thinking about the process. The focus in the first part of this exploration will be on how small proteins fold without worrying about the complexities introduced by the larger multidomain entities. The discussion will then broaden to encompass proteins involved in signaling especially those centrally involved in the diseases of body and mind. A larger picture will then emerge – one in which there are proteins that possess large unstructured regions; of proteins that simultaneously fold and bind; of proteins whose native state is not at a global minimum in free energy, and of proteins that utilize the alternative β-sheet structures as part of their normal physiological function.

16.2 Protein Folding Can Be Described in Terms of an Energy Landscape Dominated by a Folding Funnel

The way in which a nascent protein folds into its functional, tertiary-folded form can be envisioned in terms of an energy landscape picture. Each point in the energy landscape would represent a possible conformation of the protein.

Similar conformations would be found near one another, and dissimilar ones further apart. The folding process is depicted as a trajectory connecting many points on the energy landscape denoting the sequence of small conformational changes that the protein undergoes as it folds. As shown in Fig. 16.1, a folding trajectory starts out at a denatured state located at the top of the landscape at a high energy and ends at the native state located at the bottom of the landscape at a low energy.

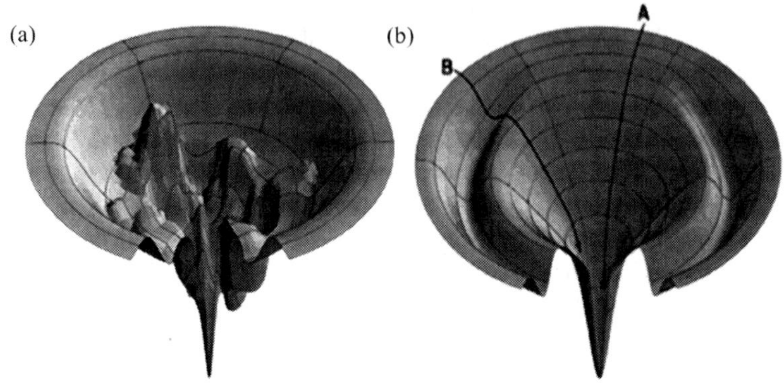

Fig. 16.1 Energy landscapes and folding funnels. (**a**) A rugged energy landscape; (**b**) an energy landscape possessing a trough. Pathway A avoids the trough and is a smooth one while pathway B has a different starting point and must surmount the kinetic barrier created by the trough. Reprinted by permission from Macmillan Publishers Ltd.: [Nature Structural Biology] Dill KA and Chan HS 1997 *Nat. Struct. Biol.* 4: 10

As a protein folds into its physiologically functional three-dimensional shape, it goes through an ensemble of intermediate shapes and structures. Thermal fluctuations are essential. They enable the protein to stochastically sample many different conformations of the polypeptide chain and escape from small local minima. At each step, a specific set of residues will come into contact with one another. Conformations more native like will have a lower energy and hence will be more stable than those less native like. Not all residues will contribute equally. Rather, a few key residues will nucleate and stabilize the formation of a native-like transition state and steer the folding in the direction of the native state.

Recall from thermodynamics that the change in Gibbs free energy, dG, is given by adding the changes in two opposing terms. One of these is the change in enthalpy, dH and the other the change in entropy, dS:

$$dG = dH - TdS$$

where T is the temperature. Favorable reaction steps are those that lower the free energy

$$dG < 0$$

These lead to the population of the native state, which lies at a (global) minimum in the free energy and is therefore thermodynamically stable. Since each point on the landscape represents a state, the width of the landscape is proportional to the number of states, that is, it is directly related to the entropy. The vertical axis in this kind of description would represent the sum of all contributions to the free energy except for the configuration entropy of the protein.

The horizontal axes of the landscape give the values of the various degrees of freedom, coordinates such as the dihedral angles. Since there are too many coordinates to depict individually, one or two coordinates or combinations thereof are selected that capture the essential behavior of the protein as it folds. As its name suggests, the folding funnel is broad at the top and narrow at the bottom. This arises because of the systematic depletion of configuration entropy as the energy is reduced. At the high energies and open character of the proteins, there are many different ways of folding the chain to arrive at the given high energy. As the protein folds toward its native low-energy state, there are fewer and fewer ways of achieving that particular energy and those configurations that do lie near one another.

16.3 Some Landscapes Are Smooth While Others Are Rugged

The amount of time required for a protein to fold into its native state is an important aspect of the process. This is referred to as a kinetic requirement. Not only must a protein fold into its native state, but also it must do so in a physiologically reasonable time interval. The speed depends critically on the topography of the potential energy surface. If the surface is studded with deep minima, and the folding trajectories pass close to them, the rate of folding will be slow. In these situations, the protein will fall into the minima and must escape before proceeding with its evolution toward its native state. The deep minima are called kinetic traps because of their slowing effects on the kinetics, or rates, of folding.

Proteins do not fold to their native state by carrying out global searches for energy minima. Instead, they fold toward the native state through sequences of small steps involving interactions between nearby residues, that is, through local interactions. These sampling mechanisms are "greedy" in nature, that is, proceed by making only small changes at each step. These changes involving mostly local interactions produce iterative improvements (energy lowering at each step) but do not examine the larger picture by searching more broadly to see if these changes are leading in the best direction to a good energy minimum. Instead, the system can easily encounter one or more kinetic traps and become stalled for appreciable amounts of time. Small single-domain proteins often have landscapes that are fairly smooth, e.g., pathway A in Fig. 16.1(b), and these proteins fold rapidly, over time frames ranging from 1 ms to 1 s. Large and complex proteins, especially those involved in signaling pathways, tend to have rugged landscapes containing minima surrounded by high barriers as depicted in Fig. 16.1(a). These require assistance in their folding as was mentioned in Section 16.1 and will be discussed in Chapter 18.

16.4 Proteins, Especially Those Involved in Signaling, Often Fold into Nonglobular, Extended Conformations

Studies of how small proteins and protein domains fold have provided fundamental insights into the process. These proteins with on the order of 100 amino acid residues or less typically fold with "two-state" kinetics. These proteins start from a stable unfolded, or denatured, state and rapidly move down a smooth folding funnel into a stable tightly folded native state. The proteins reach the native state by passing through one of an ensemble of similar transition states through a corresponding set of folding pathways. These proteins fold into small highly compact globular structures.

Not all proteins fold in the above-mentioned way. A second major way that proteins, especially large ones involved in signal transduction, fold is to assume an intrinsically unstructured form, that is, these proteins are natively unfolded. This selection is driven by the physiological roles of these proteins. Some adapt these far more open conformation in order to function as flexible scaffolds and linkers. Others do so to enable them to bind to a large number of interaction partners and fold upon binding.

Almost all of the disease-causing proteins listed in Table 16.1 are unstructured either in part or in their entirety. Proteins that have either large unstructured regions or are natively unfolded include the Aβ peptide (Alzheimer's), Tau (Alzheimer's, taupathies), α-synuclein (Parkinson's), Htt (Huntington's), and prions (SEs). Other disease-causing proteins including most of the proteins listed in Table 16.2 have native conformations dominated by well-developed secondary structure; these proteins range from all-β-sheet to mostly α-helical to mixtures of the two. Exceptions are amylin and ApoAI, both of which are natively unfolded.

A significant number of the proteins encoded by metazoan genomes are mostly unstructured or possess stretches (more than 50 amino acid residues in length) that are unstructured. More than 33% of the proteins in metazoans have partially or mostly unstructured native states. This number is far larger than the percentages of unstructured proteins in prokaryotes; these are in the range 2–9% in most organisms. This trend is consistent with the greatly increased importance of signaling in multicellular organisms. Natively unfolded conformations are especially prevalent in proteins involved in cellular signaling where the protein as part of its normal function interacts with multiple-binding partners either simultaneously or sequentially.

Two examples of how open flexible conformations are utilized in signaling are those of p27 interacting with cyclins and CDKs in cell cycle control and CREB interacting with CBP in gene transcription. A complex in which p27 inhibits the CyclinA/CDK2 complex is presented in Fig. 16.2. As can be seen, the extended unstructured conformation of p27 enables that protein to contact both cyclin A and CDK2 at multiple sites while preventing catalysis by CDK2. These actions are terminated through the following sequence of signaling events. First, resident and oncogenic nonreceptor tyrosine kinases phosphorylate p27 at Y88. This action triggers the dissociation of the 310 helix of p27 from CDK2 while other sites remain unaffected thereby preserving the tethering of p27 to the complex. CDK2 is then able to phosphorylate p27 at T187, which then leads rapidly to the ubquitination of p27 followed by degradation in the 26S proteasome. At this point the cyclin A/CDK2 complex is fully active and the cell cycle progresses to the next stage.

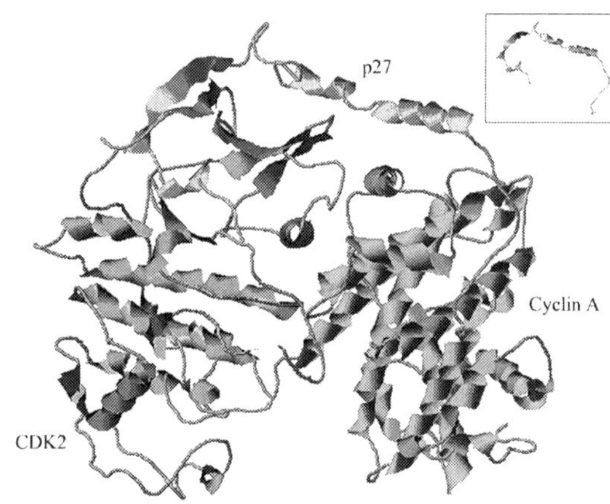

Fig. 16.2 Inhibition of cyclinA-CDK2 by p27Kip1. Insert of p27 shown in the *upper right corner* is intended as a visual guide. The figure was prepared using Jmol with atomic coordinates deposited in the PDB under accession number 1jsu

The open conformation of p27 and other similarly unstructured signaling proteins makes possible both high specificity and low affinity. These elements are crucial for the type of precisely timed signaling events such as those exquisitely choreographed with p27 as described above. Such proteins expose large surface areas permitting multiple binding partners; dissociation can be rapid, convenient binding sites are made available for phosphorylation and ubiquitination, and entry into the 26S proteasome is facilitated by the unfolded structure.

In signaling by these versatile and flexible proteins, folding and binding are coupled. In both of these operations a conformation search takes place over a funnel-dominated energy landscape. Binding by the CREB transcription factor, the second example, is illustrative. The CREB protein is regulated by phosphorylation at Ser133 by protein kinase A. The solution structure of the phosphorylated kinase-inducible domain (pKID) of CREB (which contains phosphorylated Ser133) in a complex with the KIX domain of the coactivator

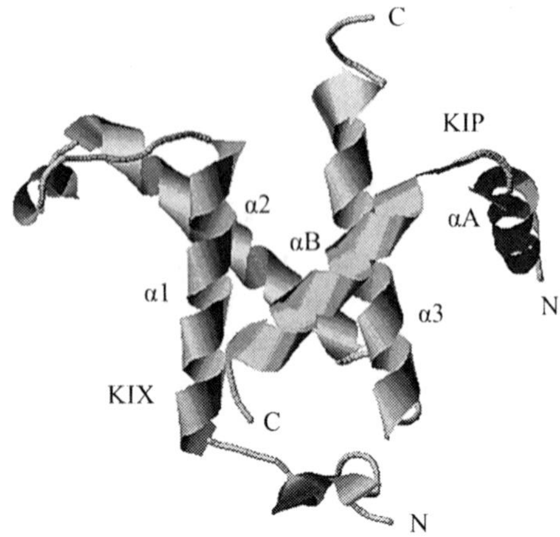

Fig. 16.3 KID domain from CREB, phosphorylated on Ser133 (pKID) in a complex with the KIX domain of CBP. The figure was prepared using Jmol with atomic coordinates deposited in the PDB under accession number 1kdx

CBP is presented in Fig. 16.3. In its free unbound form, pKID is unstructured. However, the bound form depicted in Fig. 16.3 possesses a pair of well-defined α-helices. The events underlying this transition have been studied by means of NMR spectroscopy and computer simulations. The picture that emerges is one in which the largely unfolded pKID approaches the KID and forms an initial encounter complex. This complex is formed through weak, nonspecific, and mainly hydrophobic interactions that tethers the two molecules to one another and stimulates the sequential emergence of secondary structure elements in pKID, namely, α helices A and B, resulting in a more stable binding.

16.5 Dialysis-Related Amyloidosis Is Brought on by Partial Unfolding and Aggregation of β-2 Microglobulin

Not all proteins forming amyloid fibrils and causing disease are natively unfolded. An example of an amyloid-disease-causing protein that has a compact, globular native state is β-2 microglobulin (β_2m). This is a 12-kDa subunit that together with the three-component β-, or heavy-, chain comprises the class-I major histocompatability complex (MHC-I). The β-2 microglobulin subunit maintains stability of the membrane-spanning α-chain, but these subunits are released extracellularly during MHC-I turnover and migrate to the kidneys. In diabetic individuals, they are no longer catabolized efficiently by the kidneys and as a result the concentration of β_2m in the circulation increases by more than an order of magnitude. The resulting buildup of β_2m leads to dialysis-related amyloidosis, a disease characterized by the formation of amyloid fibrils in the joints of affected individuals.

β-2 microglobulin is a small protein composed of 99 amino acid residues, which folds into a compact three-dimensional β-sandwich in which two β-sheets

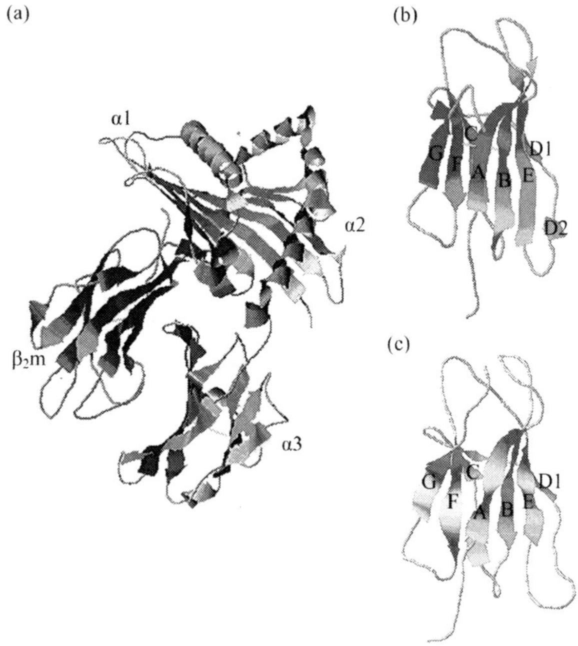

Fig. 16.4 Three-dimensional structure of β-2 microglobulin. (**a**) Crystal structure of MHC-I showing the arrangement of β_2m and the α1, α2, and α3 subunits of the heavy chain as determined through X-ray crystallography. (**b**) The bound form of the β_2m subunit in an orientation that facilitates comparison to (**c**) the solution structure of the free form of β_2m determined through NMR spectroscopy. The figure was prepared using Jmol with atomic coordinates deposited in the PDB under accession numbers 1duz for (**a**) and (**b**) and 1jnj for (**c**)

are linked by a central disulfide bond. Shown in Fig. 16.4 is the crystal structure of β_2m in a complex with the heavy-chain α subunits. Also shown in this figure is the solution structure of free monomeric β_2m. In its bound MHC-I form, the internal β-sheets and hydrophobic residues are shielded by their interactions with the subunits of the heavy chain. In addition, there is a pronounced bulge formed by the D1 and D2 β-sheet. Loops, bulges, and capping helices are some of the structural features that protect β-sheet proteins against inappropriate associations and aggregation. In its free form, the bulge is still present but the other subunits no longer provide shielding. The failure to eliminate the protein extends its residence time; the protein's hydrophobic balance is disturbed, it starts to unfold, becomes unstructured, and forms aggregates of various sizes.

One factor that helps drive the aggregation process is high concentrations of metal cations, specifically, Cu^{2+}, which are elevated in dialysis patients. These ions create transient bridges between pairs of partially unfolded β_2m molecules that stabilize the formation of a β_2m dimer. This assemblage then serves as a building block for formation of β_2m amyloid. Similarly, elevated amounts of metal cations such as Zn^{2+} and Ni^{2+} are suspected of having a facilitating role in promoting aggregation and amyloid formation in neurodegenerative diseases such as Alzheimer's disease and Parkinson's disease.

The unfolding and refolding of a protein that will aggregate can be visualized within the energy landscape picture. In this description, the protein will move along an intramolecular coordinate from its native state through one or several transition-states, and then from there along an intermolecular coordinate to a low-energy, disease-causing amyloid fibril state. An integrated energy landscape for β-2 microglobulin is presented in Fig. 16.5.

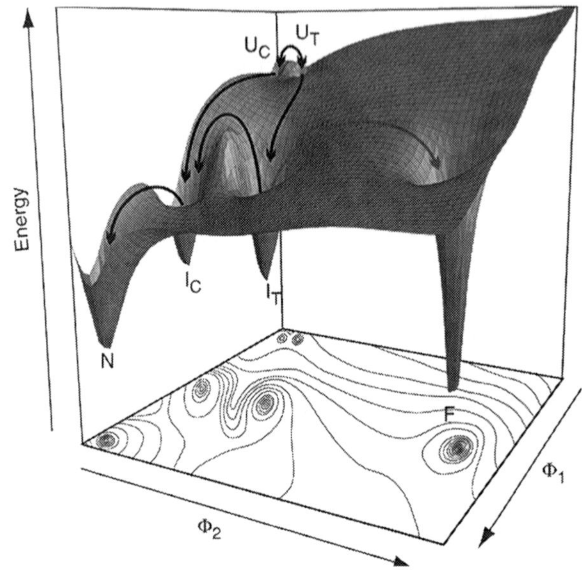

Fig. 16.5 Integrated energy landscape for free β_2m illustrating the presence of several folding and binding funnels connected by trajectories that pass over a number of transition states. The reaction coordinates Φ_1 and Φ_2 represent native-like intramolecular interactions and non-native-like intermolecular interactions. U_C and U_T denote partially unfolded states, and N, I_C and I_T and F represent the native state, two intermediate conformations, and the amyloid fibril state, respectively. Reprinted by permission from Macmillan Publishers Ltd.: [Nature Structural and Molecular Biology] Jahn TR et al. 2006 *Nat. Struct. Mol. Biol.* 13: 195

Several different mechanisms have been found that can promote oligomerization and fibrilization. One of these, tethering by metal cations, was just discussed. Another is domain swapping. To understand domain swapping consider the case of two identical proteins each consisting of two domains, A and B, separated by a flexible loop. In domain swapping, domain A from one protein occupies the place of domain A in the second protein. If similar domain pair can be swapped between the second and a third protein, and so on, then a fibril may be formed through this mechanism. A well-studied example of a protein exhibiting domain swapping is ribonuclease A, the protein whose study by Anfinsen and others led to the thermodynamic hypothesis. In the more recent studies, this protein has been shown to exhibit domain swapping in an open-ended way that promotes fibril growth.

16.6 β Cell Failure and Amyloid Formation in Type 2 Diabetes Is Brought on by Amylin Misfolding and Aggregation

Amylin, or human islet amyloid polypeptide (hIAPP), is a 37 amino acid hormone coproduced and secreted along with insulin by pancreatic β cells. Recall from Chapter 3 that a key initiating event in the progression of type 2 diabetes is insulin resistance leading to compensatory secretion by islet β cells. Amylin and insulin are secreted together by islet β cells. When the β cells become overtaxed, amylin overproduction leads to extracellular aggregation and fibril/amyloid formation. Like other amyloisoses, the buildup of amyloid fibrils is triggered by partial unfolding of amylin leading to its reassembly into the characteristic cross-β oligomers and higher-order aggregates.

Several environmental factors have been identified that stimulate the unfolding, refolding, and assembly of amylin. These include (i) catalysis through association with acidic membrane phospholipids, and (ii) an islet β cell microenvironment rich in redox signaling and stress elements. The presence of these elements in the local microenvironment helps catalyze the unfolding of amylin as well as its subsequent refolding and assembly into oligomers and amyloid fibrils.

The presence of oligomers and fibrils of misfolded amylin in association with the plasma membrane triggers a cascade of signaling events into the cell leading to death of the β cells through apoptosis. The key events following the extracellular buildup of misfolded amylin are, first, membrane damage leading to, second, the development of ER stresses brought on by an increased release of Ca^{2+} from intracellular stores and an increased Ca^{2+} influx from outside the cell as a result of increased plasma membrane permeability, accompanied by an increase in cellular ROS production, and resulting in, third, apoptosis.

16.7 Some Proteins Have Native States That Are Metastable and Not at a Global Minimum in the Free Energy

Serpins help regulate complement cascades, angiogenesis, inflammation, tumor metastasis, and apoptosis. Members of this large family include antithrombin which regulates blood coagulation, antitrypsin which mediates inflammation, and PAI-1, which is involved in tissue remodeling. Misfolded members of this family tend to polymerize and cause diseases collectively referred to as serpinopathies. α_1-Antitrypsin is synthesized by cells in the liver and by macrophages. Mutant forms of α_1-antitrypsin form polymers in the liver and cause cirrhosis. α-Antitrypsin misfolding and polymerization reduces the amount of α_1-antitrypsin available for control of proteolytic activity in the lungs. This failure leads to emphysema. A second way that polymers contribute to the lung disease is by acting as a pro-inflammatory agent. In this role, they act as a low-level chemoattractant that recruits neutrophils to the alveoli.

Under normal conditions, when harmful mutations are not present and polymerization does not take place, the serpins function as molecular mousetraps. Serpins are composed of three β-sheets and an exposed highly mobile, reactive center loop (RCL). The RCL has within it a peptide sequence that functions as the bait for the mousetrap. The mouse, the serine or cysteine protease, treats the bait as a regular substrate and binds to it. In response to the docking of the substrate, the serpin undergoes a large conformational change that clamps down and captures the protease. The new serpin conformation is a highly stable one, and the end result is a highly stable, serpin-protease complex. The conformational changes and resulting stabilization occur in these situations prior to completion of the catalytic cycle and for that reason this mechanism is referred to as suicide inhibition.

All of the proteins discussed so far have native states that are thermodynamically stable. These states correspond to global free energy minima. Serpins depart from this rule. Their native states are metastable states that function as intermediates along the folding pathway. This property and the attendant conformational flexibility are keys for the mousetrap but render the serpins susceptible to misfolding and polymerization inducing mutations that result in the serpinopathies.

An even more unusual situation is presented by an extracellular bacterial protease called α-lytic protease. This enzyme is synthesized as a pro-enzyme. The pro-region functions as a folding catalyst that lowers the transition state thereby enabling the protease to rapidly fold into its native state. Like the serpins the native state is not global free energy minimum. Instead, the native state is metastable. One it reaches this state, the pro-region is cleaved and removed. A large kinetic barrier, not present along the pro-region-assisted folding pathway, is present on the unfolding pathway and prevents unfolding. The native state in this situation is extremely long-lived even through the unfolded state lies at a lower free energy because of the high kinetic barrier.

16.8 β-Sheet Conformational Variations Underlie the Prion Strains and Disease Potential

Transmissible spongiform encephalopathies (TSEs) are a collection of neurodegenerative diseases of the central nervous system brought on by small proteinaceous infective particles, or prions, a term coined by Stanley Prusiner in 1982. Prions are misfolded forms of otherwise nondisease-causing protein with the striking property that they can trigger the conversion of the nondisease-causing form of the protein to the disease-causing one. Like many of the neuro-degenerative disease causing proteins listed in Table 16.1, the nonprion form is natively unfolded. The prion form, in contrast, is β-sheet-rich and forms amyloid fibrils. In a TSE, the brain appears spongy and hence the name for this class of diseases. Several prion diseases have been identified in mammals. These are listed in Table 16.3. The first entry in this table is bovine spongiform encephalopathy, also known as mad cow disease, a disorder that has received great public attention during the past few years with an underlying public-health question as to the likelihood of transmission across species barriers.

Table 16.3 Prion diseases (spongiform encephalopathies) and their mammalian hosts

Prion disease	Mammalian host
Bovine spongiform encephalopathy (BSE)	Cattle
Chronic wasting disease (CWD)	Deer, elk
Creutzfeldt-Jakob disease (CJD)	Humans
Gerstmann-Sträussler-Scheinker syndrome (GSS)	Humans
Kuru	Humans
Scrapie	Sheep, goats

Insight into the normal and abnormal functions of these proteins has been gained from studies of prions in the budding yeast, *Saccharomyces cerevisiae*, and the filamentous fungus, *Podospora anserina*. The two most studied yeast proteins are the prion forms of Sup25 and Ure2, while the highly studied fungal counterpart is known as Het-s. The Sup35 and Ure2 proteins function in yeast as translation termination factors and as a transcription activator, respectively; the fungal protein Het-s has a role in programmed cell death. These prions do not cause disease, but rather function as proteinaceous agents of inheritance. Like their mammalian prion-forming counterparts, the proteins exits in two distinct conformations, one that does not form amyloid fibrils and one that does. The latter form is capable of inducing the conversion of the natively unstructured nonprion form to the β-sheet-rich prion form that is propagated to offspring.

Het-s is a 289 amino acid residue long protein. The Het-s prion core consists of the C-terminal prion domain composed of residues 218–289. This domain folds into four β-strands arranged as a pair of β-strand-turn-β-strand windings. This is the basic infective unit capable of transforming the nonprion form of Het-s into the prion form. An example of how this basic unit can polymerize into a fibril is presented in Fig. 16.6. In this figure, a short amyloid fibril composed of five Het-s monomers is shown.

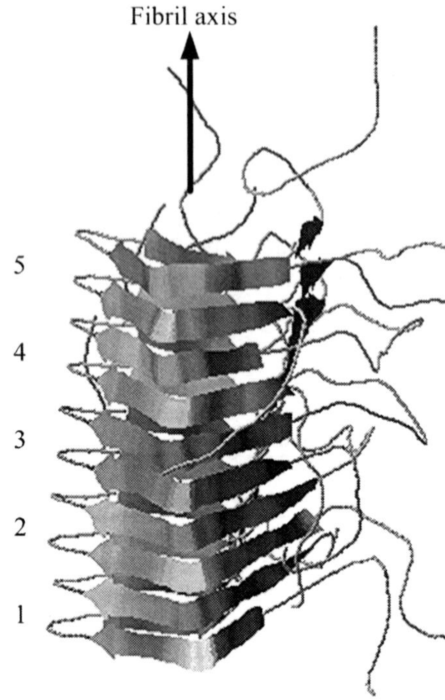

Fig. 16.6 Amyloid fibrils of the fungal prion protein Het-s (218–289) determined by means of solid-state nuclear magnetic resonance (NMR). The figure was prepared using Jmol with atomic coordinates deposited in the PDB under accession number 2NMR

Each Het-s monomer contributes two windings with three β-strands per winding. The overall structure is that of a left-handed β-solenoid. The structure is a stable one; each monomer possesses compact hydrophobic core, a number of stabilization-promoting salt-bridges, and an extensive network of hydrogen bonds. Similar triangular solenoidal structures are expected to be formed by human prions.

The atomic structure of a cross-β spine of amyloid fibrils from Sup35 is presented in Figure 16.7. This repeating structure is formed by the seven-residue sequence GNNQQNY rich in glutamine (Q) and asparagine (N) amino acids. This sequence forms a double β-sheet, each sheet erected from parallel segments exactly in register. Two sets of interactions generate this tight structure. First, an extensive network of hydrogen bonds is formed between main chain amides and between polar side chains of residues belonging to adjacent β strands within a β-sheet. This arrangement of residues was given the name "polar zipper" by Max Perutz. Second, the two β-sheets are then held together by van der Waals forces between polar side chains on β strands belonging to adjacent β-sheets oriented antiparallel to one another. These side chains are inter-digitized with one another. In this arrangement, hydrogen bonds and water molecules are absent and the interface is tight and dry (Fig. 16.7).

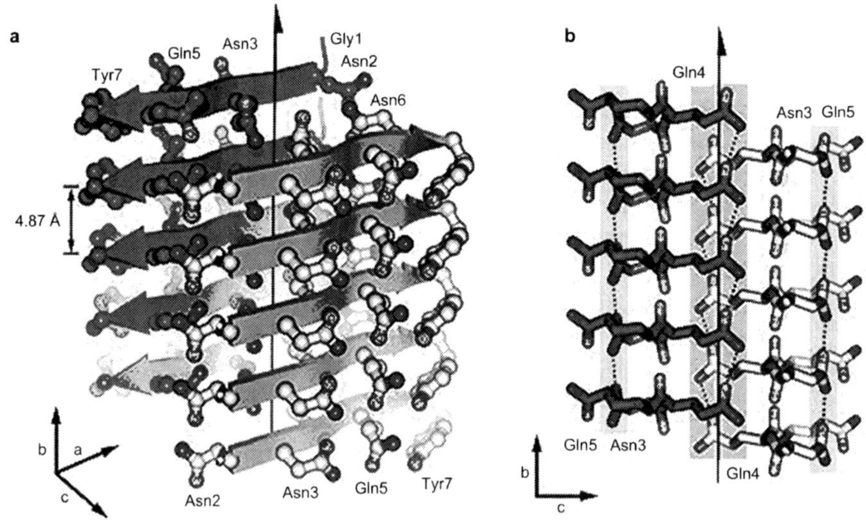

Fig. 16.7 Amyloid fibril structure for a seven-residue fragment of the yeast Sup35 prion. Reprinted by permission from Macmillan Publishers Ltd.: [Nature] Nelson R et al. 2005 *Nature* 435: 773

16.9 Strains and Transmissibility

In spongeform encephalopathies, a pathogenic form of the prion (PrP) protein, denoted as PrP^{Sc}, causes the normal, nondisease-causing form, PrP^{C} to misfold into the PrP^{Sc} form which then forms variously sized aggregates. The misfolded, β-strand-dominated forms thus act as seeds for transformation and aggregation. Unlike proteins with well-defined functional roles in the cell that provide strong structural constraints, the misfolded proteins are highly polymorphic in their secondary and super-secondary structure and aggregate morphology.

One of the most important consequences of this intrinsic variability is that multiple strains are produced. These represent different conformations of the misfolded protein or protein fragment. In some cases these arise from different fragment sequences and in other instances from fragments of the same amino acid sequence. Studies of yeast prions strains provide clear evidence for strains and further reveal that these differences give rise to differences in incubation period, patterns of distribution, toxicity, and in the ability of the prion to cross a particular species (transmission) barrier.

There is some evidence that strains and the propagation of misfolding and aggregation is not limited to prions but instead may be a more general biophysical property of proteins. The best examples pointing in this direction are provided by the Alzheimer's Aβ protein and for some of the amyloid-forming apolipoproteins listed in Table 16.2. Studies of Aβ (1–40) show that a variety of morphologies can be generated. The morphological differences arise from variations in molecular structure arising from alterations in growth conditions. The differences are self-propagating, and produce variations in toxicity that are dependent on both agent and host.

The other example of prion-like behavior is that of amyloid-forming apolipoproteins serum amyloid A and apolipoprotein AII. Both of these apolipoproteins can serve as proteinaceous agents either transmitting the disease to, or accelerating its development in, a second host. Serum amyloid A is synthesized in the liver. It is strongly induced and secreted in response to inflammatory stimuli, functioning as the major component in the acute phase response during which it associates with HDL, replacing ApoAI as its main apolipopotein. It is the precursor of the amyloid A (AA) protein. The latter is produced as an N-terminal 76 amino acid residue cleavage product of SAA. Under chronic inflammatory conditions such as those existing in atherosclerosis and rheumatoid arthritis, SAA levels increase by several orders of magnitude leading to a buildup of AA fibrils in kidneys, liver, and spleen, resulting in AA amyloidosis. The second disease in this group is systemic senile amyloidosis brought on by deposits of amyloid fibrils of misfolded ApoAII. This protein, like the prions and amyloid β, exhibits strain effects.

A second protein with misfolded forms that are a major contributor to neurodegenerative diseases is α-synuclein. This protein is natively unfolded, and in solution monomers of this protein are highly dynamic. But, like the other amyloid-forming proteins discussed in this chapter, α-synuclein forms oligomers and fibrillar structures that are highly ordered. This is perhaps the most surprising result of the various studies of amyloidogenic proteins. Rather than forming amorphous unstructured aggregates, all of these proteins assemble into highly ordered entities in which the entropic penalty is compensated for by a greater gain in stability; that is, by a compensatory decrease in enthalphy. Like the Aβ protein, the α-synuclein structures are highly polymorphic with variations in secondary structure among the fibrils. The fibril-forming α-synuclein monomers possess a β-strand-rich central region flanked by a disordered N-terminus and a flexible C-terminus. As is the case for the other fibril-forming proteins, there are differences among the fibrils in β-sheet organization and side-chain packing.

16.10 General Observations on How Proteins Fold into Alternative Disease-Causing Structures Characterized by Cross-β-Sheets

As discussed in the preceding sections there are two routes to forming fibrils and aggregates. The first route begins with a folded protein that partially unfolds into an intermediate conformation. Partial unfolding is then followed by refolding into an alternative structure composed of β strands organized into motifs such as β-sheet-turn-β-sheet, with a propensity to form higher-order structures, forming β-sheets and β-sheet sandwiches, and so on. The other route has as its starting point a natively unfolded conformation and these proteins then partially fold into an intermediate conformation and from there form β-sheet-dominated secondary and higher-order structures.

The alternative β-stranded monomeric and dimeric structures are the starting point, or seeds, for a variety of larger structures ranging from small soluble oligomers to large highly ordered multimeric aggregates to amyloid fibrils and still larger fibers. Alternatively, the β-strand-seeds may expose highly reactive, or "sticky", strands and/or surfaces that lead along different pathways to inappropriate interactions and to a buildup of large unstructured aggregates containing heterogeneous mixes of proteins. Variously sized structures may form in extracellular spaces, or within the cells, or concurrently in both places. The preeminent question posed by this wide spectrum of misfolded protein-containing structures is the proper identification of the form(s) and interactions responsible for Alzheimer's disease, Parkinson's disease, and the other neurodegenerative disorders listed in Table 16.1.

The key to the question of the identity of the toxic species as well as earlier posed questions regarding the transmissibility of disease-causing prions across species barriers is in understanding the biophysical properties of the oligomers and amyloid fibrils and the cellular responses they evoke. Core biophysical methods used in their study are, as always, X-ray crystallography and nuclear magnetic resonance (NMR) spectroscopy. These methods have been applied to amyloids since the pioneering X-ray work of Astbrury, Dickinson, and Bailey in 1935 and Eanes and Glenner in 1968. The study of amyloids is particularly challenging due to the difficulty in growing crystals of these substances, and due to the need to examine the dynamic behavior of the molecules in their functional environment or the equivalent. For these reasons each structure illuminated by these biophysical techniques represents a notable achievement.

These studies have shown that the ability to aggregate and form fibrils is a general biophysical property of polypeptide chains and not an exclusive consequence of specific amino acid sequences. The mature fibrils are themselves not toxic. Rather, the smaller oligomeric aggregates are the toxic species, containing exposed hydrophobic side chains whereas the mature fibrils are far more stable and bury these highly reactive elements. Because the larger fibrillar aggregates sequester potentially harmful misfolded chains, they actually function in a protective role.

These studies further show that there are many different ways of creating regular structures. For example, when viewed at the atomic level it becomes clear that there are many different kinds of steric zippers. Small differences in seed structures are preserved leading to corresponding differences in the final assemblages and, most importantly, these small structural differences influence the reactions and toxicity of the species with some forms far more toxic than others.

The actions taken by the small mass oligomers that render them toxic to neurons involve interactions at synaptic and intracellular membranes. These interactions involve membrane insertion and/or interactions with receptors and other membrane proteins. These interactions result in increased permeability of the membrane leading to a loss of ion homeostasis. Especially significant are those changes that lead to increases in intracellular calcium levels and increases in ROS production. The specific interactions, downstream signaling events, and cellular responses will be examined in the chapters covering each of the "big four" neurodegenerative disorders – Alzheimer's disease, Parkinson's disease, Huntington's disease, and Lou Gehrig's disease (ALS).

Further Reading

Anfinsen and the Thermodynamic Hypothesis

Anfinsen BC 1973 Principles that guide the folding of protein chains. *Science* 181: 223–230.

Hydrophobicity

Chandler D 2005 Interfaces and the driving force of hydrophobic assembly. *Nature* 437: 640–647.

Cheung MS, Garcia AE and Onuchic JN 2002 Protein folding mediated by solvation: water expulsion and formation of the hydrophobic core occur after the structural collapse. *Proc. Natl. Acad. Sci. USA* 99: 685–690.

Harano Y and Kinoshita M 2005 Translational-entropy gain of solvent upon protein folding. *Biophys. J.* 89: 2701–2010.

Frustration

Bryngelson JD et al. 1995 Funnels, pathways, and the energy landscape of protein folding: a synthesis. *Proteins: Struct. Funct. Genet.* 21: 167–195.

Energy Landscapes and Folding Funnels

Dill KA and Chan HS 1997 From Levinthal to pathways to funnels. *Nat. Struct. Biol.* 4: 10–19.

Fersht AR 2000 Transition-state structure as a unifying basis in protein-folding mechanisms: contact order, chain topology, stability, and the extended nucleus mechanism. *Proc. Natl. Acad. Sci. USA* 97: 1525–1529.

Lindorff-Larsen K et al. 2005 Protein folding and the organization of the protein topology universe. *Trends Biochem. Sci.* 30: 13–19.

Jahn TR and Radford SE 2005 The Yin and Yang of protein folding. *FEBS J.* 272: 5962–5970.

Intrinsically Unstructured Proteins

Dyson HJ and Wright PE 2005 Intrinsically unstructured proteins and their functions. *Nat. Rev. Mol. Cell Biol.* 6: 197–208.
Galea CA et al. 2007 Role of intrinsic flexibility in signal transduction mediated by the cell cycle regulator, $p27^{Kip1}$. *J. Mol. Biol.* 376: 827–838.
Levy Y, Wolynes PG and Onuchic JN 2004 Protein topology determines binding mechanism. *Proc. Natl. Acad. Sci. USA* 101: 511–516.
Shoemaker BA, Portman JJ and Wolynes PG 2000 Speeding molecular recognition by using the folding funnel: the fly-casting mechanism. *Proc. Natl. Acad. Sci. USA* 97: 8868–8873.
Sugase K, Dyson HJ and Wright PE 2007 Mechanism of coupled folding and binding of an intrinsically disordered protein. *Nature* 447: 1021–1025.

TTR Amyloidoses

Damas A and Saraiva MJ 2000 TTR Amyloidosis – structural features leading to protein aggregation and their implications on therapeutic strategies.*J. Struct. Biol.*130: 290–299.
Sekijima Y et al. 2005 The biological and chemical basis for tissue-selective amyloid disease. *Cell* 121: 73–85.
Teixeira PF et al. 2006 Endoplasmic reticulum stress associated with extracellular aggregates: evidence from transthyretin deposition in familial amyloid polyneuropathy. *J. Biol. Chem.* 281: 21998–22003.

Dialysis-Related Amylodosis

Eakin CM, Berman AJ and Miranker AD 2006 A native to amyloidogenic transition regulated by a backbone trigger. *Nat. Struct. Mol. Biol.* 13: 202–208.
Jahn TR et al. 2006 Amyloid formation under physiological conditions proceeds via a native-like folding intermediate. *Nat. Struct. Mol. Biol.* 13: 195–201.
Sambashivan S et al. 2005 Amyloid-like fibrils of ribonuclease A with three-dimensional domain-swapped and native-like structure. *Nature* 437: 266–269.
Richardson JS and Richardson DC 2002 Natural β-sheet proteins use negative design to avoid edge-to-edge aggregation. *Proc. Natl. Acad. Sci. USA* 99: 2754–2759.

Amylin and Type 2 Diabetes

Demuro A et al. 2005 Calcium dysregulation and membrane disruption as a ubiquitous neurotoxic mechanism of soluble amyloid oligomers. *J. Biol. Chem.* 280: 17294–17300.
Hayden MR et al. 2005 Type 2 diabetes as a conformational disease. *J. Pancreas (Online)* 6: 287–302.
Zhao H, Tuominen EKJ, and Kinnunen PKJ 2004 Formation of amyloid fibers triggered by phosphatidylserine-containing membranes. *Biochemistry* 43: 10302–10307.
Gellermann GP et al. 2005 Raft lipids as common components of human extracellular amyloid fibrils. *Proc. Natl. Acad. Sci. USA* 102: 6297–6302.
Engel MFM et al. 2006 Islet amyloid polypeptide inserts into phospholipid monolayers as monomer. *J. Mol. Biol.*356: 783–789.

Folding of Serpins and α-Lytic Protease

Jaswal SS et al. 2002 Energetic landscape of α-lytic protease optimizes longevity through kinetic stability. *Nature* 415: 343–346.
Jaswal SS et al. 2005 Comprehensive analysis of protein folding activation thermodynamics reveals a universal behavior violated by kinetically stable proteases. *J. Mol. Biol.* 347: 355–366.
Lomas DA and Mahadeva R 2002 α_1-Antitrypsin polymerization and the serpinopathies: pathobiology and prospects for therapy. *J. Clin. Invest.* 110: 1585–1590.
Whisstock JC and Bottomley SP 2006 Molecular gymnastics: serpin structure, folding and misfolding. *Curr. Opin. Struct. Biol.* 16: 761–768.

Prions

Govaerts C et al. 2004 Evidence for assembly of prions with left-handed β-helices into trimers. *Proc. Natl. Acad. Sci. USA* 101: 8342–8347.
Nelson R et al. 2005 Structure of the cross-β spine of amyloid-like fibrils. *Nature* 435: 773–778.
Prusiner SB 1982 Novel proteinaceous infectious particles cause scrapie. *Science* 261: 136–144.
Ritter C et al. 2005 Correlation of structural elements and infectivity of the HET-s prion. *Nature* 435: 844–848.
Wasmer C et al. 2008 Amyloid fibrils of the Het-s (218–289) prion form a β solenoid with a triangular hydrophobic core. *Science* 319: 1523–1526.

Strains and Transmissibility

Heise H et al. 2005 Molecular-level secondary structure, polymorphism, and dynamics of full-length α-synuclein fibrils studied by solid-state NMR. *Proc. Natl. Acad. Sci. USA* 102: 15871–15876.
Jones EM and Surewicz WK 2005 Fibril conformation as a basis for species- and strain-dependent seeding specificity of mammalian prion amyloids. *Cell*121: 63–72.
King CY and Diaz-Avalos R 2004 Protein-only transmission of three yeast prion strains. *Nature*428: 319–323.
Krishnan R and Lindquist SL 2005 Structural insights into a yeast prion illuminate nucleation and strain diversity. *Nature* 435: 765–772.
Lundmark K et al. 2002 Transmissibility of systemic amyloidosis by a prion-like mechanism. *Proc. Natl. Acad. Sci. USA* 99: 6979–6984.
Meyer-Luehmann M et al. 2006 Exogenous induction of cerebral β-amyloidogenesis is governed by agent and host. *Science* 313: 1781–1784.
Petkova AT et al. 2005 Self-propagating, molecular-level polymorphism in Alzheimer's β-amyloid fibrils. *Science* 307: 262–265.
Tanaka M et al. 2004 Conformational variations in an infectious protein determine prion strain differences. *Nature* 428: 323–328.
Tessier PM and Lindquist S 2007 Prion recognition elements govern nucleation, strain specificity and species barriers. *Nature* 447: 556–561.
Zhang B et al. 2008 Fecal transmission of AA amyloidosis in the cheetah contributes to high incidence of disease. *Proc. Natl. Acad. Sci. USA* 105: 7263–7268.

General Principles

Astbury WT, Dickinson S and Bailey K 1935 The X-ray interpretation of denaturation and the structure of the seed globulins. *Biochem. J.* 29: 2351–2360.

Eanes ED and Glenner GG 1968 X-ray diffraction studies of amyloid filaments. *J. Histochem. Cryochem.* 16: 673–677.

Bucciantini M et al. 2002 Inherent toxicity of aggregates implies a common mechanism for protein folding diseases. *Nature* 416: 507–511.

Sawaya MR et al. 2007 Atomic structures of amyloid cross-β spines reveal varied steric zippers. *Nature* 447: 453–457.

Chapter 17
Alzheimer's Disease

Alzheimer's disease (AD) is the leading cause of dementia in the elderly. The primary brain regions affected by the disorder are the neocortex and hippocampus. It begins with deficits in recent memory, progresses to a more general loss of memory and personality alterations, and culminates in total loss of intellectual capabilities. It affects over 5 million Americans today and has surpassed diabetes as a cause of death.

Alzheimer's disease was first described during a lecture by Alois Alzheimer in 1906. The name "Alzheimer's disease" was introduced in 1910 by Alzheimer's coworker and mentor Emil Kraepelin. As noted by Alzheimer and his colleagues, AD can be identified by the presence of two kinds of lesions – amyloid plaques and neurofibrillary tangles. Amyloid plaques are deposits of what is now known to be the amyloid β (Aβ) protein, forming insoluble aggregates in the extracellular spaces. These are usually surrounded by abnormal cells and cell structures, apoptotic cells, and elements of the immune system. The second kind of abnormality occurs within cells. Neurofibrillary tangles are amorphous collections of protein called tau that normally associates with microtubules. In Alzheimer's this protein dissociates from the microtubules and aggregates into insoluble filaments, the tangles.

By the 1960s and 1970s, it was recognized that the disease onset was associated in some way with failures in excitatory neurotransmission. There are two main excitatory neurotransmitters in the central nervous system – acetylcholine (ACh) and glutamate. The "cholinergic hypothesis" was presented in 1982. It grew out of studies emphasizing the former and was based on observations that (i) acetylcholine has a role in learning and memory, and in particular blocking its release led to memory impairments, and (ii) there are deficits in AD patients in the enzyme that catalyzes the synthesis of ACh, namely, choline acetyltransferase (ChAT). Since that time it has become increasingly apparent that the primary effects of cholinergic depletion are on attention, and is not a primary causative factor in AD nor does it correlate well with its clinical manifestations. The focus has shifted since these early groundbreaking observations on the cholinergic system to the glutamergic system and to pathologies derived from misfolded amyloid β.

M. Beckerman, *Cellular Signaling in Health and Disease*,
Biological and Medical Physics, Biomedical Engineering,
DOI 10.1007/978-0-387-98173-4_17,

The cholinergic hypothesis has been largely supplanted by the amyloid cascade model of how Alzheimer's disease develops. A key event in the development by Hardy and Higgins in 1992 of the amyloid cascade hypothesis was the discovery by Glenner and Wong in 1984 that the extracellular amyloid deposits were composed of the 39–43 amino acid long Amyloid Aβ peptide. The amyloid cascade hypothesis posited that misfolded and aggregated clumps of this protein trigger a cascade of events leading eventually to neural dysfunction and death. This hypothesis has undergone considerable revision since its initial formulation in keeping with a plethora of subsequent discoveries. The emphasis is now more on the synapse and less on space plaque, and on how the (glutamergic) synapse becomes dysfunctional during the early phases of the disease. The updated and expanded model attempts to relate the normal function of the protein to the development of the disease state, includes neuroinflammation, and incorporates the neurofibrillary tangles into the overall picture of the disease progression. That is the model that will be presented in this chapter.

The neurons most strongly affected by the disorder are those using acetylcholine and glutamate as neurotransmitters. Also affected are serotonin- and norepineprine-releasing neurons, but to a lesser extent. Synaptic transmission is impaired in these neurons and cell death is a common outcome. The disorder begins in the hippocampus where it impairs short-term memory formation. It spreads from the hippocampus to other limbic regions such as the amygdala and entorrhinal cortex, and from these regions propagates to higher associative areas of the brain producing losses in memory, judgment, and reasoning, and changes in behavior.

The causative agent in Alzheimer's disease is misfolded forms of the amyloid β protein. Like the other amyloidogenic proteins discussed in the last chapter, this misfolded protein aggregates into variously sized particles ranging from small oligomers to the large extracellular amyloid plaques. Although the amyloid plaques are one of the striking molecular level signs of the disease, it is the small oligomers that are believed to be the toxic agents that trigger development of the disease. These particles are found in intracellular and extracellular locations in equilibrium. The various stages in the progression of the disease from its earliest onset to its final outcome are illustrated in Fig. 17.1. As shown in this figure, the first element in the induction of Alzheimer's disease is altered amyloid β metabolism; this involves (1) an imbalance between generation and removal through degradation and clearance and (2) a lethal mix of folding physics and redox chemistry.

17.1 Generation of the Amyloid β Protein

The Aβ protein is derived from a larger β-amyloid precursor protein (APP) through a series of cleavage operations. These are mediated by enzymatic complexes referred to as α-secretases, β-secretases, and γ-secretases. The presenilins are part of the γ-secretase complex, which has three other members – Nicastrin,

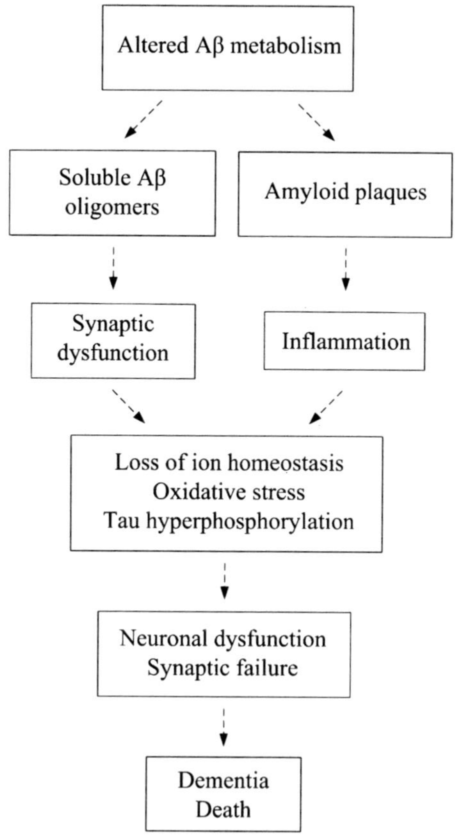

Fig. 17.1 Revised and updated formulation of the amyloid cascade hypothesis

Aph-1, and Pen-2. As illustrated in Fig. 17.2, the presenilins are 8-pass transmembrane proteins proteolytically cleaved into two chains. A pair of aspartate residues positioned in opposition to one another is crucial for presenilin's γ-secretase actions. Notch and APP pass through the gap formed by the TM segments containing the aspartates and are cleaved. The presenilins carry out their catalytic activities together with the single-pass Nicrastrin protein, the 7-pass Aph-1 protein, and the 2-pass Pen-2 protein, which help to assemble and stabilize the complex.

The APP is cleaved twice, one in its extracellular domain just outside the membrane and the other near the middle of the intermembrane region (Fig. 17.2). In the first step, the APP proteins are cleaved at one of two alternative extracellular locations termed the α- and β-sites. The γ-secretases are responsible for subsequent cleavages within the transmembrane segment. In the case of cleavage at β sites, but not the α-sites, the products are 40 and 42 amino acid residue forms of the Aβ amyloid protein. The 42 residue form is favored by the mutated Presenilins, and is especially prone to aggregation. The α-secretases and β-secretases responsible for cleaving the APPs at the α- and

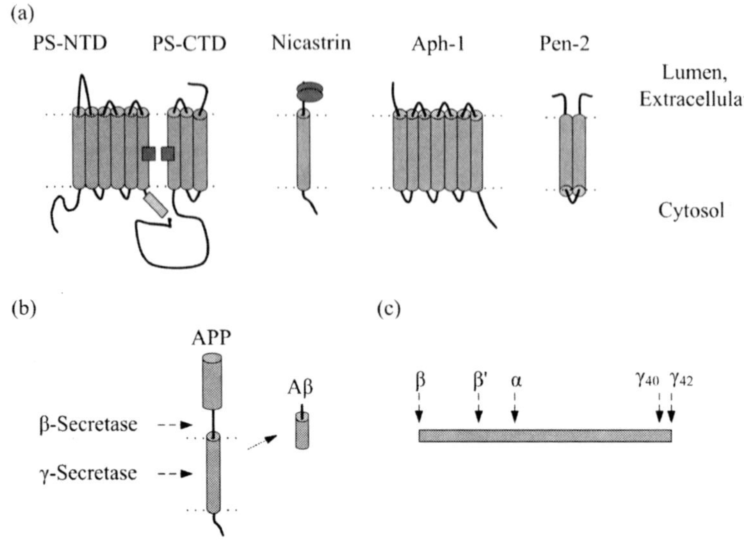

Fig. 17.2 γ-Secretase complex and generation of the Aβ through secretase cleavages. (**a**) Topology of the transmembrane helices belonging to the Presenilin N- and C-terminal domains (PS-NTD and PS-CTD), Nicastrin, Aph-1, and Pen-2. (**b**) Location of the β-secretase and γ-secretase cleavage sites, and (**c**) the resulting amyloid β proteins γ40 and γ42

β-sites are members of two other families of proteases. One or more members of the "a disintegrin and metalloprotease" (ADAM) family are the α-secretases, while aspartyl proteases belonging to the membrane-associated aspartyl protease (memapsin) family serve as the β-secretases.

Cleavage of the APP at the α-secretase site produces an N-terminal fragment referred to as sAPPα and a C-terminal C83 fragment. The latter is processed further by the γ-secretase resulting in formation of a P3 fragment. As can be seen in Fig. 17.2, the α site lies within the regions encompassed by the Aβ peptide (between resides 16 and 17 lying just outside the membrane) and serves as an alternative to β-secretase associated forms.

The enzymes responsible for the α-secretase activity are members of the ADAM (a disintegrin and metalloproteinase) family. These are single-pass transmembrane proteins with an N-terminal signal peptide, followed by a pro-domain, a metalloprotease domain, a cysteine-rich disintegrin EFG-like domain, a TM segment, and a C-terminal cytosolic region. Three members of this family acting as a team have prominent roles in APP metabolism – ADAM17, also called TACE, ADAM10, and ADAM9. TACE (tumor necrosis factor-α converting enzyme) is responsible for freeing TNF-α from its membrane bound, pro-TNF precursor protein. It cleaves several other membrane-bound proteins in addition to TNF-α and APP. Under some conditions and cell types, sAPPα is constitutively expressed. ADAM9 and to a lesser extent ADAM10 seem to be the main family members involved in these actions as well as in regulated formation of sAPPα. ADAM proteases release extracellular portions, or ectodomains, of membrane-bound proteins, which are then free to carry out their signaling roles. TACE and ADAM10, and to some extent ADAM9 too, cleave and thereby activate multiple signaling proteins. Among the proteins activated by these sheddases are cytokines, growth factors, interleukin receptors, adhesion molecules, Notch, and APP.

The β-site APP cleavage enzyme 1 (BACE1) protein is the only β-secretase found to date. This transmembrane protein is an aspartyl protease belonging to the pepsin family. It possesses pre- and pro-domains followed by a catalytic domain, transmembrane helix, and cytoplasmic segment. BACE1 is the rate-limiting enzyme in the formation of Aβ peptides and is tightly regulated at the transcription, translation, and trafficking levels. Its promoter contains binding sites for a number of transcription factors. Prominent among these are sites for NF-κB and PPARγ. As has been found in many setting the actions of NF-κB can be either repressive or stimulatory depending on the presence or absence of cellular stresses. Under healthy, nonstressed conditions, NF-κB represses transcription of BACE1 but when the cells are continually exposed to stressful conditions such as those present in AD, NF-κB promotes BACE1 gene expression. PPARγ activity is thought to repress BACE1 as well. Activated microglia and astrocytes secrete pro-inflammatory cytokines and these signaling molecules suppress PPARγ leading to an increased production of BACE1. A positive feedback loop may occur in which a resulting increased Aβ plaque buildup stimulates further inflammatory response leading to a greater production of BACE1 and hence more Aβ. BACE and this loop will be examined in more detail in the sections on the contribution to AD arising from inflammatory response to amyloid plaque buildup.

17.2 Removal Through Degradation and Clearance

Amyloid β generation and clearance are complementary actions. Under healthy conditions, clearance is sufficiently robust and in a balance that prevents large deposits of amyloid β from developing. This protein is synthesized throughout the body. Its import into the brain and export from the brain are mediated by receptor for advanced glycation end-products (RAGE) and by the low-density lipoprotein receptor-related protein (LRP), respectively (Fig. 17.3(a) and (b)). Two chaperone proteins facilitate these movements – apolipoprotein E (ApoE) and α2 macroglobulin (α2 M). RAGE is a multiligand receptor belonging to the Ig superfamily. It is responsible for transporting unbound circulating Aβ peptides across the blood-brain barrier into the brain. This action is countered by LRP, which rapidly transports back out of the brain across the BBB into the circulation. This receptor is a member of the LDL family and functions as a

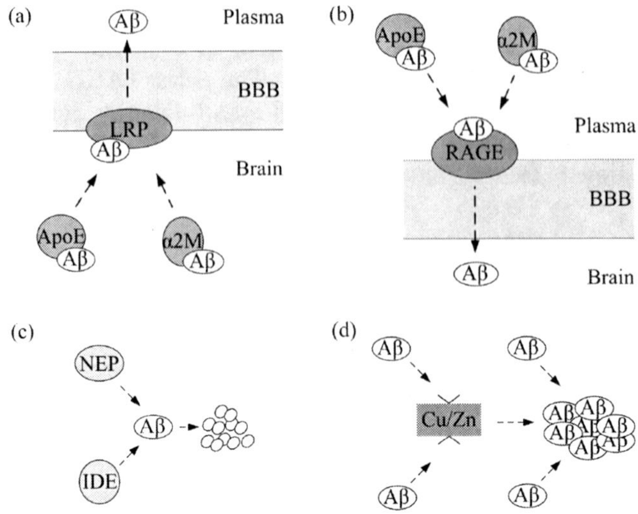

Fig. 17.3 Removal of amyloid β from the brain. Efflux and Influx of amyloid β across the blood-brain barrier (BBB). (**a**) Export of amyloid β from the brain to the circulation mediated by LRP, and (**b**) Import of amyloid β into the brain from circulation mediated by RAGE. Degradation and aggregation/sequestration. (**c**) Proteolytic degradation by neprilysin (NEP) and by insulin-degrading enzyme (IDE). (**d**) Aggregation catalyzed in part by metal cations such as copper and zinc

scavenging and signaling receptor. Under normal conditions clearance by LRP is far more rapid than influx by RAGE, and the concentration of amyloid-β peptides in the brain is kept at a low level.

Export across the blood-brain barrier is one of two mechanisms for removing amyloid-β peptides from the brain. The other is degradation. Two zinc-dependent metalloproteases – neprilysin (NEP) and insulin-degrading enzyme (IDE) – are responsible for degrading Aβ (Fig. 17.3(c) and (d)). These enzymes are found in multiple locations, in extracellular spaces either free or attached to the outer surface of the plasma membrane, and inside neurons in the cytosol or within certain organelles. Insulin-degrading enzyme targets short peptides. These include the amyloid-β peptide and, as its name suggests, insulin, amylin, and insulin-like growth factors. The targeting of both amyloid-β peptide and insulin provides a potential link between Type 2 diabetes and Alzheimer's disease that will be examined later in this chapter. At this stage in the discussion the main point as illustrated in the figure is the existence of an alternative path that removes and sequesters amyloid-β peptides through aggregation. Misfolded Aβ(1–40) and especially Aβ(1–42) resist degradation by aggregating, and this activity is enhanced by associations with heavy metals such as zinc and copper. These cations coordinate pairs of Aβ peptides thereby facilitating aggregation into protease-resistant amyloid plaques.

17.3 Folding Physics, Metal Homeostasis, and Redox Chemistry

As discussed in the previous sections, the most toxic form of the secretase-generated N-terminal fragments of the APP protein is Aβ(1–42). Using solid-state NMR, Lührs co-workers determined the three-dimensional structure of Alzheimer's amyloid-β(1–42) fibrils. Their results are presented in Fig. 17.4. Residues 1–17 are hydrophilic and disordered; their role will be discussed shortly. The fibril-generating core consists of the hydrophobic segment containing residues 18–26 that comprise the β1 strand and 31–42 that form the β2 strand.

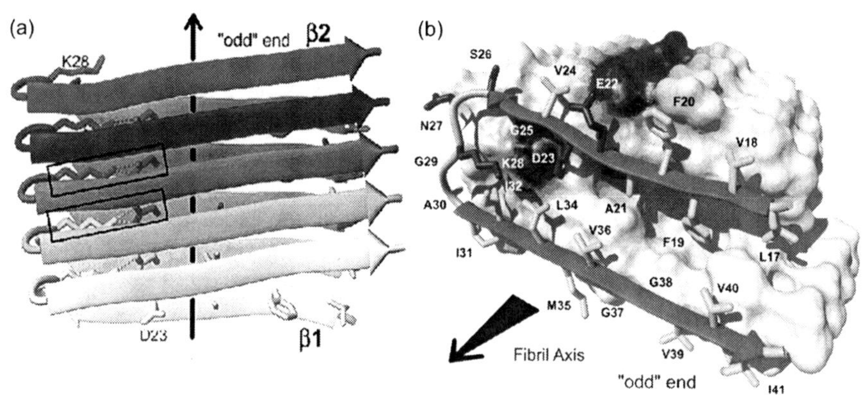

Fig. 17.4 Three-dimensional structure of the amyloid-β (1–42) fibril. The *arrow* denotes the fibril axis. (**a**) Overall structure stacked β-strand-turn-β-strand. (**b**) Detailed view of a core β-strand-turn-β-strand motif. Reprinted by permission from Lührs T et al. 2005 *Proc. Natl. Acad. Sci. USA* 102: 17342, copyright 2005 National Academy of Sciences, USA

Pore-forming toxins are the most common type of bacterial toxin. In this type of attack, the bacteria secrete proteins that form pores in the target cell's outer membrane leading to an increase in ion permeability and a loss of ion homeostasis. The toxins are secreted in water-soluble form, bind target cell membranes and/or receptors, and autonomously oligomerize into amphipathic structures that insert into membranes to form a pore. Sizes of the oligomeric complexes and the corresponding dimensions of the pore opening vary from about seven subunits for a small 20 Å wide pore by α-hemolysin (α-toxin) to 40 subunits arranged in a large ring to create a 250 Å wide pore by cholesterol-dependent cytolysins (CDCs).

Like the bacterial toxins, oligomers of amyloid-generating proteins can interact with membranes and alter their permeability to ions. Interestingly, the structures formed by amyloid β(1–40), α-synuclein, and others are somewhat pore-like, but are fairly small, typically built from 2-mers and 3-mers and far more dynamic and variable than their bacterial counterparts. These assemblages too are able to generate single ion currents and destabilize ion homeostasis but as will be discussed in the following sections and chapters in a number of different but equally potent ways.

The amino acid sequence comprising the N-terminal disordered segment of the Aβ peptide contain four residues – three histidines and one tyrosine – that bind heavy metals, most notably, Fe^{3+}, Cu^{2+}, and Zn^{2+}. Concentrations of these transition metals in the brain increase with age and are part of the pathology of Alzheimer's disease. The capacity to take up these metals is shared by APP as well as Aβ and one likely role for this protein is an antioxidant, neuroprotective one; it sequesters heavy metals thereby maintaining heavy metal homeostasis and preventing them from catalyzing ROS potentially neurotoxic reactive oxygen species. In this regard, the large extracellular deposits, the amyloid plaque, may serve as protective role, consistent with the view introduced in the last chapter that small oligomers are the toxic form.

Small oligomers of metal-bound amyloid-β peptide associate with membranes. When this happens the tendency is for Aβ to fold/refold into an α-helical form and insert into the membrane while other oligomers retain the unstructured plus β strand form and remain on the membrane surface. These membrane-associated amyloid-β peptide oligomers catalyze the reduction of Cu(II) to Cu(I) and the subsequent generation of H_2O_2 which occurs in the presence of molecular oxygen and reducing agents such as cholesterol. The production of hydrogen peroxide, in turn, leads to lipid peroxidation through generation of OH• by means of the Fenton reaction

$$M^{n+} + H_2O_2 \rightarrow M^{(n+1)+} + OH^- + OH\bullet$$

and to oxidative damage to the peptide leading to further oligomerization and enhancement of toxicity.

17.4 Normal Physiological Function of the Aβ Protein at the Synapse

Several different APP fragments are generated by the combined actions of α-, β-, and γ-secretases. Under normal conditions fragments such as sAPPα and Aβ help preserve neural health. The Aβ protein operates in a negative feedback loop that protects glutamergic neurons against excitotoxicity. Neural activity increases Aβ activity, which then depresses neural excitation thereby prevents excessive stimulation of the neurons. This aspect in its normal and pathological manifestations will be examined further in the next section.

The Aβ protein is expressed at high levels following neural injury. The occurrence of trauma and injury increases the risk of a person developing Alzheimer's disease. Chronic activation of this regulatory pathway either through injury or through genetic mutations leads to a pathological state characterized by synaptic failure and neural dysfunction. In the case of genetic mutations to either the APP or the presenilins, it leads to increased generation of the misfolded $A\beta_{42}$ fragment, which has a far greater tendency to form long-lived aggregates than either the $A\beta_{38}$ or $A\beta_{40}$ peptides.

Amyloid precursor protein intracellular domain (AICD) is generated through actions of γ-secretase on the APP at a number of sites. Cuts made at the so-called γ-cleavage sites (at residues 38, 40, 42 C-terminal to the β-cleavage site), ε-cleavage site (49), and ζ-cleavage site (46) all situated within the trans-membrane portion of the APP liberate both Aβ and the C-terminal segment, the AICD. These operations and the resulting actions of the fragments resemble the situation generated by the presenilins on the Notch protein. In the case of Notch, the released C-terminal fragment referred to as the Notch intracellular domain (NICD) translocates to the nuclear where it functions as a transcription factor. The actions taken in the case of the APP are slightly different. The AICD influences transcription by activating the Fe65 adaptor protein and this dimer interacts with other transcription coactivators such as the histone acetyltransferase Tip60. Genes influenced include those promoting developmental and injury-related alterations in actin cytoskeleton structure.

17.5 Action of the Aβ Oligomers at the Synapse – Aberrant LTD

The region just below the postsynaptic membrane is the postsynaptic density (PSD). The PSD is a disc-like structure that sits opposite the active zone in the presynaptic terminal. It is highly enriched in structural (cytoskeleton) and regulatory (receptor, ion channel, scaffolding, signaling) proteins. Several different kinds of glutamergic receptors are found in the PSD of excitatory synapses in the CNS. Two of them are ion channels (ionotropic). These are the α-amino-3-hydroxy-5-methyl-4-isoxazolepropionate (AMPA) receptors, or GluRs, and the *N*-methyl-D-aspartate (NMDA) receptors, or NRs. The third kind of glutamerigc receptor belongs to the G-protein coupled receptor superfamily. These are the metabotropic mGluRs.

The three kinds of glutamergic receptors work together (Fig. 17.5(a) and (b)). High levels of NMDA receptor stimulation triggers a signaling cascade resulting in biosynthesis, transport, and insertion of additional AMPA receptors, and to an increase in the number of dendritic spines. The nascent receptors are transported from the ER to the Golgi and from there travel along microtubule rail system; they then interact with the actin cytoskeleton and other elements of the PSD. These actions result in a stronger response to the arrival of neurotransmitter and thus potentiate the signal. Conversely, repeated low levels of NMDA receptor stimulation produce the converse biophysical reaction. The result is depression of the synaptic response to neurotransmitter arrival. Over time, AMPA are internalized and there is a loss of dendritic spines. The former process in which synaptic transmission is enhanced is called long-term potentiation (LTP) while the second kind of modification where the

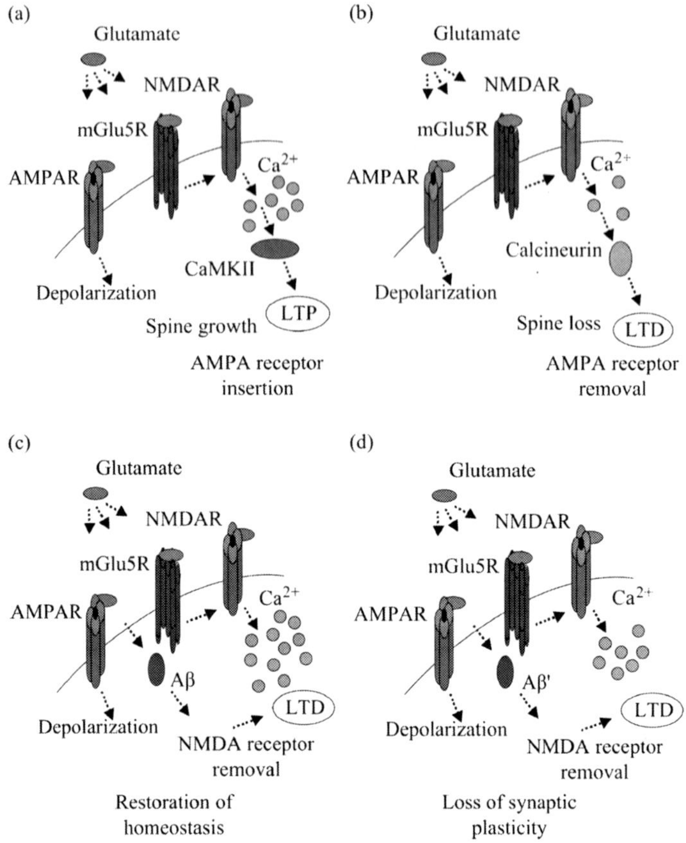

Fig. 17.5 Large amount of calcium entry stimulates the serine/threonine protein kinase CaMKII signaling leading to an increase in NMDA receptors whereas low levels of calcium entry activate the serine/threonine protein phosphatase calcineurin CaN

strength of the postsynaptic response to neurotransmitter release is reduced is called long-term depression (LTD).

In more detail, NMDA receptors are both voltage-dependent and glutamate-dependent. These receptors work in the following way. At the membrane's resting potential, the NMDA receptor ion channel is closed; it is physically blocked by an Mg^{2+} ion that sits inside the opening. When the membrane is depolarized sufficiently, the Mg^{2+} block is relieved, and once glutamate is bound, Ca^{2+} ions can pass through the channel and enter the cell. The voltage dependence is important for operation of the channel. No single synapse by itself can depolarize the postsynaptic membrane sufficiently to relieve the magnesium block. Rather, many synaptic inputs acting in close spatial and temporal proximity to one another are needed to adequately depolarize the postsynaptic membrane. When depolarization and ligand binding occurs within an appropriate time window, the NMDA channel opens. Thus, NMDA receptors integrate together multiple signals, and determine whether the requisite conditions for synaptic modification have been satisfied or not and are responsible for synaptic plasticity.

At many postsynaptic terminals only NMDA receptors are active during synapse development. These synapses remain silent at resting potentials because of the Mg^{2+} block, but are capable of transmitting signals when the membrane is depolarized by one means or another. In response to entry of calcium into the terminal through the NMDA receptors, AMPA receptors are inserted into the postsynaptic membrane and become active. The presence of active AMPA receptors allows the synapses to respond rapidly to glutamate and generate stronger responses at the postsynaptic terminal to the release of glutamate. AMPA receptors and AMPA receptor trafficking have a prominent role in learning and memory formation.

One of the very earliest effects of amyloid β oligomers on brain function is an impairment of synaptic function. This involves a loss of synaptic plasticity leading to dendritic spine loss and the death of the affected neurons. The steps leading to this loss of synaptic function are related to a protective function normally carried out by amyloid precursor protein and its cleavage products at glutamergic synapses situated in the hippocampus and elsewhere in the brain. These proteins protect against excessive excitation and signaling through glutamergic receptors. It accomplishes this by stimulating signaling pathway leading to the internalization of NMDA receptors. If excessive excitation and calcium entry occurs this negative feedback loop is activated resulting in fewer NMDA receptors and a restoration of homeostasis (Fig. 17.5(c)).

Perturbations to the normal expression levels and functioning of the Aβ protein affect synaptic plasticity through the above-mentioned pathway resulting in impaired learning and memory. In particular, excessive levels and misfolded forms of Aβ peptides that have aggregated into small oligomers trigger the internalization of NMDA receptors even when the amount of excitatory stimulation is not excessive (Fig. 17.5(d)). This continual activity by the modified Aβ proteins leads to the removal of first NMDA receptors and then AMPA receptors from the membrane followed by loss of dendritic spines. These events derail synaptic plasticity and replace it with a persistent depression of synaptic transmission as perhaps the earliest stage in the disease.

17.6 The Local Microenvironment Contains Neurons, Astrocytes, and Microglia

The local microenvironment has an important role in AD just as it does in atherosclerosis, cancer, and other age-related diseases. Neurons in the central nervous system are in communication with two other types of cells – astrocytes and microglia. Astrocytes (also called astroglia or macroglia) carry out multiple supportive roles in the brain, functioning as an interface between the microvasculature and the neurons and as local communication elements within the microenvironment. They supply neurons with nutrients and protection; they regulate the chemical and fluid milieu, maintain ion and neurotransmitter homeostasis in and near the synaptic cleft, and help regulate synaptic plasticity. They are able to carry out these multiple roles through their positioning in between neurons and the vasculature with end feet making near contact with both. They are able to sense the amount of synaptic activity and in response balance vasodilation and vasoconstriction so that blood supply matches demand.

Astrocytes are excitable. They express voltage-gated ion channels and neurotransmitter receptors on their surface. In contrast to neurons, they do not generate action potentials but instead propagate calcium waves and other chemical signals within the cells. The positioning of astrocytes end feet strategically across from synaptic clefts gives rise to the notion of a tripartite synapse composed of a presynaptic terminal, postsynaptic terminal, and astrocyte process. Astrocytes remain in two-way communication with neurons at these sites. They are responsive to neurotransmissions across the synaptic cleft, and send out chemical messages to the pre- and postsynaptic terminals.

The other major component of the local microenvironment of the neurons is the microglia. These are the immune system cells of the brain. They are phagocytes derived from the same bone marrow progenitors as macrophages and dendritic cells. They migrate to the central nervous system during development prior to formation of the blood-brain barrier. They differentiate into microglia and function as sentinels of the brain continuously monitoring their microenvironment for signs of injuries such as stroke and tumor formation, bacterial and viral pathogens, and toxins. Once damage is perceived they migrate to site of the damage where they engulf and remove tissue debris.

17.7 Microglia Respond to Amyloid Plaque Buildup by Mounting an Inflammatory Response

Microglia utilize pattern recognition receptors (PRRs) to detect the presence of unwanted biomaterials in the local environment. As discussed in Chapter 7 the receptors bind macromolecules expressed on the surface of pathogens that distinguish them from the host. These macromolecules form pathogen-associated molecular patterns (PAMPs). Several families of PRRs are expressed on microglia. One of these is the Toll-like receptor (TLR) family. These cytokine receptors mediate the body's innate immunity to microbial pathogens ranging from viruses to parasites. These PRRs and others react to deposits of fibrillar Aβ as though they were pathogens thereby triggering a chronic inflammatory response to the continual presence and buildup of these amyloids. Another key family of pattern recognition receptors is the scavenger receptor (SR) family. These receptors not only recognize pathogens such as bacteria but also oxidized LDL (OxLDL), as was discussed in Part I, and fibrillar Aβ.

Binding of extracellular amyloid deposits to pattern recognition receptors on microglia sets off a chain of signaling events that produces a toxic milieu of inflammatory cytokines and reactive oxygen species. Amyloids containing fibrillar Aβ in the case of Alzheimer's disease, prions, or α-synuclein as in Parkinson's disease are each capable of producing this type of response in which the microglia converge to the site of the accumulation and become activated. Binding of the amyloids to pattern recognition receptors is the first step in a signal transduction cascade that upregulates transcription of genes encoding inflammatory cytokines, most notably, tumor necrosis factor α (TNFα) and IL-1β. Also upregulated are genes encoding enzymes that stimulate the production of nitric oxide. The cytokines and reactive oxygen species are then secreted from the micoglia.

In phagocytes such as microglia, NADPH oxidase (NOX) complex catalyzes the generation of superoxide anions (O_2^-) from molecular oxygen (O_2) using NADPH as the electron donor. The superoxide anions, along with more potent reactive oxygen species derived from them, are used to kill the invading micro-organisms. The superoxide is highly damaging to neurons and also serves as a microglial intracellular signal to stimulate cytokine production.

The primary NOS enzyme in the response to amyloid buildup is iNOS. This family member, in contrast to the others, does not require calcium – calmodulin binding whereas both eNOS and nNOS are calcium-dependent enzymes. Inducible nitric oxide synthase is upregulated in activated microglia by the downstream signal transduction pathway activated in response to amyloid binding at the cell surface. The iNOS promoter is a complex one enabling multiple signaling pathways to converge onto the site and stimulate iNOS gene transcription. Binding sites are present for STATs, IFN-1, CREB/ATF, C/EBP, AP-1, and NF-κB/Rel proteins. This organization allows for the induction of iNOS gene expression by injures, infections, inflammation, and other cellular challenges. The production of NO by iNOS differs from that derived from either eNOS or nNOS in that it is far more intense and longer lasting.

Although NO may be thought of as a free radical, it is also an important signaling molecule and is utilized as a negative feedback regulator of NMDA activity in neurons. This aspect was discussed briefly in Chapter 8. Whenever NMDA receptors are overstimulated, excessive amounts of calcium enter the cell and calcium homeostasis is disrupted, large numbers of free radicals are produced, and oxidative phosphorylation is impaired. To prevent this happening, nNOS molecules are tethered in the vicinity of the NMDA receptors, and NO production is stimulated when calcium enters the cell and binds to calmodulin. S-nitrosylation of critical cysteine residues located in the regulatory subunits of the NMDA receptors then throttles back the calcium influx through the receptors.

Under normal healthy conditions, the production of superoxide anions by NADPH oxidase and the generation of nitric oxide by the NOS enzymes are protective actions. However, under chronic amyloidogenic conditions these actions become harmful and help drive the disease. The primary way these two classes of activities become dangerous and toxic is through the combination of superoxide and NO to form peroxynitrite, or $ONOO^-$. These free radicals are generated rapidly when both superoxide and nitric oxide are present. Peroxynitrite is able to pass through membranes and attack resident macromolecules, leading to apoptosis when present at low concentrations and necrosis when present at high concentrations. In the nucleus, it causes DNA damage leading to activation of poly(ADP-ribose) polymerase-1 (PARP-1).

Poly(ADP-ribose) polymerase, or PARP, is a DNA damage response enzyme. This enzyme facilitates the repair of DNA damage by marking locations of the damage. In the event that the damage is severe, it triggers either apoptotic or necrotic cell death depending on the severity of the damage. Under nonstressed conditions in the absence of DNA damage, PARP associates with DNA but is not catalytically active. DNA damage, most notably single-strand breaks, serves as a signal that turns on the enzyme's catalytic activities. Upon activation, PARP uses NAD^+ to synthesize and covalently attach either linear or multibranched chains of ADP ribose (ADPr) to substrate chromatin-associated proteins and to itself. The synthesis of ADP ribose polymers is an energy-consuming process. It uses a high energy bond (free energy of −8.2 kcal/mol) formed between nicotinamide and the ribose moiety of NAD^+ to catalyze the synthesis of the polymers of repeating ADPr units. These chains can have as many as 200 units. In the process, a large number of NAD^+ molecules, an essential cofactor in ATP synthesis, is consumed and energy stores are depleted. In another set of toxic actions, peroxynitrite enters mitochondria and attacks components of the electron transport chain, most notably, complex I in Parkinson's disease and Complex IV in Alzheimer's disease.

17.8 Inflammatory and Synaptic Cytokines Are Released by Microglia and Astrocytes

Inflammatory cytokines of several different types are usually secreted together to marshal needed resources required to deal with injuries and infections. Microglia and astrocytes respond in this manner to the buildup of amyloid deposits in the extracellular spaces. They are recruited to the sites of amyloid buildup and take up the unwanted material thereby acting in a protective role. Microglia secrete inflammatory mediators, that is, "there's something amiss that needs fixing so come here and help" cytokines such as IL-1β, IL-6, and TNFα and migration-stimulating chemokines such as IL-8 and MCP-1 that guide astrocytes to the site of amyloid-β buildup (Fig. 17.6(a)). As was the case for microglia, scavenger receptors expressed on the surface of the astrocytes bind the Aβ fibrils. In response to Aβ binding, the astrocytes secrete a mix of cytokines and chemokines similar to that sent out by the microglia. They also secrete the excitatory neurotransmitter glutamate and growth factors (Fig. 17.6(b)).

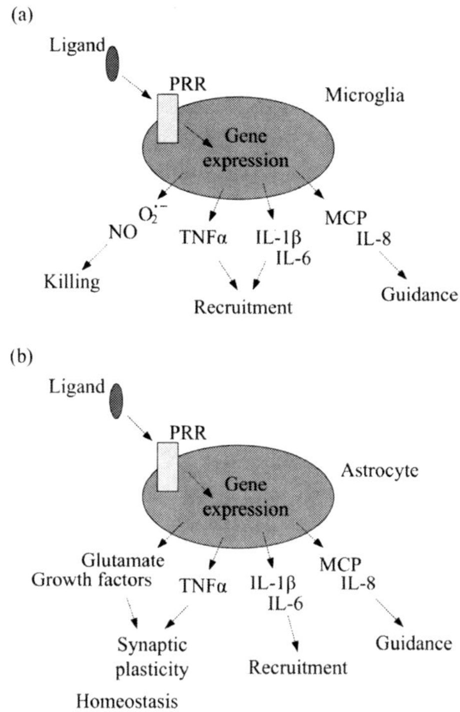

Fig. 17.6 Neuroinflammatory and synaptic actions of microglia (**a**) and astrocytes (**b**) in the healthy brain

TNFα has a dual role in the brain. In its role as a component of the body's immune system, it functions as an important pro-inflammatory cytokine. It also functions in the nervous system as a modulator of synaptic transmission. The TNF receptors can be divided into two groups according to the kind of adapter proteins that are used. Members of the first group of TNF receptors (TNFR2) have TRAF-binding motifs in their cytoplasmic domain that mediates binding to TRAFs, while members of the second group (TNFR1) have death domain (DD) motifs and bind to DD-bearing adapters. Receptors belonging to the second group are called death receptors because they convey cell suicide instructions to the recipient. Both types are expressed on hippocampal neurons.

Under normal, healthy, and noninflammatory conditions, TNFα is secreted constitutively by astrocytes and functions as a gliotransmitter. Binding of TNFα to TNFR1s on neurons triggers a signaling cascade resulting in the regulation of AMPA receptor trafficking and through that regulatory role maintains a balance between excitatory (glutamergic) and inhibitory ($GABA_A$) activity. This balancing mechanism, known as synaptic scaling, is yet another component in the regulation of neural excitability and prevention of excitotoxicity. In mediating this activity, the TNFα signaling may alter the mix of subunits comprising the active AMPA receptor by favoring the exocytosis of GluR1 subunits over that of GluR2 subunits. The Glu2 receptor subunit prevents permeation by Ca^{2+} and by favoring exocytosis of GluR1 subunits instead of GluR2s it stimulates an increased entry of calcium into the cell.

TNFα does not operate alone but instead works together with several other gliotransmitters to maintain a healthy neural environment. The arsenal of modulatory gliotransmitters includes glutamate, D-serine, and ATP in addition to TNFα. Astrocyte end feet are positioned to monitor neurotransmitter flow across the synaptic cleft and when necessary release glutamate. The glutamate binds and activates extrasynaptic NMDA receptors thereby activating a signaling pathway that relays signals to CREB that shut down excessive excitatory synaptic transmission. D-serine binds to the glycine site on NMDA receptors and functioning as a receptor coagonist regulates synaptic plasticity. ATP released by astrocytes is converted to adenosine, which then binds A1 receptors expressed on presynaptic terminals and P2X receptors on postsynaptic terminals resulting in a modulation of neural excitability.

In developing AD and other neurodegenerative situations, TNFα and glutamate secretions from microglia and astrocytes increase considerably. The resulting signaling events amplify glutamate excitotocxicity through alterations in the balance of excitation versus inhibition and changes in AMPA Ca^{2+} permeability. TNFα together with activation of NMDA receptors leads to increased Ca^{2+} entry, and also to a greater iNOS production and peroxynitrite buildup all leading to excitotoxicity and neuronal death. In addition, there is an inhibition of glial glutamate transporters on astrocytes; these transporters help maintain healthy synapses through their uptake and clearance of excess glutamate from the synaptic cleft.

17.9 Tau Hyperphosphorylation and Formation of the Tangles

Neurofibrillatory tangles, or NFTs, are intracellular aggregates of misfolded Tau proteins. In healthy neurons, Tau stabilizes microtubules by serving as ties for the microtubule rails over which cargo is transported out to the axons. In developing AD, Tau proteins dissociate from the microtubules and form the tangles. This pathological condition results in the clogging of axons and impaired transport of cargo, and contributes to synaptic failure and neural death. This impairment occurs in several other neurodegenerative diseases and not just in AD. The so-called taupathies include AD, Down's syndrome, frontotemporal dementia with parkinsonism linked to chromosome 17 (FTDP-17), and Pick's disease.

Tau proteins are natively unfolded. By assuming an extended shape, the Tau proteins are able to establish contact with multiple tubulin dimers thereby serving as rail ties. They have few hydrophobic amino acids and an excess of positively charges both properties supporting an unfolded native state. The transition from the extended native state to a more folded shape with at least some β-sheet and α helical secondary structure elements present occurs in the transition to polymeric aggregates from tubulin ties. As shown in Fig. 17.7, Tau proteins possess two hydrophobic segments, one in the N-terminal region and

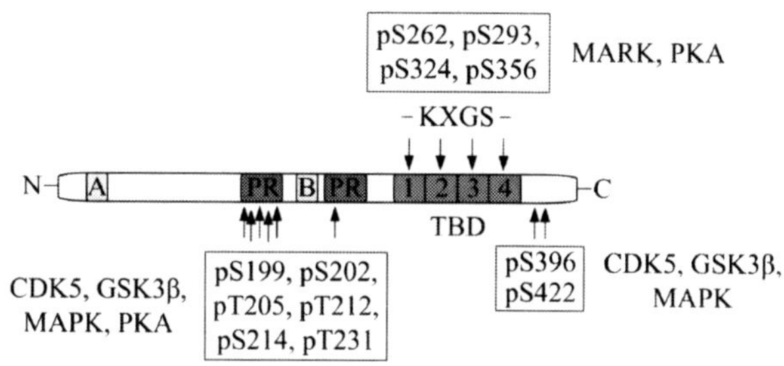

Fig. 17.7 Structure of Tau. There are six Tau isoforms in humans. Depicted is the largest one, 441 amino residues long. Regions A and B are hydrophobic; there are two proline-rich regions (PRs) and four tubulin-binding repeats (TBDs) labeled 1–4. CDK5, cyclin-dependent kinase 5; GSK3β, glycogen synthase kinase 3β; MARK, Microtubule-affinity-regulating kinase; MAPK, mitogen activated protein kinase; PKA, cAMP-dependent protein kinase

the other near the center of the sequence. It has four repeats functioning as tubulin-binding segments characterized by the presence of a KXGS motif and located in the C-terminus, and a pair of proline-rich regions in the center. The most prominent of the secondary structure elements are a type 2 polyproline helix (PPII) established through phosphorylation of serine and threonine residues located in the proline-rich regions and a β-sheet element associated with a critical hexapeptide (VQIVYK) element located in the third tubulin-binding repeat.

Tau contains multiple (80) phosphorylation sites, about 18% of the amino acid residues are either serines or threonines. Some of the most prominent of these sites are shown in Fig. 17.7 along with the attendant protein kinases. Many of these residues are phosphorylated as part of Tau's normal function as a regulator of microtubule stability and growth and as a regulator of fast axonal transport. However, in AD at least 30 serine/threonine residues are modified to produce aberrantly hyperphosphorylated tau proteins that with the addition of these extra negative charges detach from the microtubules. These aberrantly phosphorylated proteins are found in the NFTs. Two classes of kinases phosphorylate Tau – proline directed and nonproline directed. The former group includes GSK3 and the p38 and JNK MAP kinases while the latter group encompasses PKA, PKC, and CamKII. Phosphorylation of some of these sites, most notably S231 by GSK3β and S293 and S356 by MARK, substantially decreases microtubule binding. Phosphorylation at T231, S396, and S422 by CDK5, GSK3β, and/or MAPK is strongly associated with Tau paired filament formation.

The stages in the development of a taupathy are illustrated in Fig. 17.8. The addition of negative charges through phosphorylation triggers the dissociation of tau from the microtubules. This activity along with caspase-induced cleavage of N-terminal or C-terminal segments promotes Tau misfolding and oligomer-

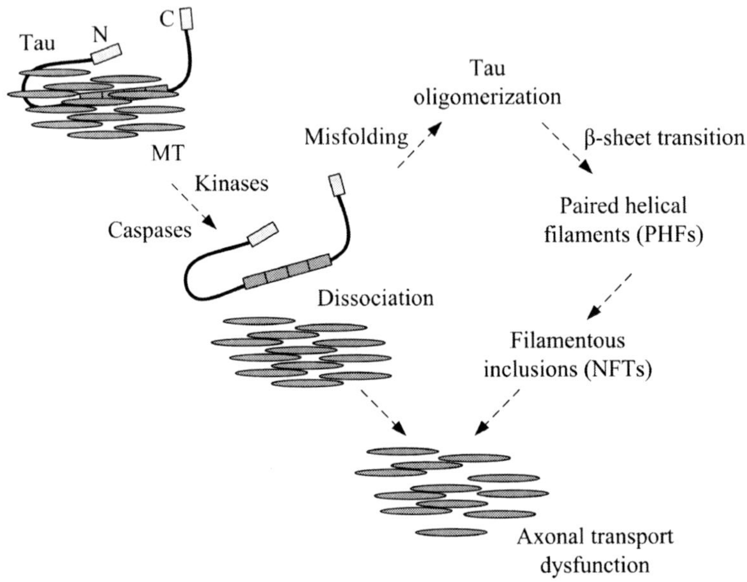

Fig. 17.8 Stages in the development of NFTs and the deleterious effects of Tau hyperphosphorylation/cleavage and NFT formation on axonal transport

ization. These stages in which small mass oligomers are generated are followed by a transition to the β-sheet secondary structure. The folding alteration from a random coil to a β-sheet form is mediated by a key sequence motif, ^{306} VQI-VYK^{311}, in the third tubulin-binding repeat that seeds the transition. This step leads to the formation of paired helical filaments and eventually to formation of the neurofibrillay tangles.

The amyloid β peptide and Tau may operate synergistically through positive feedback to drive Alzheimer's disease. Stresses brought on by oligomers and aggregates of misfolded Aβ may stimulate activity by stress kinases and caspases 3 and 6. These stresses persist; the kinases phosphorylate Tau and the caspase cleave its N- and C-terminal segments leading to Tau fragments with an increased propensity for misfolding and aggregation. The resulting malfunctions in axonal transport, including that of APP, in turn, generate further stresses and Aβ-driven damage. The small effects accumulate until the synapses fail and the neurons die.

Further Reading

Bartus RT et al. 1982 The cholinergic hypothesis of geriatric memory dysfunction. *Science* 217: 408–414.
Glenner GG and Wong 1984 Alzheimer's disease: initial report of the purification and characterization of a novel cerebrovascular amyloid protein. *Biochem. Biophys. Res. Commun.* 120: 885–890.
Hardy JA and Higgins GA 1992 Alzheimer's disease: the amyloid cascade hypothesis. *Science* 256: 184–185.
Kraepelin E 1910 *Lehrbuch der Psychiatrie*. Leipzig: Barth.
Selkoe DJ 2001 Alzheimer's disease: genes, proteins, and therapy. *Physiol. Rev.* 81: 741–766.

Amyloid β and Its Metabolism

Allinson TMJ et al. 2003 ADAMs family members as amyloid precursor protein α-secretases. *J. Neurosci. Res.*74: 342–352.
Cao X and Südhof TC 2004 Dissection of the amyloid-β precursor protein-dependent transcriptional transactivation. *J. Biol. Chem.* 279: 24601–24611.
Huovila APJ et al. 2005 Shedding light on ADAM metalloproteinases. *Trends Biochem. Sci.* 30: 413–422.
Sisoda SS and St. George-Hyslop PH 2002 γ-secretase, Notch, Aβ and Alzheimer's disease: where do the presenilins fit in? *Nat. Rev. Neurosci.* 3: 281–290.
Spasic D et al. 2006 Presenilin-1 maintains a nine-transmembrane topology throughout the secretory pathway. *J. Biol. Chem.* 281: 26569–26577.
Takasugi N et al. 2003 The role of presenilin cofactors in the γ-secretase complex. *Nature* 422: 438–441.

Amyloid β Removal

Iwata N et al. 2001 Metabolic regulation of brain Aβ by neprilysin. *Science* 292: 1550–1552.
Qiu WQ and Folstein MF 2006 Insulin, insulin-degrading enzyme and amyloid-β peptide in Alzheimer's disease: review and hypothesis. *Neurobiol. Aging* 27: 190–198.
Zlokovic BV 2004 Clearing amyloid through the blood-brain barrier. *J. Neurochem.* 89: 807–811.

Folding Physics, Metal Homeostasis, and Redox Chemistry

Bush AI 2003 The metallobiology of Alzheimer's disease. *Trends Neurosci.* 26: 207–214.
Curtain CC et al. 2001 Alzheimer's disease amyloid-β binds copper and zinc to generate an allosterically ordered membrane-penetrating structure containing superoxide dismutase-like subunits. *J. Biol. Chem.* 276: 20466–20473.
Lührs T et al. 2005 3D structure of Alzheimer's amyloid-β (1–42) fibrils. *Proc. Natl. Acad. Sci. USA*102: 17342–17347.
Quist A et al. 2005 Amyloid ion channels: a common structural link for protein-misfolding disease. *Proc. Natl. Acad. Sci. USA* 102: 10427–10432.

Amyloid β at the Synapse

Cao X and Südhof TC 2004 Dissection of the amyloid-β precursor protein-dependent transcriptional transactivation. *J. Biol. Chem.* 279: 24601–24611.
Cleary JP et al. 2005 Natural oligomers of the amyloid-β protein specifically disrupt cognitive function. *Nat. Neurosci.* 8: 79–84.
Hsieh H et al. 2006 AMPAR removal underlies Aβ-induced synaptic depression and dendritic spine loss. *Neuron* 52: 831–843.
Kamenetz F et al. 2003 APP processing and synaptic function. *Neuron* 37: 925–937.
Shankar GM et al. 2007 Natural oligomers of the Alzheimer amyloid-β protein induce reversible synapse loss by modulating an NMDA-type glutamate receptor-dependent signaling pathway. *J. Neurosci.* 27: 2866–2875.
Shankar GM et al. 2008 Amyloid-β protein dimers isolated from Alzheimer's brains impair synaptic plasticity and memory. *Nat. Med.* (in press).
Snyder EM et al. 2005 Regulation of NMDA receptor trafficking by amyloid-β. *Nat. Neurosci.* 8: 1051–1058.

Local Microenvironment

Abbott NJ, Rönnbäck L and Hansson E 2006 Astrocyte-endothelial interactions at the blood-brain barrier. *Nat. Rev. Neurosci.* 7: 41–53.
Nimmerjahn A, Kirchhoff F and Helmchen F 2005 Resting microglial cells are highly dynamic surveillants of brain parenchyma in vivo. *Science* 308: 1314–1318.
Volterra A and Meldolesi J 2005 Astrocytes, from brain glue to communication elements: the revolution continues. *Nat. Rev. Neurosci.* 6: 626–640.

Microglia and Astrocytes in AD

Akiyama H, et al. (Neuroinflammatory Working Group) 2000 Inflammation and Alzheimer's disease. *Neurobiol. Aging* 21: 383–421.
Andrabi SA et al. 2006 Poly(ADP-ribose) (PAR) polymer is a death signal. *Proc. Natl. Acad. Sci. USA* 103: 18308–18313.
Block ML, Zecca L, Hong JS 2007 Microglia-mediated neurotoxicity: uncovering the molecular mechanisms. *Nat. Rev. Neurosci.* 8: 57–69.
Combs CK et al. Inflammatory mechanisms in Alzheimer's disease: inhibition of β-amyloid-stimulated pro-inflammatory responses and neurotoxicity by PPARγ antagonists. *J. Neurosci.* 20: 558–567.
Moncada S and Bolaños JP 2006 Nitric oxide, cell bioenergetics and neurodegeneration. *J. Neurochem.* 97: 1676–1689.
Pickering M, Cumisky D and O'Conner JJ 2005 Actions of TNFα on glutamergic synaptic transmission in the central nervous system. *Exp. Physiol.* 90: 663–670.
Qin LY et al. 2002 Microglia enhance β-amyloid peptide-induced toxicity in cortical and mesencephalic neurons by producing reactive oxygen species. *J. Neurochem.* 83: 973–983.
Yu SW et al. 2006 Apoptosis-inducing factor mediates poly(ADP-ribose) (PAR) polymer-induced cell death. *Proc. Natl. Acad. Sci. USA* 103: 18314–18319.

Tau and the Tangles

Augustinack JC et al. 2002 Specific tau phosphorylation sites correlate with severity of neuronal cytopathology in Alzheimer's disease. *Acta Neuropathol.* 103: 26–35.
Avila J 2006 Tau phosphorylation and aggregation in Alzheimer's disease pathology. *FEBS Lett.* 580: 2922–2927.

Ballatore C, Lee VMY and Trojanowski JQ 2007 Tau-mediated neurodegeneration in Alzheimer's disease and related disorders. *Nat. Rev. Neurosci.* 8: 663–672.
Gamblin TC et al. 2003 Caspase cleavage of tau: linking amyloid and neurofibrillary tangles in Alzheimer's disease. *Proc. Natl. Acad. Sci. USA* 100: 10032–10037.
Mukrasch MD et al. 2007 The "jaws" of the Tau-microtubule interaction. *J. Biol. Chem.* 282: 12230–12239.
von Bergen M et al. 2001 Assembly of τ protein into Alzheimer's paired helical filaments depends on a local sequence motif (^{306}VQIVYK^{311}) forming β structure. *Proc. Natl. Acad. Sci. USA* 97: 5129–5134.

Chapter 18
Chaperones, Endoplasmic Reticulum Stress, and the Unfolded Protein Response

Failures to properly fold and dispose of damaged proteins are central to most if not all neurodegenerative disorders. The buildups discussed in the case of Alzheimer's disease, of extracellular and intracellular aggregates and toxic oligomers, are not restricted to that disorder but are a common thread running through all the major neurodegenerative disorders and minor ones too.

Newly formed polypeptide chains exit the ribosome in unfolded states. The environment surrounding them upon leaving is a crowded one. Other polypeptides are leaving ribosomes from nearby sites, and cells contain large numbers of macromolecules in various states of folding and unfolding. Hydrophobic patches exposed on the not yet folded polypeptide chains promote inappropriate clumping together of chains and aberrant folded forms. Cells rely on an extensive network of molecular chaperones that assist in folding, in refolding, and in removal of aberrantly folded proteins to maintain protein homeostasis. Chaperones are present at exit sites from ribosomes; they are found in the cytosol, and in the lumen of the endoplasmic reticulum.

Molecular chaperones serve as the first layer in a multitiered defense system in the cell that maintains protein homeostasis. This network will be examined in the first part of the chapter. If a protein becomes malformed and cannot be rescued by the chaperones it is targeted for proteolytic destruction by the 26S proteasome. The endoplasmic reticulum (ER), in particular, monitors the folding of signaling proteins destined for secretion or insertion in the plasma membrane. If ER misfolding occurs, this system directs the protein toward the proteasome. Folding in the ER will be explored in the second half of the chapter. There is a third system that back up the ubiquitin-proteasome system if it becomes overloaded or impaired for some reason. That system sequesters badly folded proteins, and ships them for lysosomal degradation in a process called autophagy. That system will be discussed in the next chapter.

M. Beckerman, *Cellular Signaling in Health and Disease*,
Biological and Medical Physics, Biomedical Engineering,
DOI 10.1007/978-0-387-98173-4_18,

18.1 The Cellular Complement of Molecular Chaperones

The three most prominent families of molecular chaperones in mammalian cells are the chaperonins, the Hsp70 family, and the Hsp90 family (Table 18.1). All of these chaperones work in concert with a number of cochaperones and with each other. Members of the HSP60 (chaperonin) family of molecular chaperones form

Table 18.1 Three families of molecular chaperones in mammalian cells

Chaperone family	Chaperone	Cochaperones
Chaperonins		
– Class I	Hsp60	Hsp10
– Class II	TriC/CCT	GimC
Hsp70	Hsc70	J-proteins (Hsp40), Hsp110, Grp170, Bag1, HspBP1 NEFs
Hsp90	Hsp90	HOP, p23, Aha, p50, Immunophilin

TRiC/CCT, TCP-1 ring complex/cytosolic chaperonin containing TCP-1; Hsc70, heat shock protein cognate 70; Grp170, glucose-regulated protein.

folding (Anfinsen) cages enabling proteins to fold and recover from folding errors in a chamber protected from the stressful cellular environment. There are two kinds of chaperonins. Type-1 chaperonins are found in mitochondria and chloroplasts, and they require cochaperones that form lids over the folding chambers. The folding chambers are built from 16 identical subunits arranged in two rings one stacked on top of the other to form "Anfinsen cages." Type-2 chaperonins such as TrRiC are found in the cytosol; they do not require a cochaperone to form a lid, and their rings are constructed from eight different subunits.

Chaperones directly participate in the folding process, supplying energy through ATP hydrolysis to the substrate protein. The additional energy supplied to the protein enables that protein to surmount energy barriers and escape from kinetic traps. The protein is able to fold far more rapidly into its native state since it can avoid getting stuck in nonoptimal and nonfunctional states for long periods of time. The subunits that form the folding cage repeatedly bind and release the substrate protein. They disrupt the misfolded structures that are kinetically trapped and release the protein in a less folded configuration through the use of mechanical stretching forces. At the end of each round of binding and release, the protein is free to continue to fold and search for its low-energy native state.

Members of the Hsp70 family of molecular chaperones assist nascent polypeptide chains in their folding. Eukaryotic members of this family bind hydrophobic patches on newly synthesized chains; they help transport proteins across organelle membranes and into the endoplasmic reticulum and mitochondria. These chaperones work in concert with Hsp60 family members, handing off the proteins to TRiC chaperones for later stages of folding and prevention of aggregation. The Hsp70 substrate binding and release cycle is involved in switching between low-affinity ATP-bound and high-affinity ADP-bound states. The cycle is regulated by J-domain proteins (JDPS), or alternatively, J-proteins as in Table 18.1, that direct the Hsp70 chaperones to their substrates, and by nucleotide exchange factors that control the duration of the substrate binding.

Members of the Hsp90 family assist in signaling by maintaining the functional integrity of proteins. These chaperones often work in association with Hsp70 family members and several other small ancillary proteins to prevent aggregation and mediate refolding. The Hsp90 chaperones form complexes with signal molecules, and then help in their translocation to the correct subcellular compartment and assembly into signaling complexes and signaling nodes. Members of the Hsp90 family of molecular chaperones are among the most abundant proteins in the cell. They account for 1–2% of the total cellular protein even under nonstressful conditions. Proteins targeted by Hsp90 typically have low-energy conformations that are only marginally stable. These signaling proteins have difficulty in remaining in their physiologically competent states and require the assistance of Hsp90 to stabilize them.

18.2 Hsp70 Structure and Function

Heat shock protein 70 (Hsp70) assists newly synthesized proteins in their folding to the native stage; it prevents aggregation and, if aggregation occurs, it helps solubilize and refold the clumped proteins. Cochaperones belonging to the JDP family assist in the cycle and together with the substrate lower the activation energy for hydrolysis. The cochaperone assists in substrate binding by Hsp70. Once bound the substrate and cochaperone stimulate hydrolysis of ATP leading to the closing of the substrate binding cavity about the substrate. In the next step, the ADP molecule is released to be rapidly replaced by another ATP molecule leading to opening of the substrate binding cavity, and the cycle is ready to be repeated.

Hsp70 possesses two domains, an N-terminal ATPase domain and a C-terminal substrate binding domain. The three-dimensional structure of the ATPase domain reveals its organization into two globular subdomains, I and II, each of which are further partitioned into A and B regions (Fig. 18.1). This arrangement creates four distinct lobes that surround and bind the ATP molecule along with an Mg^{2+} ion and a pair of K^{+} ions. The C-terminal domain contains two four-stranded β-sandwiches possessing a set of loops that angle upward and along with the β-sheets bind the substrate. The cavity formed by these structures and the two nearby α-helices, together with hydrophobic residues that line the cavity, is able to recognize exposed hydrophobic patches on a wide variety of misfolded proteins.

There are three families of Hsp70-related proteins. The Hsps discussed above belong to the DnaK subfamily, so named because they are found in all

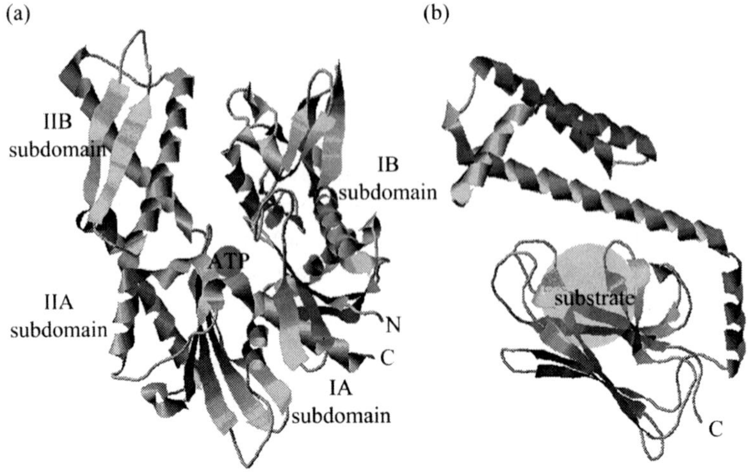

Fig. 18.1 Three-dimensional structure of heat-shock protein 70 (Hsp70). (**a**) N-terminal ATPase domain from bovine Hsc70; (**b**) C-terminal substrate-binding domain from *E. coli* DnaK. The figure was prepared using Jmol with atomic coordinates deposited in the PDB under accession numbers 1hpm (**a**) and 1dkx (**b**)

kingdoms of life including the bacterial DnaKs. A second family of Hsp70-like chaperones includes Hsp110. This subfamily of chaperones is found exclusively in eukaryotes. These proteins are similar to Hsp70, the main difference being the addition of a segment situated in-between the β-sandwich peptide-binding motif and the α-helical lid domain, and an expanded C-terminal domain. These proteins function as holdases rather than as foldases. The third subfamily of Hsp70s is the 170-kDa glucose-related proteins, or Grp170 subfamily, found in the ER. Like other ER chaperones, these proteins are upregulated by ER stresses. They too have an extended segment situated in-between the β-sandwich peptide-binding motif and the α-helical lid domain. Members of the Hsp110 and Grp170 subfamilies are designed to bind and hold partially unfolded polypeptide chains. The Hsp110s assist Hsp70s in the cytosol while the Grp170s assist the ER stress-response chaperone BiP/Grp78 (to be discussed later in this chapter).

18.3 Hsp90 Structure and Function

The Hsp90 chaperones associate with a number of cochaperones and operate as homodimers. Shown below in Fig. 18.2 is the structure of Hsp90 in a dimeric complex with its cochaperone p23.

The Hsp90 molecular chaperone contains three domains, an N-terminal ATP-binding domain, a middle domain that participates in ATP and client protein binding, and a C-terminal dimerization domain (Fig. 18.2(a)). In its

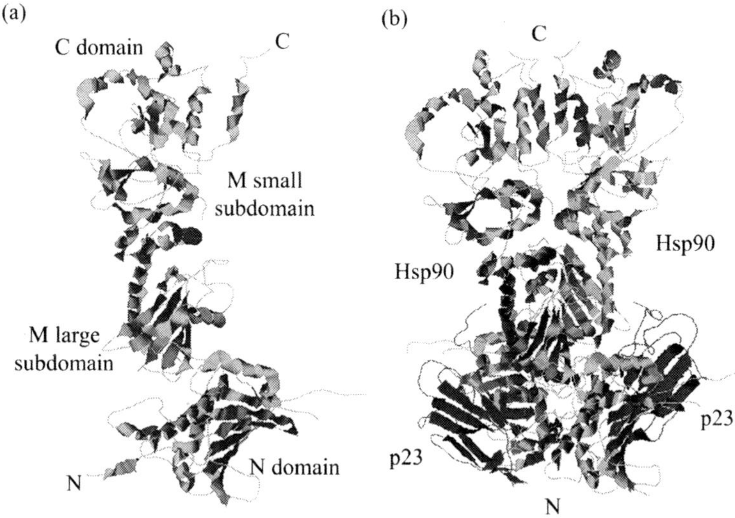

Fig. 18.2 Three-dimensional structure of heat-shock protein 90 (Hsp90) in a dimeric complex with its cochaperone p23. (**a**) Hsp90 monomer illustrating the positions of the N-terminal domain (N domain), large and small middle (M large and M small) subdomains, and C-terminal domain (C domain). (**b**) Hsp90-p23 dimer. The figure was prepared using Jmol with atomic coordinates deposited in the PDB under accession number 2cg9

active form, the Hsp90 molecular chaperone dimerizes and associates sequentially with a number of cochaperones. Both Hsp90 and Hsp70 possess EEVD sequences in their N-terminal. These motifs serve as binding sites for cochaperones possessing a tetratricopeptide repeat (TPR) clamp that recognizes the EEVD motif. The cochaperone Hsp70–Hsp90 organizing protein (HOP) binds this sequence, and as its name suggests mediates the folding handoff from Hsp70 and Hsp40 chaperoning to Hsp90 and its cohorts. The p23 cochaperone acts later in the Hsp90 ATPase cycle than HOP. As shown in Fig. 18.2(b), p23 binds to a site in the C-terminal domain while another cochaperone, activator of Hsp90 ATP 1 (Aha1) that participates still later in the cycle, binds to a site in the middle domain.

The stages of the Hsp90 ATP hydrolysis cycle are depicted schematically in Fig. 18.3. The cycle begins with the handoff of the proteins substrate (referred to in the preceding paragraph as the "client") from the hsp70/hsp40 system. As mentioned above the cochaperone HOP mediates the transfer. The client binds in a cleft formed by the Hsp90 dimer. In the next stage, the p23 cochaperone along with an immunophilin displaces HOP and binds to the hsp90 dimer. This stage along with ATP binding induces large conformational changes in the chaperone complex leading to its closed conformation. In the third stage of the cycle, the cochaperone Aha1 replaces p23 binding and stimulates hydrolysis leading to opening of the dimer and release of the client and cochaperones. The chaperones and client are then set for another round of binding and release.

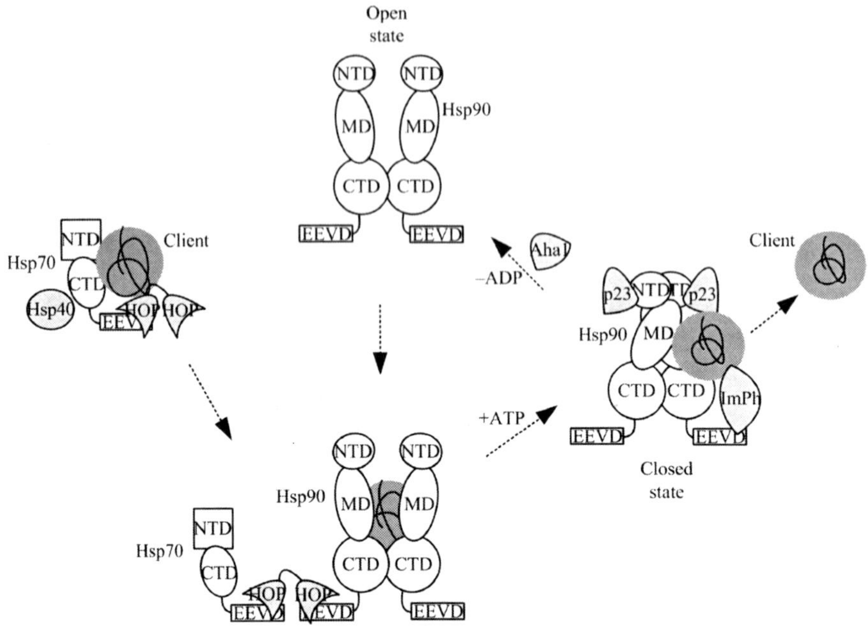

Fig. 18.3 The ATP-dependent Hsp90 chaperone cycle. The cycle begins with Hsp70 and Hsp40 binding. This chaperone duo hands off the client protein to Hsp90. Aided by the Hsp90 cochaperones (Table 18.1), the client-bound Hsp90 undergoes a series of conformational changes leading to the closed state and then to the open state in which the client is released. If folding is not completed, the client undergoes another cycle of binding and release

18.4 Heat Shock Factor 1 Is a Master Regulator of Protein Homeostasis

Heat shock factor 1 (HSF1) is the guardian of the proteome. It is a transcription factor with a domain organization consisting of an N-terminal winged helix-loop-helix binding domain along with nuclear localization signal sequences and a C-terminal transcription activation domain (Fig. 18.4). Three sets of hydrophobic repeats provide binding sites for converging activating and repressing signals, while further regulatory activities are mediated by the presence of several serine phosphorylation sites.

Under low-stress conditions, HSF1 is kept in an inactive monomeric state through binding to Hsp90 and its co-chaperones p23 and immunophilin. It detaches from its Hsp90 negative regulators in response to the buildup of ubiquitinated proteins indicative of protein folding-related stresses. Once freed from its sequestration eEF1A along with heat shock RNA 1 (HSR1) stimulates HSF1 relocation to the nucleus, trimerization, and binding to DNA heat-shock response elements (HREs) in heat-shock promoters. In unstressed cells, eEF1A is tied to translation and cytoskeleton management.

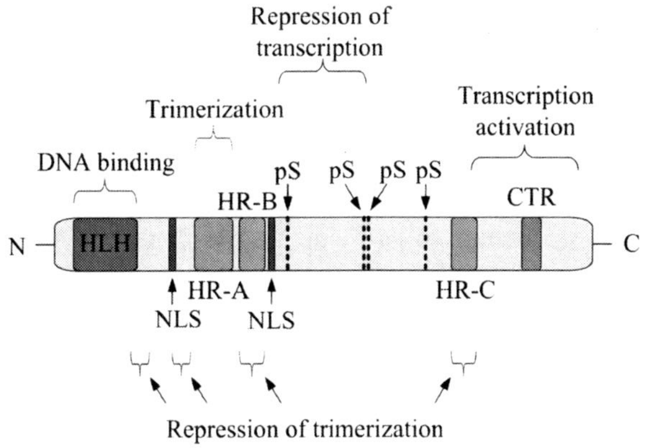

Fig. 18.4 Domain organization of the heat-shock factor 1 (HSF1) protein. Abbreviations: HLH, winged helix-loop-helix; NLS, nuclear localization signal; HR, hydrophobic repeat; CTR, C-terminal repeat of vertebrate Hsf1/Hsf2; pS, phosphorylation sites at Ser230, Ser303, Ser307, and Ser363

When translation is shut down in response to protein folding-related stresses, eEF1A becomes available to interact with HSF1. The second partner in this activity, HSR1, is not completely understood at this point of time. It is believed to function in some fashion as a sensor of certain stresses and becomes activated through conformational changes in their presence. Once at promoters, additional stress-related factors such as ROS may potentiate transcription by extending its strength and duration, and when necessary this transcription factor can trigger a massive upregulation of heat-shock proteins.

18.5 Folding, Processing, and Maturation of Membrane and Secreted Proteins

The overall organization of a folded membrane protein is remarkably similar to that of a cytosolic protein – the interior is nonpolar and tightly packed, and composed largely of hydrogen-bonded α-helices and β-sheets. These secondary structure elements span the hydrocarbon core of the membrane bilayer. As a result, these elements tend to be longer than their cytosolic counterparts. Also, their outer surfaces, which are in contact with the core, are more hydrophobic.

The process of folding membrane and secreted proteins begins in mammals with the transport of nascent polypeptide chains exiting the ribosome across the ER membrane. The translocation of the chain is mediated by the ribosome-translocon complex, which forms an opening in the membrane to allow for insertion of the polypeptide chain. The proteins are inserted into ER membrane by this complex, and then folded and processed in a series of stages starting in the endoplasmic reticulum and continuing in the Golgi apparatus.

Newly synthesized proteins are processed, subjected to quality control with respect to their folding, and then shipped to their cellular destinations. Prosthetic groups – sugars and lipids – are added to proteins destined for insertion in membrane to enable them to attach to the membranes. These modifications are made subsequent to translation, in several stages, as the proteins pass through the ER and Golgi apparatus. The overall process resembles an assembly line that builds up the proteins, folds them, inserts them into membranes, sorts them, labels them with targeting sequences, and ships them out to their cellular destinations (Fig. 18.5).

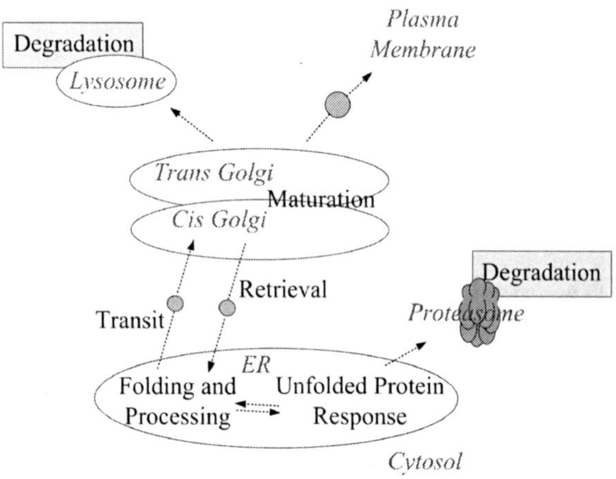

Fig. 18.5 Schematic view of the folding, processing, and maturation of membrane and secreted proteins. This series of processing stages includes degradation pathways that eliminate malformed proteins

The Golgi apparatus consists of a stacked system of membrane-enclosed sacs called cisternae. Some of the polysaccharide modifications needed to make glycoproteins are either made or started in the rough ER. Proteins, especially signaling proteins destined for export (secretion) from the cell or for insertion into the plasma membrane, are sent from the rough ER to the smooth ER where they are encapsulated into transport vesicles pinched off from the smooth ER. The transport vesicles are then sent to the Golgi for further processing and eventual shipping to their cellular destinations. The Golgi apparatus takes the carbohydrates and attaches them as oligosaccharide side chains to some of these proteins to form glycoproteins and to complete modifications started in the rough ER. Both proteins and lipids are modified in the Golgi. Other proteins, synthesized as inactive precursor molecules, are processed to produce activated forms in the Golgi. Modified proteins are enclosed in transport vesicles, pinched off from the Golgi, and shipped to destinations such as the plasma membrane and the extracellular matrix.

One of the types of protein modification made in the ER is the formation of disulfide bonds. The process of folding proteins and forming disulfide bonds in the ER is referred to as oxidative protein folding. Disulfide bond formation requires a redox reaction and an electron acceptor. The key players in this step are Ero1p and protein disulfide isomerase (Pdi1p). In this process illustrated in Fig. 18.6, Ero1p uses a flavin-dependent reaction to pass electrons to molecular oxygen. In a series of disulfide exchange reactions, Ero1p oxidizes Pdi1p, which then oxidizes the substrate protein.

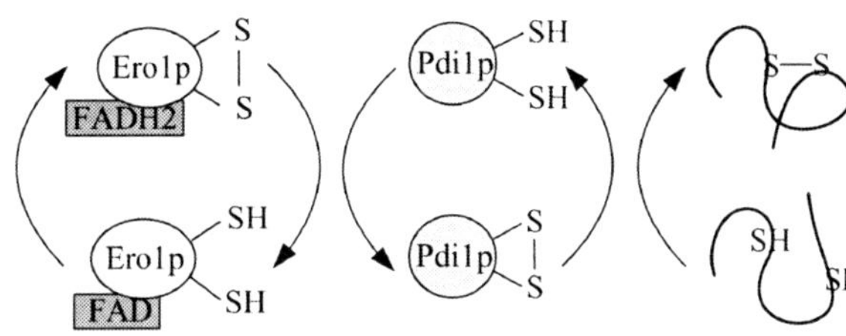

Fig. 18.6 Oxidative protein folding. In a series of oxidation reactions, the substrate protein is converted from a reduced (SH) to an oxidized (S) form

18.6 N-Linked Glycan Processing

Most proteins destined for insertion in the plasma membrane contain covalently linked oligosaccharides that extend out from their extracellular side. These proteins are referred to as glycoproteins. Their posttranslational modifications are started in the ER and finished in the Golgi apparatus. There are two forms of modification, N-linked and O-linked. In N-linked glycoproteins, a carbohydrate, or glycan, is added to a side chain NH_2 group of an asparagine amino acid residue. In O-linked glycoproteins, the oligosaccharide chain is appended to a side chain hydroxyl group of a serine or threonine amino acid residue.

Carbohydrates are hydrophilic so addition of the N-linked glycans increases solubility of the proteins; and their additions influence protein folding and stability of the proteins. In addition, they facilitate intracellular transport and targeting and finally they assist in the signaling roles of the mature proteins. In the first step of the process, a N-glycan unit is added to the newly synthesized protein. This module has a branched oligosaccharide structure consisting of three glucoses, nine mannoses, and two *N*-acetyl-glucosamines ($Glc_3Man_9GlcNAc_2$) arranged in the manner shown in Fig. 18.7. This structure is then trimmed to the needed final form.

Proteins involved in the trimming and folding operations of N-glycans are modified by ER-resident glycanases glucosidase I (GI) and glucosidase II (GII), and these operations are assisted by two molecular chaperones, calnexin (CNX) and calreticulin (CRT), which interact with the newly synthesized glycoproteins. In the first step, the oligosaccharide $Glc_3Man_9GlcNAc_2$ is attached to asparagine residues containing the consensus sequence Asn-Xxx-Ser/Thr. The newly synthesized polypeptide chain is scanned by a multisubunit complex known as oligosaccharyltransferase (OST) for the consensus sequences. The OST complex then catalyzes the transfer of the core sugar block in a step-by-step manner to create the branched structure depicted in Fig. 18.7.

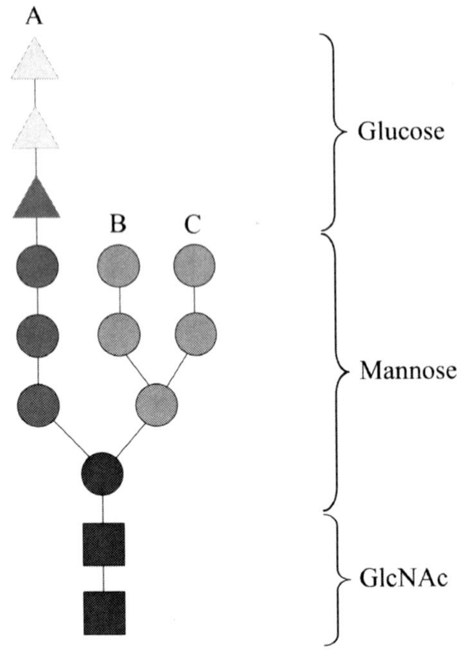

Fig. 18.7 N-linked oligosaccharide unit containing 14 sugars – three glucoses, 9 mannoses, and two *N*-acetyl-glucosamines arranged into a three-branched (A, B, C) structure

Simultaneous trimming and folding occur next. In the first step, Glucosidases I and II (GI and GII) trim two of the glucose residues residing on branch A. The modified protein chain then enters the so-called calnexin-calreticulin (CNX/CRT) cycle (Fig. 18.8). These ER chaperones bind to the $Glc_1Man_9GlcNAc_2$ unit. Folding including disulfide bond formation is followed by dissociation. If the protein has folded correctly, it is packaged in a transport vesicle and

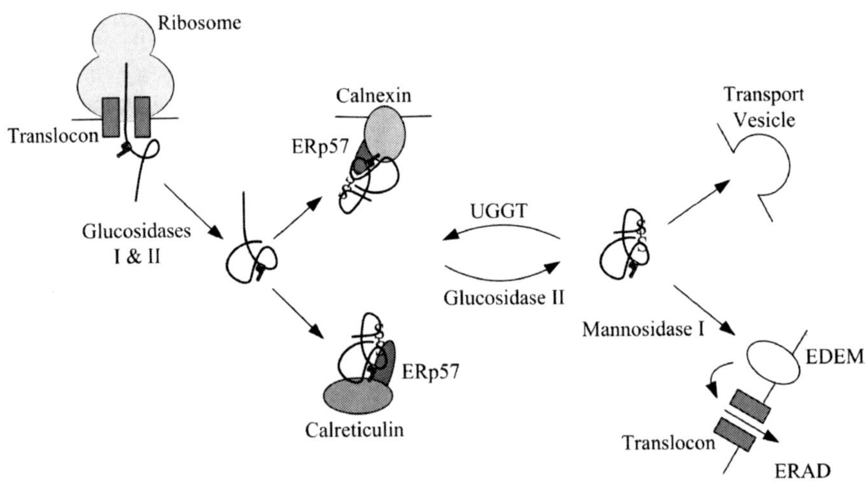

Fig. 18.8 The calnexin-calreticulin folding cycle

sent to the Golgi. If the protein has not folded correctly, the protein chain is reglycosylated by uridine diphosphate (UDP)-glucose: glycoprotein glucosyl transferase (UGGT), which functions as a folding sensor and which adds back sugars permitting reassociation and folding.

18.7 The Unfolded Protein Response

The unfolded proteins response (UPR) is active in cells that secrete large numbers of proteins. A prominent example is the pancreatic β cell that secretes insulin. The continued secretion of signaling proteins places a burden on the endoplasmic reticulum because of the high workload and because this organelle is responsible for preparing proteins for exposure to the extracellular milieu. Such proteins utilize disulfide bond formation and glycosylation, and these finishing or maturation operations are carried out in the ER's membraneous network. Failure to carry out these functions adequately because of a buildup in unfolded, misfolded, and/or unfinished proteins is termed ER stress.

In order to deal with changing numbers of proteins requiring folding assistance from the ER, signals are sent to both the transcription and translation machinery to coordinate their behavior. These actions stimulate increases in the production of chaperones to deal with an increased load, and slow protein synthesis to avoid excessive accumulation of unfolded proteins in the ER while the transcription machinery is accessed and produces the additional chaperone mRNAs.

The UPR involves several activities. Specifically, (1) signals are sent to increase expression of genes involved in ERAD to remove un- and misfolded proteins either to the ubiquitin-proteasome system (UPS) (to be discussed later in this chapter) or by means of autophagy; (2) signals are sent to the transcriptional machinery that upregulate genes involved in alleviating the stresses such as chaperones and antioxidants; (3) signals are sent to the translational machinery in order to reduce the folding burden placed on the organelle; and (4) if the stresses are severe and persistent so that protein homeostasis cannot be restored, signals are sent to trigger apoptosis leading to destruction of the overburdened cell.

The above-mentioned recovery measures are orchestrated beautifully by the three sensors – inositol requiring enzyme 1 (IRE1), PKR-like ER kinase (PERK), and activating transcription factor 6 (ATF6). In the absence of ER stresses, these proteins are inactive residents of the ER membrane. Once they sense excessive stresses, they coordinately mediate the four sets of unfolded protein responses. The domain organizations of the three sensors are depicted in Fig. 18.9. IRE1 and PERK are Type 1 transmembrane proteins; their N-terminals reside in the ER lumen and their C-terminals in the cytosol or nucleoplasm. The third sensor and responder, ATF6, is a Type 2 transmembrane protein; its N-terminal is situated in the cytosol or nucleoplasm and its C-terminal extends out into the ER lumen.

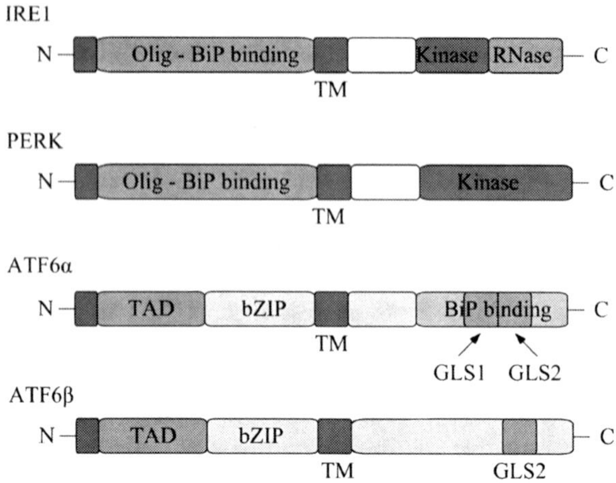

Fig. 18.9 Domain structure of the three sensors – IRE1, PERK and ATF6 (α and β isoforms)

The luminal domains of these proteins function as an ER stress sensor. As indicated in Fig. 18.9, one of the functions of these domains is to bind BiP, a resident ER chaperone belonging to the Hsp70 family. This protein has several roles in the ER. One of these is to seal the Sec61 translocon against unwanted entry of ions. A second is as a master regulator of IRE1, PERK, and ATF6. BiP binds unfolded proteins. When these are not present in significant amounts, BiP is able to bind the three sensors and maintain them in an inactive state. This inhibition is relieved by a buildup in unfolded proteins. These become bound by BiP, which then no longer binds and inhibits the sensors.

The IRE1 protein is a single-pass resident of the ER membrane containing three domains. Its N-terminal luminal domain (LD) is followed by a transmembrane segment, a cytosolic serine-threonine kinase domain, and a cytosolic C-terminal RNase domain. Direct detection of unfolded proteins by the LD is accompanied by dimer- and oligomerization, autophosphorylation, and activation of the RNase domain, which is then able to recruit its downstream effector, the mRNA for Hac1 in the case of yeast and X-box binding protein 1 (XBP1) in mammals. This messenger RNA molecule is constitutively expressed but remains untranslated due to the presence of an inhibitory intron. These mRNAs are cleaved in two locations by IRE1 resulting in removal of the inhibitory intron and formation of a translationally competent mRNA. Once translated the Hac1/XBP1 proteins translocate to the nucleus where they function as transcription factor that upregulates UPR target genes.

Once activated by the presence of unfolded proteins, PERK phosphorylates the α subunit of the translation initiation factor eIF2 on Ser51 resulting in inhibition of eIF2B. This action has two consequences. The first is that it transiently attenuates global mRNA translation. The second action is that it promotes translation of a small subset of mRNAs, most notably, the ATF4 mRNA resulting in an increased production of the ATF4 transcription factor.

The ATF4 transcription factor then activates transcription of genes involved in amino acid metabolism and transport, in antioxidant factors and chaperones, and the transcription factor CHOP, a key inducer of apoptosis. PERK also phosphorylates Nrf2 leading to its dissociation from KEAP, which is then able to translocate to the nucleus and stimulate transcription of phase II detoxification enzyme encoding genes.

The third sensor, ATF6, is activated through regulated intramembrane proteolysis. Under high levels of ER stress, ATF6 is transported to the Golgi where it is proteolytically cleaved twice to free the cytoplasmic/nucleoplasmic portion. It undergoes regulated intramembrane proteolysis (RIP) by the same proteases, site-1 protease (S1P) and site-2 protease (S2P), that activate the cholesterol sensor SREBP. This freed segment then translocates to the nucleus where it binds to cAMP response elements (CRE) and ER stress-response elements I and II (ERSE I and II) and stimulates transcription of genes encoding CHOP, ER chaperones, and folding enzymes. Translocation to the Golgi is mediated by two Golgi localization sequences, GLS1 and GLS2, and through association with the lectin CRT.

The three ER stress response pathways coordinately regulate cellular metabolism in their maintenance of protein homeostasis. They are able to sense excessive metabolic load and impaired redox status through their effects on protein folding. They continually adjust protein, amino acid, lipid, and membrane metabolism. In addition, they activate central inflammatory responses through their regulation of NF-κB and JNK signaling. These activities may well have a crucial role in the metabolic syndromes. For example, obesity-related buildup of cholesterol and fatty acids in the ER leads to activation of XBP1. This transcription factor then increases membrane biogenesis through its ability to upregulate phospholipid synthesis and restore a healthy phospholipid-to-cholesterol ratio. It further acts to maintain the correct ratio by antagonizing SREBP and reducing cholesterol synthesis. In adipose tissue and liver subjected to obesity-related conditions, IRE1 signals to the JNK pathway which as discussed in Part I can phosphorylate the IRS proteins and contribute to insulin resistance. Furthermore, IRE1 and also PERK can activate NF-κB and regulate inflammation through their varied effects on the IκB inhibitory proteins.

Cells handle protein overload in multiple ways. One of these, a rapid means of relieving stress, is called cotranslocational degradation. The purpose of cotranslocational degradation is to increase the decay rate of mRNAs thereby reducing the load placed on the ER. This can be accomplished by means of IRE1's RNase activity by not only targeting XBP1 but also other membrane- and ribosome-associated mRNAs. Another way of achieving this is through the actions of a DnaJ family cochaperone called $p58^{IPK}$, which recruits HSP70 to the cytosolic opening in the ER translocon, which then assists in the extraction of the misfolded or stalled chain from the ribosomal machinery and directs it toward proteosomal destruction.

18.8 ERAD and the Sec61 Translocon

Endoplasmic reticulum-associated protein degradation, or ERAD, is the process whereby misfolded proteins are extracted from the ER, ubiquitinated, and degraded by the 26S proteasome. The first step is the removal of misfolded proteins and unassembled protein chains from the repeating calnexin/calreticulin cycle. Referring back to Fig. 18.8, this step is mediated by α-mannosidase I, an enzyme that trims the protein to a Man_8 form thereby reducing the affinity of UGGT for the protein and promoting binding and sequestration by α-mannosidase I-like protein EDEM.

The primary components of the ERAD machinery are (i) ER residents that detect misfolded and damaged proteins and direct them to protein channels for passage out of the ER into the cytosol; (ii) ER channel proteins such as Sec61 and associated factors that allow passage through the ER membrane; and (iii) cytosol escort proteins that chaperone the damaged proteins from the ER membrane to the 26S proteasome for degradation.

The Sec61 translocon, the key element in the extraction and translocation machinery, is a heterotrimeric protein-conducting channel. Unlike channels that allow passage of small molecules and ions across lipid membranes, the protein-conducting channel both enables passage from one side of the channel to the other and insertion of the protein into the lipid bilayer. The Sec61 translocon is a heterotrimeric assembly. The largest component is the α-subunit containing 10 transmembrane helices. The other two chains, β and γ, are far smaller and contain just one membrane-spanning helix each. The structure of SecY, the bacterial and archaeal homolog of Sec61, illustrating the organization of these three subunits is presented in Fig. 18.10. The protein forms an hourglass-shaped, water-filled channel with a plug formed in its center. The two halves of the hourglass are separated by a ring of hydrophobic residues that form a collar while a gate for lateral movement of the protein chains into the lipid bilayer is shaped by two of the transmembrane helices.

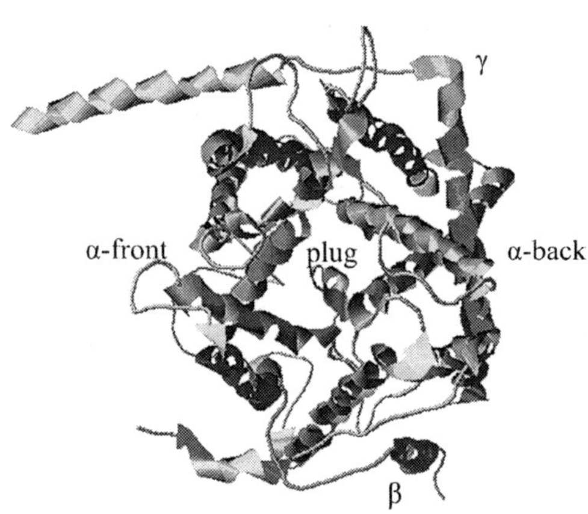

Fig. 18.10 The SecY protein-conducting channel from the archaeon *Methanococcus jannaschii* at 3.2Å. The figure was prepared using Jmol with atomic coordinates deposited in the PDB under accession number 1rhz

18.9 The p97 Motor Protein Is a Molecular Chaperone Required for ERAD

Ubiquitinated proteins undergoing removal from the ER membrane and transport to the 26S proteasome for degradation are chaperoned by two complexes. The first of these complexes to act is the p97-Ufd1-Npl4 complex, and the second is the Rad23-Dsk2 complex. The p97 motor protein is also known as valosin-containing protein (VCP) and in yeast it is referred to as Cdc48. This abundant and versatile protein belongs to the ATPase associated with various cellular activities, or AAA, family of ATPases. These proteins are involved in cellular processes ranging from membrane fusion to DNA replication to proteolysis and degradation.

Their defining feature of p97 is the presence of one or two ATPase domains, approximately 230–250 amino acid residues in size along with a sequence called a second region of homology (SRH) motif and Walker A and B motifs. The p97 structure is presented in Fig. 18.11. As can be seen, the protein has an N domain and two ATPase domains. The N domain consists of two subdomains, Na and

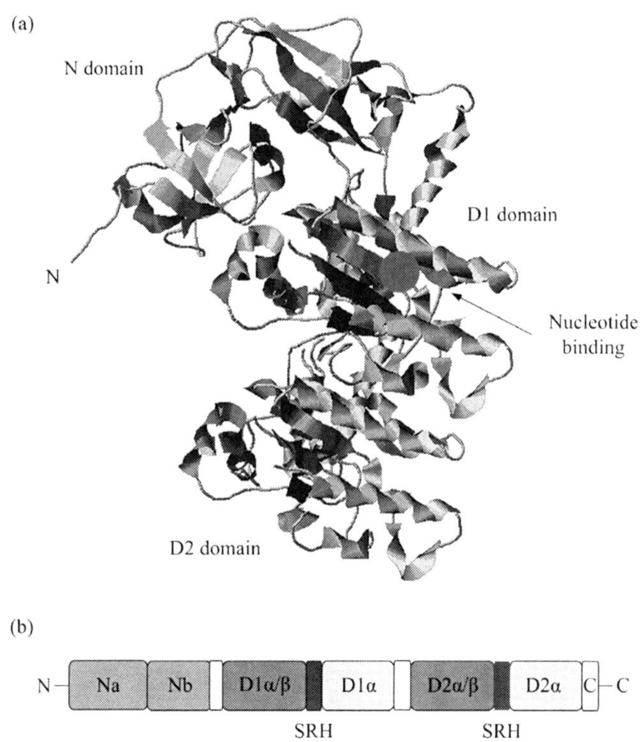

Fig. 18.11 Structure of the p97 monomer. The figure was prepared using Jmol with atomic coordinates deposited in the PDB under accession number 1r7r

Nb. These structures serve as p97's primary interface to its binding partners. These subdomains are linked to the D1 and D2 domains that lie C-terminal to it. Each of the two ATPase domains, D1 and D2, consist of a subdomain that assumes an α/β fold and contains the Walker A and B nucleotide binding and hydrolysis motifs, and a subdomain that is α-helical in form. The α/β and α subdomains are separated from one another by the SRH motif.

The p97 motor proteins assemble into homohexameric rings. This arrangement enables the enzyme to use the energy of ATP hydrolysis to unfold and extract proteins from membranes, unwind DNA, and disassemble proteins complexes. The homohexameric ring contains a small pore in its center. In taking this form, the two ATPase domains are stacked one on top of the other. The sixfold symmetric arrangement of p97 subunits is depicted in Fig. 18.12. In this figure, the N-terminal domain binds the p47 adapter module through interactions with its ubiquitin-binding regulatory X (UBX) domain. The p97 protein has a variety of binding partners and the particular selection of binding partner to a large extent determines its particular cellular role. In mediating the extraction of proteins from the ER and chaperoning their passage to the 26S proteasome during ERAD p97 binds Ufd1 and Npl4.

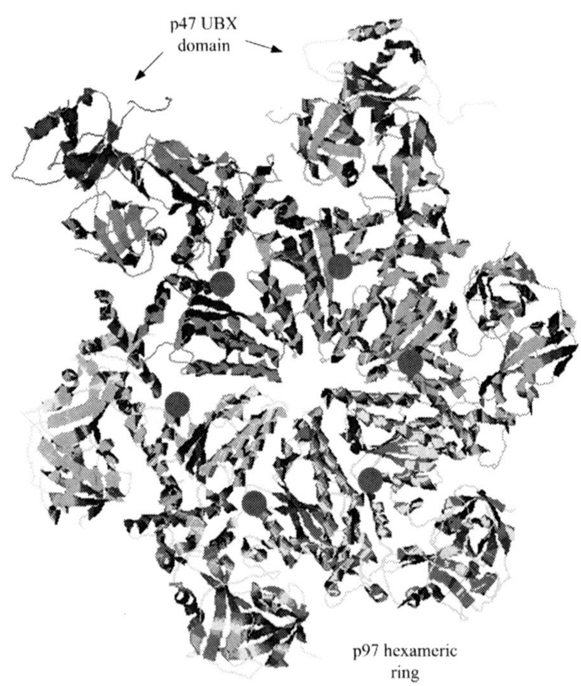

Fig. 18.12 The p97 molecular machine complexed with p47 UBX domain. Shown are p47 UBX together with pore-containing ring structure formed by the p97 N and D1 domains. The figure was prepared using Jmol with atomic coordinates deposited in the PDB under accession number 1s3s

Repetitive cycles of ATP hydrolysis are used by these molecular machines such as p97 to generate mechanical pulling forces that extract misfolded proteins from the ER and unfold them. The rate of unfolding depends upon the

ATPase hydrolysis rate and the number of residues that can be pulled through the machine for each ATP molecule hydrolyzed. The process occurs in a sequential and stochastic annealing-like manner. The unfolding rate depends at least in part upon unfolding energy landscape of the substrate; some proteins can be effectively unfolded by a particular AAA ATPase while others cannot. Instead, these others will refold and escape because of a too slow pulling rate, and as a result will require a different enzyme.

As noted earlier, p97 interacts with a number of different proteins through its N-terminal domain. In its role as an ERAD chaperone, p97 forms a complex with the Ufd1 and Npl4 heterodimers. These proteins belong to the UBX family of substrate-recruiting cofactors. Another component of this complex is Ufd2. This latter member is an E4 ubiquitin-elongating factor. This escort complex then interacts with a pair of polyubiquitin-binding enzymes which in yeast are called Dsk2p and Rad23p. These proteins possess N-terminal ubiquitin-like domains that bind the 19S cap of the proteasome, and these proteins mediate the handoff to the proteasome.

Two additional factors bind p97 and have a role in proteasomal targeting. One of these is named phospholipase A2 activating protein (PLAA) in humans and Doa1/Ufd3 in yeast. The Doa1 protein possesses three domains, an N-terminal WD40 domain, a centrally situated ubiquitin binding domain, and a C-terminal Cdc48/p97 binding domain. It and the second additional factor, the deubiquitinating enzyme Otu1, are involved in maintaining ubiquitin homeostasis in the cell. The Out1 protein targets polyubiquitinated K48-linked ubiquitin chains. It hydrolyzes isopeptide bonds formed between the C-terminus of one ubiquitin and a lysine residue of the other. The protein contains an N-terminal UBX domain that interfaces with Chc48 and a catalytic domain.

Further Reading

Aberrant Protein Folding and Disease

Bucciantini M et al. 2002 Inherent toxicity of aggregates implies a common mechanism for proteins misfolding diseases. *Nature* 416: 507–511.

Dobson CM 2001 The structural basis of protein folding and its link with human disease. *Phil. Trans. R. Soc. Lond. B* 356: 133–145.

Haass C and Selkoe DJ 2007 Soluble protein oligomers in neurodegeneration: lessons from the Alzheimer's amyloid β-peptide. *Nat. Rev. Mol. Cell Biol.* 8: 101–112.

Hsp70

Easton DP, Kaneko Y and Subjeck JR 2000 The Hsp110 and Grp170 stress proteins: newly recognized relatives of the Hsp70s. *Cell Stress Chaperones* 5: 276–290.

Mayer MP and Bukau B 2005 Hsp70 chaperones: cellular functions and molecular mechanism. *CMLS Cell. Mol. Life Sci.* 62: 670–684.

Hsp90

Ali MMU et al. 2006 Crystal structure of an Hsp90-nucleotide-p23/Sba1 closed chaperone complex. *Nature* 440: 1013–1017.
Shiau AK et al. 2006 Structural analysis of *E. coli* Hsp90 reveals dramatic nucleotide-dependent conformational changes. *Cell* 127: 329–340.

Hsf1

Pirkkala L, Nykänen P and Sistonen L 2001 Roles of the heat shock transcription factors in regulation of the heat shock response and beyond. *FASEB J.* 15: 1118–1131.
Shamovsky I et al. 2006 RNA-mediated response to heat shock in mammalian cells. *Nature* 440: 556–560.
Voellmy R 2004 On mechanisms that control heat shock transcription factor activity in metazoan cells. *Cell Stress Chap.* 9: 122–133.

Chaperones

Miller EJ, Meyer AS and Frydman J (2006) Modeling of possible subunit arrangements in the eukaryotic chaperonin TRiC. *Prot. Sci.* 15: 1522–1526.
Pratt WB and Toft DO 2003 Regulation of signaling protein function and trafficking by the Hsp90/Hsp70-based chaperone machinery. *Exp. Biol. Med. (Maywood)* 228: 111–133.
Young JC et al. 2004 Pathways of chaperone-mediated protein folding in the cytosol. *Nat. Rev. Mol. Cell Biol.* 5: 781–791.

Membrane and Secreted Proteins

Schnell DJ and Hebert DN 2003 Protein translocons: multifunctional mediators of protein translocation across membranes. *Cell* 112: 491–505.
Tu BB and Weissman JS 2002 The FAD and O_2-dependent reaction cycle of Ero1-mediated oxidative protein folding in the endoplasmic reticulum. *Mol. Cell* 10: 983–994.
Tu BB and Weissman JS 2004 Oxidative protein folding in eukaryotes: mechanisms and consequences. *J. Cell Biol.* 164: 341–346.
White SH and Wimley WC 1999 Membrane protein folding and stability: physical principles. *Annu. Rev. Biophys. Biomol. Struct.* 28: 319–365.

N-Linked Glycans

Helenius A and Aebi M 2001 Intracellular functions of N-linked glycans. *Science* 291: 2364–2369.
Ruddock LW and Molinari M 2006 N-glycan processing in ER quality control. *J. Cell Sci.* 119: 4373–4380.

The UPR

Aragon T et al. 2009 Messenger RNA targeting to endoplasmic reticulum stress signaling sites. *Nature* 457: 736–740.
Gregor MF and Hotamisligil GS 2007 Adipocyte stress: the endoplasmic reticulum and metabolic disease. *J. Lipid Res.* 48: 1905–1914.
Hetz C et al. 2006 Proapoptotic BAX and BAK modulate the unfolded protein response by a direct interaction with IRE1α. *Science* 312: 572–576.
Hollien J and Weissman JS 2006 Decay of endoplasmic reticulum-localized mRNAs during the unfolded protein response. *Science* 313: 104–107.
Kang SW et al. 2006 Substrate-specific translocational attenuation during ER stress defines a pre-emptive quality control pathway. *Cell* 127: 999–1013.
Korennykh AV et al. 2009 The unfolded protein response signals through higher-order assembly of Ire1. *Nature* 457: 687–693.
Marcinak SJ and Ron D 2006 Endoplasmic reticulum stress signaling in disease. *Physiol. Rev.* 86: 1133–1149.
Oyadomari S et al. 2006 Cotranslational degradation protects the stressed endoplasmic reticulum from protein overload. *Cell* 126: 727–739.
Ron D and Walter P 2007 Signal integration in the endoplasmic reticulum unfolded response. *Nat. Rev. Mol. Cell Biol.* 8: 519–527.
Schröder M and Kaufman RJ 2005 ER stress and the unfolded proteins response. *Mut. Res.* 569: 29–63.
Yoshida H et al. 2001 XBP! mRNA is induced by ATF6 and spliced by IRE1 is response to ER stress to produce a highly active transcription factor. *Cell* 107: 881–891.
Zhang K and Kaufman RJ 2008 From endoplasmic-reticulum stress to the inflammatory response. *Nature* 454: 455–462.
Zhou J et al. 2006 The crystal structure of human IRE1 luminal domain reveals a conserved dimerization interface required for activation of the unfolded proteins response. *Proc. Natl. Acad. Sci. USA*103: 14343–14348.

ERAD and Sec61

Hessa T et al. 2005 Recognition of transmembrane helices by the endoplasmic reticulum translocon. *Nature* 433: 377–381.
van den Berg B et al. 2003 X-ray structure of a protein-conducting channel. *Nature* 427: 36–44.

p97

Boyault C et al. 2006 HDAC6-p97/VCP controlled polyubiquitin chain turnover. *EMBO J.* 25: 3357–3366.
Richly H et al. 2005 A series of ubiquitin binding factors connects CDC48/p97 to substrate multiubiquitylation and proteasomal targeting. *Cell* 120: 73–84.
Rumpf S and Jentsch S 2006 Functional division of substrate processing cofactors of the ubiquitin-selective Cdc48 chaperone. *Mol. Cell* 21: 261–269.

The p97 Chaperone Machine

Davies JM, Brunger AT and Weis WI 2008 Improved structures of full-length p97, an AAA ATPase: implications for mechanisms of nucleotide-dependent conformational changes. *Structure* 16: 715–725.

DelaBarre B et al. 2006 Central pore residues mediate the p97/VCP activity required for ERAD. *Mol. Cell* 22: 451–462.
Shcherbik N and Haines DS 2007 Cdc48p$^{Npl4p/Ufd1p}$ binds and segregates membrane-anchored/tethered complexes via a polyubiquitin signal present on the anchors. *Mol. Cell* 25: 385–397.

AAA ATPases

Hanson PI and Whiteheart SW 2005 AAA+ proteins: have engine, will work. *Nat. Rev. Mol. Cell Biol.* 6: 519–529.
Martin A, Baker TA and Sauer RT 2008 Protein unfolding by a AAA+ protease is dependent on ATP-hydrolysis rates and substrate energy landscapes. *Nat. Struct. Mol. Biol.* 15: 139–145.
Meusser B et al. 2005 ERAD: the long road to destruction. *Nat. Cell Biol.* 7: 766–772.

p97 to the Proteasome

Messick TE et al. 2008 Structural basis for ubiquitin recognition by the Otu1 ovarian tumor domain protein. *J. Biol. Chem.* 283: 11038–11049.

Chapter 19
Parkinson's Disease

Parkinson's disease is the leading disease affecting motor function and the second most common neurological disorder overall. This movement disorder is named for the English physician James Parkinson who first described the disease as "shaking palsy" in 1817. As described by Parkinson, the disease is characterized by a slowing of movement (bradykinesia), muscle rigidity, resting tremor, and postural righting defects. In PD, there is a degeneration of neurons in a region of the midbrain called the substantia nigra pars compacta (SNc). These neurons use dopamine as their neurotransmitter, sending their axons to neurons in the striatum responsible for initiating and controlling movement. The lack of adequate dopamine signaling results in failed motor control, and leads in its later stages to cell death and eventually to falling and dementia. The disease is not limited to nigrostriatal dopamine area; but rather, other, non-dopamine areas are involved, as well. Parkinson's disease affects 1% of the population at age 65 and up to 5% at age 85.

The formation of insoluble protein clumps in certain classes of cells in the brain is not restricted to Alzheimer's disease, but rather is encountered in a host of neurological disorders. One of the most commonly encountered examples is Parkinson's disease. In this neurological disorder, inclusion bodies called Lewy bodies form in the above-mentioned dopamergic cells. These intracellular protein aggregates were discovered by Friedrich Lewy in 1912. Since then similar aggregates have been found in two other neurological disorders – dementia with Lewy bodies and multiple system atrophy. In multiple system atrophy, aggregates form in oligodendrocytes and these are referred to glial cytoplasmic inclusions.

The first clues as to the molecular underpinnings for the disease emerged in 1997 with two sets of findings. First, the appearance of PD in certain individuals was associated with mutations in α-synuclein, a small, 140 amino acid residue protein enriched in the presynaptic terminals of neurons. Second, it was established that the α-synuclein molecules were the main component of the Lewy bodies. Lewy bodies are spherically shaped, 5–25 μm in diameter, structures. They contain a dense core of filamentous and granular material surrounded by

M. Beckerman, *Cellular Signaling in Health and Disease*,
Biological and Medical Physics, Biomedical Engineering,
DOI 10.1007/978-0-387-98173-4_19, © Springer Science+Business Media, LLC 2009

a filamentous halo. They were found to be composed of neurofilaments, ubiquitin, heat-shock proteins and, as just mentioned, α-synuclein.

Three PD-causing mutations have been found. The first, the one discovered in 1997, involves substitution of threonine for alanine at position 53 (A53T). In the second, found within the next year, was that of a proline substitution for alanine at position 30 (A30P) and the third E46K. In addition to these mutations, a locus triplication has been found to lead to familial inherited PD, thereby pointing to the amount of α-synuclein residing in the cell as a factor.

In discussing PD, the emphasis placed in the earlier examination of AD on events taking place at the postsynaptic side of the synaptic cleft will be replaced by a focus on activities occurring at presynaptic terminals and the axonal transport system that moves materials between these terminals and the soma. A key factor that will emerge and be explored in detail is "garbage collection"; that is, the bringing together and disposal of misfolded and nonfunctional proteins and organelles. A hint of that is already present in the observation that ubiquitin is a component of the Lewy bodies. It becomes yet clearer when other mutated proteins found to give rise to rare familial forms of PD are examined.

The best characterized of the Parkinson's disease-causing mutated proteins and their functions are listed in Table 19.1. This group of proteins is a remarkable one for several reasons. Some of these proteins are exceptionally abundant in neurons, comprising more than 1% of the protein complement. They are encountered in other neurodegenerative disorders, not just Parkinson's disease, and are connected in a not yet understood way to an increased risk for certain cancers. α-Synuclein is a natively unfolded protein and is believed to be a central player in PD.

Table 19.1 Mutated proteins implicated in the pathogenesis of Parkinson's disease

Locus	Protein	Function
PARK1 – 4q21	α-Synuclein	Presynaptic protein
PARK2 – 6q25-27	Parkin	E3 ubiquitin ligase
PARK5 – 4p14	UCHL1	Deubiquitinating enzyme
PARK6 – 1p35-36	PINK1	Serine-threonine kinase
PARK7 – 1p36	DJ-1	Oxidative stress response regulator
PARK8 – 12p11.2	LRRK2	Mixed lineage kinase
PARK13 – 2p12	Omi/HtrA2	Stress-induced serine protease

Mutated forms of the six other proteins listed in Table 19.1 give rise to various familial, that is, inherited, forms of PD. Parkin accounts for 10–20% of these cases; LRRK2 for approximately 5–10%, PINK1 about 5%, and the others on the order of 1% or less. The domain structure of each of these proteins is depicted in Fig. 19.1. Included in this figure are a sampling of the best characterized mutations associated with familial PD. Most of the cases of PD are sporadic; that is, they arise spontaneously, without any obvious cause. The

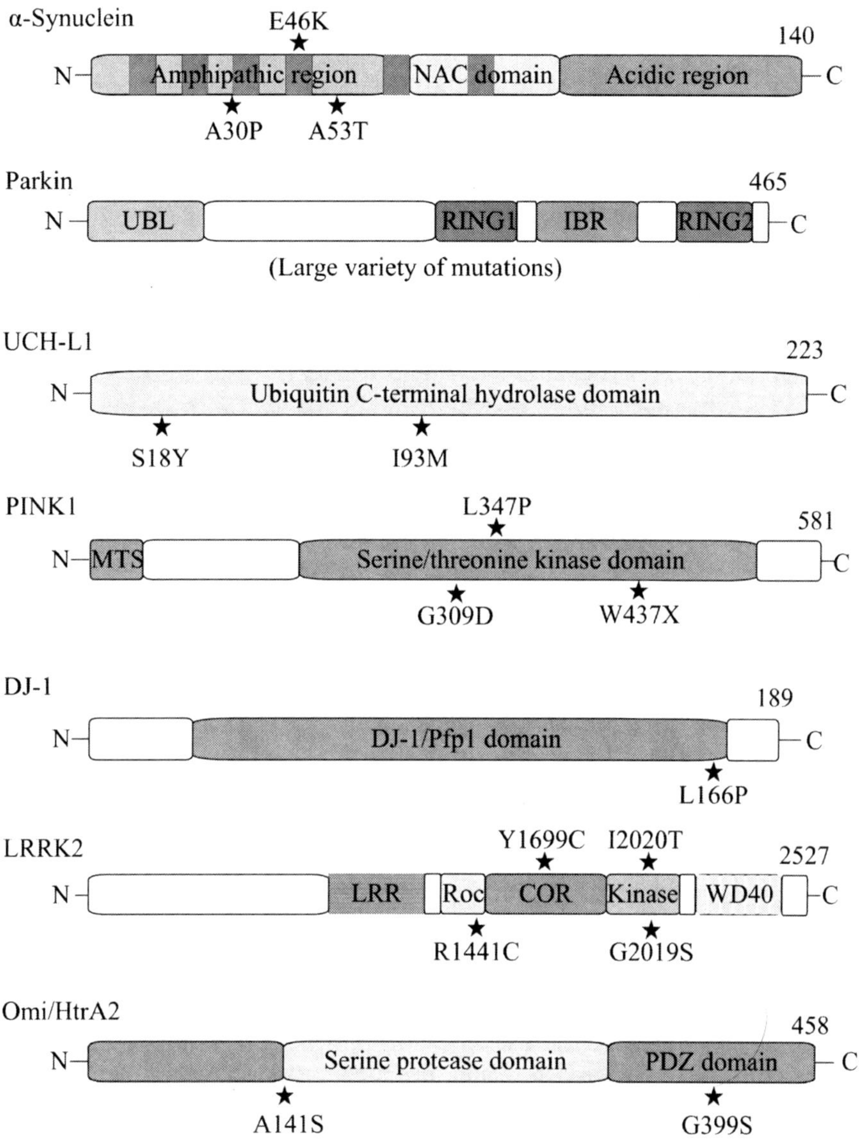

Fig. 19.1 Domain structures of the seven proteins whose genetic mutations give rise to familial forms of Parkinson's disease. Major disease-causing mutations are indicated next to star symbols. Abbreviations are defined in the text

hope in finding these rare PD-inducing mutations is that they serve as a beacon, illuminating the nature of the disease and pointing the ways toward effective treatments. Each of these proteins and the pathways they uncover will be examined in this chapter.

19.1 α-Synuclein Is a Presynaptic Protein

The domain structure of α-synuclein protein is depicted in Fig. 19.1. As can be seen, it consists of three domains. It has an N-terminal amphipathic region in which are embedded a series of KTVEGV repeats that mediates binding to the negatively charged membranes. It possesses a central region nonamyloid component (NAC) domain that contains a stretch of hydrophobic residues responsible for its tendency to polymerize into filaments, and this domain is followed by a C-terminal acidic region. α-Synuclein is a highly flexible, natively unfolded protein that upon association with phospholipid-containing (negatively charged) membranes develops amphipathic α-helical secondary structure in its N-terminal while the C-terminal regions remain free and unstructured.

α-Synuclein is a secretory vesicle protein abundantly expressed in presynaptic terminals of neurons in the central nervous system where it helps protect nerve terminals against injury. It does so at least in part by acting as a chaperone that complements the actions of cysteine string protein α (CSPα) in folding and refolding of v-SNARE and t-SNARE proteins centrally involved in neurotransmitter release and vesicle recycling. α-Synuclein may function as a negative regulator of vesicle priming, a step involving conformational changes in the SNARE proteins that prepares secretory vesicles for release of their contents. Other functions have been suggested for α-synuclein, and its precise role(s) in neurons remains to be determined.

The cysteine string protein α to which it associates is a synaptic cochaperone that assists in the localization and exocytosis in presynaptic terminals of secretory vesicles containing neurotransmitters and peptides such as dopamine. CSPα helps direct chaperone complexes to the machinery responsible for vesicle exocytosis where they help preserve the correct folding of proteins. Processes influenced either directly or indirectly by the CSPαs include vesicle filling and docking, and SNARE protein and Ca^{2+} channel functioning. Cysteine string protein α possesses an N-terminal J domain and as a result belongs to the J-domain family of cochaperones that includes DnaJ, Hsp40, and Hsc70. The J-domains of these proteins enable them to interact with and regulate the activities of members of the Hsp70 family of chaperones. The second main structural feature of CSPα is its centrally located cysteine-rich region, or cysteine string domain that lends its name to the proteins. Residues in this region are extensively palmitoylated, and this domain enables the protein to attach to secretory vesicles. At these sites, CSPα along with Hsc70 helps maintain correct folding and assembly.

19.2 Abnormalities and Toxicity Result from α-Synuclein Misfolding and Aggregation

Misfolded α-synuclein forms aggregates, and is the main constituent of Lewy bodies. α-Synuclein is associated with dysfunctions in vesicle transport; in particular, it disrupts vesicle trafficking through the ER to the Golgi apparatus by interfering with members of the Rab family of GTPases that coordinate the docking and fusion of cargo-carrying vesicles to Golgi membranes. Most significantly, one of the earliest signs of disease that appears prior to a buildup of Lewy bodes is the accumulation of clusters of vesicles indicative of emerging dysregulation and malfunctions in vesicle trafficking and dopamine handling. The propensity of α-synuclein in its α helical form to associate with phospholipid-enriched membranes, its tendency to misfold and oligomerize, and its ability to disrupt normal vesicle trafficking all contribute to vesicle clustering.

Paralleling the situation encountered elsewhere, not all disease-causing mutations are equal. The missense mutations shown in Fig. 19.1 have somewhat different effects on α-synuclein aggregation. The A53T mutation promotes formation of fibrils while the A30P mutation does not but instead enhances oligomerization. Another disease-causing alteration is the aforementioned triplication in the gene encoding α-synuclein. This leads to overexpression of α-synuclein and illustrates the dependence of aggregation upon expression levels and concentration.

19.3 Oxidative Damage Is a Cause of α-Synuclein Aggregation and PD

Parkinson's disease affects dopamergic neurons in the substantia nigra, which sends projections to neurons in the dorsal motor striatum. Atrophy of these connections eliminates signaling to the striatum leading to tremors and other physical manifestations of Parkinson's disease. The dopamergic neurons release dopamine from their presynaptic terminals and also take up excess dopamine from the synaptic spaces via the dopamine transporter (DAT). Dopamergic neurons are particular susceptible to oxidative stress due to the production of reactive oxygen species during dopamine metabolism. The excess dopamine taken up by the DAT is sequestered in synaptic vesicles and if this system is overwhelmed, the dopamine remains in the cytosol or is concentrated in mitochondria and causes damage to complex I of the electron transport chain. Damage to complex I generates a further buildup of ROS and results in mitochondrial dysfunction. Environmental factors such as pesticides can act as accelerants. Pesticides such as rotenone are taken up by the DAT and once inside the neurons cause further damage to complex I.

The clumping together of α-synuclein into fibrils and incorporation of these aggregates into Lewy bodies take place in a series of intermediate stages. Dopamine interacts with α-synuclein; it forms α-synuclein-dopamine adducts that inhibit fibril formation and growth of Lewy bodies, and extends the half-life of early intermediates called protofibrils. Small protofibrils of α-synuclein are then able to form pores in the small synaptic vesicles containing dopamine. As a result of this permeabilization, dopamine is released into the cytosol increasing its ability to damage complex I. Mitochondrial complex I inhibition may be the main causal event in sporadic PD. This induces the clumping of α-synuclein and leads to the death of dopamergic neurons. In addition, α-synuclein binds fatty phospholipids and is involved in brain lipid metabolism. Its impairment can lead to alterations in mitochondrial lipid composition, most notably, reductions in cardiolipin content, further impairing mitochondrial function.

19.4 Parkin Is an E3 Ubiquitin Ligase

The discovery of parkin's involvement in Parkinson's disease came shortly after the discovery that mutated forms of α-synuclein were centrally involved in at least some forms of the disorder. The parkin E3 ubiquitin ligase is associated with a form of Parkinson's disease known as autosomal recessive juvenile parkinsonism (AR-JP). This form of Parkinson's disease significantly lacks the Lewy bodies – they are absent in most AR-JP neurons. This absence supports the notion that normal parkin is needed for Lewy body formation and, furthermore, Lewy bodies may serve a protective function in which misfolded proteins are sequestered and rendered harmless.

Parkin is an E3 ubiquitin ligase that utilizes several E2s in assembling polyubiquitin chains. Recall that ubiquitin can be conjugated on any of a number of lysine residues among which are K48 and K63. The outcome in the case of K48 ubiquitination is degradation of the tagged proteins by the 26S proteasome. In situations where K63 is the site for ubiquitination, the chain serves as a platform for assembly of a signaling complex. Parkin can mediate both K48 and K63 polyubiquitination. One of parkin's substrates synphilin-1, a protein that interacts with α-synuclein, is found in Lewy bodies, and provides a link between parkin and α-synuclein. Under most circumstances, parkin catalyzes the polyubiquitination of synphilin-1 on K63 and this activity promotes Lewy body formation.

A key substrate of parkin with respect to AR-JP is the parkin-associated endothelin receptor-like receptor (Pael-R). This transmembrane protein is an orphan G-protein-coupled receptor situated in the ER lumen. It is highly expressed in dopamergic neurons and is involved in dopamine metabolism. When overexpressed in cells, the receptor misfolds and accumulates in Lewy bodies. Misfolded Pael-R can be toxic. Misfolded Pael-R triggers an ERAD unfolded proteins response. When not handled adequately by the UPR, it accumulates, induces ER stress, and induces cell death. Parkin functions at least in part as an endoplasmic reticulum E3 ligase. Mutated forms of parkin may drive AR-JP by its failure to ubiquitinate misfolded Pael-R and tag it for proteasomal degradation. S-nitrosylation of parkin can also derail its E3 ubiquitin ligase activity. This susceptibility provides a potential link between redox signaling imbalances and sporadic Parkinson's disease.

19.5 Protein Carbonylation and UCH-L1

In cells where the redox balance is disturbed, excessive oxidation of proteins can take place. One of the most damaging of these modifications is protein carbonylation in which proteins are modified through the addition of carbonyl groups to specific side chains located in their active center. This is an important process because protein carbonylation is both damaging and irreversible thereby necessitating proteasomal degradation to prevent excessive buildup of these oxidized proteins.

The 20S proteasome is the catalytic core of the proteasomal degradation machine. The 26S proteasome has in addition to the 20S core unit two 19S regulatory units containing the machinery for ATP-hydrolysis and ubiquitin recognition. The 20S proteasome carries out the degradation of misfolded, heavily oxidized and cross-linked proteins in an ATP- and ubiquitin-independent way. Protein carbonylation leads to partial unfolding and exposure of hydrophobic patches. If these proteins are not removed early in the oxidation and cross-linking stages, they can become resistant to proteasomal degradation; they may form insoluble aggregates, and jam the proteasomal machinery. Protein carbonylation becomes more prevalent as cells age and has a role in aging as well as in age-related neurodegenerative disorders.

One of the proteins with a propensity to become carbonylated is Ubiquitin C-terminal hydrolase L1 (UCH-L1). A mutated form of UCH-L1 known as I93M UCH-L1 is associated with a rare familial form of PD. This protein is one of the most abundant proteins encountered in cells, accounting for more than 1% of the total soluble protein in neurons. This protein functions as a deubiquitinating enzyme and possibly as a ubiquitin ligase and as a stabilizing agent. Both the mutated form and carbonylated form become insoluble and exhibit aberrant interactions with other proteins. One of the targets of altered UCH-L1 is tubulin, and the interactions between the two promote tubulin polymerization. Oxidized modifications of UCH-L1 are encountered not only in sporadic PD but also in sporadic AD.

19.6 PINK1 Is a Neuroprotective Serine/Threonine Kinase

The PTEN induced putative kinase 1 (PINK1) is a serine/threonine kinase with two domains – an N-terminal mitochondria-targeting motif and a C-terminal serine/threonine kinase domain. The normal function of this kinase is a neuroprotective one carried out by phosphorylating a number of key substrates and ensuring mitochondrial function during periods of increased oxidative stress. One of its substrates is the fusion/fission machinery that plays a critical role in maintaining mitochondrial morphology and function within the network of mitochondria distributed throughout the soma, axons, and dendrites. These activities are required for synaptic plasticity and neural survival.

In the second group of activities are prosurvival actions taken at mitochondria that raise the barrier to mitochondrial-induced apoptosis in the presence of oxidative stress. One of the proteins phoshorylated by PINK1 is the tumor necrosis factor receptor-associated protein 1, or Trap1. This protein is a mitochondrial member of the Hsp90 family of molecular chaperones (to be discussed in the next chapter). When activated by upstream stress-responsive kinases such as PINK1, Trap1 binds cyclophilin D and inhibits its proapoptotic activities through a folding–refolding mechanism.

A second mitochondrial substrate of PINK1 is the proapoptotic serine protease, HtrA2. This protein is released from mitochondria in response to proapoptotic stimuli; once it is outside in the cytosol, it binds and inhibits members of the inhibitor of apoptosis protein (IAP) protein family, which would otherwise bind and inhibit caspase apoptotic effectors. In the absence of proapoptotic factors triggering its release from the mitochondria, HtrA2 functions as a prosurvival factor. Stress-induced PINK1 and p38 induced phosphorylation activate the protease, which then degrades misfolded proteins through its ability to recognize unstructured hydrophobic patches. Thus, PINK1 protects dopamergic neurons against oxidative stress by enhancing mitochondrial effectiveness and shifting the apoptotic tipping point toward survival.

19.7 DJ-1 Protects Against Oxidative Stress

DJ-1 is a 189 amino acid residue long protein that folds into a compact structure consisting of six-stranded β-sheet sandwiched by eight α-helices, a β-hairpin at one end, and a three-stranded antiparallel β-sheet at the other end. The protein is found in several locations in the cell – in the cytosol, in the nucleus, and in mitochondria, but it is the mitochondrial sites that are perhaps most interesting. Several possible functions for this protein have been uncovered. The three exposed cysteine residues shown in Fig. 19.2 are believed to be of special

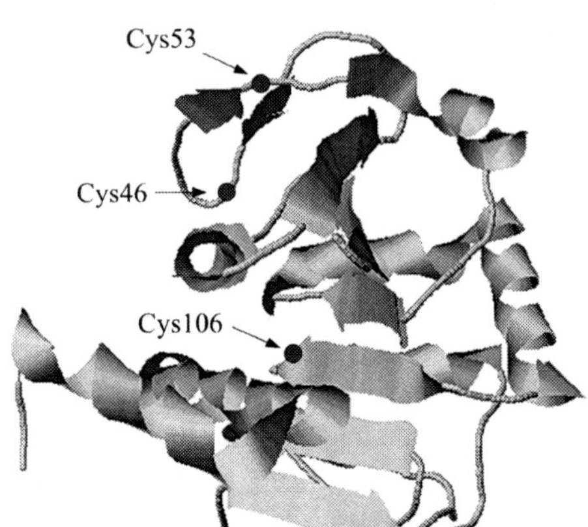

Fig. 19.2 Three-dimensional structure of DJ-1 showing locations of the three critical cysteine residues. The figure was prepared using Jmol with atomic coordinates deposited in the PDB under accession number 1p5f

significance. In particular, the DJ-1 protein scavenges hydrogen peroxide through oxidation at Cys106. This is a particularly important function in dopamergic neurons because of their increased susceptibility to oxidative stresses. The other two residues, Cys46 and Cys53, are targets of S-nitrosylation.

Mutations in the PARK7, DJ-1 gene are associated with autosomal recessive early-onset Parkinson's disease. The most prominent of these mutations, L166P, abolishes the ability of DJ-1 to associate into dimers, the physiologically active form of the protein. When mutated in this way, the protein, instead, loses stability and unfolds.

In addition to its role as an antioxidant sensor and scavenger, DJ-1 protects cells against oxidative damage by upregulating the synthesis of the front-line antioxidant glutathione. It accomplishes this in two distinct ways. First, it stimulates the transcription and catalytic activity of glutamate cysteine ligase (GCL), the rate-limiting enzyme in glutathione synthesis. Second, in its oxidized form DJ-1 interacts with the α-synuclein mutant A53T, activates the Hsp70 molecular chaperone, and decreases α-synuclein aggregation and toxicity.

19.8 LRRK2 Is a ROCO Family Member and Mutations in This Protein Are Most Strongly Associated with PD

Leucine-rich repeat kinase 2 is a member of the ROCO (Ras in complex proteins [Roc] with a C-terminal of Roc domain) family of complex proteins. Its domain structure is depicted in Fig. 19.1. As is the case for other members of this family of proteins, LRRK2 contains not only a RasGTPase domain called Roc but also a kinase domain characteristic of MAPKKKs. The Roc and kinase domains are accompanied by a series of leucine-rich repeats situated N-terminal to the Roc domain. A COR (domain C-terminal to Roc) domain is located in between the Roc and kinase domains and a number of WD40 repeats lie C-terminal to the kinase domain. Mutations associated with Parkinson's disease are situated in all domains of this protein. These have been identified as giving rise in 5–6% of the cases to familial PD and in 1–2% of the cases to sporadic PD.

The Roc domain forms homodimers with a Roc domain from a dimerization partner. The Roc domain is composed of two subdomains linked together by a bent α helix to form body, neck, and head regions (Fig. 19.3). In forming the active homodimer, the head of one domain lies near the body of the other domain in a domain swapping configuration. The dimer has a pair of active sites with residues contributed from both partners to each site. The evidence obtained to date is suggestive that the GTPase domain regulates the kinase domain with the COR domain situated in between the two acting as a molecular hinge.

The mechanism of activation is, as follows. Dimer formation of the two Roc domains brings the kinase domains together to form a dimer. This mirror action leads to autophosphorylation and activation of the kinase domains. Hydrolysis by an as yet unknown GAP causes the dissociation of the GTPase homodimer

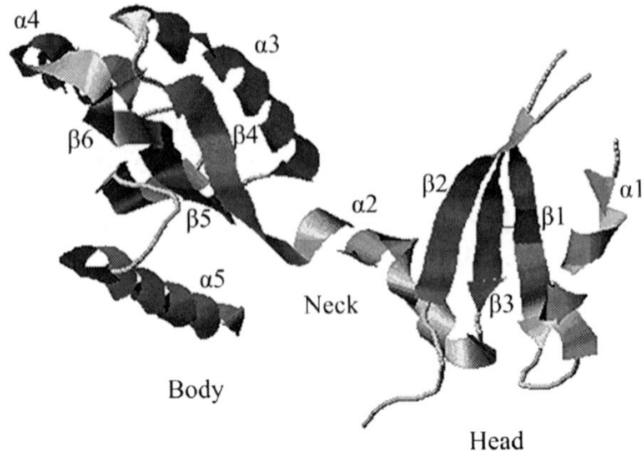

Fig. 19.3 Structure of the Roc domain of the ROCO protein kinase, LRRK2, as revealed by X-ray crystallography. The head subdomain is composed of β1, α1, β2, and β3. The body subdomain is comprised of β4, α3, β5, α4, β6, and α5 while the neck linker region consists of a partially disordered α2 helix. The figure was prepared using Jmol with atomic coordinates deposited in the PDB under accession number 2zej

thereby shutting down the kinase domain's homodimerization. Mutations associated with PD are situated in the interface between the GTPase dimer partners. These mutations are likely to impair its ability to be turned off in a timely fashion as is the situation for Ras in the presence of oncogenic mutations. In the case of Roc, malfunctions in GTP hydrolysis leads to excessive activation of LRRK2 that promotes neurodegeneration.

19.9 HtrA2/Omi Removes Misfolded Proteins

The last entry in Table 19.1 is HtrA2/Omi, a mitochondrial serine protease. This protein is released from mitochondria under apoptotic conditions. Once released into the cytosol, it targets members of the inhibitor of apoptosis (IAP) family of caspase negative regulators thereby freeing the caspases to trigger apoptosis. This early finding, which led to the assessment of HtrA2 as a proapoptotic protein, has been substantially revised with several more recent findings. These newer discoveries point to a fundamental, protective role for the protease consistent with similar functions observed for bacterial family members, which have chaperone and protease capabilities.

The three-dimensional structure of HtrA2 is shown in Fig. 19.4. As can be seen, HtrA2 has a well-defined secondary structure that consists of 7 α helices and 19 β-strands organized into two tightly folded domains – an N-terminal serine protease domain and a C-terminal PDZ domain. These two domains are connected by a flexible linker. Recall that PDZ domains are protein–protein interaction modules. There are several kinds of PDZ domains, and the one

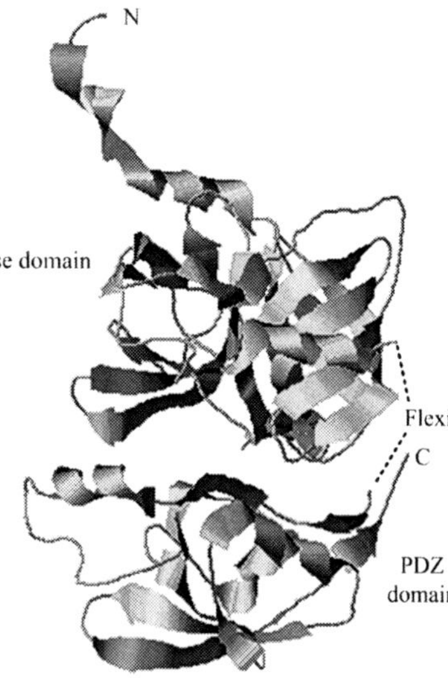

Fig. 19.4 Three-dimensional structure of Omi/HtrA2. *Dashed lines* show the location of the missing flexible linker connecting the N- and C-terminal domains. The figure was prepared using Jmol with atomic coordinates deposited in the PDB under accession number 1lcy

possessed by HtrA2 closely resembles those possessed by its bacterial homologues. In HtrA2, the PDZ domain recognizes exposed hydrophobic patches on proteins, of the type resulting from protein misfolding.

The functional form of HtrA2 is that of a trimer in which the PDZ domains face outward and the three protease domains are in the center, mediate subunit-subunit binding, and are catalytically inactive. Ligand binding to a PDZ domain triggers a set of conformational changes leading to activation of the proteases. These enzymes processively cleave the substrates into small equally sized pieces thereby maintaining the health of the mitochondria.

19.10 The Pathway Is Illuminated

The importance of protecting against oxidative damage and attendant protein misfolding is highlighted by the existence of multiple layer of protection. First, as discussed in the last chapter, molecular chaperones aid in folding, refolding, and directing proteins that cannot be repaired and refolded to proteasomal degradation. Second the ubiquitin-proteasome system removes damaged proteins, thereby maintaining protein homeostasis. Third, the aggesome-autophagy-lysosomal pathway cleans up residual damaged and no-longer-needed cellular materials. Malfunctions and overloads in these pathways are one of two main components of the pathway to Parkinson's disease. The other part of the route, the production line, has at its core oxidative damage, misfolded α-synuclein, and mitochondrial damage.

The picture that emerges from studies of familial PD and the six mutations that have been well characterized is presented in Fig. 19.5. In this figure, there is

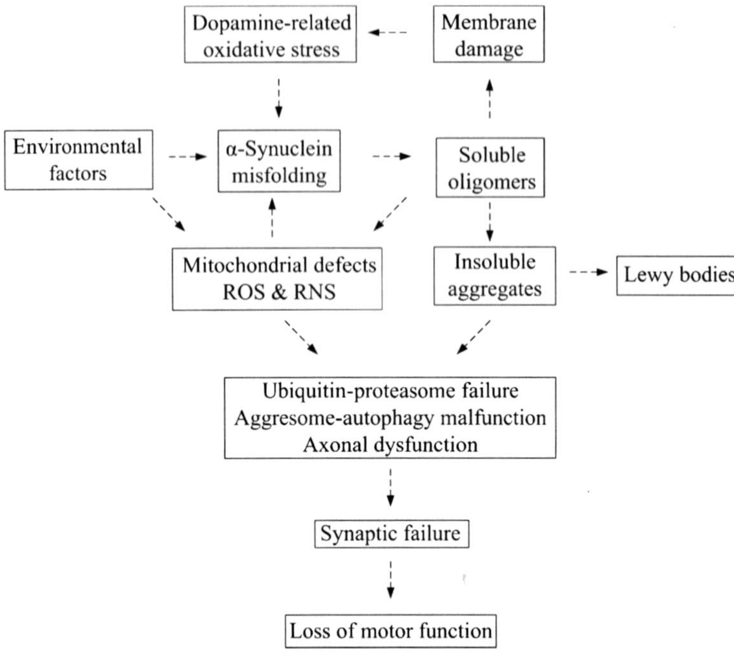

Fig. 19.5 The Parkinson's disease pathways

a core production process involving α-synuclein, the mitochondria, and the ubiquitin-proteasome system. Several factors can cause α-synuclein to misfold. There is evidence for an increased incidence of PD arising from pesticides. That heavy metals can catalyze protein misfolding was discussed earlier with regard to Amyloid β. Oxidative damage is a key contributor to the entire ensemble of neurodegenerative disorders. Dopamine metabolism as well as mitochondrial defects, most notably, to Complex I of the electron transport chain, contributes to α-synuclein misfolding.

Two positive feedback loops drive α-synuclein misfolding and buildup. The first of these is anchored by dopamine vesicles and metabolism. The buildup of misfolded α-synuclein leads through α-synuclein oligomerization to increased damage to dopamine-bearing vesicles, generation of oxidants, and further α-synuclein misfolding. The other loop has at its center the mitochondria with S-nitrosylation and ROS production enhancing α-synuclein misfolding.

19.11 Proteasome Organization

The ubiquitin-proteasome system (UPS) functions as a third multiply connected node in the PD pathway depicted in Fig. 19.5. Failures in the upstream production core lead to shortfalls in ATP generation by the mitochondria and to excessive production of insoluble aggregates. Both the loss of ATP and the buildup of insoluble aggregates impact the UPS; the former is required for degradation and the latter can easily jam the UPS thereby preventing its proper operation. The third, backup system, the autophagic-lysosomal railway can become overloaded and fail from lack of sufficient ATP and from transport blockages.

The 20S and 26S proteasomes are chamber-shaped proteolytic machines that degrade misfolded and oxidized proteins. The 20S core unit, as depicted in Fig. 19.6, is composed of four rings stacked one on top of the other. Each ring is composed of seven different subunits. The two inner rings consist of β-subunits and the two outer rings contain α-subunits. The β-subunits comprise the catalytic core of the proteasome while the α-subunits have a structural role. Three catalytic sites are positioned on the inner surface of the catalytic rings; these are responsible for the proteolytic degradation of proteins that enter the chamber into a set of short peptide fragments.

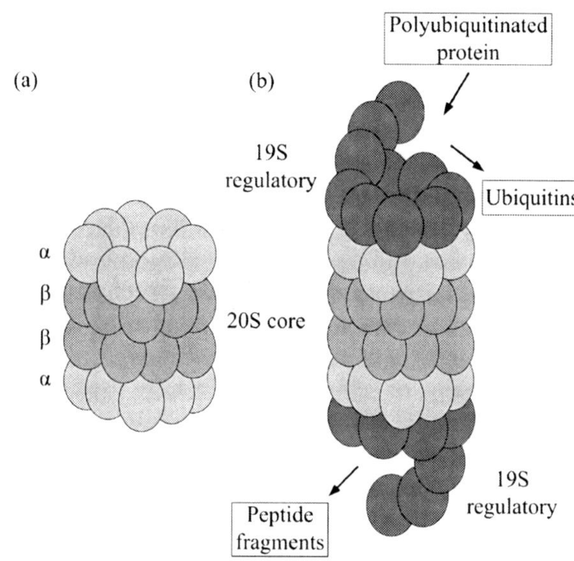

Fig. 19.6 Arrangement of the heptameric α and β rings and 19S cap structure to form the (**a**) 20S proteasome and (**b**) 26S proteasome

The 26S proteasome contains a pair of 19S regulatory units in addition to the 20S catalytic core. These units assist in unfolding proteins so they may more easily fit into the chamber and remove ubiquitin so that it may be recycled. These processes are carried out in an ATP-dependent manner. Partially unfolded oxidized proteins have exposed hydrophobic patches. They can be degraded in an ATP- and ubiquitin-independent manner by the 20S proteasome, but heavily oxidized proteins may fragment, become cross linked, aggregate, and block the proteasomes leading to a buildup of unfolded and damaged proteins in the cell.

19.12 Cellular Garbage Collection and the Aggresomal – Autophagic Railway

If the first two lines of defense, chaperone-mediated refolding and ubiquitin-proteasome-mediated degradation, fail to maintain protein homeostasis against misfolded protein buildup, an additional response that may be called cellular garbage collection is initiated. In this response, the misfolded proteins that tend to aggregate into insoluble clumps through hydrophobic interacts are collected together and sequestered in dynamic compartments called aggresomes.

Aggresomes form at and about microtubule (MT) organizing centers (MTOCs). These centers are located at or near centromeres; the aggregates are transported to these central locales along microtubule rails by minus-end directed dynein motor proteins. Aggresomes are characterized by a cage of intermediate filament (IF) proteins, most notably, vimentin. The aggresomes are dynamic entities and situated near and recruit components of the ubiquitin-proteasome system and molecular chaperones to deal with the misfolded proteins.

The ubiquitin-proteasome system can become overwhelmed by buildups in misfolded and aggregated proteins. When this happens, the removal of misfolded proteins is increasingly handled through aggregation of those proteins into large assemblages and their subsequent collection and degradation through autophagic pathways. The actions of the two routes of disposal are coordinated. Impairments in UPS lead to the collection of the damaged proteins and small proteins into aggresomes, inclusion bodies (IBs), and other aggregating structures. Large numbers of these structures can in themselves become disruptive, and to prevent this from happening the buildups are accompanied by increased activity of autophagic pathways.

Autophagy, or self-eating, is the term used to describe the process whereby cellular components are degraded in lysosomes. Cytosolic materials, proteins, parts of organelles such as the ER, the Golgi, and the nucleus, and even entire organelles, for example, mitochondria are disposed of through autophagic transport and digestion in lysosmes. This process maintains cellular homeostasis in the face of environmental challenges and changing requirements on cellular resources. There are three kinds of autophagic processes – macroautophagy, microautophagy, and chaperone-mediated autophagy. The predominant form found in mammalian cells is macroautophagy, hereafter called autophagy, in which proteins and organelles are packaged in double-membraned vesicles called autophagosomes and transported to lysosomes.

Autophagy occurs constitutively in almost all mammalian cells, and is rapidly upregulated in response to cellular demands. It is adjusted according to the nutrient and growth demands through the insulin-Akt-TOR pathway. Under plentiful nutrient supply, this pathway negatively regulates expression of autophagy-related (Atg) proteins centrally involved in vesicle nucleation and elongation, docking and fusion, steps required to form the autophagosomes.

This disposal route is rapidly upregulated in response to starvation conditions and also when there is a buildup of misfolded-protein aggregates and the ubiquitin-proteasome system is not able to handle the load. Failure to clear the buildup in cellular garbage in neurons can by itself trigger neurodegeneration. An example of how this might happen is through impairment in the MT transport. An indication of this happening during neurodegeneration is the presence of inclusion bodies. These differ from aggresomes in that they are dispersed through the cytoplasm.

19.13 Histone Deacetylase 6 Mediates Transport Along the Disposal Railway

The histone deacetylase-6 (HDAC6) protein has a key role in the pathway leading from aggregates to aggresomes to autophagic clearance in lysosomes. Histone deacetylases, or HDACs, and their roles in the epigenetic regulation of gene expression were discussed in depth in Chapter 12. The deactetylase HDAC6 is a highly unusual member of this class of enzymes. It does not deacetylate histones but rather functions as a cytoplasmic regulator of protein homeostasis with a major role in the clearance of misfolded and aggregated proteins. It is a microtubule-associated protein that regulates microtubule acetylation, associates with dynein motor complexes, and has an ubiquitin-binding motif that enables it to bind mono- and polyubiquitinated proteins. By possessing this set of properties, HDAC6 is able to function as an adapter between ubiquitinated proteins and microtubule-associated dynein motor complexes. These proteins are then transported and delivered to collection sites to form large structures such as aggresomes and inclusion bodies (IBs) that contain the misfolded and ubiquitinated proteins thereby isolating them and rendering them relatively harmless.

The UPS is controlled by p97/VCP, a caspase-like AAA ATPase. This enzyme dissociates complexes formed between VDAC6 and ubiquitinated proteins; it regulates ubiquitin turnover, and the cellular fate of the misfolded and ubiquitinated proteins. Another key protein in the management of protein homeostasis is heat shock factor 1 (HSF1). This protein is a transcription factor that stimulates the upregulation of heat shock proteins (HSPs) in response to a buildup in misfolded proteins. This protein was discussed previously in Chapter 18 along with the HSPs. Recall that the HSPs help to refold the proteins and direct those that cannot be repaired to proteasomes and other collection and disposal sites.

The control system operates in the following manner (Fig. 19.7). The HSF1 protein is kept in an inactive state through its binding to Hsp90. To become activated it must dissociate from Hsp90, and then undergo phosphorylation and trimerization. Binding of HDAC6 to ubiquitinated proteins triggers the dissociation of HSF1 from Hsp90. The p97/VCP protein is a key intermediary. HDAC6 forms a complex with p97/VCP, which is disrupted by the binding of ubiquitinated proteins to HDAC6. This may be thought of as a sensing and triggering stage resulting in release of the p97/VCP protein that then stimulates the freeing of HSF1 from it binding to Hsp90 through its ATPase function. The HSF1 protein then promotes the increased expression of HSPs, and this process shuts down once the number of ubiquitinated proteins is reduced and the p97/VCPs become bound to HDAC6.

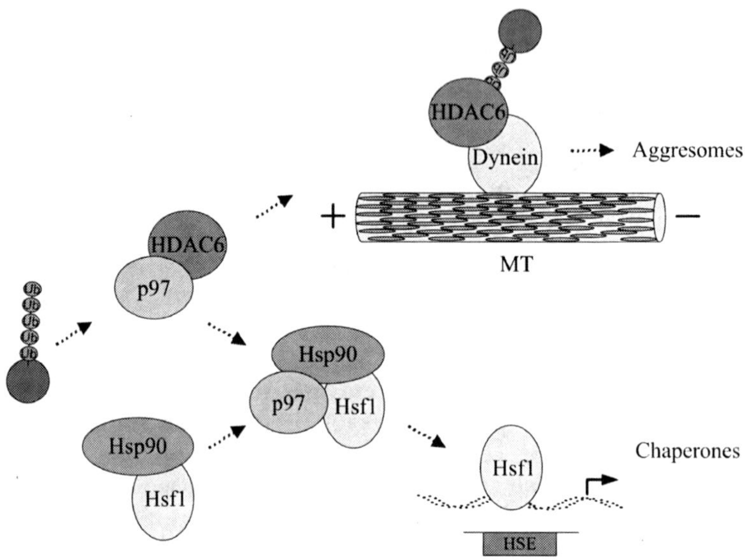

Fig. 19.7 Dual actions by the Hsfl-Hsp90 and p97-HDAC6. Ubiquitinated proteins trigger the translocation of Hsfl to the nucleus where it stimulates transcription of heat shock proteins while HDAC6 facilitates transport along microtubules of the misfolded and damaged proteins to aggresomes

For example, misfolded DJ-1 is a parkin substrate. Misfolded DJ-1 proteins are K63-polyubiquitinated by parkin using as its E2 the UbcH13/Uev1a heterodimer. The K63-polyubiquitinated DJ-1 proteins are then recruited to aggresomes. The K53 polyubiquitin tag serves as a platform and signal for HDAC6. This enzyme modifies DJ-1 so that it may bind to dynein motors and undergo transport to the aggresomes.

Further Reading

Forster E and Lewy FHG 1912 Paralysis agitans, in *Pathologische Anatomie. Handbuch der Neurologie*(ed. Lewandowsky M). Berlin: Springer Verlag, pp 920–933.

Parkinson J 1817 *An Essay on the Shaking Palsy*. London: Sherwood, Neely & Jones.

Polymeropoulos MH et al. 1997 Mutation in the α-synuclein gene identified in families with Parkinson's disease. *Science* 276: 2045–2047.

Spillantini MG et al. 1997 α-Synuclein in Lewy bodies. *Nature* 388: 839–840.

α-Synuclein

Chamberlain LH and Burgoyne RD 2000 Cysteine-string protein: the chaperone at the synapse. *J. Neurochem.* 74: 1781–1789.
Chandra S et al. 2005 α-Synuclein cooperates with CSPα in preventing neurodegeneration. *Cell* 123: 383–396.
Fernandez-Chacón et al. 2004 The synaptic vesicle protein CSPα prevents presynaptic degeneration. *Neuron* 42: 237–251.
Larsen KE et al. 2006 α-Synuclein overexpression in PC12 and chromaffin cells impairs catecholamine release by interfering with a late step in exocytosis. *J. Neurosci.* 26: 11915–11922.

α-Synuclein Pathology

Cabin DE et al. 2002 Synaptic vesicle depletion correlates with attenuated synaptic responses to prolonged repetitive stimulation in mice lacking α-synuclein. *J. Neurosci.* 22: 8797–8807.
Cooper AA et al. 2006 α-Synuclein blocks ER-Golgi traffic and Rab1 rescues neuron loss in Parkinson's models. *Science* 313: 324–328.
Gitler AD et al. 2008 The Parkinson's disease protein α-synuclein disrupts cellular Rab homeostasis. *Proc. Natl. Acad. Sci. USA* 105: 145–150.
Soper JH et al. 2008 α-Synuclein-induced aggregation of cytoplasmic vesicles in *Saccharomyces cerevisiae*. *Mol. Biol. Cell* 19: 1093–1103.

Oxidative- and α-Synuclein-Driven Damage

Conway KA et al. 2001 Kinetic stabilization of the α-synuclein protofibril by a dopamine-α-synuclein adduct. *Science* 294: 1346–1349.
Dawson TM and Dawson VL 2003 Molecular pathways of neurodegeneration in Parkinson's disease. *Science* 302: 819–822.
Ellis CE et al. 2005 Mitochondrial lipid abnormality and electron transport chain impairment in mice lacking α-synuclein. *Mol. Cell. Biol.* 25: 10190–10201.
Goedert M 1999 Filamentous nerve cell inclusions in neurodegenerative diseases: tauopathies and α-synucleinopathies. *Phil. Trans. R. Soc. Lond. B*354: 1101–1118.
Lashuel HA et al. 2002 α-Synuclein, especially the Parkinson's disease-associated mutants, form pore-like annular and tubular protofibrils. *J. Mol. Biol.* 322: 1089–1102.
Lotharius J and Brundin P 2002 Pathogenesis of Parkinson's disease: dopamine, vesicles and α-synuclein. *Nat. Rev. Neurosci.* 3: 1–10.
Volles MJ and Lansbury PT Jr 2002 Vesicle permeabilization by protofibrillar α-synuclein is sensitive to Parkinson's disease-linked mutations and occurs by a pore-like mechanism. *Biochemistry* 41: 4595–4602.

Parkin

Chung KKK et al. 2004 S-nitrosylation of parkin regulates ubiquitination and compromises parkin's protective function. *Science* 304: 1328–1331.
Imai Y et al. 2001 An unfolded putative transmembrane polypeptide, which can lead to endoplasmic reticulum stress, is a substrate of parkin. *Cell* 105: 891–902.

Lim KL et al. 2005 Parkin mediates nonclassical, proteasomal-independent ubiquitination of synphilin-1: implications for Lewy body formation. *J. Neurosci.* 25: 2002–2009.

Palacino JJ et al. 2004 Mitochondrial dysfunction and oxidative damage in *parkin*-deficient mice. *J. Biol. Chem.* 279: 18614–18622.

UCH-L1

Choi J et al. 2004 Oxidative modifications and down-regulation of ubiquitin carbonyl-terminal hydrolase L1 associated with idiopathic Parkinson's and Alzheimer's disease. *J. Biol. Chem.* 279: 13256–13264.

Dale-Donne I et al. 2006 Protein carbonylation, cellular dysfunction, and disease progression. *J. Cell. Mol. Med.* 10: 389–406.

Kabuta T et al. 2008 Aberrant molecular properties shared by familial Parkinson's disease-associated mutant UCH-L1 and carbonyl-modified UCH-L1. *Hum. Mol. Genet.* 17: 1482–1496.

Nyström T 2005 Role of oxidative carbonylation in protein quality control and senescence. *EMBO J.* 24: 1311–1317.

Shringarpure R et al. 2003 Ubiquitin conjugation is not required for the degradation of oxidized proteins by the proteasome. *J. Biol. Chem.* 278: 311–318.

PINK1

Clark IE et al. 2006 *Drosophila pink1* is required for mitochondrial function and interacts genetically with *parkin*. *Nature* 441: 1162–1166.

Park J et al. 2006 Mitochondrial dysfunction in *Drosophila PINK1* mutants is complemented by *parkin*. *Nature* 441: 1157–1161.

Plun-Favreau H et al. 2007 The mitochondrial protease HtrA2 is regulated by Parkinson's disease-associated kinase PINK1. *Nat. Cell Biol.* 9: 1243–1252.

Poole AC et al. 2008 The PINK1/parkin pathway regulates mitochondrial morphology. *Proc. Natl. Acad. Sci. USA* 105: 1638–1643.

Pridgeon JW et al. 2007 PINK1 protects against oxidative stress by phosphorylating mitochondrial chaperone TRAP1. *PLoS Biol.* 5: 1494–1503.

Valente EM et al. 2004 Hereditary early-onset Parkinson's disease caused by mutations in *PINK1*. *Science* 304: 1158–1160.

Yang Y et al. 2008 PINK1 regulates mitochondrial dynamics through interaction with the fusion/fission machinery. *Proc. Natl. Acad. Sci. USA*105: 7070–7075.

DJ-1

Andres-Mateos E et al. 2007 *DJ-1* gene deletion reveals that DJ-1 is an atypical peroxiredoxin-like peroxidase. *Proc. Natl. Acad. Sci. USA* 104: 14807–14812.

Bonifati V et al. 2003 Mutations in the DJ-1 gene associated with autosomal recessive early-onset parkinsonism. *Science* 299: 256–259.

Carnet-Avilés RM et al. 2004 The Parkinson's disease protein DJ-1 is neuroprotective due to cysteine-sulfinic acid-driven mitochondrial localization. *Proc. Natl. Acad. Sci. USA*101: 9103–9108.

Zhou W and Freed CR 2005 DJ-1 up-regulates glutathione synthesis during oxidative stress and inhibits A53T α-synuclein toxicity. *J. Biol. Chem.* 280: 43150–43158.

LRRK2

Bosgraaf L and van Haastert PJM 2003 Roc, a Ras/GTPase domain in complex proteins. *Biochem. Biophys. Acta* 1643: 5–10.
Deng J et al. 2008 Structure of the ROC domain from Parkinson's disease-associated leucine-rich repeat kinase 2 reveals a dimeric GTPase. *Proc. Natl. Acad. Sci. USA* 105: 1499–1504.
West AB et al. 2007 Parkinson's disease-associated mutations in *LRRK2* link enhanced GTP-binding and kinase activities to neuronsal toxicity. *Hum. Mol. Genet.* 16: 223–232.

Omi/HtrA2

Plum-Favreau H et al. 2007 The mitochondrial protease HtrA2 is regulated by Parkinson's disease-associated kinase PINK1. *Nat. Cell Biol.* 9: 1243–1252.
Strauss KM et al. 2005 Loss of function mutations in the gene encoding Omi/HtrA2 in Parkinson's disease. *Hum. Mol. Genet.* 14: 2099–2111.

Mitochondrial Dysfunction and Oxidative Stress

Abou-Sleiman PM, Muqit MMK and Wood NW 2006 Expanding insights of mitochondrial dysfunction in Parkinson's disease. *Nat. Rev. Neurosci.* 7: 207–219.
Lin MT and Beal MF 2006 Mitochondrial dysfunction and oxidative stress in neurodegenerative diseases. *Nature* 443: 787–795.

The UPS

Bence NF, Sampat RM and Kopito RR 2001 Impairment of the ubiquitin-proteasome system by protein aggregation. *Science* 292: 1552–1555.
Ciechanover A and Brundin P 2003 The ubiquitin proteasome system in neurodegenerative diseases: sometimes the chicken, sometimes the egg. *Neuron* 40: 427–446.
McNaught KSP et al. 2001 Failure of the ubiquitin-proteasome system in Parkinson's disease. *Nat. Rev. Neurosci.* 2: 589–594.

Aggresome-Autophagy

Hara T et al. Suppression of basal autophagy in neural cells causes neurodegenerative disease in mice. *Nature* 441: 885–889.
Komatsu M et al. 2006 Loss of autophagy in the central nervous system causes neurodegeneration in mice. *Nature* 441: 880–884.
Ravikumar B et al. 2004 Inhibition of mTOR induces autophagy and reduces toxicity of polyglutamine expansions in fly and mouse models of Huntington disease. *Nat. Genet.* 16: 585–595.

HDAC6

Bennett EJ et al. 2005 Global impairment of the ubiquitin-proteasome system by nuclear or cytoplasmic proteins aggregates precedes inclusion body formation. *Mol. Cell* 17: 351–365.

Boyault C et al. 2007 HDAC6 controls major cell response pathways to cytotoxic accumulation of protein aggregates. *Genes Dev.* 21: 2172–2181.

Iwata A et al. 2005 HDAC6 and microtubules are required for autophagic degradation of aggregated Huntingtin. *J. Biol. Chem.* 280: 40282–40292.

Kawaguchi Y et al. 2003 The deacetylase HDAC6 regulates aggresome formation and cell viability in response to misfolded protein stress. *Cell* 115: 727–738.

Olzmann JA et al. 2007 Parkin-mediated K63-linked polyubiquitination targets misfolded DJ-1 to aggresomes via binding to HDAC6. *J. Cell Biol.* 178: 1025–1038.

Pandey LB et al. 2007 HDAC6 rescues neurodegeneration and provides an essential link between autophagy and UPS. *Nature* 447: 859–863.

Chapter 20
Huntington's Disease and Amyotrophic Lateral Sclerosis

Huntington's disease is a fatal disorder affecting neurons in the brain resulting in motor dysfunction and progressing to cognitive decline, dementia, and death some 15–20 years after onset of the illness. This disease was first described by George Huntington in 1872. The primary targets are spiny neurons situated in the striatum that release the neurotransmitter γ-aminobutyric acid (GABA) but in later stages of the illness populations of neurons in the hippocampus and cortex are affected.

The causal agent of the disease is mutated forms of Huntingtin (Htt), a large, highly flexible, and most likely natively unfolded signaling protein composed of 3144 amino acid residues. The most prominent feature of this protein from the disease viewpoint is the presence of large tract of polyglutamine (PolyQ) repeats in the N-terminal region. In individuals with Huntington's disease, there are too many of these repeats. As will be discussed shortly, in the first part of this chapter, the Htt protein misfolds and acquires toxic gain-of-function properties along with loss-of-function defects.

Amyotrophic lateral sclerosis (ALS) was first described 140 years ago by the French pathologist Jean-Martin Charcot. His explorations extended over several years starting in the period 1865–1869. He gave the disease the name amytrophic lateral sclerosis in 1874 to emphasize its two primary features – lateral sclerosis and amyotrophia. ALS is a disease affecting upper and lower motor neurons. The upper neurons involved in this disease are located in the motor cortex; these cells convey signals and control lower motor neurons situated in the brainstem/spinal cord which, in turn, convey signals to the muscles. In ALS, both kinds of neurons degenerate and lose axons resulting in the "myelin pallor" observed in the lateral portion of the spinal cord by Charcot (feature 1) The retraction of presynaptic axons from muscles results in their atrophy, and to paralysis, respiratory failure, and death (feature 2).

Amyotrophic lateral sclerosis is referred to as Lou Gherig's disease in the United States and as Charcot's disease elsewhere. It is an adult-onset, fatal disorder primarily affecting adults 45–60 years old. Most cases are sporadic (sALS), but roughly 10% are inherited (fALS). About 20% of the inherited (fALS) cases involve mutations to the superoxide dismutase 1 (SOD1) gene. In

M. Beckerman, *Cellular Signaling in Health and Disease*,
Biological and Medical Physics, Biomedical Engineering,
DOI 10.1007/978-0-387-98173-4_20, © Springer Science+Business Media, LLC 2009

contrast to the other protein-folding diseases, aggregates do not develop in all cases pointing to a nonaggregate mechanism underlying the toxicity. A diverse set of more than 100 mutations in the gene encoding SOD1 have been found and these produce a variety of alterations in the protein. A central question has been identifying the common factor(s) that make the mutant gene product toxic. To answer this, one must look at how the protein folds and how metalloenzymes are formed. This question and its possible answers will be the starting point for an examination of ALS in the second part of this chapter.

20.1 Huntington's Disease Is an Expanded PolyQ Repeat Disorder

Nine disorders involving inappropriate expansions of polyQ tracts have been uncovered to-date. A summary of these diseases is presented in Table 20.1. Each disease is brought on by expanded CAG repeats in genes that encode the otherwise unrelated proteins listed in the last column of the table. In all situations, proteins containing fewer than 38 repeats function normally while those containing more than 41 promote disease. The proteins subjected to the poly-

Table 20.1 Polyglutamine (CAG) repeat disorders

Disease	Gene	Protein	Function
Spinal and bulbar muscular atrophy (SBMA)	AR	Androgen receptor	Steroid hormone receptor
Huntington's disease (HD)	HTT	Huntingtin	Signaling, transport, transcription
DentatoRubral and PallidoLuysian atrophy (DRPLA)	DRPLA	Atrophin-1	Transcription
Spino-cerebellar ataxia 1	ATXN1	Ataxin-1	Transcription
Spino-cerebellar ataxia 2	ATXN2	Ataxin-2	RNA metabolism
Spino-cerebellar ataxia 3	ATXN3	Ataxin-3	De-ubiquitination
Spino-cerebellar ataxia 6	CACNA1A	α1A-VDCC subunit	P/Q-type VDCC subunit
Spino-cerebellar ataxia 7	ATXN7	Ataxin-7	Transcription
Spino-cerebellar ataxia 17	TBP	TATA-binding protein	Transcription

glutamine expansions misfold into aggregation- and inappropriate-association prone conformations, acquire and generate new functions and lose old ones. The increased number of repeats results in formation of soluble oligomers, buildup of insoluble aggregates, and formation of inclusion bodies.

Neurons in several areas of the brain are affected in these diseases. Prominent among these regions are spinal motor nuclei, brainstem nuclei, basal ganglia, and cerebellum. The clinical features of each of these diseases are distinct, however, and depend crucially on the repeat length and the structural properties and associated normal cellular functions of the protein. Understanding the proteins' normal cellular function is turning out to be critical not only for loss-of-function mechanisms but also for the gain-of-function ones.

The proteins responsible for these diseases are all involved in signaling either directly or indirectly through a support function. As is the case for all neurodegenerative disorders one of the paramount questions that needs to be answered for each disease is why that particular set of cells is affected and not others, particularly in cases where the responsible protein is broadly expressed in other neural regions, and even throughout the body and not just in the brain. That is, there is a clear dependence upon cellular signaling context, and this needs to be understood too.

The first entry in the Table 20.1 is (X-linked) spinal and bulbar muscular atrophy, also known as Kennedy disease. This disease is characterized by muscle weakness and wasting in arms and legs, and spreads to (bulbar) muscles in the face and throat. This disorder is caused by expanded polyQ tracts in the androgen receptor (AR), a member of the nuclear receptor family that upon ligand binding translocates to the nucleus where it stimulates gene transcription. Normal forms of the AR contain from 11 to 35 CAG triplets. Like the other entries in the table, this tract is expanded in disease-causing forms, which contain from 40 to 62 copies of the triplet.

All of the diseases with the exception of SBMA (Kennedy's disease) are inherited in an autosomal dominant fashion. Kennedy's disease, the first PolyQ disease to be discovered (1991), is inherited in an X-linked autosomal recessive manner. One of the most striking features of these diseases is the dynamic character of the PolyQ expansions – they tend to increase in size from generation to generation and these changes correlate with a growing severity of the illnesses in successive generations of offspring. Another general property of these diseases is the presence of inclusion bodies in the nucleus; these are termed neuronal intranuclear inclusions, or NIIs.

Spino-Cerebellar ataxia 1 (SCA1) is caused by expanded polyglutamine tract in the N-terminal of human Ataxin-1 protein. The disease begins in adulthood with a progressive loss of balance and coordination, advancing in time to swallowing and breathing difficulties. The primary neural targets are cerebellar Purkinje neurons, the inferior olive, brain-stem nuclei, and spinocerebellar tracts.

The Ataxin-1 protein is roughly 800 amino acid residues long. In addition to its PolyQ tract, it possesses a C-terminal AXH domain involved in transcription and RNA binding as well as several other cellular activities. AXH domain is important for the pathology illustrating that regions of the protein outside the polyQ tract along with native function of the wt protein can be important contributors to the pathology. For example, SCA1 depends in part on interactions between Ataxin-1 and the 14-3-3 protein and this interaction, in turn, is controlled by Akt-mediated phosphorylation of Ataxin-1 on S776.

20.2 The Structure of the Huntingtin Protein Is That of a Multipurpose Signaling Organizer

Huntingtin is a scaffolding protein. It is localized in synaptic terminals where it mediates a number of signaling activities. It interacts with cytoskeleton proteins and with synaptic vesicle protein required for exocytosis and endocytosis. The organization of the Huntingtin protein is depicted in Fig. 20.1. The PolyQ, or CAG tract, begins at about residue 15. As shown in the figure, this variable section is immediately followed by a polyproline (PolyP) tract and then by a series of HEAT repeats that spans most of the protein.

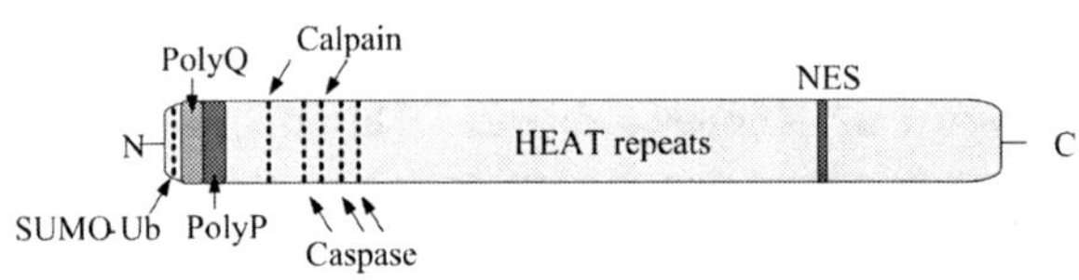

Fig. 20.1 Domain structure of the Huntingtin protein. Calpain cleavage sites are located at amino acid residues 469 and 536; caspase cleavage sites are situated at 513 (caspase 3), 552 (caspase 3), and 589 (caspase 6). NES: nuclear export sequence

Recall from Chapter 11 that a HEAT repeat is a short helix-loop-helix hairpin-shaped segment. These are arranged one after another to form long bent rod-like helical structures. Prominent examples of HEAT repeat proteins are the ATM and ATR proteins crucially involved in DNA damage signaling and checkpointing. The HEAT repeats in these proteins serve as flexible scaffolds, or platforms, for attachment of signaling and structural proteins. The Huntingtin PolyP stretch mediates protein–protein interactions, most notably those involving WW and SH3 domains that bind PP sequences. The PolyP stretch helps sequester vesicle-associated proteins in IBs and modulates aggregation and toxicity. Embedded in the HEAT repeat region are a number of caspase cleavage sites, and also a pair of calpain cleavage sites. The extreme N-terminus contains a site for posttranslational attachment of SUMO or ubiquitin groups and the C-terminal region contains a NES and a NLS (not shown).

20.3 Synaptic Terminal Interactions Occur

One of the earliest, if not the earliest, effect of expanded full-length Huntingtin occurs in presynaptic terminals. Recall that calcium influx through voltage-dependent calcium channels in response to membrane depolarization stimulates neurotransmitter release in a process mediated by SNARE proteins. Mutant full-length Huntingtin interferes with this process; in particular it disrupts the actions of the SNARE protein family member syntaxin 1A, which has an inhibitory role, and with other components of the neurotransmitter release machinery. These actions result in increased calcium influx into the presynaptic terminals, loss in calcium homeostasis, and increased neurotransmitter release from presynaptic vesicles. Activated proteolytic enzymes such as the caspases and calpains generate Huntingtin fragments that further disrupt the release machinery in the presynaptic terminals.

Huntingtin proteins, especially those with expanded N-terminal PolyQ tracts, interact with a number of substrates involved with calcium signaling and homeostasis. On the postsynaptic side, two of the best characterized of these are the PSD95 scaffolding protein that binds and regulates calcium entry through NR1/NR2B NMDA receptors, and the Htt-associated protein 1 (HAP1) that binds and regulates calcium release through the Type 1 inositol 1,4,5-triphosphate receptor ($InsP_3R1$). The first of these interactions results in an increased influx of Ca^{2+} into the cell through the NMDA receptors in response to glutamate binding (and membrane depolarization). The second type of interaction potentiates the release of Ca^{2+} ions into the cytosol from intracellular stores in response to $InsP_3$ binding.

The overall result of these interactions is disturbed Ca^{2+} signaling in the cytosol that spreads to the mitochondria. This leads to mitochondrial calcium overload. The deranged calcium signaling activates proteolytic enzymes such as the caspases and the calcium-dependent calpains. Apoptosis becomes more likely and N-terminal Htt cleavage products are generated in greater numbers.

Several caspase and calpain cleavage sites are present on Htt, and when cleavage occurs at these sites variously sized N-terminal fragments are generated. Calpains are Ca^{2+}-dependent cysteine proteases. When activated in synaptic terminals these proteases target critical signaling elements thereby permitting synaptic remodeling critical for synaptic plasticity and normal physiological functions such as memory formation. Prominent calpain substrates are PSD-95, glutamate receptors, and structural proteins (e.g., spectrins, neurofilaments, and tau). Under normal operating conditions, calpain activity is maintained at a low, controlled level, but Ca^{2+} overload triggers calpain hyperactivation as part of a protective feedback response aimed at substrates that promote calcium entry.

Mutant Htt destabilizes calcium homeostasis leading to activation of the caspases and calpians. The N-terminal fragments that are generated are toxic to varying degrees depending on the length of the fragment and the size of the polyQ tract. The fragments generated in this fashion are capable of attachment to membranes, and aggregate into cytoplasmic and nuclear inclusions, the NIIs mentioned earlier in the chapter. These cleavage activities, particularly those of the calpains, and antagonists of calcium channels, present a possible target for therapeutic intervention, the goal being to prevent sustained, elevated calcium entry as this is one of the very earliest events in the disease progression.

20.4 Impaired Fast Axonal Transport Happens

The buildup of Huntingtin aggregates impedes fast axonal transport; that is the rapid movement of cargo along microtubule tracks. Recall that two types of motor proteins – dynein and kinesin – mediate the movements of cargo. These transport operations both anterograde out from the cell body along the axons, and retrograde back from the ends of the axons to the cell body are essential; interruptions and impairments reduce the ability of the neurons to carry out their signaling roles. Huntingtin in its normal form is involved in these transport actions through interactions with HAP1 and motor protein subunits. Mutant Huntingtin and especially mutant N-terminal fragments interfere with the normal functioning of the dynein and kinesin motor proteins. The result is impaired axonal transport of organelles and signaling proteins.

One of the signaling molecules affected by mutant-Huntingtin-impaired fast axonal transport is brain-derived neurotrophic factor (BDNF), a protein that promotes neural survival. Under healthy conditions, Huntingtin associates with the motor proteins enhancing the velocity of vesicles and diminishing their pause time. In Huntington's disease situations, mutant Huntingtin interferes with the normal operations of the motor proteins resulting in a diminished transport of crucial signaling proteins such as BDNF. Reductions in this trophic factor contribute to diminished survival of the affected neurons.

The Huntingtin-interacting protein 1 (HIP1) is an adaptor protein that participates in clathrin-mediated receptor uptake and recycling of AMPA (and NMDA) receptors. A key structural feature of HIP1 is the presence of an adaptor-protein 180 (AP180) NH2-terminal homology (ANTH) domain capable of simultaneously binding phosphatidylinositol-bearing lipids such as PtdIns(4,5)P2 and clathrin as part of the AP-2 endocytosis complex. Both HAP1 and HIP1 contain coiled-coil domains and interact with each other, thereby providing a link between endocytic activity and microtubule-mediated vesicle transport. In addition to acting as a member of an endocytic adaptor complex, HIP1 shuttles between the plasma membrane and the nucleus where it associates with several types of nuclear receptors including the androgen receptor and functions as a transcription coregulator.

20.5 Zippers, Aggregation, Fibrils, Inclusion Body Formation, and Toxicity

The Huntingtin gene is unstable. The tendency is for the number of PolyQ repeats to increase from generation to generation. Along with that increase, the disease itself increases in severity since Huntingtin lethality is positively correlated with the length of its PolyQ tract. The mechanism of repeat instability is tied to slippage in the lagging strand during DNA replication and repair. Secondary structure elements, that is, loops or hairpins, that may form are a key factor along with the DNA repair processes brought on by gaps that can occur as a result of the hairpin formation.

The Huntingtin protein with its PolyQ tract of regularly spaced glutamines is able to bind other polyQ-containing proteins, most notably transcription factors, through a polar zipper mechanism. A widespread example of a zipper mechanism is the leucine zipper found in a variety of transcription factors. In this type of structure, first presented in 1988, hydrophobic leucine residues are positioned at every seventh amino acid position in a chain. The chain folds into a short α-helix characterized by 3.5 residues per turn. A pair of these chains then forms the leucine zipper in which leucines from one chain in contact with the leucines from the partner chain to form a strong, highly stable coiled-coil structure.

The polar zipper structure, based on repeating glutamines, was introduced by Max Perutz in 1994. In this kind of structure, the chains form β-sheets strongly held together by hydrogen bonds between their main-chain and side-chain amides. In the case of the expanded PolyQ Huntingtin protein, the proteins misfold into β-sheets with ~20 residues per turn that oligomerize and form fibrillar structures, characterized by Perutz as water-filled nanotubes. Because of their increased stability, these structures are resistant to degradation and clearance by the ubiquitin-proteasome system.

An emerging consensus is that like the other neurodegenerative diseases, small oligomers are likely to be more toxic than the large aggregates and nuclear or cytoplasmic inclusions. These larger deposits may form with the assistance of molecular chaperones such as TriC. Recall from Chapter 18 that molecular chaperones such as Hsp70 and TRiC assist in the folding of polypeptide chains in the cytosol. Hsp70 acts first to assist in folding nascent chains as they exit the translation apparatus. The TRiC chaperone then takes over to further mediate the proper folding of the protein. These chaperones limit the toxicity of improperly folded proteins. In the case of the Huntingtin protein, these chaperones influence how these proteins aggregate. They direct the assembly of PolyQ-expanded Huntingtin toward their assembly into relatively benign ~500 kDa assemblages and away from forming more toxic, smaller ~200 kDa assemblages.

The various stages of aggregate buildup are depicted schematically in Fig. 20.2. As shown in this figure, the mutant Huntingtin proteins misfold into β-sheets and helices. These aggregate to form small oligomers, which can aggregate into yet larger structures such as fibrils and fibers. Chaperones promote their aggregation into the larger and mostly benign structures and eventually inclusion bodies form in the cytoplasm and especially in the nucleus. If these buildups continue they may disrupt normal cellular functions but otherwise the consensus in the field is that they are protective, a means of isolating potentially harmful structures.

There is some uncertainty as to the relative importance with regard to toxicity from full-length versus proteolytically generated fragments of the Huntingtin proteins. The underlying picture is one in which a progressive

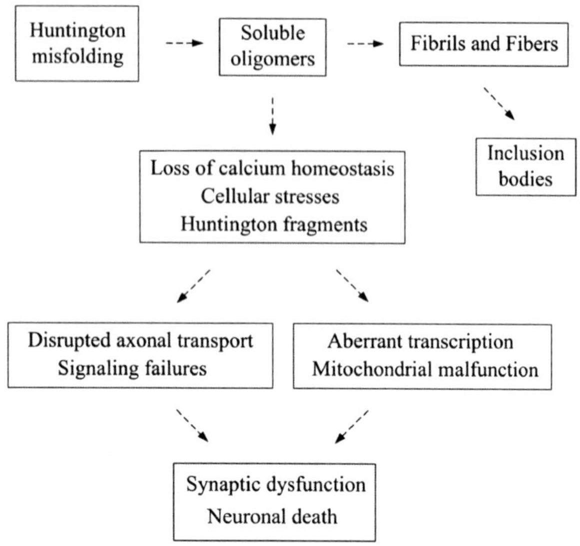

Fig. 20.2 Huntington's disease progressions

buildup of aggregates over time causes stresses that stimulate the activation of caspases and other proteolytic enzymes. These cleave the Huntingtin protein into variously sized fragments at, for example, the cleavage sites indicated in Fig. 20.1. These fragments tend to form inappropriate associations with cellular components leading to additional cellular stresses and damage. The resulting stages of Huntington disease progression triggered by the mutant and misfolded Huntingtin will now be examined.

20.6 The Ubiquitin-Proteasome System Regulates Synaptic Transmission and This Function Is Impaired by Mutant Htt

Three factors specific to neurons help explain why these cells and not others expressing the same aberrant proteins are affected. The first of these is the postmitotic character of neurons. This allows for the slow buildup of defects that would be slight and nonfatal over short time-scales, but in the case of neurons leads to the onset and progression of fatal disorders over time-scales of months and years. The second key factor is the high metabolic load and attendant oxidative stresses in neurons. These have already been examined in the context of Alzheimer's disease and Parkinson's disease. The third, which will be examined in more detail in this chapter, is the existence in neurons, and not other cell types, of processes, the axons and dendrites, with their long and narrow shape, and having at their ends rich signaling machineries. Maintenance of protein homeostasis in such a structure is a grave logistical challenge; it is met extraordinarily well over the first part of life but less so as one ages.

As seen in the last chapter the ubiquitin-proteasome system is a critical part of a neuron's protein homeostasis machinery. It is an important regulator of signaling at both pre- and postsynaptic terminals. At presynaptic terminals, the several families of E3 ubiquitin ligases along with a large family of deubiquitinating enzymes (DUBs) maintain ubiquitin homeostasis and regulate the timing and strength of neurotransmitter release. They do so by ubiquitinating and degrading proteins that control neurotransmitter-vesicle priming, fusion, and recycling.

At postsynaptic terminals, the UPS functions as a critical regulator of synaptic remodeling and plasticity. This system targets a number of key regulators of synaptic plasticity such as the PSD-95 scaffolding protein and the turnover of AMPA and NMDA receptors associated with it. Preserving a balance between synthesis and turnover of these receptors maintains synaptic plasticity viability and prevents excessive synaptic activity.

Mutant Htt proteins impede the UPS in pre- and postsynaptic terminals. In Huntington's disease, the proteasome is unable to digest polyQ filaments and release polyQ peptides. Instead, these structures block proteasomes thereby preventing the degradation of their normal substrates. These events occur early in the development of HD, before inclusion bodies form. The UPS is an ATP-dependent process. One of the drivers for impairment of the UPS is the failure to properly transport mitochondria to locations where they are needed resulting in an inadequate ATP supply for proteasomes. Mutant Htt impairs this transport, as well. The overall result of these shortcomings is the progressive loss of protein homeostasis leading first to synaptic dysfunction and then to axonal degeneration.

20.7 Impaired Transcription: CBP and PGC-1 – and Mitochondrial Dysfunction

In addition to acting in the axoplasm, mutant Htt accumulates in the nucleus where it alters normal gene transcription. Nuclear targets of expanded Htt include the CREB-binding protein (CBP) and PPARγ-coactivator 1α (PGC-1α). Recall that CBP is a histone acetyltransferase. This protein binds mutant Huntingtin and aggregates along with it in the nucleus. The amount of CBP available for its normal HAT activities is reduced as a consequence of this sequestration. Histone acetylation is reduced, and the resulting hypoacetylation leads to altered gene transcription patterns. These alterations may contribute to the repeat instability discussed in Section 20.2.

The disruptions of normal gene transcription are enhanced by mutant Htt fragments. Soluble monomers and small soluble oligomers of these aberrant proteolytically cleaved forms fold in a way that enables them to bind PolyQ sequences in CBP and in the TATA box binding protein (TBP). These alterations are a major contributor to cellular toxicity along with Htt's interference with the transcriptional coactivator PGC-1α.

PGC-1α has a key role in mitochondrial biogenesis and function and in regulation of ROS. This transcription coactivator forms transcription activating complexes with CREB, TBP, TAF4, and TFIID at PGC-1α promoters where it helps transcribe genes encoding ROS handling superoxide dismutase 1 and 2 (SOD1, 2), and glutathione peroxidase (GPx1) and the adaptive thermogenesis mediator uncoupling protein 1 (UCP-1). Mutant Huntingtin binds and interferes with components of this complex. The result is impaired transcription of detoxification enzymes and a buildup of ROS. The increases in ROS together with excessive buildups of calcium result in mitochondrial dysfunction.

Another transcription factor targeted by mutant Huntingtin is Sp1. This transcription factor regulates a variety of genes through its ability to recruit the general transcription factor TFIID to promoters. The latter is a complex containing TBP and a number of TBP-associated factors, or TAFs. Mutant Htt proteins bind and disrupt the normal functions of Sp1 and TAFII130. These actions alter the patterns of gene expression of Sp1 target genes in a cell-context-dependent manner. The overall result of these deleterious activities is synaptic dysfunction and neuronal death as indicated in Fig. 20.2.

20.8 Structure and Folding of the Superoxide Dismutase Protein SOD1

Turning now to ALS, Superoxide dismutase 1 (SOD1) is a small, 153 amino acid long metalloenzyme, formerly called Cu,Zn-superoxide dismutase. This enzyme is an important member of the cell's redox control machinery, being responsible for catalyzing the conversion of superoxide to hydrogen peroxide. To become enzymatically active, the SOD1 monomeric apoprotein must acquire a copper ion and a zinc ion and form an intramolecular disulfide bond. These posttranslational modifications stabilize the catalytically active, dimeric form of the enzyme. Copper is delivered to SOD1 by the copper chaperone for superoxide dismutase (CCS). There are relatively few proteins requiring copper as a cofactor and this delivery mechanism is an effective one, serving also in a protective role against toxic effect of copper ions. In contrast, there is an abundance of zinc-binding proteins. Perhaps as a result no zinc chaperone has been found; rather it appears that zinc arrives where it is needed through simple diffusion.

The three-dimensional structure of the SOD1 holodimer as determined by means of X-ray crystallography is presented in Fig. 20.3. The SOD1 monomer folds into flattened Greek-key β-barrel erected out of eight antiparallel β-strands. In this figure, the four cysteine residues present in an SOD1 monomer are highlighted. Two of these, Cys57 and Cys146, form the intramolecular disulfide bond while the other two residues, Cys6 and Cys111, remain unaffected.

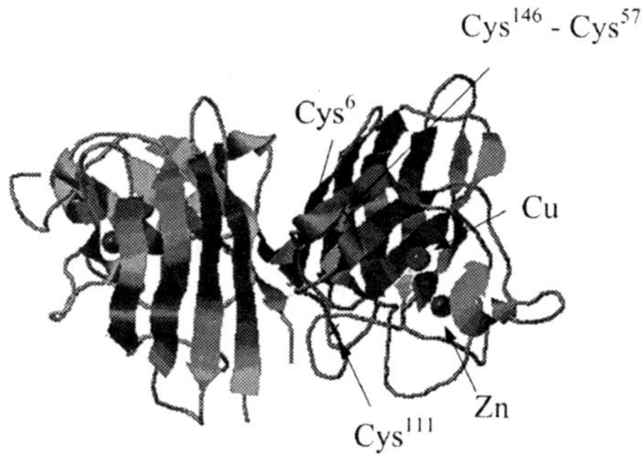

Fig. 20.3 Crystal structure of the human SOD1 holodimer determined by means of X-ray crystallography. Small xs mark the locations of the four crucial cysteine residues while balls denote the locations of the Cu and Zn atoms. The figure was prepared using Jmol with atomic coordinates deposited in the PDB under accession number 1hl5

20.9 SOD1 Mutations and Aggregation

More than 100 ALS-causing single point mutations in SOD1 have been identified. Many of the mutations seen in fALS appear to disrupt the folding of the apoSOD1 monomers. As can be seen in Fig. 20.3, the SOD1 monomer is largely composed of β-strands. The mutations destabilize the caps that form at the extrema of the folded protein and protect the β-sheet structures from undesirable side reactions. As a result the proteins become "sticky". This is similar to the pathology seen in β_2m (see Chapter 17) where the protective role of the edge strands becomes compromised leading to aggregation and other defects.

A number of general observations can be made from an examination of the mutations. One of these is that the main result of the aggregation-stimulating mutations seems to be a gain-of-toxic activity rather than a loss of enzymatic activity. Another observation is that mutations that prevent monomers from acquiring their metal cofactors have a greater destabilizing effect than mutations that leave this capability alone. Monomers lacking their Cu and Zn ions are more prone to unfold and refold into aggregation-prone conformations. Disulfide bonds are another major factor. Incorrect disulfide bonds do not lead to dimerization but instead promote multimeric aggregation from destabilized monomeric folding intermediates.

Yet another theme that has emerged is that of oxidative damage leading to misfolding and aggregation. The central observation here is that fully metalized, properly disulfide-linked SOD1 is extraordinarily stable. In order to dispose of the enzyme by proteasomes when no longer needed, it first has to be destabilized. This is accomplished through a negative feedback loop in which a buildup in hydrogen peroxide leads to a loss of metal ions and destabilization as a preliminary step in the pathway to its degradation. The resulting metal-free monomers are sensitive to oxidative misfolding and aggregation and, as was the case for many of the mutated forms, might serve as seeds for aggregate development. Oxidative stress-induced aggregation thus joins the other mechanisms just discussed as a possible unifying theme linking sALS to fALS.

One other observation is important, this one pertaining to the folding of SOD1 into its native state. The disease-causing mutations in SOD1 are distributed throughout the proteins and not all produce the same effect on stability. What these mutations have in common is a tendency to alter the folding kinetics. Small single-domain proteins such as SOD1 should fold fairly rapidly into its native state. That point was emphasized in Chapter 16. In contrast to this expectation, mutant SOD1 proteins take orders of magnitude longer to find their native state than the 1 sec fast folding time frame that is characteristic of small proteins. The energy landscapes for these variants are rugged and the folding pathways are populated by kinetic traps. The formation of a variety of folding intermediates and apo monomers may provide ample opportunity for the sampling of aggregation-prone conformers.

20.10 Impaired Fast Axonal Transport and Retraction of Axons from Synapses

One of the main factors that ties specificity of these diseases to neurons is the extreme polarity of these cells. They possess processes – axons and dendrites – that extend over large distances requiring active transport of organelles, vesicles, cytoskeleton components, and signaling proteins between the cell body and the terminal regions. Interruptions and malfunctions in this machinery contribute substantially to impaired transport, synaptic failure, and neurodegeneration in Alzheimer's disease and Huntington's disease, and there is emerging evidence for a similar involvement in ALS.

The active transport is carried out by molecular motor complexes that travel along tracks built from actin filaments and microtubules. Three kinds are motor complexes are utilized. Myosin motors travel along actin filaments while kinesin and dynein motor complexes travel along microtubules. Kinesins mediate the anterograde transport from the cell body out along axons and dyneins move cargo in the opposite or retrograde direction back to the cell body. Similar complexes carry out comparable functions in dendrites.

In general, microtubule-mediated transport is used to move cargo over long distances, while actin filaments mediate transport locally over short distances. In fast axonal transport (FAT), cargo is transported along microtubule tracks at rates up to 1 μ/sec. Slow axonal transport takes place at speeds 100 times lower, on the order of 1 mm/day. Mitochondria are transported by FAT, but at reduced rates, punctuated by stops and reverses designed to uniformly distribute mitochondria throughout the axon. Motor neurons are particularly susceptible to disruption in these transport systems because they have long axons.

Mutated SOD1 interferes with fast axonal transport. In particular, it interacts with the dynein complex and disrupts retrograde transport thereby reducing its efficiency. Trophic factor delivery to the cell body is impaired and SOD1 is mislocated to mitochondria. Both anterograde and retrograde transport are interfered with, and mitochondrial FAT is impacted leading to a failure to uniformly distribute these energy-bearing organelles. Thus, both function and localization of mitochondria are affected leading to degradation of axonal transport, synaptic function, and cellular survival.

In more detail, one of the clear indications of fALS and also sALS in postmortem and biopsy samples is abnormal mitochondria. Gross alterations are observed in their structure, number, and localization. SOD1 is found mostly in the cytosol. However, a fraction of the mutated SOD1 forms localize to the mitochondria. They can be found attached to the cytoplasmic face of the outer mitochondrial membrane, in the intramembrane spaces, and within the mitochondrial matrix. This relocation is specific to mutated SOD1 in the affected neural populations. The mutated SOD1 proteins are prone to aggregate and undergo inappropriate interactions. However, the nature of the interactions and mechanisms linking the accumulation of SOD1 to the mitochondrial pathologies are as yet unknown.

20.11 A Model for Amyotrophic Lateral Sclerosis

The focus in the preceding discussion has been on mutant SOD1 and how it can lead amyotrophic lateral sclerosis. Mutations in genes that encode other proteins and when mutated can also produce fALS have been found. However, studies of these rare instances so far have not yielded major new insights into the disease and what can be done to stop its progression. Furthermore, superoxide dismutase 1 may well be the main culprit in both fALS and sALS. Evidence in support of this is provided by clinical studies showing that the two forms of the disease are indistinguishable. They affect the same populations of neurons and in both forms the neurons contain ubiquitin-bearing aggregates. The stages of ALS as they are currently understood in terms of misfolded and aggregated SOD1 are depicted in Fig. 20.4.

As was the case for Huntington's disease as represented by Fig. 20.2, the key starting point for ALS is the progressive deterioration of the synaptic terminals on both sides of the synaptic cleft arising in this case from misfolded superoxide dismutase. A number of defects are produced ranging from loss of calcium homeostasis and excessive quantities of neurotrasmitters released from the presynaptic terminals to blockage of vesicle and organelle transport and damage to mitochondria in dendrites and axons. Among the consequences include inadequate energy production and distribution, excitotoxicity, garbage collection breakdown, and loss of synaptic plasticity. These would then lead to axon retraction. In humans, motor neuron axons can exceed 1 m in length, and

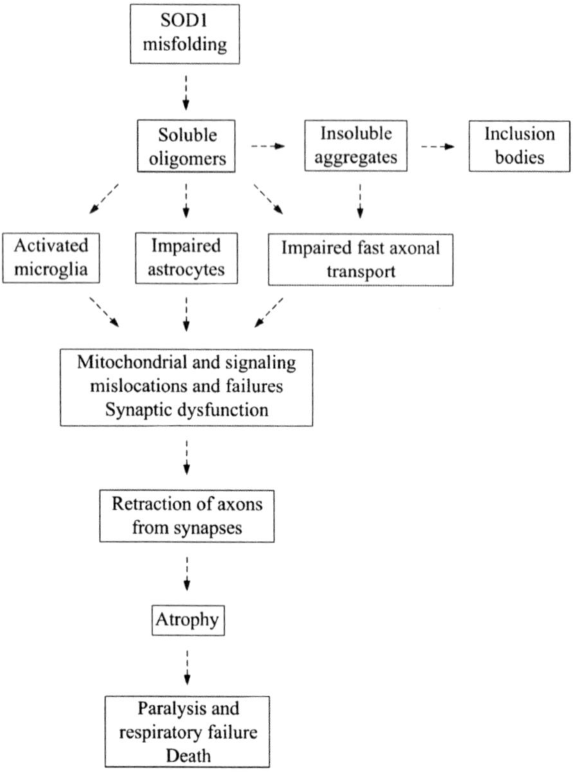

Fig. 20.4 Amyotrophic lateral sclerosis disease progressions

because of this are particularly vulnerable to breakdowns and blockages in the microtubule railways and their motors that deliver cargo.

In amyotrophic lateral sclerosis, several cell types contribute to the disease progression. As was the case for other neurodegenerative disorders neurons, astrocytes, and activated microglia send and receive signals and at some point produce a neuroinflammatory microenvironment in which agents such as nitric oxide and TNFα are secreted. In the SOD1-centric picture presented here, a key trigger would be the secretion of mutated SOD1 oligomers from neurons plus the expression in the glia of mutated SOD1 forms.

20.12 Acceleration of ALS Through Interactions Between Neurons and Other Cellular Residents of Its Microenvironment

Recall from Chapter 17 that astrocytes support and protect neurons. They supply neurons with nutrients and protection. They regulate the chemical and fluid milieu, maintain neurotransmitter homeostasis in and near the synaptic cleft, and help regulate synaptic plasticity. In carrying out these roles, astrocytes located in different parts of the brain encounter different kinds of neurons and express appropriately different sets of proteins. The astrocytes involved in protecting the motor neurons express the glutamate uptake transporter EAAT2. The uptake of glutamate prevents excessive glutamate buildup in and around the synaptic cleft from the glutamate-releasing neurons that make synaptic contact with motor neurons that in turn drive muscle activity. Another way these astrocytes protect the motor neurons is by regulating their expression of the AMPA receptor subunit GluR2. This subunit controls the permeability of the AMPA receptor to Ca^{2+} ions. When present, the receptors are impermeable to these ions, but this property is lost if GluR2s are not incorporated into the receptors. In this case, large number of GluR2-lacking AMPA receptors can produce excitotoxicity.

These two protective activities are compromised in the presence of mutated SOD1. The mutated forms of SOD1 interfere with the ability of the astrocytes to express EAAT2 and regulate the motor neuron's expression of GluR2. The mechanistic steps linking mutated SOD1 to these negative consequences are not yet known. One of the key steps is the release of soluble toxic factors by the astrocytes; possible agents range from cytokines to ROS to secreted mutated SOD1.

The other major cellular resident in the local microenvironment of the neurons is the microglia. These cells are smaller than astrocytes; hence, the name microglia while astrocytes are referred to as macroglia. Recall that microglial cells are phagocytes that migrate to the central nervous system during development and become the immunological sentinels of the brain. They are mobile and continuously monitor the microenvironment for signs of injuries such as stroke and tumor formation, bacterial and viral pathogens, and toxins. Once damage is perceived, they migrate to site of the damage where they engulf and remove tissue debris.

SOD1 functions not only as a superoxide dismutase but also as a signaling protein functioning as a component of a feedback control system that regulates microglial NADPH oxidase (Nox) activity in response to ROS levels. As discussed previously, microglia utilize Nox as part of their normal cellular role as phagocytes. This multicomponent enzyme complex facilitates the killing of invading microbes by generating superoxide in response to inflammatory stimuli.

Recall that the small GTPase, Rac1, functions as a regulatory subunit of the NADPH oxidase complex. SOD1 carries out its regulatory role by binding Rac1 and stabilizing its activity by inhibiting GAP-mediated hydrolysis. This inhibition is terminated by elevated amounts of H_2O_2 relative to superoxide levels. Under these oxidized conditions, SOD1 decouples from Rac1, which then deactivates upon conversion to the GDP-bound form leading to reductions in Nox activity. This normal regulatory circuit is depicted in Fig. 20.5(a,b). Some ALS-mutated SOD1 forms increase the activity of this ROS-generating enzymatic complex when they bind to Rac1. These mutated SOD1 proteins maintain the Rac1-GTP bound state even in the presence of elevated H_2O_2 levels leading to a sustained production of superoxide by the Nox enzymes as illustrated in Fig. 20.5(c,d). This sustained production of ROS helps create a

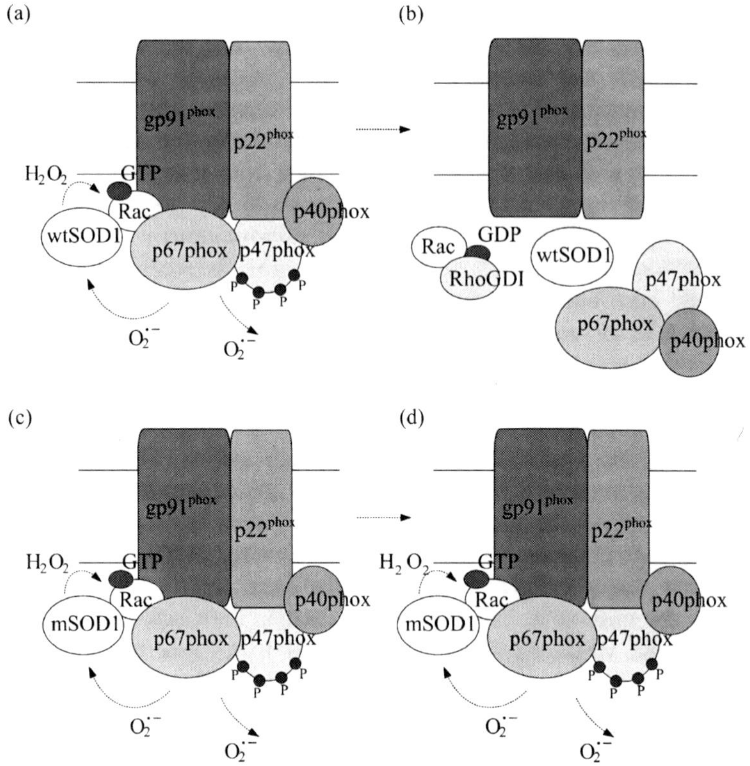

Fig. 20.5 Regulation of the Nox complex by SOD1. (**a**) and (**b**) Normal regulation by wtSOD1; (**c**) and (**d**) abnormal regulation by mutant SOD1 (mSOD1)

hostile local microenvironment that may drive the disease progression. Although extent to which this model is applicable to sALS is an open question at this time, the observation is a significant one and serves as yet another warning that redox balance and cell signaling are deeply intertwined in multiple ways and simplistic approaches to manipulating them will not work.

20.13 PolyQs, Mutant SOD1, and Impaired ERAD

The prompt removal of misfolded proteins maintains protein homeostasis and health. Several systems including ERAD/proteosomal and aggresomal pathways have key roles in ensuring that buildups of malfunctioning and potentially toxic proteins do not occur. These aspects were discussed in the last two chapters; they are central to the pathogenesis of Parkinson's disease and are believed to have a role in amyotrophic lateral sclerosis and polyQ disorders such as Huntington's disease.

ATXN3, or SCA3, is the causative agent in spinocerebellar ataxia type 3, the most widely encountered autosomal dominantly inherited ataxia. This protein contains an N-terminal Josephin domain (JD), a pair of ubiquitin-binding motifs, a series of polyQ repeats, and a C-terminal variable domain. It functions as a deubiquitinating enzyme that cleaves UB-48 and Ub-63 chains. Under normal conditions it helps regulate the retrotranslocation of ERAD substrates and serves a protective role, especially against polyQ toxicity. However, when the number of expanded polyQ repeats exceeds 50, the enzyme misfolds and because of its abnormal conformation it disrupts the normal interactions between p97, Ufd1, and Npl4 resulting in impaired ERAD.

A similar set of events occurs in Huntington's disease. In this disorder, the ERAD route central to the ER stress response is affected negatively. The misfolded Htt polyQ generates ER stresses and, in response, stress proteins such as BiP, PDI, and CHOP are upregulated. However, and most importantly, the Htt-expanded polyQ fragments entrap essential components of the ERAD; in particular, the fragments attach to the p97/Npl4/Ufd1 complex causing the ERAD to malfunction. As a result misfolded proteins accumulate in the ER, and protein homeostasis is lost.

An impaired ERAD is also implicated in ALS. Mutated and misfolded SOD1 interacts with Derlin-1, a component of the ERAD machinery, and, as a consequence, misfolded proteins are not exported out of the ER to the proteasome. Derlin-1 is a four-pass ER membrane protein with both N- and C-terminals located in the cytosol. Mutated SOD1 interacts with the C-terminal portion of Derlin-1 and inhibits proper function of the ERAD. The resulting buildup in ER stresses activates ASK1. This ER stress-activated pathway is anchored by the ER stress sensor IRE1, which signals to and activates ASK1 via the adapter protein TRAF2. The ASK protein, in turn, activates JNK and p38 MAP kinases and, under the influence of sustained mutant SOD1 signaling, may stimulate the expression of pro-apoptotic Bcl2 proteins thereby increasing the likelihood that the affected cells undergo apoptosis.

20.14 Mutations in Genes Other Than That for SOD1 Can Cause fALS

The transactive response (TAR) DNA-binding protein 43 (TDP-43) is a 414 amino residue long protein widely expressed throughout the body. It contains two RNA recognition motifs RRM1 and RRM2 in its central region and a C-terminal glycine-rich domain that mediates protein–protein interactions. The involvement of TDP-43 in neurodegeneration was discovered in 2006. These finding were followed in 2008 by the discovery of a cluster of disease-causing mutations, all but one in the C-terminal region and the lone exception in

RRM1. The existence of disease-related mutations implies that TDP-43 has an active role in the disease progression. Further evidence for a causal role was supplied by yeast data showing that the C-terminal RNA recognition motifs are centrally involved in the pathology, and toxicity is not just a consequence of increased cellular stress brought on by accumulation of misfolded TDP-43.

TDP-43 has been studied previously with respect to its function as a transcription repressor and as a splicing factor involved in exon skipping. In particular, it was found to be a component of a complex involved in the disease-associated skipping of exon 9 of the cystic fibrosis conductance regulator (CFTR) gene. Other roles have since then been identified for this protein. The protein shuttles between nucleus and cytoplasm and, by interacting with components of the pRb pathway, influences cell survival.

C-terminal TDP-43 fragments that are proteolytically cleaved, hyperphosphorylated, and ubiquitinated accumulate in cytoplasmic inclusions. These inclusions are a hallmark of frontotemporal lobar dementia and also familial and sporadic forms of amyotrophic lateral sclerosis. Frontotemporal lobar dementia (FTLD), as its name suggests, is a group of neurodegenerative diseases that affects neurons and glia in the frontal and temporal lobes of the brain. These dementias are second only to Alzheimer's disease in their frequency of occurrence in individuals under the age of 65. At a molecular level the FTLD group can be divided into two subgroups, those with tau-containing inclusions (taupathies) and those with TDP-43-containing inclusions (TDP-43 proteinopathies). The latter is collectively designated as FTLD-U disorders and the former group, which includes disorders such as FTLD with Pick bodies, as FTLD taupathies.

Mutations in another RNA-binding protein, fused in sarcoma (FUS), alternatively called translocated in liposarcoma (TLS), can cause fALS. In the fALS cases involving mutated FUS, cytoplasmic inclusions are again formed, but these now contain FUS and lack any TDP-43 content. Similarly, TDP-43 inclusions are absent in those instances of familial ALS in which there are SOD1 mutations. Several functional roles have been ascribed to FUS. These include transcription, RNA processing, and mRNA transport into dendrites. FUS is translocated into dendritic spines in a Ca^{2+}-dependent manner, by myosin V motors, upon metabotropic glutamate receptor-5 (mGLUR5) activation, and once there regulates spine morphology. One common feature, shared by both TDP-43 and FUS, is their roles in RNA metabolism. Mechanistic details regarding TDP-43 and FUS yet to be uncovered may well produce a model that strongly resembles that of SOD1 in its depiction of the steps leading to neuronal malfunction and death, but involving different combinations of proteins, mutation-specific misfolding, and protein aggregation. The particular failure modes (points of failure) leading to impaired transport, synaptic dysfunction, and neuronal death may vary but all will ultimately produce the same negative outcome.

20.15 Interlocking Signaling Networks Underlie Health and Disease

Figure 20.6 summarizes what appears to be happening across the entire spectrum of age-related diseases. For simplicity the figure includes just two states, both highly stable and dynamically maintained. One of these, the healthy state, dominates during most of life. That state is represented by two concentric circles, an inner one denoting the interlocking networks of intracellular signaling networks that regulate cellular life and an outer one signifying the signaling system belonging to the microenvironment. This outer system encompasses the extracellular matrix and its cellular constituents which, depending on the particular tissue, will be a mix of macrophages, fibroblasts, stem cells, microglia, astrocytes, and so on responsible for maintaining a healthy microenvironment. The arrows connecting the two circles denote two-way signaling in which each system influences and assists the other. The disease state is represented by another pair of concentric circles. It is not a stable state and is seldom populated. Perturbations in the form of injuries and illnesses can produce a transition to the disease state but these states are short lived and the body recovers rapidly.

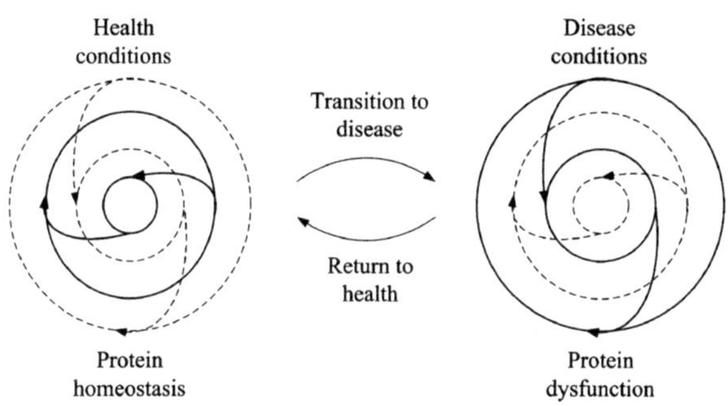

Fig. 20.6 Health and disease conditions. Under both healthy and disease conditions, there are ensembles of healthy and disease states. Under healthy condition, the healthy states, denoted by heavy *circles* and *arrows*, are stable and the disease states as indicated as *dashed circles* and *arrows* are transiently populated. The opposite holds for disease conditions. The rotation in the symbolic depiction of healthy and disease states is meant to indicate that the signaling circuitry has been altered in an essential way

Under the healthy conditions, protein homeostatis is maintained, but over time this may be lost. Factors that have been found to contribute to protein dysregulation include somatic mutations, excessive oxidation, calcium dysregulation, and heavy metal buildup. These lead to aberrant posttranslational modifications, protein misfolding, accumulation of insoluble aggregates, and persistant inflammation. Over time alterations occur in the intracellular signaling networks and in those belonging to the microenvironment; the disease state becomes the stable state, and the healthy state becomes at best metastable.

The two-condition figure is an obvious simplification, and a more detailed representation would include an ensemble of conditions in-between the two shown. The goal of a therapeutic intervention is, from this viewpoint of this picture, to drive the individual's health toward the left, that is, toward a more healthy condition. Advances are being made in this direction. Examples where this goal is furthered include gene therapy and the use of stem cells. Progress in terms of drug treatments is also being made with the recognition that not all mutations are equal at the biophysical and biochemical levels, and also with the recognition that in many instances single nucleotide polymorphisms have an effect, and must be taken into account in devising a treatment strategy. However, a deeper understanding of the signaling networks that sustain health and disease states is lacking.

The stability of the disease state in the presence of sustained disease conditions manifests itself as difficulties in finding effective and long-lasting drug treatments. Magic bullets designed to attack a discovered potential weak point in the signaling networks that sustain the disease state are countered over time, and the drugs are rendered weakly and temporarily effective, at best.

Somehow, the logic, however excellent, is being defeated. Some of the factors responsible for this were mentioned in the last paragraph. A true goal as recongnized by all is to devise ways to restore proper function and this depends on knowing what each specific disease-causing mutation to a given protein is doing.

It is also crucial to understand far better what is happening at the tissue level. It is at this level that multiple interlocking positive and negative feedback circuits establish themselves, and along with epigenetic factors stabilize the state, whether it be a healthy one or a disease-promoting one.

Further Reading

Expanded PolyQ Tract Disorders

Chai Y et al. 2004 Poly-ubiquitin binding by the polyglutamine disease protein Atanin-3 links its normal function to protein surveillance pathways. *J. Biol. Chem.* 279: 3605–3611.

Lam YC et al. 2006 ATAXIN-1 interacts with the repressor Capicua in its native complex to cause SCA1 neuropathology. *Cell* 127: 1335–1347.

Lim J et al. 2008 Opposing effects of polyglutamine expansion on native protein complexes contribute to SCA1. *Nature* 452: 713–718.

Taroni F and DiDonato S 2004 Pathway to motor incorrdination: the inherited ataxias. *Nat. Rev. Neurosci.* 5: 641–655.

Tsuda H et al. 2005 The AXH domain of Ataxin-1 mediates neurodegeneration through its interaction with Gfi-1/senseless proteins. *Cell* 122: 633–644.

Synaptic Terminal Interactions

Romero E et al. 2008 Suppression of neurodegeneration and increased neurotransmission caused by expanded full-length Huntingtin accumulating in the cytoplasm. *Neuron* 57: 27–40.

Swayne LA et al. 2005 Crosstalk between Huntingtin and syntaxin 1A regulates N-type calcium channels. *Mol. Cell. Neurosci.* 30: 339–351.

Loss of Calcium Homeostasis, Proteases, and Fragments

Gafni J et al. 2004 Inhibition of calpain cleavage of Huntingtin reduces toxicity. *J. Biol. Chem.* 279: 20211–20220.
Graham RK et al. 2006 Cleavage at the caspase-6 site is required for neuronal dysfunction and degeneration due to mutant Huntingtin. *Cell* 125: 1179–1191.
Lunkes A et al. 2002 Proteases acting on mutant Huntingtin generate cleaved products that differentially build up cytoplasmic and nuclear inclusions. *Mol. Cell* 10: 259–269.
Panov AV et al. 2002 Early mitochondrial calcium defects in Huntington's disease are a direct effect of polyglutamines. *Nat. Neurosci.* 5: 731–736.
Romero E et al. 2008 Suppression of neurodegeneration and increased neurotransmission caused by expanded full-length Huntingtin accumulating in the cytoplasm. *Neuron* 57: 27–40.
Tang TS et al. 2003 Huntingtin and Huntingtin-associated proteins 1 influence neuronal calcium signaling mediated by inositol-(1,4,5) triphosphate receptor type 1. *Neuron* 39: 227–239.
Wellington CL et al. 2002 Caspase cleavage of mutant Huntingtin precedes neurodegeneration in Huntington's disease. *J. Neurosci.* 22: 7862–7872.
Zeron MM et al. 2002 Increased sensitivity to *N*-methyl-D-aspartate receptor-mediated excitotoxicity in a mouse model of Huntington's disease. *Neuron* 33: 849–860.

Impaired Axonal Transport

Gauthier LR et al. 2004 Huntingtin controls neurotrophic support and survival of neurons by enhancing BDNF vesicular transport along microtubules. *Cell* 118: 127–138.
Gervais FG et al. 2002 Recruitment and activation of caspase-8 by the Huntingtin-interacting protein Hip1 and a novel protein Hippi. *Nat. Cell Biol.* 4: 95–105.
Gunawardena S et al. 2003 Disruption of axonal transport by loss of Huntingtin or expression of pathogenic PolyQ proteins in *Drosophila*. *Neuron* 40: 25–40.
Mills IG et al. 2005 Huntingtin interacting protein 1 modulates the transcriptional activity of nuclear hormone receptors. *J. Cell Biol.* 170: 191–200.
Mishra SK et al. 2001 Clathrin- and AP-2 binding sites in HIP1 uncover a general assembly route for endocytic accessory proteins. *J. Biol. Chem.* 276: 46230–46236.
Szebenyi G et al. 2003 Neuropathogenic forms of Huntingtin and ardrogen receptor inhibit fast axonal transport. *Neuron* 40: 41–52.

Zippers, Oligomers, Aggregates, and IBs

Arrasate M et al. 2004 Inclusion body formation reduces levels of mutant Huntingtin and the risk of neuronal death. *Nature* 431: 805–810.
Behrends C et al. 2006 Chaperomin TRiC promotes the assembly of PolyQ expansion proteins in nontoxic oligomers. *Mol. Cell* 23: 887–897.
Landschultz WH, Johnson PF and McKnight SL 1988 The leucine zipper – a hypothetical structure common to a new class of DNA-binding proteins. *Science* 240: 1759–1764.
Perutz MF 1994 Glutamine repeats as polar zippers: their possible role in inherited neurodegenerative diseases. *Proc. Natl. Acad. Sci. USA* 91: 5355–5358.
Perutz M 1999 Glutamine repeats and neurodegenerative diseases: molecular aspects. *Trends Biochem. Sci.* 24: 58–63.
Perutz MF et al. 2002 Amyloid fibers are water-filled nanotubes. *Proc. Natl. Acad. Sci. USA* 99: 5591–5595.
Wetzel R 2002 Ideas of order for amyloid fibril structure. *Structure* 10: 1031–1036.

Impaired UPS

Bennett EJ et al. 2005 Global impairment of the ubiquitin-proteasome system by nuclear or cytoplasmic protein aggregates precedes inclusion body formation. *Mol. Cell* 17: 351–365.
Bennett EJ et al. 2007 Global changes to the ubiquitin system in Huntington's disease. *Nature* 448: 704–708.
Colledge M et al. 2003 Ubiquitination regulates PSD-95 degradation and AMPA receptor surface expression. *Neuron* 40: 595–607.
Fonseca R et al. 2006 A balance of protein synthesis and proteasome-dependent degradation determines the maintenance of LTP. *Neuron* 52: 239–245.
Orr AL et al. 2008 N-terminal mutant Huntingtin associates with mitochondria and impairs mitochondrial trafficking. *J. Neurosci.* 28: 2783–2792.
Venkatraman P et al. 2004 Eukaryotic proteasomes cannot digest polyglutamine sequences and release them during degradation of polyglutamine-containing proteins. *Mol. Cell* 14: 95–104.
Zhao Y et al. 2003 The ubiquitin proteasome system functions as an inhibitory constraint on synaptic strengthening. *Curr. Biol.* 13: 887–898.

Imapired Transcription

Cui LB et al. 2006 Transcriptional repression of PGC-1α by mutant Huntingtin leads to mitochondrial dysfunction and neurodegeneration. *Cell* 127: 59–69.
Dunah AW et al. 2002 Sp1 and TAFII130 transcriptional activity disrupted in early Huntington's disease. *Science* 296: 2238–2243.
Jung J and Bonini N 2007 CREB binding protein modulates repeat instability in a Drosophila model for PolyQ disease. *Science* 315: 1857–1859.
Li SH and Li XJ 2004 Huntingtin-protein interactions and the pathogenesis of Huntington's disease. *Trends Genet.* 20: 146–154.
Schaffar G et al. 2004 Cellular toxicity of polyglutamine expansion proteins: mechanism of transcription factor deactivation. *Mol. Cell* 15: 95–105.
St-Pierre J et al. 2006 Suppression of reactive oxygen species and neurodegeneration by the PGC-1 transcription coactivators. *Cell* 127: 397–408.
Weydt P et al. 2006 Thermoregulatory and metabolic defects in Huntington's disease transgenic mice implicates PGC-1α in Huntington's disease neurodegeneration. *Cell Metab.* 4: 349–362.
Zuccato C et al. 2001 Loss of Huntingtin-mediated BDNF gene transcription in Huntington's disease. *Science*. 293: 493–498.
Zuccato C et al. 2003 Huntingtin interacts with REST/NRSF to modulate the transcription of NRSE-controlled neuronal genes. *Nat. Genet.* 35: 76–83.

SOD1 Structure, Folding, and Aggregation

Bruns CK and Kopito RR 2007 Impaired post-translational folding of familial ALS-linked Cu,Zn superoxide dismutase mutants. *EMBO J.* 26: 855–866.
Deng HX et al. 2006 Conversion to the amyotrophic lateral sclerosis phenotype is associated with intermolecular linked insoluble aggregates of SOD1 in mitochondria. *Proc. Natl. Acad. Sci. USA* 103: 7142–7147.
Lindberg MJ et al. 2005 Systematically perturbed folding patterns of amyotrophic lateral sclerosis (ALS)-associated SOD1 mutants. *Proc. Natl. Acad. Sci. USA* 102: 9754–9759.

O'Halloran TV and Culotta VC 2000 Metallochaperones, an intracellular shuttling service for metal ions. *J. Biol. Chem.* 275: 25057–25060.

Rakhit R et al. 2002 Monomeric Cu,Zn-superoxide dismutase is a common misfolding intermediate in the oxidative models of sporadic and familial amyotrophic lateral sclerosis. *J. Biol. Chem.* 279: 15499–15504.

Impaired Fast Axonal Transport

Hafezparast M et al. 2003 Mutations in dynein link motor neuron degeneration to defects in retrograde transport. *Science* 300: 808–812.

Liu J et al. 2004 Toxicity of familial ALS-linked SOD1 mutants from selective recruitment to spinal mitochondria. *Neuron* 43: 5–17.

Vale RD 2003 The molecular motor toolbox for intracellular transport. *Cell* 112: 467–480.

Vijayvergiya C et al. 2005 Mutant superoxide dismutase 1 forms aggregates in the brain mitochondrial matrix of amyotrophic lateral sclerosis mice. *J. Neurosci.* 25: 2463–2470.

Williamson TL and Cleveland DW 1999 Slowing of axonal transport is a very early event in the toxicity of ALS-linked SOD1 mutants to motor neurons. *Nat. Neurosci.* 2: 50–56.

ALS Microenvironment – Astrocytes

Di Giorgio FP et al. 2007 Non-cell autonomous effect of glia on motor neurons in an embryonic stem cell-based ALS model. *Nat. Neurosci.* 10: 608–614.

Nagai N et al. 2007 Astrocytes expressing ALS-linked mutated SOD1 release factors selectively toxic to motor neurons. *Nat. Neurosci.* 10: 615–622.

Rothstein JD et al. 2005 β-Lactam antibiotics offer neuroprotection by increasing glutamate transporter expression. *Nature* 433: 73–77.

Van Damme P et al. 2007 Astrocytes regulate GluR2 expression in motor neurons and their vulnerability to excitotoxicity. *Proc. Natl. Acad. Sci. USA* 104: 14825–14830.

ALS Microenvironment – Microglia

Boillée S, Vande Velde C, and Cleveland DW 2006 ALS: a disease of motor neurons and their nonneuronal neighbors. *Neuron* 52: 39–59.

Boillée S et al. 2006 Onset and progression in inherited ALS determined by motor neurons and microglia. *Science* 312: 1389–1392.

Block ML, Zecca L and Hong JS 2007 Microglia-mediated neurotoxicity: uncovering the molecular mechanisms. *Nat. Rev. Neurosci.* 8: 57–69.

Harraz MM et al. 2008 SOD1 mutations disrupt redox-sensitive Rac regulation of NADPH oxidase in a familial ALS model. *J. Clin. Invest.* 118: 659–670.

Lobsiger CS and Cleveland DW 2007 Glial cells as intrinsic components of non-cell-autonomous neurodegenerative disease. *Nat. Neurosci.* 10: 1355–1360.

Impaired ERAD

Duennwald ML and Lindquist S 2008 Impaired ERAD and ER stress are early and specific events in polyglutamine toxicity. *Genes Dev.* 22: 3308–3319.

Mao Y et al 2005 deubiquitinating function of ataxin 3: insights from the solution structure of the Josephin domain. *Proc. Natl. Acad. Sci. USA* 102: 12700–12705.

Nishitoh H et al 2008 ALS-linked mutant SOD1 induces ER stress- and ASK1-dependent motor neuron death by targeting Derlin-1. *Genes Dev.* 22: 1451–1464.
Warrick JM et al 2005 Ataxin-3 suppresses polyglutamine neurodegeneration in Drosophila by a ubiquitin-associated mechanism. *Mol. Cell* 18: 37–48.

TDP-43 and FUS

Ayala YM et al 2008 TDP-43 regulates retinoblastoma protein phosphorylation through the repression of cyclin-dependent kinase 6 expression. *Proc. Natl. Acad. Sci. USA* 105: 3785–3789.
Johnson BS et al 2008 A yeast TDP-43 proteinopathy model: exploring the molecular determinants of TDP-43 aggregation and cellular toxicity. *Proc. Natl. Acad. Sci. USA* 105: 6439–6444.
Kwiatkowski TJ et al 2009 Mutations in FUS/TLS gene on chromosome 16 cause familial amyothrophic lateral sclerosis. *Science* 323: 1205–1208.
Kwong LK et al 2008 TDP-43 proteinopathies: neurodegenerative protein misfolding diseases without amyloidosis. *Neurosignals* 16: 41–51.
Neumann M et al 2006 Ubiquitinated TDP-43 in frontotemporal lobar degeneration and amyotrophic lateral sclerosis. *Science* 314: 130–133.
Vance C et al 2009 Mutations in FUS, an RNA processing protein, cause familial amyotrophic lateral sclerosis type 6. *Science* 323: 1208–1211.

Index

O

P

T

U

V

W

Lightning Source UK Ltd.
Milton Keynes UK
UKOW040215020213

205739UK00001B/36/P

9 780387 981727